The Utilization of Bioremediation to Reduce Soil Contamination: Problems and Solutions

edited by

Václav Šašek

Institute of Microbiology,
Academy of Sciences of the Czech Republic,
Prague, Czech Republic

John A. Glaser

U.S. Environmental Protection Agency,
National Risk Management Research Laboratory,
Cincinnati, Ohio, U.S.A.

and

Philippe Baveye

Cornell University,
Ithaca, New York, U.S.A.

Kluwer Academic Publishers

Dordrecht / Boston / London

Published in cooperation with NATO Scientific Affairs Division

Proceedings of the NATO Advanced Research Workshop on
The Utilization of Bioremediation to Reduce Soil Contamination: Problems and Solutions
Prague, Czech Republic
14–19 June 2000

A C.I.P. Catalogue record for this book is available from the Library of Congress.

ISBN 1-4020-1141-5 (HB)
ISBN 1-4020-1142-3 (PB)

Published by Kluwer Academic Publishers,
P.O. Box 17, 3300 AA Dordrecht, The Netherlands.

Sold and distributed in North, Central and South America
by Kluwer Academic Publishers,
101 Philip Drive, Norwell, MA 02061, U.S.A.

In all other countries, sold and distributed
by Kluwer Academic Publishers,
P.O. Box 322, 3300 AH Dordrecht, The Netherlands.

Printed on acid-free paper

The Utilization of Bioremediation to Reduce Soil Contamination: Problems and Solutions

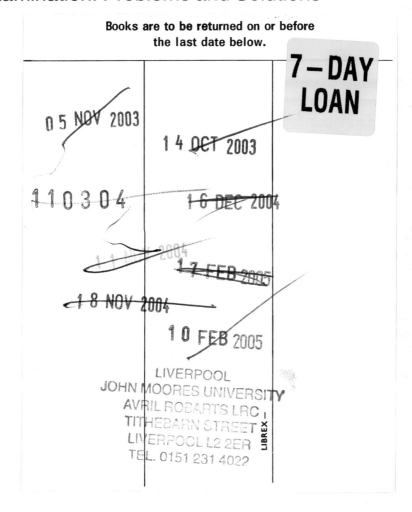

Books are to be returned on or before
the last date below.

7 – DAY
LOAN

0 5 NOV 2003

1 4 OCT 2003

110304

1 6 DEC 2004

1 7 FEB 2005

1 8 NOV 2004

1 0 FEB 2005

NATO Science Series

A Series presenting the results of scientific meetings supported under the NATO Science Programme.

The Series is published by IOS Press, Amsterdam, and Kluwer Academic Publishers in conjunction with the NATO Scientific Affairs Division

Sub-Series

I. **Life and Behavioural Sciences**	IOS Press
II. **Mathematics, Physics and Chemistry**	Kluwer Academic Publishers
III. **Computer and Systems Science**	IOS Press
IV. **Earth and Environmental Sciences**	Kluwer Academic Publishers
V. **Science and Technology Policy**	IOS Press

The NATO Science Series continues the series of books published formerly as the NATO ASI Series.

The NATO Science Programme offers support for collaboration in civil science between scientists of countries of the Euro-Atlantic Partnership Council. The types of scientific meeting generally supported are "Advanced Study Institutes" and "Advanced Research Workshops", although other types of meeting are supported from time to time. The NATO Science Series collects together the results of these meetings. The meetings are co-organized bij scientists from NATO countries and scientists from NATO's Partner countries – countries of the CIS and Central and Eastern Europe.

Advanced Study Institutes are high-level tutorial courses offering in-depth study of latest advances in a field.
Advanced Research Workshops are expert meetings aimed at critical assessment of a field, and identification of directions for future action.

As a consequence of the restructuring of the NATO Science Programme in 1999, the NATO Science Series has been re-organised and there are currently five sub-series as noted above. Please consult the following web sites for information on previous volumes published in the Series, as well as details of earlier sub-series.

http://www.nato.int/science
http://www.wkap.nl
http://www.iospress.nl
http://www.wtv-books.de/nato-pco.htm

Series IV: Earth and Environmental Sciences – Vol. 19

TABLE OF CONTENTS

---

[*] Initially given at the Workshop as a short volunteered presentation

Part III. Ecotoxicology and Toxicity Monitoring of Bioremediation Measures

Part IV. Application of Bioremediation to Environmental Problems

PREFACE

Increasingly, modern society relies on a striking array of organic chemicals and the quantities of these compounds used on a yearly basis are staggering. As a direct consequence of their industrial, agricultural or domestic usage, occasionally as a result of accidents or negligence, ever-increasing amounts of these chemicals are released into the environment. There, they present risks to ecosystems and human health, since many organic pollutants are toxic to living organisms even in minute quantities. As a result, pollution in the atmosphere, seas and oceans, surface waters, and subsurface porous formations (soils, aquifers) by organic chemicals has become a cause of major societal concern in industrialized countries during the past forty years.

Since the seventies, public officials and the private sector have responded to this concern with a vigorous remediation effort. Various strategies have been adopted, largely based at first on *ex situ* treatments. For contaminated sediments, for example, they include incineration or disposal in landfills. "Pump-and-treat" (or "soil washing") procedures have also been developed, in which the polluted groundwater is pumped to the soil surface, treated to reduce the contaminant(s) concentration to an environmentally-acceptable level, and re-injected in the aquifer some distance away. The aboveground treatment of the pumped groundwater relies on physical, chemical or, occasionally, biological processes.

In the early eighties, interest shifted from above- to below-ground (or, more generally, from outside to inside polluted ecosystems) with the proposal that remediation might be carried out more directly in situ by fostering the growth of microorganisms capable of degrading targeted organic pollutants. These microorganisms may be indigenous to the polluted ecosystem, or may be injected specifically to carry out the in situ bioremediation. Compared with most physical or chemical procedures (like incineration, vitrification, or extraction), biologically mediated treatment strategies present the highly desirable feature of allowing, in principle, the complete eradication or mineralization of organic xenobiotics. In addition, they have generally lower capital and operational costs. These different advantages, combined with the enthusiasm of scientists, convinced a number of entrepreneurs to start bioremediation companies, or to reorient existing businesses in this promising direction.

Unfortunately, since that "pioneer" time in the mid-eighties, few of the original expectations seem to have been met by practitioners, and the initial euphoria has drastically subsided. The rate of adoption of in situ bioremediation strategies has been stagnating in recent years. In 1992, bioremediation was adopted in approximately 9 % of all Superfund remedial actions in the United States, with a little less than half of these actions taking place *in situ*. Two years later, in 1994, again at Superfund sites in the US, bioremediation was the technology of choice in 10 % of the cases. In situ bioremediation was used in 4% of all cases, i.e., at a mere 25 sites. Statistics are not available for more recent years, but it is generally agreed that the market share of bioremediation (*ex situ* and *in situ*) has only marginally increased since 1994, in the US as well as in Europe.

Some reasons for the disappointing rate of adoption of *in situ* bioremediation strategies are common to all the remediation strategies at the low end of the cost spectrum. Delays, whether because of disincentives to initiate remediation or because of bureaucratic inefficiency, substantially weaken the market for inexpensive, effective remediation technologies. Specifically, however, the key reasons for the stagnation of the bioremediation market (and for the ensuing bankrupcy of many start-up bioremediation companies over the last few years) seem to be the frequent inability of *in situ* bioremediation strategies to decrease contaminant concentrations below regulatory levels in a timely fashion. This may be caused by extreme pollution at the sites, with contaminant concentrations at or above levels that are toxic to microorganisms. However, even when conditions seem optimal for biodegradation of the pollutants to occur, the rate of the process often drops rapidly to inconsequential levels, allowing unacceptably high amounts of the contaminant(s) to remain in the system. In addition, in some well-documented cases where bioremediation had successfully decreased pollutant concentrations below regulatory levels, the clean-up turned out to be only temporary. After a few months, the contaminants were once again present in ground- and surface water at environmentally unacceptable concentrations.

In retrospect, it seems clear that bioremediation, *in situ* or *ex situ*, was promoted in the mid-eighties at a time when the basic knowledge needed to make it work in the field was woefully inadequate or even was entirely lacking. As is often the case under similar circumstances, and in spite of the fact that some of the required basic information was slowly but surely being produced by scientists, bioremediation lost much of its appeal and popularity in the nineties, and suffered a severe backlash, particularly in the US. This pendulum swing, in turn, contributed to distort somewhat in the minds of many the true usefulness of bioremediation strategies.

In that context, a NATO Advanced Research workshop was organized and held in Liblice (Czech Republic) in June 2000, to set the record straight. Specifically, this ARW was meant to provide to a group of specialists the opportunity to assess as accurately and objectively as possible the state of the science related to bioremediation, to point out areas where further research is vital, and to document and evaluate in detail a number of practical situations where bioremediation either failed or successfully met its intended objectives.

Even before the ARW was publicized, the organizers began receiving what eventually became a flood of applications, from several hundred individuals, practitioners as well as scientists. This enthusiastic response alone evinced that, paraphrasing Mark Twain, news of the demise of bioremediation may have been exaggerated, and that there was still a widespread belief that bioremediation remained a viable and promising technology. Of these hundreds of applicants, 47 participants and 11 observers were selected, representing 23 countries (Austria, Belarus, Belgium, Bulgaria, Canada, Czech Republic, Estonia, France, Germany, Greece, Israel, Italy, Latvia, Lithuania, The Netherlands, Norway, Poland, Portugal, Slovak Republic, Slovenia, Ukraine, United Kingdom and United States).

Among the many topics addressed during the workshop, three were of particular significance. The first concerns the development of suitable toxicological tests to be used in conjunction with bioremediation strategies. Traditional reliance on chemical analysis to understand the direction and extent of treatment in a bioremediation process has been found to be inadequate. Whereas the goal of bioremediation is toxicity reduction, few direct, reliable measures of this process are as yet available. Toxicity measures in aquatic systems have been developed in recent years. However, equivalent methods for use in soils and sediments are direly needed. The efficacy of each method must be evaluated for application to soil. Test reliability, difficulties in the interpretation of the results, and lack of correlation among methods are the major hurdles limiting at present the utility of toxicology tests to determine suitable targets for bioremediation operations. Discussion of this concern led to heated debate among workshop participants.

Another area of intense discussion was the assessment of market forces contributing to the acceptability of bioremediation. Perspectives of the US and Czech Republic markets provided significant understanding of this poorly recognized factor.

Finally, another important component of the meeting was a series of lectures and lively exchanges devoted to practical applications of different bioremediation technologies. The range of subjects covered a wide spectrum, encompassing emerging technologies as well as actual, full-scale operations. Examples discussed included landfarming, biopiling, composting, phytoremediation and mycoremediation. Each technology was explored for its utility and capability to provide desired treatment goals. Advantages and limitations of each technology were discussed. The concept of natural attenuation was also critically evaluated since in some cases where time to remediation is not a significant factor, it may be an alternative to active bioreme-diation operations.

All of these themes are covered in the present book, which emanates from the 2000 NATO ARW and covers much of the same ground. Yet, in the intervening two years, the scope and content of the chapters have evolved in many cases compared to the lectures presented in Liblice, in part because of the insightful comments of reviewers and of the improved understanding of a number of issues as a direct result of the ARW itself. The result is a unique book that provides apparently for the first time an up-to-date, comprehensive assessment of bioremediation strategies, which we hope will be invaluable to scientists and practitioners alike.

Publication of this book would not have been possible without the extremely helpful assistance of Professor A. Kotyk, who polished the English of the various chapters authored by non-native English speakers, and of Dr. J. Cudlín and Dr. Č. Novotný, who prepared the final, camera-ready version of the text. Sincere gratitude is also expressed toward Dr. Alain Jubier, director of NATO's Scientific Affairs Division, and to his administrative assistant, Mrs. Lynn Campbell-Nolan, for their infaillible and stimulating support.

The editors:
Václav Šašek
John A. Glaser
Philippe Baveye

Contributors

Altman, D.J.
    Westinghouse Savannah River Company, Bldg 704-8T, Aiken, SC 29808, USA
Angelini, G.
    Institute of Biochemistry and Plant Physiology, Nuclear Chemistry, National
    Council of Research, Area della Ricerca di Roma, 00016 Monterotondo Scalo, Italy
Alessandrelli, S.
    Institute of Biochemistry and Plant Physiology, Nuclear Chemistry, National
    Council of Research, Area della Ricerca di Roma, 00016 Monterotondo Scalo, Italy
Anoliefo, O.
    National Agricultural Research Foundation, Institute of Kalamata, Lakonikis 85,
    24100 Kalamata, Greece
Bailey, M.J.
    Institute of Virology and Environmental Microbiology, Centre for Ecology and
    Hydrobiology, Natural Environment Research Council, Mansfield Road, Oxford
    OX1 3SR, UK
Baláž, Š.
    Department of Pharmaceutical Sciences, College of Pharmacy, North Dakota State
    University, Sudro Hall 108, Fargo, ND-58105, USA
Baldrian, P.
    Institute of Microbiology, Academy of Sciences of the Czech Republic, Vídeňská
    1083, CZ-142 20 Prague 4, Czech Republic
Baltrènas, P.
    Vilnius Gediminas Technical University, Environmental Protection Department,
    Sauletekio al. 11 – 2306, 2040 Vilnius, Lithuania
Barclay, M.
    Institute of Virology and Environmental Microbiology, Centre for Ecology and
    Hydrobiology, Natural Environment Research Council, Mansfield Road, Oxford
    OX1 3SR, UK
Baveye, P.
    Bradfield Hall, Cornell University, Ithaca, New York 14853, USA
Bhatt, M.
    UFZ – Umweltforschungszentrum Leipzig-Halle GmbH, Permosestrasse 15, 04318
    Leipzig, Germany
Boller, V.
    IFA-Tulln, Konrad Lorenz Str. 20, A-3430 Tulln, Austria
Braun, R.
    IFA-Tulln, Konrad Lorenz Str. 20, A-3430 Tulln, Austria
Braun-Lullemann, A.
    Institut für Forstbotanik, Universität Göttingen, Busgenweg 2, 37077 Göttingen,
    Germany
Burkhard, J.
    Institute of Chemical Technology, Technická 3, CZ-16620 Prague 6, Czech
    Republic

Cerniglia, C.E.
National Center for Toxicological Research, 3900 NCTR Rd., Jefferson, AR 72079, USA

Chang, Y.-J.
Center for Environmental Biotechnology, University of Tennessee, 10515 Research Drive, Suite 300, Knoxville, TN 37932-2575, USA

Chial, B.
Centro de Ciencias del Mar y Limnologia, Universidad de Panama City, Panama

Chromá, L.
Institute of Chemical Technology, Technická 3, CZ-16628 Prague 6, Czech Republic

Day, J.C.
Institute of Virology and Environmental Microbiology, Centre for Ecology and Hydrobiology, Natural Environment Research Council, Mansfield Road, Oxford OX1 3SR, UK

Dela Cruz, M.A.
IFA-Tulln, Konrad Lorenz Str. 20, A-3430 Tulln, Austria

Demnerová, K.,
Department of Biochemistry and Microbiology, Faculty of Food and Biochemical Technology, Institute of Chemical Technology, Technická 3, CZ-16628 Prague 6, Czech Republic

Dercová, K.
Department of Biochemical Technology, Faculty of Chemical Technology, Slovak University of Technology, Radlinskeho 9, 812 37 Bratislava, Slovak Republic

Dubourguier, H.-C.
Institut Supérieur d'Agriculture, 41, rue du Port, 59046 Lille Cedex, France

Eggen, T.
Centre for Soil and Environmental Research, Fredrik A. Dahl vei 20, N-1432 Aas, Norway

Ehaliotis, C.
National Agricultural Research Foundation, Institute of Kalamata, Lakonikis 85, 24100 Kalamata, Greece

Esposito, D.
Institute of Biochemistry and Plant Physiology, Nuclear Chemistry, National Council of Research, Area della Ricerca di Roma, 00016 Monterotondo Scalo, Italy

Gabriel, J.
Institute of Microbiology, Academy of Sciences of the Czech Republic, Vídeňská 1083, CZ-142 20, Prague 4, Czech Republic

Gadd, G.M.
Department of Biological Sciences, University of Dundee, Dundee, DD1 4HN, Scotland, UK

Gan, Y.-D. M.
Center for Environmental Biotechnology, University of Tennessee, 10515 Research Drive, Suite 300, Knoxville, TN 37932-2575, USA

Geyer, R.
   Center for Environmental Biotechnology, University of Tennessee, 10515 Research
   Drive, Suite 300, Knoxville, TN 37932-2575, USA, and UFZ Centre for
   Environmental Research Leipzig-Halle, Germany
Giardi, M.T.
   Institute of Biochemistry and Plant Physiology, Nuclear Chemistry, National
   Council of Research, Area della Ricerca di Roma, 00016 Monterotondo Scalo, Italy
Giardi, P.
   Institute of Biochemistry and Plant Physiology, Nuclear Chemistry, National
   Council of Research, Area della Ricerca di Roma, 00016 Monterotondo Scalo, Italy
Glaser, J.A.
   US EPA, National Risk Manag. Res. Lab., 26 W. Martin Luther King Dr.,
   Cincinnati, OH 45268, USA
Goncharova, N.
   Institute of Radiobiology af the Academy of Sciences of Belarus, Plant
   Radiobiology Laboratory, Kuprevich Str., 2200141 Minsk, Belarus
Groudev, S.
   University of Mining and Geology, Sofia 1100, Bulgaria
Harris, J.
   Institute of Water and Environment, Cranfield University, Silsoe, Bedfordshire
   MK45 4DT, UK
Hazen, T.C.
   Institute of Water ans Environment, University of California, Lawrence Berkeley
   National Laboratory, 1 Cyclotron Road, MS 70A-3317, Berkeley, CA 94720, USA
Hüttermann, A.
   Institut für Forstbotanik, Universität Göttingen, Busgenweg 2, 37077 Göttingen,
   Germany
Ignatavičius, G.
   Vilnius Gediminas Technical University, Environmental Protection Department,
   Sauletekio al. 11 – 2306, 2040 Vilnius, Lithuania
Isikhuemhen, O.S.
   North Carolina Agricultural and Technical University, Greensboro, NC 27411,
   USA
Jerger, D.
   Shaw Environmental & Infrastructure, Inc., 312 Directors Drive, Knoxville, TN
   37923, USA
Joseleau, J.-P.
   CERMAV-CNRS, BP 53, 38041 Grenoble cedex09, France
Kahru, A.
   Institute of Chemical Physics and Biophysics, Laboratory of Molecular Genetics,
   Akadeemia tee 23, Tallinn EE0026, Estonia
Kardimaki, A.
   National Agricultural Research Foundation, Institute of Kalamata, Lakonikis 85,
   24100 Kalamata, Greece
Kästner, M.
   UFZ – Umweltforschungszentrum Leipzig-Halle GmbH, Permosestrasse 15, 04318
   Leipzig, Germany

Kislushko, P.
Institute of Plant Protection, Lesnaya Str. 9, Priluki, 223011, Belarus

Knowles, C.J.
Institute of Virology and Environmental Microbiology, Centre for Ecology and Hydrobiology, Natural Environment Research Council, Mansfield Road, Oxford OX1 3SR, UK

Koller, G.
UFZ – Umweltforschungszentrum Leipzig-Halle GmbH, Permosestrasse 15, 04318 Leipzig, Germany

Kučerová, P.
Institute of Chemical Technology, Technická 3, CZ-16628 Prague 6, Czech Republic

Lednická, D.
Faculty of Science, University of Ostrava, 30. dubna.22, CZ-70103 Ostrava 1, Czech Republic

Leigh, M.B.
University of Oklahoma, Department of Botany and Microbiology, Norman, OK 73019, USA

Loibner, A.P.
IFA-Tulln, Konrad Lorenz Str. 20, A-3430 Tulln, Austria

Lytle, C.A.
Center for Environmental Biotechnology, University of Tennessee, 10515 Research Drive, Suite 300, Knoxville, TN 37932-2575, USA

Macek, T.
Institute of Organic Chemistry and Biochemistry, Academy of Sciences of the Czech Republic, Flemingovo nám. 2, CZ-16620 Prague 6, Czech Republic

Macková, M.
Institute of Chemical Technology, Technická 3, CZ-16628 Prague 6, Czech Republic

Majcherczyk, A.
Institut für Forstbotanik, Universität Göttingen, Busgenweg 2, 37077 Göttingen, Germany

Malachová, K.
Faculty of Science, University of Ostrava , 30. dubna 22, CZ-70103 Ostrava 1, Czech Republic

Malachowska-Jutsz, A.
Silesian Technical University, Environmental Biotechnology Dept., ul. Akademicka 2, 44-101 Gliwice, Poland

Maliszewska-Kordybach, B.
Institute of Soil Science and Plant Cultivation, Department of Soil Science and Land Reclamation, Czartoryskich 8, 24-100 Pulawy, Poland

Manko, T.
Institute for Ecology of Industrial Areas, 6 Kossutha St., Katowice, Poland

Margonelli, A.
Institute of Biochemistry and Plant Physiology, Nuclear Chemistry, National Council of Research, Area della Ricerca di Roma, 00016 Monterotondo Scalo, Italy

Matějů, V.
ENVISAN-GEM, Inc., Biotechnological Division, Radiová 7, CZ-102 31 Prague 10, Czech Republic

McIntyre, T.
Environment Canada, 351 St. Joseph Blvd, 18th Floor, Hull, Quebec, K1A OH3, Canada

Meckenstock, R.U.
Center for Applied Geosciences, Eberhard-Karls-University of Tübingen, Willhelmstr. 56, 72076 Tübingen, Germany

Miksch, K.
Silesian Technical University, Environmental Biotechnology Dept., ul. Akademicka 2, 44-101 Gliwice, Poland

Möder, M.
UFZ – Umweltforschungszentrum Leipzig-Halle GmbH, Permosestrasse 15, 04318 Leipzig, Germany

Molitoris, H.P.
Botanical Institute, University of Regensburg, Universitätsstrasse 31, D-93040 Regensburg, Germany

Nerud, F.
Institute of Microbiology, Academy of Sciences of the Czech Republic, Vídeňská 1083, CZ-142 20, Prague 4, Czech Republic

Novotný, Č.
Institute of Microbiology, Academy of Sciences of the Czech Republic, Vídeňská 1083, CZ-142 20, Prague 4, Czech Republic

Pace, E.
Institute of Biochemistry and Plant Physiology, Nuclear Chemistry, National Council of Research, Area della Ricerca di Roma, 00016 Monterotondo Scalo, Italy

Papadopoulou, K.
National Agricultural Research Foundation, Institute of Kalamata, Lakonikis 85, 24100 Kalamata, Greece

Pazlarová, J.
Institute of Chemical Technology, Technická 3, CZ-16628 Prague 6, Czech Republic

Patel, M.
Department of Biosciences, S.P. University, Vidyanagar-388120, India

Peacock, A.M.
Center for Environmental Biotechnology, University of Tennessee, 10515 Research Drive, Suite 300, Knoxville, TN and Microbial Insights, Inc., 2340 Stock Creek, Blvd., Rockford, TN 37853-3044, USA

Perdih, A.
University of Ljubljana, Faculty of Chemistry and Chemical Technology, Askerceva 5, Ljubljana, Slovenia

Persoone, G.
Laboratory for Environmental Toxicology and Aquatic Ecology, Ghent University, J. Plateaustraat 22, 9000 Ghent, Belgium

Podgornik, H.
   University of Ljubljana, Faculty of Chemistry and Chemical Technology, Askerceva 5, Ljubljana, Slovenia

Põllumaa, L.
   Institute of Chemical Physics and Biophysics, Laboratory of Molecular Genetics, Akadeemia tee 23, Tallinn EE0026, Estonia

Rawal, B.
   Botanical Institute, University of Regensburg, Universitätsstrasse 31, D-93040 Regensburg, Germany

Richnow, H.H.
   Centre for Environmental Research Leipzig-Halle (UFZ), Permoserstr. 15, 04318 Leipzig, Germany

Ruel, K.
   CERMAV-CNRS, BP 53, 38041 Grenoble cedex09, France

Šašek, V.
   Institute of Microbiology, Academy of Sciences of the Czech Republic, Vídeňská 1083, CZ-142 20, Prague 4, Czech Republic

Schwittzguébel, J.-P.
   Laboratory for Environmental Biotechnology (LBE), Swiss Federal Institute of Technology Lausanne (EPFL), CH-1015 Lausanne, Switzerland

Sims, J.L.
   Utah Water Research Laboratory, 8200 Old Main Hill, Utah State University, Logan, UT 84322-8200, USA

Sims, R.C.
   Utah Water Research Laboratory, 8200 Old Main Hill, Utah State University, Logan, UT 84322-8200, USA

Steer, J.
   School of Health and Bioscience, University of East London, Stratford London E15 4LZ, UK

Szolar, O.
   IFA-Tulln, Konrad Lorenz Str. 20, A-3430 Tulln, Austria

Tabouret, A.
   Institute of Virology and Environmental Microbiology, Centre for Ecology and Hydrobiology, Natural Environment Research Council, Mansfield Road, Oxford OX1 3SR, UK

Tandlich, R.
   Department of Pharmaceutical Sciences, College of Pharmacy, North Dakota State University, Sudro Hall 108, Fargo, ND-58105, USA

Thompson, I.P.
   Institute of Virology and Environmental Microbiology, Centre for Ecology and Hydrobiology, Natural Environment Research Council, Mansfield Road, Oxford OX1 3SR, UK

Tien, A.J.
   Westinghouse Savannah River Company, Bldg 704-8T, Aiken, SC 29808 USA

Ulfig, K.
   Institute for Ecology of Industrial Areas, 6 Kossutha, Katowice, Poland

Vaněk, T.
Institute of Organic Chemistry and Biochemistry, Academy of Sciences of the Czech Republic, Flemingovo nám. 2, 166 10 Prague 6, Czech Republic

Vrana, B.
Department of Biochemistry and Microbiology, Faculty of Food and Biochemical Technology, Institute of Chemical Technology, Technická 3, CZ-16628 Prague 6, Czech Republic

White, D.C.
Center for Environmental Biotechnology, University of Tennessee, 10515 Research Drive, Suite 300, Knoxville, TN 37932-2575, USA

Worsztynowicz, A.
Institute for Ecology of Industrial Areas, 6 Kossutha St., Katowice, Poland

Zervakis, G.
National Agricultural Research Foundation, Institute of Kalamata, Lakonikis 85, 24100 Kalamata, Greece

PART I

BIOMASS ESTIMATION TECHNIQUES

RAPID DETECTION/IDENTIFICATION OF MICROBES, BACTERIAL SPORES, MICROBIAL COMMUNITIES, AND METABOLIC ACTIVITIES IN ENVIRONMENTAL MATRICES

D. C. WHITE [1], A. M. PEACOCK[1,2], R. GEYER[1,3], Y.-J. CHANG[1], YING-DONG M. GAN[1] and C. A. LYTLE[1]

[1]Center for Environmental Biotechnology, University of Tennessee, 10515 Research Drive, Suite 300, Knoxville, TN 37932-2575, USA, Phone 865- 974-8001; FAX 865-974-8027, Dwhite1@utk.edu

[2]Microbial Insights, Inc., 2340 Stock Creek, Blvd., Rockford, TN 37853-3044, USA

[3]UFZ Centre for Environmental Research Leipzig-Halle, Germany

Abstract

Modern molecular methods such as lipid biomarker analysis (LBA) and PCR-Denaturing Gradient Gel Electrophoresis (DGGE) of rDNA or functional genes have enabled sensitive and quantitative assessment of *in situ* microbial communities that is independent of isolation and culture of the microbes. Application of artificial neural network analysis (ANN) to biomarker data has demonstrated a powerful means to characterize soil, marine and deep subsurface microbial communities. This analysis has provided accurate detection of bioavailability, bioremediation effectiveness, and rational end points for remediation based on the community microbial ecology. Sequential high pressure/temperature extraction greatly shortens sample-processing time and increases the recovery of lipid biomass from soils. Sequential extraction by supercritical CO_2 and/or accelerated solvent extraction (ACE) lyses cells and allows sequential recovery of neutral lipids, then polar lipids, and facilitates recovery of DNA. The extracted residue is then derivatized *in situ* and re-extracted with supercritical CO_2 or hexane with ACE to yield dipicolinic acid originating from bacterial spores and ester-linked hydroxy fatty acids derived from the lipopolysaccharide of Gram-negative bacteria. Once extracted, components in each fraction are separated by capillary high performance liquid chromatography (HPLC) with solvent systems compatible with electrospray ionization and analyzed by tandem quadrupole mass spectrometry (HPLC/ESI/MS/MS) and capillary gas chromatography/mass spectrometry (GC/MS). HPLC utilizing electrospray ionization (ESI) provides the potential to analyze intact lipid biomarkers and provides greater insights into the viable biomass, community composition and physiological status than with previous techniques. Use of HPLC/ESI/MS/MS enables analysis of respiratory quinones, diacylglycerols, and sterols from the neutral lipids. Also, the structures of intact phospholipids that have been collisionally dissociated can be determined at sensitivities (for synthetic molecules) of ≈ 90 attomol/μL (concentration of total lipids in a single *E. coli*). Moreover HPLC/ESI/MS/MS also enables analysis/detection of the isoprenoid tetraethers of the Archaea, ornithine lipids of Gram-negative bacteria signaling for bio-available phosphate, and the lysyl esters of phosphatidyl glycerol of Gram-positive bacteria. This expanded LBA not only facilitates recovery of DNA for highly specific PCR amplification but provides deeper insight into predicting *in situ*

V. Šašek et al. (eds.),
The Utilization of Bioremediation to Reduce Soil Contamination: Problems and Solutions, 3–19.
© 2003 *Kluwer Academic Publishers. Printed in the Netherlands.*

metabolic activities without disturbance artifacts. In addition, the analysis is based on the responses to microniche environments—the Holy Grail of microbial ecology.

1. Introduction

Analysis of the cellular lipids provides a convenient quantitative way to gain insight into critical attributes of microbial communities without isolation or culture of the microbes. Lipids are cellular components recoverable by extraction with organic solvents. The extraction provides both a purification and concentration. Lipids are an essential component of the membrane of all cells and play a role as storage materials. The lipid biomarker analysis (LBA) provides quantitative insight into three important attributes of microbial communities and facilitates recovery of the DNA for increased specificity.

1.1 VIABLE BIOMASS

The determination of the total phospholipid ester-linked fatty acids (PLFA) provides a quantitative measure of the viable or potentially viable biomass. Viable microbes have an intact membrane that contains phospholipids (and PLFA). Cellular enzymes hydrolyze (release) the phosphate group from phospholipids within minutes to hours of cell death resulting in the formation of diacylglycerols [1]. A careful study of subsurface sediment showed the viable biomass determined by PLFA was equivalent (but with a much smaller standard deviation) to that estimated by intercellular ATP, cell-wall muramic acid, and very carefully done acridine orange direct counts (AODC)[2].

1.2 COMMUNITY COMPOSITION

The analysis with lipid biomarkers provides a quantitative definition of the microbial community structure. Specific groups of microbes often contain unusual lipids [3]. For example, specific PLFA are prominent in the hydrogenase-containing *Desulfovibrio* sulfate-reducing bacteria, whereas the *Desulfobacter* types of sulfate-reducing bacteria contain distinctly different PLFA [4, 5]. Patterns of the prominent PLFA from isolated microbes after growth on standardized media are used to differentiate over 8,000 species of organisms with the Microbial Identification System (MIDI, Newark DE)[6]. Hierarchical cluster analysis of the patterns of phospholipid fatty acids shows similarities between species of isolated methane oxidizing and sulfate-reducing bacteria that almost exactly parallel the phylogenetic relationships based on the sequence similarities of 16S rRNA [7, 8]. Hierarchical cluster analyses of PLFA patterns of the total microbial communities quantitatively define relatedness of different microbial communities. This hierarchical cluster analysis of PLFA patterns has been done with deep subsurface sediments in which the microbial communities of permeable strata are different from surface soils, clay aqualudes, and drilling fluids used to recover the samples [9].

Analysis of other lipids, such as sterols (for the microeukaryotes – algae, protozoa) [10], glycolipids (phototrophs, Gram-positive bacteria), or the hydroxy fatty acids in the lipopolysaccharide (LPS) lipid A (Gram-negative bacteria) [11] can provide an even more detailed community structure analysis.

1.3 NUTRITIONAL STATUS

The formation of poly β-hydroxyalkanoic acid (PHA in bacteria) [12], or triacylglycerol (in microeukaryotes) [13] relative to the PLFA provides a measure of the nutritional status. Bacteria grown with adequate carbon and terminal electron acceptors form PHA when they cannot divide because some essential component is missing (phosphate, nitrate, trace metal, etc.). Furthermore, specific patterns of PLFA can indicate physiological stress [14]. Exposure to toxic environments can lead to minicell formation and a relative increase in specific *trans*-monoenoic PLFA compared to the *cis* isomers. It has been shown that for increasing concentrations of phenol toxicants, the bacterium *Pseudomonas putida* forms increasing proportions of *trans*-PLFA [15].

1.4 CONCOMITANT DNA EXTRACTION

This powerful quantitative assessment method developed over the past 20 years to define the viable biomass, community structure and nutritional/physiological status of environmental microbial communities based on LBA can be expanded to include analysis of DNA. The DNA probe analysis offers powerful insights because of the exquisite specificity in the detection of genes for enzyme processes and/or for 16S rDNA for organism identification at the kingdom, family, genus or species levels. Extraction of DNA from soil requires lysis of the cells prior to obtaining the DNA [16]. Recent evidence indicates that the lipid extraction used for LBA also liberates the cellular DNA [17]. The combined analysis using lipid biomarkers and PCR of rDNA with subsequent separation of the amplicons by denaturing gradient gel electrophoresis (DGGE) has proved especially powerful in the analysis of microbial communities impacted by pollution [18]. One problem with DNA analysis is that it is sometimes difficult to achieve quantitative results. The concomitant DNA/lipid analysis readily provides quantitative recoveries independent of the ability to isolate or culture the microbes, and the presence of intact cellular membranes containing polar lipids provides an accurate measure of the microbial biomass.

As part of the Department of Energy Natural and Accelerated Bioremediation Research Program (NABIR), our laboratory has been able to compare the relative effectiveness of phospholipid fatty acid analysis (PLFA) vs the polymerase chain reaction (PCR) based techniques, terminal Restriction Fragment Length Polymorphism (tRFLP) and Denaturant Gradient Gel Electrophoresis (DGGE), for predicting the impact of chromium on the microbial community at a contaminated site. Contamination at the site ranged from 50 to 200,000 ppm [19, 20]. As a consequence of the relative lack of toxicity of Cr^{III}, the microbial populations were relatively diverse over this wide range. By using artificial neural network (ANN) analysis, the impact of Cr on the microbial community down to about 10^2–10^3 ppm Cr was detected with PLFA analysis. t-RFLP, while also able to pick up the impact of the Cr, only did so at $>10^3$ ppm Cr [21]. Preliminary analysis showed the same sensitivity of PLFA analysis in predicting effects of Cr toxicity on the microbial community in vadose-zone sediment microcosms [22].

As a membrane marker, PLFA analysis picks up the response of not only the *in situ* community gene pool but also the individual physiological response of cells to shifts in the local microniche ecosystem processes. It is this extra response that gives PLFA its advantage over the DNA based techniques, especially when analysis of the whole community response is the goal. PLFA is not as effective as DNA analysis when looking for one specific organism. It is analogous to determining the shape of a cathedral from one brick. DNA-based techniques show you the shape of the individual bricks (mathematically the DNA sequence is noncompressible) but does not provide much data on the community. PLFA, on the other hand, provides an idea of how the whole community (the "cathedral") is shaped. As such, PLFA seems to provide a "holistic" answer to mapping community change, and one that takes into account any perturbation that may occur. Enhancement of lipid biomarker analysis as reported herein not only greatly increases the functional insights into the community dynamics by expanding the lipids that can be readily analyzed but provides this analysis more rapidly and at greater sensitivity as well as at a lower cost.

2. Methods

One of the drawbacks of the classical PLFA analysis is the extraction process by which lipids are recovered from environmental samples. Classical room temperature/pressure extraction takes 8-12 h to allow emulsions to settle and requires careful analytical technique by a skilled operator. Once extracted, the lipids are then separated with bulk elution on a silicic acid column into the three fractions of neutral lipid, glycolipid and polar lipid. Each fraction is then transmethylated for analysis by GC/MS [3, 23].

2.1 A RAPID, POTENTIALLY AUTOMATABLE, EXTRACTION SYSTEM

"Flash" sequential extraction/fractionation of neutral lipids with supercritical CO_2 (SFECO$_2$), or with hexane utilizing accelerated solvent extraction (ASE) followed ASE extraction of polar lipids with chloroform methanol with no carryover [24, 25]. The neutral lipids after SFECO$_2$ or ASE are analyzed for respiratory quinones and, after derivatization, for diacylglycerol and sterols. ASE of polar phospholipids is then performed on the residue of the neutral lipid extraction with one-phase aqueous chloroform/methanol/aqueous buffer and deposited in a fraction collector for analysis. Next the residue of the two sequential extractions is split and a portion hydrolyzed in acid and the hydroxy fatty acids of the lipopolysaccharide (LPS-HFA) are extracted and analyzed by GC/MS. The other portion is derivatized *in situ* and the 2,6-dipicolinic acid (DPA) derivative extracted with the neutral lipids; intact polar lipids are analyzed by HPLC/ES/MS/MS.

2.1.1 SFECO$_2$ Extraction of Neutral Lipids
Nivens and Applegate working in this laboratory [26] have shown that CO_2 when rendered supercritical at temperatures > 31.1°C and pressures greater than 72.85 atm, extracted neutral lipids (sterols, diacylglycerols, respiratory quinones) and pollutants/contaminants like polynuclear aromatic hydrocarbons, pesticides, and

petroleum hydrocarbons and facilitated the recovery of DNA amplifiable by PCR. Neutral lipids are also quantitatively recoverable in hexane with ASE [25].

2.1.2 Accelerated Solvent Extraction of Polar Lipid

The second phase in the rapid recovery is to use the ASE of the residue after $SFECO_2$ cell lysis and neutral lipid extraction. The yield and lipid patterns are comparable to the classical Bligh and Dyer one-phase chloroform/methanol/buffer solvent extraction at room temperature and pressure [3, 23]. The Dionex ASE-200 Accelerated Extraction system with 2 cycles, 80° C and 1200 PSI produced recovery of 3 fold more PLFA from *Bacillus* sp. spores and 2-fold more PLFA from *Aspergillus niger* spores than did the standard one-phase extraction system [24]. Preliminary studies [23] showed that increases in pressure to 55 MPa at 80° C in a hand operated ISCO SFX apparatus further increased the recovery of lipids from *Bacillus* spores. The lipids were detectable after a minute exposure to high pressure/temperature solvent.

2.2 RECOVERY OF DPA AND LPS-HFA AFTER *IN SITU* DERIVATIZATION

The chelator, 2,6-dipicolinic acid (DPA), is specific for all known bacterial spores and is primarily responsible for their remarkable resistance to heat and dryness. A combined derivatization/extraction procedure that quantitatively forms the dipentafluoroproponyl ester of spore 2,6-DPA *in situ* and allows extraction has been developed [27]. The dipentafluoroproponyl ester of spore 2,6-DPA is sensitively detected by either HLC/ES/MS/MS or gas chromatography/negative ion chemical ionization mass spectrometry (GC/NICIMS) at levels of several spores.

2.3 HPLC/ES/MS/MS OF NEUTRAL LIPID

One of the primary considerations in the bioremediation or the immobilization of U and Pu contaminants is the ability to biologically remove high potential electron donors such as oxygen and nitrate, thereby decreasing the redox potential. This results in a reduction of the nuclides to insoluble compounds following interactions with metal- or sulfate-reducing bacteria or by geochemical processes. Estimating the terminal electron acceptor in the subsurface microbial microniche has proved especially difficult as the recovery of unaltered samples or the placement of microsensors from the subsurface has generally been unsatisfactory. Gram-negative bacteria form respiratory ubiquinones (UQ) when the terminal electron acceptor is oxygen [28]. Under anaerobic growth conditions Gram-negative bacteria may form respiratory naphthoquinones or no quinones. Gram-positive bacteria form respiratory naphthoquinones when grown aerobically. Respiratory ubiquinones are found at concentrations about 200 times lower than the PLFA or about 0.5 μmole/g dry weight [29]. Eukaryote mitochondria contain UQ-10 (80 carbon side chain). Gram-negative bacteria contain isoprenologues from UQ-4 to UQ-14 [30]. A new technology utilizing atmospheric pressure photo-ionization enables sensitive detection of both menaquinone, desmethylquinone and UQ isoprenologues with the HPLC/APPI/MS/MS at sensitivities of 1.5 fmol/μL, the equivalent of 10^4 exponentially growing aerobic *E. coli* [25, 31]. With the UQ isoprenologues the HPLC/ES/MS/MS analysis is about 10-times more sensitive [32].

8

Two other important neutral lipids require derivatization of their primary alcohol moieties to be charged and thus are good candidates for electrospray ionization. The electrospray ionization requires coulombic repulsion to generate the microdroplets containing a single molecular species and thus requires a charge to be present or be induced on the analyte molecule. Primary alcohols have been particularly difficult to electrospray successfully. Two alcohols, sterols and diacylglycerols are especially valuable indicators of the microbial community composition and physiological status. Sterols are valuable in assessing microeukaryotes. In work done in this laboratory analysis of sterols allowed the diets of filter feeding marine invertebrates to be assessed [10]. With cell death the bacterial phospholipids rapidly form diacylglycerols by the action of endogenous and exogenous phospholipases [1]. The ratio diacylglycerols/PLFA is a valuable indicator of conditions promoting cell lysis in the subsurface region[33]. Recently a derivatizing agent was developed for primary alcohols based on the formation of ferrocenoyl carbamates from ferrocenoyl azide that have sub-femtomolar sensitivities in ES/MS [34]. Fig. 1 illustrates the mass spectrum of the ferrocenyl derivative of cholesterol after HPLC/ESI/MS:

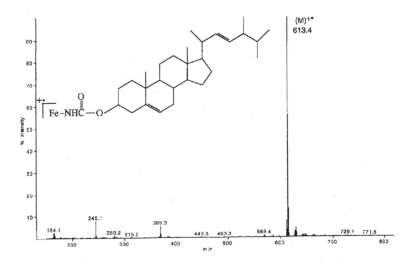

Figure 1. Mass spectrum of the ferrocenyl derivative of cholesterol after HPLC/ESI/MS.

The primary alcohol of the 1,2-diacylglycerols can also be derivatized with another charge-promoting moiety. This derivatization of uncharged alcohols based on the formation of ES-active *N*-methylpyridyl alcohol salt. The *N*-methylpyridyl salt is formed from the diacylglycerol by direct reaction with 2-fluoro-1-methylpyridinium *p*-toluenesulfonate [35]. The *N*-methylpyridyl ether salt will give an $(M + 92)^+$ (where M is the mass of the digacylglycerol or sterol) [35].

2.4 HPLC SEPARATION OF POLAR LIPIDS

There is a rich literature on the HPLC separation of lipids [36]. Unfortunately the columns and elution gradients used on normal-phase HPLC columns that give excellent separation of polar lipid classes are incompatible with the conditions for ES. For optimal ES the analytes must either be charged in solution, or readily ionized electrolytically in the ES capillary and the presence of nonvolatile ions in the eluent and certain solvents greatly compromises the robustness of the ES source. To provide a negative charge on the intact phospholipids for ready ionization the pH should be above 9.0. However, at this pH the silicic acid based column material used in the normal phase separation of polar lipids dissolves. Use of a post column addition of ammonium acetate (which is volatile) achieves the pH but generates an emulsion that compromises the ES source. The use of polar solvents with chloroform greatly depressed ionization in the ES inlet. Therefore, reverse phase HPLC with a resistant C-18 column material provided base line separation of intact polar lipids in a methanol solvent with 0.002 % piperidine as base with a post column addition of 0.2% piperidine in methanol. This has proved a significant advance in providing rapid (< 8 min) separation of intact polar lipids with ES compatible solvents isocratically with no re-equilibration [37].

2.4.1 Bioavailable Phosphate

Other phospholipids provide significant insight into the nutritional/physiological states of the microbial communities. Bioavailable phosphate has been a particularly difficult parameter to estimate [38] and a wide variety of extractants have been utilized to quantify soil phosphorus status. These methods vary in chemical aggressiveness. Methods utilizing water, acid or alkali do not accurately reflect the bioavailable P as reflected in plant productivity. Estimates of mineralization have utilized a sequence of bicarbonate extracts over time to gain a truer insight into the extent of mineralization and immobilization processes [39, 40]. An alternative to chemical extractants is the application of isotopic techniques. A long series of experiments are reviewed in Frossare et al, [41] who came to the conclusion that great caution should be taken before extrapolating the quantity of isotopically exchangeable phosphate to isotopic exchange times longer than a minute. Suffice it to say the estimation of bioavailable P in soils and sediments is highly problematic. The HPLC/ES/MS can provide a biological sensor for phosphate bioavailability. Sedimentary universally distributed Gram-negative soil organisms like *Pseudomonas* form acyl-ornithine lipids when grown with limited bioavailable phosphate [42]. The acyl-ornithine lipids replace phospholipids as the predominating polar lipids in the cell membranes.

2.4.2 Amino Acid PG Esters

Another phospholipid class lipid is the lysyl ester of PG that is formed by micrococci when grown at acid pH [43]. Both the acyl ornithine lipids and lysyl-PG should be readily analyzed by HPLC/ES/MS/MS and the separation conditions are currently being developed in our laboratory.

2.4.3 Archaea

Another important group of organisms are the Archaea. The Archaea contain isoprenoid glyceroldialkyl diethers (GDE) with an archaeol core or diglycerol dialkyl glycerol tetraethers (GDGT) with a caldarcheol core and a large variety of carbohydrate, phosphate or both as polar head groups [44]. There are considerably more complex intact polar lipid components in archeal isolates than in eubacterial isolates so the polar head groups are usually removed by hydrolysis and the glycerol diether or bi-diethers analyzed. Some hydroxylated ethers can be lost in this hydrolysis. The ether cores of GDE and GDGT characteristic of the Crenarchaeota are analyzed by HPLC/atmospheric pressure chemical ionization(APCI)/MS/MS [45].

2.4.4 Extraction Recovery, Acid Methanolysis/Derivatization of Poly β-hydroxy-alkonate (PHA)

The detection of the bacterial polyester storage product is a measure of bacterial unbalanced growth which occurs when bacteria have an adequate carbon and terminal acceptor supply but lack some essential trace nutrient such as phosphate, nitrogen or trace elements [12]. The most sensitive and specific assay for PHA, developed in this laboratory, unfortunately requires hydrolysis of the polyester with derivatization to a volatile ethyl ester for GC/MS [46]. Unless great care is taken the esterification often leads to losses. In addition, PHA in its most common form is remarkably insoluble and maximal recovery at room pressure requires boiling chloroform. PHA precipitates on glass when the solvent cools and the transfer to the GC/MS is often not quantitative. Preliminary experience has shown that much more PHA is recovered by the "flash" high temperature/pressure extraction method proposed herein. We propose *in situ* transesterification to form the isopropanol ester of the β-hydroxyacid monomer subunits of the PHA in the $SFECO_2$ chamber prior to "flash high-temperature/high-pressure extraction. We propose to transesterify the PHA with isopropanol, recover the β-hydroxyalkanyl isopropanol ester and then derivatize the primary alcohol with the ester directly with 2-fluoro-1-methylpyridinium *p*-toluenesulfonate to form the ES active *N*-methylpyridyl ester salt. The *N*-methylpyridyl ether salt will give an $(M +92)^+$ (where M is the mass of the β-hydroxy acid monomers of the PHA hydroxyalkanyl isopropyl ester) derivative [35]. Measuring PHA is particularly important as we have shown that bacteria in the rhizosphere attached to fine roots show no evidence of unbalanced growth (PHA/PLFA < 0.02), whereas bacteria from the surrounding soil show evidence of unbalanced growth (PHA/PLFA >0.1), [47].

2.4.5 ES/MS/MS Analysis of Polar Lipids

The HPLC/ES/MS/MS greatly facilitates ultrasensitive analysis of intact polar lipids. In situations where ultrasensitive analysis of phospholipids is required it is possible to analyze the PG and PE and ignore the PC. This is important as nearly all the trace contaminants in chromatographic-grade redistilled solvents are in PC of human origin.

2.4.6 ES/MS/MS

Tandem mass spectrometry greatly increases the sensitivity and specificity of this environmental analysis system without any increase in the analysis time [48-51]. Neutral loss scans are particularly useful for the rapid screening of targeted lipid ions as demonstrated by Cole and Enke [51] who showed that the PE and PG could be readily detected using a neutral loss of m/z = 141 and 154, respectively, by FAB. The neutral loss scan for PE can be performed at LC/ES/MS/MS in the positive mode, whereas PG has to be done by Precursor ion scan of m/z = 153 in negative mode. In the negative ion mode, it proved possible to detect the position of each ester-linked fatty acid, *i.e.* the fatty acid component at the *sn*1 position had 20 % of the abundance of the fatty acid at the *sn*2 position of the glycerol. Tandem mass spectrometry, ES/MS/MS, provides great advantages in the structural analysis of phospholipids. Product ion spectra and multiple reaction monitoring (MRM) were used to investigate PG containing a 16:0 fatty acid at the *sn*1 position and an 18:0 fatty acid at the *sn*2 position. Figure 2A shows the ES spectrum for 1 ppm PG when scanning from m/z = 110-900; Fig. 2B shows the product ion spectra for m/z = 747 with collision-induced dissociation (CID) in the second quadrupole in the presence of 0.37 Pa Ar yielding the product ions that were analyzed by scanning over m/z = 110-900 in the third quadrupole. MS/MS decreases chemical noise in the product ion spectra thereby increasing the signal-to-noise (s/n) ratio and the resulting sensitivity. Comparing the s/n ratio of the up-front CID of the *sn*1 and *sn*2 fatty acids in the upper panel of Fig. 2 A to the product ion spectra from CAD in the second quadrupole, the product ion spectrum for PG is more sensitive by a factor of 50 (Fig. 2 B lower panel). By scanning over a narrower range, sensitivity can be increased. When acquiring product-ion spectra, the LOD and LOQ were experimentally determined to be 446 amol/μL and 1.3 fmol/μL, respectively. If the third quadrupole is scanned for the m/z = 281 product of m/z = 745, a roughly 5-fold additional gain in sensitivity was achieved. This is multiple reaction monitoring (MRM), m/z = 747 → m/z = 281 and represents the most sensitive application of this HPLC/ES/MS/MS.

We have demonstrated limits of detection of 90 amol/μL using a single molecular species of phosphatidyl glycerol, [sn1 palmitic acid m/z = 255, sn2 oleic acid 18:1 m/z = 284, parent ion m/z = 747] as the parent ion selected CID by monitoring m/z = 284 negative ions. That is essentially equivalent to the total phospholipid in a single *E. coli* cell!

In the positive ion mode, intact molecular ions are seen [51]. The introduction of ethyl vinyl ether into the Q-2 chamber at a pressure of 3.5–4.0 mTorr resulted in the addition of m/z 26, 31, 44, 57 and/or 72 to the polar head group positive ions of the phospholipids. Each phospholipid class resulted in a unique neutral gain upon reaction with ethyl vinyl ether due to differences in the polar head groups. For some phospholipids these neutral gain positive ion spectra can sometimes be more intense than the with the neutral loss negative ion spectra [51]. The CID with Ar in Q$_2$ readily forms negative ions in MS/MS but to determine the position of unsaturation, branching and other structures transmethylation and GC/MS are much more specific.

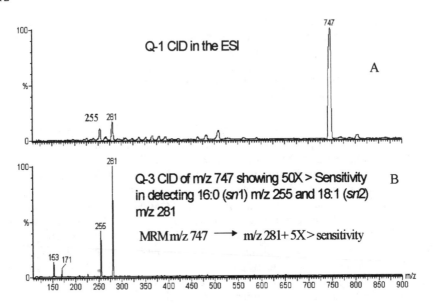

Figure 2. A. ES spectra for 1 ppm PG 16:0/18:1 when scanning from m/z = 110-900. B. ES production spectra for m/z = 747 of 1 ppm PG 16:0/18:1 when scanning from m/z = 110-900. This panel shows the fragmentation of the molecular ion at m/z = 747 detected in Q-1 (panel A) in the CID chamber with the negative ion spectra of the constituents fatty acids 16:0 at m/z 255- from sn1 and, 18:2 at m/z 281- from sn2, and the head group at m/z = 153-, 171- with an intensity > 50 times that seen in Q-1. Using MRM of m/z 747- on Q-1 'm/z = 281 further increases the intensity > 5-folds.

Such details of the phospholipid fatty acid structure including position and chirality of methyl branching, unsaturation, hydroxylation, cyclopropyl ring formation all have great significance for identification and physiological status [3, 15].

3. Expanded Lipid Biomarker Analysis

Figure 3 represents a diagram of the possibilities for lipid analysis. We have included the several GC/MS analyses as they continue to be useful. They include standard PLFA by GC/MS that has been expanded by additional steps to provide dimethyl acetals derived from plasmalogens by mild acid methanolysis (which indicate an anaerobic environment and components of other lipids released from lipids by strong acid hydolysis in a relatively facile highly useful analysis [52].

The sequential "flash" extraction begins with 1. The ASE/(hexane) or SFECO$_2$ extraction of the neutral lipid. The respiratory benzoquinones are analyzed directly and the diglycerides, and sterols derivatized with charged derivatives. 2. ASE recover the polar lipids. A portion of the intact lipids would be separated by capillary HPLC for ES/MS/MS analysis. A second portion of the polar lipid would be transesterified and the methyl esters of the PLFA analyzed by GC/MS, mild acid methanolysis to

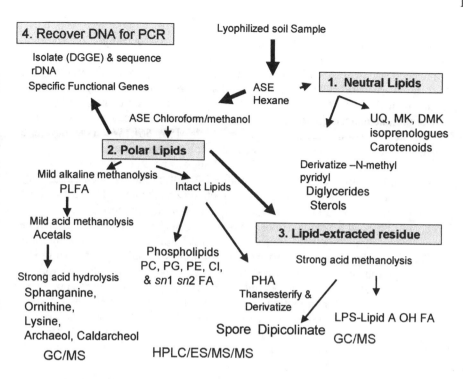

Figure 3. Schematic for expanded lipid biomarker analysis.

release the vinyl ethers of the plasmalogens as dimethyl acetals, and a variety of amino acids carbohydrates, archeols and sphanganine for GC/MS. A third portion of the polar lipids would be transesterified under high pressure and temperature and the hydroxyl derivatized for analysis of the PHA monomers by HPLC/ES/MS/MS. 3. The extracted residue is derivatized *in situ* and the 2,6-dipicolinic acid recovered for analysis of bacterial spores, and the ester linked 3-hydroxy fatty acids from the lipopolysaccaharides analyzed rendered extractible for analysis by GC/MS or HPCL/ES/MS/MS. The presence of tannins in soils could complicate interpretation of hydroxy fatty acid analysis after the derivatization.

4. Interpretation

The expanded LBA allows insight not only into the community composition and viable biomass but also into the physiological status of the microbial community. From the analysis of the community physiological status the microniche environment can be ascertained and metabolic activities predicted. The most difficult measurements in microbial ecology are estimations of *in situ* metabolic activity. The problem is the determination of metabolic activity of a community poised for rapid acquisition of any resources presented by disturbance. We have shown that LBA provided a quantitative means to measure the effects of disturbance in sediments where an aerobic surface overlies anaerobic sediment rich in reduced substrates. Disturbance introduces aerated water to the anaerobic sediment and a burst of activity ensues with rapid growth until a critical nutrient again limits cell division. However, carbon accumulation continues without cell division and new cell-membrane synthesis. This was readily detectable as an initial decrease in the ratio between the rates of synthesis of PHA and PLFA followed by a rebound with PHA accumulation without PLFA synthesis. The measurement of the PHA/PLFA ratio has been shown to be an exquisitely sensitive measure of the unbalanced growth engendered by disturbance in sediments [53, 54] or level of impact by pollutants on stream periphyton [55, 56]. We have developed systems in which microniche conditions and implied metabolic activity could be directly observed without disturbance artifacts. This was based on the hypothesis that responses of monocultures to specific conditions will predict responses of these organisms in communities to the same specific conditions. This requires a biofilm to be generated on a substratum in a flow-through in-line monitoring of microbial biofilm communities in a device where the biofilm is subjected to controlled laminar flow sufficient to remove pelagic microbes and provide nutrients from a dilute bulk phase. The biofilm is generated on a substratum in a flow-through system that models the slimes of periphyton on rocks in a stream. If the system is outfitted with quartz windows that can be cleaned it is possible to utilize tryptophan fluorescence of the bacterial proteins to monitor formation and biomass of the biofilm [57]. The metabolic status of naturally bioluminescent *Vibrio harveyi* can be followed by comparing the fluorescence yield per cell of monocultures to a series of sublethal impacts [58]. With the system it proved possible to clearly correlate the changes in metabolism to the antifouling and fouling-release properties of coatings [58].

With LBA not only can metabolic activities be predicted but also the longer-term responses by the total community used in estimates of bioavailability and as ecologically defensible end-points in remediation [59]. The rapid responses of periphyton to pollution abatement in surface run-off waters [55, 56] seem to be matched by responses to hydrocarbon degradation and putative return of communities to the uncontaminated status of the aquifer up-gradient of the pollution source [60].

5. Predictions

Based on expanded LBA and judicious use of the PCR of rDNA with DGGE and sequence analysis for identification of specific organisms in communities where dominance of less than 50 "species units" is sufficient to make it practical, the following predictions of consequences of microniche environments and thus metabolic activities can be made:

1. The microniche redox potential and terminal electron acceptor will be correlated to the respiratory ubiquinone/menaquinone or ubiquinone/PLFA ratio. Aerobic microniches will be found with high respiratory ubiquinone/PLFA ratios, the absence of plasmalogens, and phosphatidyl glycerol (PG) with ratios of $i15:0^*$ to $a15:0 < 0.1$

---

* Fatty acids are defined as the number of carbon atoms in the longest chain: the number of double bonds with the position of the first double bond counting from the hydrocarbon end (omega (ω) end) of the carbon chain. Double bond conformation is indicated by the suffix c for *cis* or t for *trans*. The prefix indicating methyl branching is Me. The prefix number indicates by the position of the mid chain methyl branching or br if not known. Terminal methyl branching is i for *iso* and a for *anteiso* for 2 carbons from the hydrocarbon terminal. Cyclopropane rings are indicated by the prefix cy.

characteristic for the aerobic Gram-positive cocci and the presence of actinomycete biomarkers (high 10Me17:0 and 10Me18:0). Straight-chain sphinganines from *Sphingomonas* are found in aerobic microsites [15, 32, 52].

2. Anaerobic microniches will be correlated to the plasmalogen/PLFA ratio (plasmalogens are characteristic for *Clostridium* spp.) and the isoprenoid ether lipids of the methanogenic Archaea. In the presence of sulfate the anaerobic microniches will have a ratio of i15:0 to a15:0 > 2.0 and of $17:1\omega7c$ in the phosphatidyl ethanolamine (PE) (characteristic of Gram-negative *Desulfovibrio*) and high relative proportions of 10Me16:0 (characteristic of sulfate/metal reducing bacteria) [15, 32].

3. Microniches with low pH will have elevated concentrations of lysyl esters of PG characteristic of Gram-positive *Micrococcus*-type bacteria; microniches with low bioavailable phosphate will show high ratios of acyl ornithine polar lipids characteristic of Gram-negative bacteria such as *Pseudomonas* [15, 42, 43].

4. Microniches with suboptimal growth conditions (low water activity, missing nutrients or trace components) will show high (>1) cyclopropane to monoenoic fatty acid ratios in phosphatidylglycerol (PG) and phosphatidylethanolamine (PE). They will show greater ratios of cardiolipin (CL) to PG ratios [14].

5. Microniches will show lower viable biomass but increased activity under conditions inducing high levels of microeukaryote predation as indicated by higher proportions of phospholipid polyenoic fatty acids in phosphatidylcholine (PC) and cardiolipin (CL) as well as in sterols. Viable biomass (total PLFA) will decrease but there will be an increase in the mineralization/unit viable biomass as we have shown for amphipods grazing detrital periphyton [61, 62].

6. Reductive process will increase with increases in the microniches with non-growth-limiting carbon sources and terminal electron acceptors but with limiting nitrogen or trace growth factors as reflected in PHA/PLFA >0.8. Conditions favoring cometabolic fortuitous biodegradation of refractory compounds are associated with very high PHA/PLFA ratios (>10) [63].

7. In microniches with suboptimal growth conditions (low water activity, missing nutrients or trace components), high (>1) cyclopropane to monoenoic fatty acid ratios in the PG, PE and PLFA, as well as greater ratios of cardiolipin (CL) to PG ratios will be detected [64].

8. Conditions of maximal fine root proliferation will mirror the bacterial viable biomass (PLFA) and evidence of balanced growth (PHA/PLFA <0.06), whereas conditions of fine root dieback will be associated with a much slower loss of viable bacterial biomass and higher fungal biomass expressed as increases in ergosterol and $18:2\omega6$ in PC and UQ-10 vs. bacteria which contain no sterols, little PC, $16:1\omega7c$, br 15:0 and other bacterial PLFA in PG and PE, with the range of UQ-4-13 [47].

9. Microniches with surfactants high bacteriophage multiplicities or other cell-lysis conditions can be detected as increased diacylglycerol/PLFA. Desiccation and suboptimal growth will be detected as cyclopropane/monoenoic precursors >0.1 in the PLFA or PG/PE [15, 64].

10. Microniches exposed to toxic solvents will result in ratios of *trans* to *cis* PLFA ($16:1\omega7t/16:1\omega7c$) > 0.05 . As the bacteria make membranes less permeable, they produce membranes with higher relative levels of *trans*-PLFA most prominently in PE. Gram-negative bacteria reversibly make *trans*-fatty acids as a result of changes in the environment, usually as a result of stress (*i.e.* toxicity or starvation) [15].

11. Microniches with high activity in anaerobic iron reduction contain α- and ω-diacids, dioics-chemical species as well as PLFA in PE and PG typical of sulfate-reducing bacteria.

12. Microniches exposed to oxidative biocides or reactive agents show epoxide (oxirane) PLFA in PE in conditions, *e.g.*, in mine service water and wastewater treatment facilities. They also occur in the fungal rust parasite *Pneumocystis carni*. The presence of oxirane PLFA results in Gram-negative bacteria being rendered not culturable [65].

13. Microniches exposed to high temperatures show ω-cyclohexane PLFA and short chain (<14 carbon) PLFA. Average PLFA chain length and average degree unsaturation (ACL & ADU) has been shown to increase with soil temperature but may also be related to substrate availability.

14. Metabolic stress can be reflected in iso10me16:0/i17:1ω7c in Gram-positive bacteria in PG. This indicator is similar to the cyclopropyl to monoenoic PLFA ratio in Gram-negative bacteria.

Acknowledgements

This work was supported by National Science Foundation grant DEB 9814813, EAR 9714215, and EAR 9978161 together with the Department of Energy, Office of Science, Office of Biology and Environmental Research, grant number DE-FC02-96ER62278 (Assessment Component of the Natural and Accelerated Bioremediation Research Program, NABIR).

References

1. White, D.C., Davis, W.M , Nickels, J.S., King, J.D. and Bobbie, R.J.. (1979)   Determination of the sedimentary microbial biomass by extractable lipid phosphate. *Oecologia* **40**, 51-62.
2. Balkwill, D.L., Leach, F.R., Wilson, J.T., McNabb, J.F. and White, D.C. (1988) Equivalence of microbial biomass measures based on membrane lipid and cell wall components, adenosine triphosphate, and direct counts in subsurface sediments. *Microbial Ecology* **16**, 73-84.
3. White, D.C., Stair, J.O. and Ringelberg, D.B. (1996) Quantitative Comparisons of *in situ* Microbial Biodiversity by Signature Biomarker Analysis. *J. Indust. Microbiol.* **17**, 185-196.
4. Dowling, N.J.E., Widdel, F. and White, D.C. (1986) Phospholipid ester-linked fatty acid biomarkers of acetate-oxidizing sulfate reducers and other sulfide forming bacteria. *J. Gen. Microbiol.* **132**, 1815-1825.
5. Edlund, A., Nichols, P.D., Roffey, R. and White, D.C. (1985) Extractable and lipopolysaccharide fatty acid and hydroxy acid profiles from *Desulfovibrio* species. *J. Lipid Res.* **26**, 982-988.
6. Welch, D.F. 1991. Applications of cellular fatty acid analysis. *Clinical Microbiology Reviews* **4**, 422-438.
7. Guckert, J.B., Ringelberg, D B., White, D C., Henson, R. S. and Bratina, B. J. (1991). Membrane fatty acids as phenotypic markers in the polyphasic taxonomy of methylotrophs within the proteobacteria. *J. Gen. Microbiol.* **137**, 2631-2641.
8. Kohring, L.L., Ringelberg, D.B., Devereux, R., Stahl, D.A., Mittelman, M.W. and White, D.C. (1994) Comparison of phylogenetic relationships based on phospholipid fatty acid profiles and ribosomal RNA sequence similarities among dissimilatory sulfate-reducing bacteria. *FEMS Microbiol. Letters.* **119**, 303-308.
9. White, D.C., Ringelberg, D.B., Guckert, J.B. and Phelps, T. J. (1991).. Biochemical markers for *in situ* microbial community structure. in: Fliermans, C.B. and Hazen, T.C. (Eds.), *Proceedings of the First International Symposium on Microbiology of the Deep Subsurface*. WSRC Information Services, Aiken, SC. pp. 4-45 to 4-56.
10. Canuel, E.A., Cloen, J.E., Ringelberg, D.B., Guckert, J.B.and Rau G.H.. (1995) Molecular and isotopic tracers used to examine sources of organic matter and its incorporation into the food webs of San Francisco Bay. *Limnol Oceanogr.* **40**, 67-81.
11. Parker, J.H., Smith, G.A., Fredrickson, H.L., Vestal, J.R. and White, D.C. (1982) Sensitive assay,

based on hydroxy-fatty acids from lipopolysaccharide lipid A for gram negative bacteria in sediments. - *Appl. Environ. Microbiol.* **44**, 1170-1177.

12. White, D.C. , Ringelberg, D.B. and Macnaughton S.J. (1997) Review of PHA and signature lipid biomarker analysis for quantitative assessment of *in situ* environmental microbial ecology. In (Eggink, G., Steinbuchel, A., Poirer, Y. and Witholt, B eds.) *1996 International Symposium on Bacterial Polyhydroxylalkanoates,* NRC Research Press, Ottawa, Canada, pp. 161-170

13. Gehron, M.J. and White, D.C. (1982) Quantitative determination of the nutritional status of detrital microbiota and the grazing fauna by triglyceride glycerol analysis. *J. Exp. Mar. Biol.* **64**, 145-158.

14. Guckert, J.B., Hood, M.A. and White, D.C. (1986) Phospholipid, ester-linked fatty acid profile changes during nutrient deprivation of *Vibrio cholerae*: increases in the *trans/cis* ratio and proportions of cyclopropyl fatty acids. *Appl. Environ. Microbiol.* **52**, 794-801.

15. White, D.C. (1995) Chemical ecology: Possible linkage between macro-and microbial ecology. *Oikos* **74**, 174-181.

16. Ward, D.M., Bateson, M.M.,Weller, R. and Ruff-Roberts, A.L. (1992) Ribosomal RNA analysis of microorganisms as they occur in nature. *Adv. Microbial Ecology* **12**, 219-286.

17. Kehrmeyer, S.R., Appelgate, B.M., Pinkert, H C., Hedrick, D.B., White, D.C. and Sayler, G.S. (1996) Combined lipid/DNA extraction method for environmental samples. *J. Microbiol. Methods* **25**, 153-163.

18. Stephen, J.R., Chang, Y-J., Gan, Y.D., Peacock, A., Pfiffner, S.M, Barcelona, M. J, White D.C., and Macnaughton S. J. (1999) Microbial Characterization of JP-4 fuel contaminated-site using a combined lipid biomarker/PCR-DGGE based approach. *Environmental Microbiology.* **1**, 231-241.

19. White, D.C., Macnaughton, S.J., Stephen, J.R., Almeida, J.A., Chang, Y-J., Gan, Y-D and Peacock A.. (2000) The in-situ assessment of microbial community ecology within samples from Chromium and Uranium contaminated Sites. DOE- NABIR workshop Abstracts. Reston, VA January 31-February 2, Pp. 32

20. Macnaughton, S. J., Marsh, T. L., Stephen, J. R., Long, D., Gan, Y-D., Chang, Y-J., Almeida, J. A., and White, D.C. (1999) Lipid biomarker analysis of bacterial and eukaryote communities of chromium contaminated soils. *Abst. Am Soc. Microbiol.* 1999

21. Macnaughton, S. J. (2000) Community dynamics presentation, NABIR workshop Abstracts. Reston, VA January 31-February 2, Pp. 3.

22. Brockman, F., Kieft, T. and Balkwill, D.. (2000) Vadose zone microbial community structure and activity in metal/radionuclide-contaminated Sediments. NABIR workshop Abstracts. Reston, VA January 31-February 2, Pp. 25.

23. Ringelberg, D.B., Davis, J.D., Smith, G.A., Pfiffner, S.M., Nichols, P.D., Nickels, J. B., Hensen, J.M., Wilson, J.T., Yates, M., Kampbell, D.H., Reed, H.W., Stocksdale, T.T and White D.C.. (1988) Validation of signature polarlipid fatty acid biomarkers for alkane-utilizing bacteria in soils and subsurface aquifer materials. *FEMS Microbiol. Ecology* **62**, 39-50.

24. Macnaughton, S.J., Jenkins, T.L., Wimpee, M.H., Cormier, M.R. and White, D.C. (1997) Rapid extraction of lipid biomarkers from pure culture and environmental samples using pressurized accelerated hot solvent extraction. *J. Microbial Methods* **31**, 19-27.

25. Geyer, R., Bittkau, A., Gan, Y-D., White, D. C., and Schlosser, D. 2002. Advantages of lipid biomarkers in the assessment of environmental microbial communities in contaminated aquifers and surface waters. Proceed. Third International Conference on Wate Resources and Environmental Research (ICWRER) Dresden Germany.

26. Nivens, D.E., and Applegate, B.M. (1997) Apparatus and methods for nucleic acid isolation using supercritical fluids. *U. S. Patent #* 5922536 issued July 13, 1999.

27. Cody, R. B., J. A. Laramar, G. L. Samuelson, E. G. Owen 2002. Application of the JEOL tunable-Enengy electron monochromoter: Bacterial spores, BTX and Biowepons. 50th ASMS Conference Proceedings poster 298,TPX, Orlando FL.

28. Hedrick, D.B., and White, D.C. (1986) Microbial respiratory quinones in the environment I. A sensitive liquid chromatographic method. *J. Microbiol. Methods* **5**, 243-254.

29. Hollander, R., Wolf, G. and Mannheim, W. (1977). Lipoquinones of some bacteria and mycoplasmas, with considerations of their functional significance. *Antonie van Leeuwenhoek* **43**, 177-185.

30. .Collins, M.D., and Jones, D. (1981) Distribution of isoprenoid quinone structural types in bacteria and their taxonomic implications. *Microbiol. Rev.* **45**: 316-354.

31. Lytle, C.A., G.J. Van Berkel, and D.C. White (2001) Comparison of Atmospheric Pressure Photoionization and Atmospheric Pressure Chemical Ionization for the Analysis of Ubiquinones and Menaquinones, 49th American Society for Mass Spectrometry Meeting Proceedings Chicago, IL., TPC 074, May 27-31

32. Lytle, C.A., Y-D. M. Gan, K. Salone, and DC. White (2001) Sensitive Characterization of Microbial Ubiquinones from Biofilms by Electrospray/Mass Spectrometry Environ. Microbiol. **3** (4): 265-272.

18

33. Ringelberg, D.B., Sutton, S. and White, D.C. (1997) Biomass bioactivity and biodiversity: microbial ecology of the deep subsurface: analysis of ester-linked fatty acids. *FEMS Microbiology Reviews* **20**, 371-377.

34. Van Berkel, G.J., Quirke, J.M., Tigani, R.A. and Dilley, A.S.(1998) Derivatization for electrospray ionization mass spectrometry. 3. Electrochemically ionizable derivatives. *Ann. Chem.* **70**: 1544-1584.

35. Quirke, J.M.E., Adams, C.L.and Van Berkel, G.J. (1994) Chemical derivatization for electrospray-ionization mass spectrometry. 1. alkyl-halides, alcohols, phenols, thiols, and amines. *Ann. Chem.* **66**, 1302-1315.

36. Christie, W.W. (1996) Separation of phospholipid classes by high-performance liquid Chromatography, in (Christie, W.W., Ed.), *Advances in Lipid Methodology*. The Oily Press, Dundee, Scotland, Vol. 3 Pp. 77-107.

37. Lytle, C.A, Gan, Y-D. and White, D.C. (2000) Electrospray Ionization Compatible Reversed Phase Separation of Phospholipids: Incorporating Piperidine As A Post Column Modifier for Negative Ion Detection by Mass Spectrometry. *J. Microbiol. Methods* 41: 227-234.

38. Sadusky. M.C. and Datta, S.K.DE. (1991) Chemistry of P transformations in soil. *Adv. In Soil Sci.* **16**: 1-120.

39. Condon, L.M., Tissen, H., Trasar-Cepeda, C., Noir, J. and Stewart, J.W.B. (1993) Effects of liming on organic matter decomposition and phosphorous extractability in an acid humic ranker soil from northwest Spain. *Biology and Fertility of Soils* **15**, 279-284.

40. Zhang, Y.S., Werner, W., Scherer, H.W. and Sun, X.. (1994) Effect of organic manure on organic phosphorous fractions in two paddy soils. *Biology and Fertiltity of Soils* **17**, 64-68.

41. Frossard, E., Lopez-Hernandez, D. and Brossard, N. (1996) Can isotopic exchange kinetics give valuable information on the rate of mineralization of organic phosphorous in soils? *Soil Biol. Biochem.* **28**, 857-864.

42. Minnikin, D.E. and Abdolrahimzadeh, H. (1974) The replacement of phosphatidylethanlolamine and acidic phospholipids by ornithine-amid lipid and a minor phosphorus-free lipid in *Pseudomonas fluorescence* NCMB129. *FEMS Letters* **43**, 257-260.

43. Lennarz, W.J. (1970) Bacterial lipids. In Wakil, S. (ed) *Lipid Metabolism*. Academic Press, New York, NY, pp. 155-183.

44. Kates, M. (1993) Membrane Lipids of archea. IN The biochemistry of Archea (Archaebacteria), M.Kates, D. J. Kushner, and A. T. Matheson (eds), Chapter 9, 261-296, Elsevvier, New York.NY.

45. Hopmans, E.C., S. Schouten, R.D Pancost, M.T.J. van der Meer, and J.S. Sinnlughe Damste. (2000) Analysis of intact tetraether lipids in archeal cell material and sediments by high performance liquid chromatography/atmospheric pressure chemical ionization mass soectrometry. *Rapid Comm. Mass Spectrometry* 14: 585-589.

46. Findlay, R.H., and White, D.C. (1983) Polymeric beta-hydroxyalkanoates from environmental samples and *Bacillus megaterium*. *Appl. Environ. Microbiol.* **45**, 71-78.

47. Tunlid, A., Baird, B.H., Trexler, M.B., Olsson, S., Findlay, R.H., Odham, G. and White, D.C. (1985). Determination of phospholipid ester-linked fatty acids and poly beta hydroxybutyrate for the estimation of bacterial biomass and activity in the rhizosphere of the rape plant *Brassica napus* (L.). *Canad. J. Microbiol.* **31**, 1113-1119.

48. Fang, J. and Barcelona, M.J. (1998) Structural determination and quantitative analysis of bacterial phospholipids using liquid chromatography/electrospray ionization/mass spectrometry. *J. Microbiol. Methods* **33**, 23-35.

49. Smith, P.B.W., Snyder, A. P. and Harden, C.S. (1995) Characterization of bacterial phospholipids by electrospray ionization tandem mass spectrometry. *Anal. Chem.* **67**, 1824-1830.

50. Le Quere, J.L. 1993. Tandem mass spectrometry in the structural analysis of lipids, in Christie, W.W (Ed.), *Advances in Lipid Methodology*. The Oily Press, Dundee, Scotland, Vol 2 pp. 215-245.

51. Cole, M. J. (1990) Application of triple quadrupole mass spectrometry to the direct determination of microbial biomarkers and the rapid detection and identification of microorganisms. PhD Dissertation, Department of Chemistry, Michigan State University, E. Lansing, MI.

52. Lane, J.R. W.R. Mayberry, and D.C. White. 2002. A method for the sequential fractionation and derivatization of fatty acids, aldehydes and long-chain bases from diester lipids, plasmalogens and sphingolipids in lipid extracts . in review.

53. Findlay, R.H.,Trexler, M.B.,Guckert, J.B. and White, D.C. (1990) Laboratory study of disturbance in marine sediments: response of a microbial community. *Marine Ecological Progress Series* **61**,: 121-133.

54. Findlay, R.H.,Trexler, M.B. and White, D.C. (1990). Response of a benthic microbial community to biotic disturbance. *Marine Ecology Progress Series*. **61**, 135-148.

55. Guckert, J.B., Nold, S.C., Boston, H.L. and White, D.C. (1991). Periphyton response along an industrial effluent gradient: Lipid-based physiological stress analysis and pattern recognition of microbial community structure. *Canad. J. Fish. Aquat. Sci.* **49**, 2579-2587.

56. Napolitano, G.E., Hill, W.R., Stewart, J.B., Nold, S.C. and White, D.C. (1993) Changes in periphyton fatty acid composition in chlorine-polluted streams. *J. N. Amer. Benthol. Soc.* **13**, 237-249.

57. Angell, P., Arrage, A., Mittelman, M.W and White, D.C. (1993) On-line, non-destructive biomass determination of bacterial biofilms by fluorometry. *J. Microbiol. Methods* **18**, 317-327.

58. Arrage A. A., Vasishtha, N., Sunberg, D., Baush, G., Vincent, H. L. and White, D. C (1995) On-line monitoring of biofilm biomass and activity on antifouling and fouling-release surface using bioluminescence and fluorescence measurements during laminar-flow. *J. Indust. Microbiol.* **15**, 277-282.

59. White, D.C, Flemming, C.A., Leung, K.T. and Macnaughton, S.J. (1998) *In situ* microbial ecology for quantitative appraisal, monitoring, and risk assessment of pollution remediation in soils, the subsurface and in biofilms. *J. Microbiol. Methods* **32**, 93-105.

60. Macnaughton, S.J., Stephens, J.R., Venosa., A.D.,. Davis, G. A, Chang, Y-J. and White, D. C. (1999). Microbial population changes during bioremediation of an experimental oil spill. *Appl. Environ. Micobiol.* **65**, 3566-3574.

61. Smith, G.A., J.S. Nickels, W.M. Davis, R.F. Martz, R.H. Findlay, and D.C. White. (1982) Perturbations of the biomass, metabolic activity, and community structure of the estuarine detrital microbiota: resource partitioning by amphipod grazing. *J. Exp. Mar. Biol. Ecol.* **64**: 125-143.

62. Morrison, S.J., and D.C. White (1980) Effects of grazing by estuarine gammaridean amphipods on the microbiota of allochthonous detritus. *Appl. Environ. Microbiol.* **40**: 659-671.

63. Nichols, P.D., and White, D.C. (1989) Accumulation of poly-beta-hydroxybutyrate in a methane-enriched halogenated hydrocarbon-degrading soil column: implications for microbial community structure and nutritional status. *Hydrobiologia* **176/177**, 369-377.

64. Kieft, T. L. E. Wilch, K. O'Connor, D. B. Ringelberg, and D. C. White. 1997. Survival and phospholipid fatty acid profiles of surface and subsurface bacteria in natural microcosms. *Appl. Env. Microbiol.* **63**: 1531-1542.

65. Smith, C.A, Phiefer, C.B., Macnaughton, S. J., Peacock, A., Burkhalter, R.S., Kirkegaard, R. and White, D.C. (1999) Quantitative lipid biomarker detection of unculturable microbes and chlorine exposure in water distribution system biofilms. *Water Res.* **34**: 2683-2688.

COMPARISON OF EFFECTIVE ORGANISMS IN BIOREMEDIATION PROCESSES: POTENTIAL USE OF NUCLEIC ACID PROBES TO ESTIMATE CYANIDE DEGRADATION *IN SITU*

M. BARCLAY [1,2], A. TABOURET [1], J.C. DAY [1],
I.P. THOMPSON [1], C.J. KNOWLES [2] and M.J. BAILEY [1]

[1] *NERC IVEM Centre for Ecology and Hydrology Oxford, Mansfield Rd., Oxford, OX1 3SR*

[2] *Department of Engineering Sciences, University of Oxford, Parks Road, Oxford, OX1 3PJ*

Abstract

Contamination of the environment with cyanide and its metal complexes is a major environmental problem. Fungi that degrade cyanide and metal complexes through induction of the enzyme cyanide hydratase were studied. The screening of six *Fusarium* strains demonstrated that cyanide degrading activity was inducible, a trait that is possibly common to the genus. Following sequence alignment probes were designed for cyanide hydratase (*chy*) allowing the application of molecular probes to amplify the gene (DNA), and amplification of mRNA by RT-PCR to measure degradative potential. DNA sequence analysis of the *chy* gene shows a high degree of homology between species. Interspecies variation within *chy* allows the application of tools such as RT-PCR denaturing gradient gel electrophoresis (DGGE) to determine that fungi actually degrade cyanide within a community.

1. Introduction

The application of molecular biology techniques has provided useful tools for the analysis of species diversity and function in the environment. The use of molecular probes for the detection of ribosomal nucleic acids (rRNA and rDNA) has been successfully used for the determination of phylogenetic relationships amongst microorganisms in the soil environment. Probes can also be designed for specific genes, for example functional genes involved in degradative pathways, in order to assess the genetic potential of a environment (by examining DNA) and to assess gene expression, and as a measure of degradative activity, by extracting and quantifying mRNA. The polymerase chain reaction (PCR) is a powerful tool for detecting and amplifying low concentrations of specific DNA within complex mixtures. There are a number of examples of the application of molecular probes for monitoring bioremediation as well as looking at species diversity in the environment. For example,

V. Šašek et al. (eds.),
The Utilization of Bioremediation to Reduce Soil Contamination: Problems and Solutions, 21–28.
© 2003 *Kluwer Academic Publishers. Printed in the Netherlands.*

Fleming *et al.* [1] describe the use of environmental probes for the detection and measurement of naphthalene degradative genes in PAH-contaminated gasworks soil. RT-PCR has also been applied to analyze *in situ* the expression of the catabolic *dmpN* gene in a bioreactor to monitor phenol degradation by *Pseudomonas putida* [2]. Maximum levels of gene expression coincided with maximum phenol concentrations and aeration, and subsequently decreased with decreasing phenol.

There has been relatively little work carried out on fungi even though eukaryotes have a major advantage over prokaryotic organisms due to the presence of introns within genes, facilitating differentiation between the amplification of DNA and gene transcripts. The use of RT-PCR to examine gene expression in degradative fungi has been described for a number of lignin and manganese peroxidase genes (*lip* and *mnp*) from white-rot fungi [3-5]. White-rot fungi have been widely used in bioremediation processes due to the ability of LiP and MnP enzymes to degrade polyaromatic hydrocarbons (PAHs). The most extensively studied of these fungi is *Phanerochaete chrysosporium*. The molecular genetics of the ligninolytic system of this fungus are complex and involve at least ten structurally related *lip* genes and three different *mnp* genes [6]. However, these genes are differentially regulated in response to carbon, nitrogen [7, 8], cAMP [9] and, in the case of *mnp*, manganese [8]. As a result *lip* transcript patterns in anthracene- or PCP-contaminated soils did not indicate whether gene expression was a direct result of the pollutant or was due to growth-mediated differences in nutrient deficiency [3, 10].

Use of temperature/denaturing gradient gel electrophoresis (T/DGGE) in combination with PCR (DNA amplification) and RT-PCR (RNA amplification) has also provided another approach for the analysis of genetic diversity of complex microbial populations. PCR (or RT-PCR) can be used initially to amplify the sequence of interest prior to T/DGGE. PCR amplification of a specific gene from DNA extracted from a mixed community does provide a measure of the number of individual species that have that gene; however, T/DGGE allows the separation of DNA fragments that are of the same length but with different base-pair fragments. This allows the separation of closely related sequences to provide a profile of a community for the DNA sequence in question. Whereas PCR-T/DGGE of DNA provides information on gene diversity, RT-PCR-T/DGGE of mRNA provides vital evidence of which microorganisms are transcriptionally active.

2. Application to Cyanide Biodegradation

The overall aim of this work has been the development of molecular probes for monitoring bioremediation *in situ* when applied to the model of cyanide degradation by fungi. Hydrogen cyanide is one of the most rapidly acting metabolic inhibitors known, due to its universal inhibition of respiration [11]. Despite this, cyanide occurs widely in the environment both naturally from cyanogenic plants and microorganisms, as well as the actions of industrial processes, such as the former manufacture of municipal gas, extraction of metal ores, and steel manufacture [12-15]. Cyanide complexes tightly to metals, which occur widely in industrial wastes. For example, the majority of cyanide in contaminated gasworks soil is complexed to iron.

The enzyme cyanide hydratase is responsible for the detoxication of cyanide to formamide:

$$\text{HCN} \xrightarrow[\text{cyanide hydratase}]{\text{H}_2\text{0}} \text{HCONH}_2$$

A number of plant pathogenic fungi can induce cyanide hydratase in response to cyanide. It was first characterized from *Stemphylium loti,* a pathogenic fungus of the cyanogenic plant, birdsfoot trefoil (*Lotus corniculatus* L.) [16], and has since been identified in a number of other fungi including *Fusarium solani* [17], *F. solani* IHEM 8026 [18], *F. oxysporum* CCMI 876 [19], *F. lateritium* [20] and *Gloeocercospora sorghi* [21, 22]. Two *Fusarium* isolates, *F. solani* and *F. oxysporum,* isolated from contaminated gasworks soil, have the ability to utilize metal-complexed cyanides as the sole source of nitrogen for growth [23]. Both these isolates could degrade $K_2Ni(CN)_4$ and $K_4Fe(CN)_6$ at pH 4 and 7, respectively.

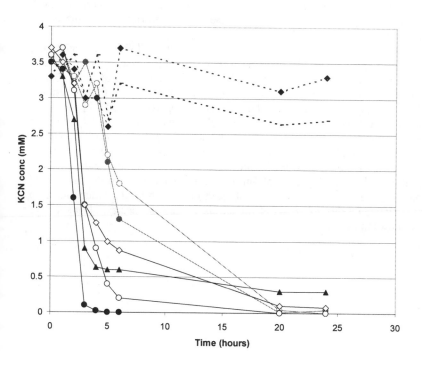

Figure 1: Biodegradation of 4 mM KCN by *F. solani* (•), *F. solani pisi* (○), *F. oxysporum* (□), *F. lateritium* (•), *F. graminearum* (O), and *F. culmorum* (□) in phosphate buffer (pH 7). Controls *Penicillium* sp. (♦), and phosphate buffer (−) alone. Cyanide analysis was carried out using the free cyanide assay of Lambert *et al.* [24].

The nucleic and amino acid sequences of the *chy* gene have been published for both *G. sorghi* and *F. lateritium* (M99044 & M99046, respectively). Sequence alignment of the cyanide hydratase nucleic acid and amino acid sequence from *F. lateritium* and *G. sorghi* has revealed that the enzymes are by 75 % identical at the level of the protein with 65 % nucleotide homology [20]. Genetic similarity between these two species has allowed the design of nucleic acid probes for PCR amplification of the *chy* gene from a number of *Fusarium* species. Of six *Fusarium* species and one strain of *Penicillium*, all *Fusarium* species grew on $K_2Ni(CN)_4$ as a sole source of nitrogen, biotransformed KCN to formamide and by PCR amplification we were able to demonstrate that the *chy* gene was present. The *Penicillium* sp. was negative in each case. The biodegradative capacity of the six *Fusarium* species investigated, with respect to KCN, is shown in Fig. 1.

Sequence alignment for these 6 isolates was obtained by standard di-deoxy methods (Ceq 2000, Beckman Coulter) following the isolation of PCR products obtained, using previously described primers. The sequence alignment of the 5' termini (<600 bp) of cyanide hydratase gene from *F. lateritium, F. solani, F. solani pisi, F. culmorum, F. graminearum* and the cyanide-degrading fungus *G. sorghi* (database number) has been compared. As anticipated, the genetic variation occurs within the two introns of the gene, therefore analysis of the transcribed (mRNA) sequence will show higher homology than complete DNA sequences.

The presence of introns within the eukaryotic genome allows differentiation of DNA and RNA amplification. This has been found to be the case with the *chy* gene from *Fusarium*. The amplified fragment of *chy* contains two introns, of approximately 50 bp each, at the 5' end of the gene. As a result, PCR amplification of DNA results in products that are approximately 600 bp in size, compared to 500 bp for RNA amplification. We have recently demonstrated that primers based on *chy* sequence consensus can be successfully used to show induction of *chy* in *F. solani* on exposure to KCN or metal-complexed cyanides, such as $K_2Ni(CN)_4$. The assumption that cyanide hydratase is induced by HCN, produced from both simple and complexed cyanides, allows the development of an effective molecular sensor for cyanide biodegradation. We have extended these studies to demonstrate the induction of *chy* expression following exposure to KCN for all six *Fusarium* species. Data for four isolates are shown in Fig. 2.

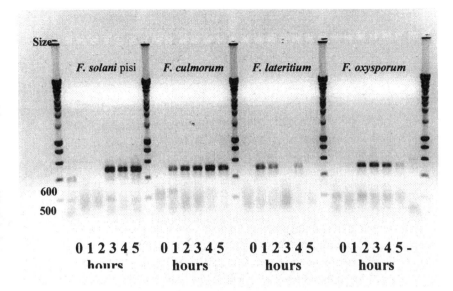

Size

F. solani pisi   F. culmorum   F. lateritium   F. oxysporum

600
500

0 1 2 3 4 5   0 1 2 3 4 5   0 1 2 3 4 5   0 1 2 3 4 5 -
hours   hours   hours   hours

Figure 2. RT-PCR of *chy* during biodegradation of 4 mM KCN in phosphate buffer (pH 7) by *F. solani pisi*, *F. culmorum*, *F. lateritium* and *F. oxysporum*. Gel shows induction of the gene in the presence of cyanide for all strains. Identical induction pattern has been seen with *F. graminearum* and *F. solani*. No RT-PCR product was detected in the absence of cyanide. No bands of 600 bp in size were detected, confirming the absence of contaminating DNA.

Species variation with the DNA sequence analysis and evidence of genetic induction of *chy* allows us to propose that methods such as TGGE/DGGE could be applied to environmental samples to resolve the activity of individual components in mixed samples of DNA. The amplified DNA is separated according to the GC content facilitating the detection of microorganisms that successfully degrade cyanide within the environment samples.

Fig. 3 shows DGGE band separation of RT-PCR of six *Fusarium* species on a 45 % constant gradient polyacrylamide gel. There is some variation between the melting points of products between species; however, only two distinct bands can be seen for the mixture of cDNA.

It is important to note that to date we have amplified and sequenced the first 600 bp of the cyanide hydratase gene. Comparison of the available full-length published nucleic acid sequences for *F. lateritium* and *G. sorghi* show that there is significantly less homology in the 3' region (600 bp) of the of the 1.2 kb–size gene. To date we have not been able to amplify this portion of the gene from the other *Fusarium* isolates using 3' primers based on *F. lateritium* sequence. We plan to obtain the full-length transcript of the gene by genome walking to gain greater insight into the diversity and functionality of this important enzyme.

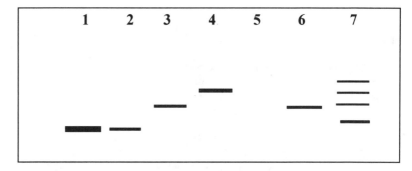

Figure 3. Diagram showing separation of GC-clamped RT-PCR products on a 45 % constant gradient DGGE gel run for 16 h at 200 V. 1: *F. culmorum.* 2: *F. graminearum.* 3: *F. solani.* 4: *F. lateritium.* 5: *F. oxysporum.* 6: *F. solani pisi.* 7: Mixed cDNA.

3. Concluding Remarks

To date we have identified six strains of *Fusarium* which each have the ability to degrade cyanide and metal-complexed cyanides, and utilize them as a source of nitrogen for growth. The design of PCR primers has allowed amplification of the *chy* gene from each of these isolates and subsequent amplification of RNA extracted from isolates grown in the presence of cyanide has shown that cyanide hydratase is induced in all *Fusarium* species tested. DNA sequence analysis shows that 600 bp of the 5' region of the gene are similar, with most of the variation falling within the introns. The targeting of the gene and mRNA transcripts has provided a powerful tool for monitoring the regulation of gene expression in the laboratory that can be applied to monitor gene expression *in situ* during bioremediation processes and the clean up of contaminated land. Application of RT-PCR and T/DGGE helps to differentiate which organisms are active within a population.

These data will allow further manipulation and bioaugmentation to optimize the bioremediation of soils and other polluted sites. This is of particular relevance as the application of nucleic acids for monitoring bioremediation as it has been successfully applied for naphthalene-degrading bacteria in PAH-contaminated gasworks sites [1], and for monitoring phenol degradation in controlled bioreactors [2]. Bogan *et al.* [6] proposed the use of specific primers for RT-PCR quantification of fungal degradative gene expression in soil, with respect to PAH degradation by *P. chrysosporium.* However, the regulation of *lip* and *mnp* transcription is complicated, due to multiple structurally related genes and complex gene regulation. The simplicity of cyanide hydratase in terms of rapid induction and sensitivity to cyanide-containing compounds and - as far as can be deduced to date – there is only one structural gene, make this an ideal model for monitoring cyanide degradation and gene expression in the environment.

References

1. Fleming, J.T., Sanseverino, J. and Sayler, G.S. (1993) Quantitative relationship between naphthalene catabolic gene frequency and expression in predicting PAH degradation in soils at manufactured town gas sites. *Environ. Sci. Technol.* **27**, 1068-1074.

2. Selvaratnam, S., Schoedel, B.A. McFarland, B.L. and Kulpa, C.F. (1995) Application of reverse transcriptase PCR for monitoring expression of the catabolic dmpN gene in a phenol-degrading sequencing batch reactor. *Appl. Environ. Microbiol.* **61**, 3981-3985

3. Lamar, R.T., Schoenike, B., Vanden Wymelenberg, A., Stewart, P., Dietrich, D.M. and Cullen, D. (1995) Quantitation of fungal mRNAs in complex substrates by reverse transcription PCR and its application to *Phanerochaete chrysosporium*-colonized soil. *Appl. Environ. Microbiol.* **61**, 2122-2126.

4. Collins, P.J., Field, J.A., Teunissen, P. and Dobson, A, D.W. (1997) Stabilization of lignin peroxidases in white rot fungi by trytophan. *Appl. Environ. Microbiol.* **63**, 2543-2548.

5. Ruiz-Duenas, F.J., Guillen, F., Camarero, S., Perez-Boada, M., Martinez, M.J. and Martinez, A.T. (1999) Regulation of peroxidase transcript levels in liquid cultures of the ligninolytic fungus *Pleurotus eryngii*. *Appl. Environ. Microbiol.* **65**, 4458-4463.

6. Bogan, B.W., Schoenike, B., Lamar, R.T. and Cullen, D. (1996). Manganese peroxidase mRNA and enzyme activity levels during bioremediation of polycyclic aromatic hydrocarbon-contaminated soil with *Phanerochaete chrysosporium*. *Appl. Environ. Microbiol.* **62**, 2381-2386.

7. Stewart, P., Kersten, P., Vanden Wymelenberg, A., Gaskell, J. and Cullen, D. (1992) Lignin peroxidase gene family of *Phanerochaete chrysosporium*: complex regulation by carbon and nitrogen limitation and identification of a second dimorphic chromosome. *J. Bacteriol.* **174**, 5036-5024.

8. Gettemy, J.M., Ma, B., Margaret, A. and Gold, M.H. (1998) Reverse transcription-PCR analysis of the regulation of the manganese peroxidase gene family. *Appl. Environ. Microbiol.* **64**, 569-574.

9. Boominathan, K. and Reddy, C.A. (1992) cAMP-mediated differential regulation of lignin peroxidase and manganese dependent peroxidase production in the white rot basidiomycete *Phanerochaete chrysosporium*. *Proc. Natl. Acad. Sci. USA* **89**, 5586-5590.

10. Bogan, B.W., Schoenike, B. Lamar, R.T. and Cullen, D. (1996). Expression of *lip* genes during growth in soil and oxidation of anthracene by *Phanerochaete chrysosporium*. *Appl. Environ. Microbiol.* **62**, 3697-3703.

11. Stryer, L. (1988) Oxidative phosphorylation. Chapter 17. *In* Biochemistry, 3rd edition, W.H. Freeman & Company, New York.

12. De-Walle, I.F. (1988) Land rehabilitation in the Netherlands. *In* Proceedings of soil: The aggressive agent. IBC Technical Services, London. **9**, 1-17.

13. Wild, S.R., Rudd, R. and Neller, A. (1994) Fate and effects of cyanide during wastewater treatment processes. *Sci. Total Environ.* **156**, 93-107.

14. Environment Resources Ltd. (1987) Problems arising from the redevelopment of gasworks and similar sites. 2nd edition, MHSO. London.

15. Thomas, A.O. and Lester, J.N. (1993) Gaswork sites as sources of pollution and land contamination: An assessment of past and present public perceptions of their physical impact on the surrounding environment. *Environmemt. Technol.* **14**, 801-814.

16. Fry, W. E. and Millar, R.L. (1972). Cyanide degradation by an enzyme from *Stemphylium loti*. *Arch. Biochem. Biophys.* **161**, 468-474.

17. Barclay, M., Tett, V.A. and Knowles, C.J. (1998). Metabolism and enzymology of cyanide/metallocyanide biodegradation by *Fusarium solani* under neutral and acidic conditions. *Enzyme Microbial Technol.* **23**, 321-330.

18. Dumestre, A., Chone, T., Portal, J.M., Gerard, M., and Berthelin, J. (1997). Cyanide degradation under alkaline conditions by a strain of *Fusarium solani* isolated from contaminated soils. *Appl. Environ. Microbiol.* **63**, 2729-2734.

19. Pereira, P.T., Arrabaca, J.D. and Amaral-Collaco, M.T. (1996). Isolation, selection and characterization of a cyanide-degrading fungus from an industrial effluent. *Internat. Biodeter. Biodegrad.* **45**, 45-52.

20. Cluness, M.J., Turner, P.D., Clements, E., Brown, D.T. and O'Reilly, C. (1993). Purification and properties of cyanide hydratase from *Fusarium lateritium* and analysis of the corresponding *chy1* gene. *J. Gen. Microbiol.* **139**, 1807-1815.

21. Wang, P. and VanEtten, H.D. (1992). Cloning and properties of a cyanide hydratase gene from the phytopathogenic fungus *Gloeocercospora sorghi*. *Biochem. Biophys. Res. Commun.* **187**, 1048-1054.

22. Fry, W.E. and Munch, D.C. (1975). Hydrogen cyanide detoxification by *Gloeocercospora sorghi*. *Physiol. Plant Pathol.* **7**, 23-33.

23. Barclay, M., Hart, A., Knowles, C.J., Meeussen, J.C.L. and Tett, V.A. (1998). Biodegradation of metal cyanides by mixed and pure cultures of fungi. *Enzyme Microbial Technol.* **22**, 223-231.
24. Lambert, J.L, Kamasary, J. and Paukskelis, J.V. (1975) Stable reagents for the colorimetric detection of cyanide by modified konig reactions. *Anal. Chem.* **47**, 916-918.

MODERN METHODS FOR ESTIMATING SOIL MICROBIAL BIOMASS AND DIVERSITY: AN INTEGRATED APPROACH

J.A. HARRIS[1] AND J. STEER[2]

1Institute of Water and Environment, Cranfield University, Silsoe
Bedfordshire MK45 4DT, UK
2School of Health and Bioscience, University of East London,
Romford Road, London E15 4LZ, UK

1. Introduction

Soil is a complex medium, undergoing constant change in its component parts: chemical, physical, and biological. These components are significantly affected by environmental factors, and anthropogenic management and influences. The microbial community of soil is a keystone of the function and structure of soil, which may be characterized from one system to the next by its size, composition and activity. The soil microbial biomass has been defined as the part of the organic matter in soil that constitutes living microorganisms smaller than 5–10 μm^3. The importance of the soil microbial biomass was summarized by Jenkinson:

"The microbial biomass accounts for only 1-3 % of soil organic C but it is the eye of the needle through which all organic material that enters the soil must pass" [1]

In other words the microbial community acts as a "funnel" in elemental recycling, and by determining its characteristics we cannot only follow the path of a particular component, or the effect of a particular treatment or management regime, but we can also obtain some sense of the "health" of the soil system of interest in relation to "pristine" environments [2]. The current debate over the best way to determine "ecosystem health" [3] provides an opportunity for the application of soil microbial community measures as a vital component of such considerations. Also, we are just beginning to understand the importance of species diversity in the functioning of ecosystems (for references see [4]). The insurance hypothesis was examined by Yachi and Loreau [4] in a mathematical model. They found that increased species richness both resulted in increased mean ecosystem productivity over time and buffered variance in productivity.

Before looking at the ways in which the soil microbial community responds to pollutants it is important to consider naturally occurring stresses on soil microorganisms. There are often large spatial variations in soils. There may also be temporal variability in the soil microbial biomass [5]. This highlights the need for a suitable control when attempting to determine the influence of an additional factor such as a pollutant. Options used in different studies include:

- Identification of a suitable control (for experimentation) or reference site (for monitoring).
- Long-term field observations, in order to account for seasonal fluctuations.

V. Šašek et al. (eds.),
The Utilization of Bioremediation to Reduce Soil Contamination: Problems and Solutions, 29–48.
© 2003 *Kluwer Academic Publishers. Printed in the Netherlands.*

- Laboratory experiments - pollutant added in the lab under controlled conditions (artificial) but no problem of a control.

Once soil has been collected from the field it can be incubated under optimum conditions for several days after which the microbial biomass and activity will approach a base level for that soil. For a review of micro- and mesocosm experimental design see [6].

Brookes [7] has suggested some criteria for judging the utility of particular microbiological measurements as indicators in soil pollution monitoring.

- Accuracy and precision of measurement in wide range of soils and under a variety conditions.
- Ease and economy (especially if there are large numbers of samples)
- Availability of reference/control measurements.
- Sensitivity - being sensitive enough to indicate pollution yet robust enough not to give false alarms.
- Scientific validity – with some theoretical basis for the measurement.

The use of two or more independent measures may be necessary in many situations, particularly when soils of different types are being compared. In such cases it is best to use measures where interrelationships under nonpolluted conditions are known.

2. Types of Information

The three main conceptual categories to be determined are as follows:
- **Size** – the total mass of the *viable* microbial community, usually expressed in units of carbon, but also as other essential elements or component parts.
- **Composition** – the *numbers* and *abundances* of particular species or functional groups.
- **Activity** – the *metabolic turnover* of the biomass, from internal metabolism to conversion of nutrient pools.

Some studies have focussed on the importance of the physical arrangement of the community, but this takes us beyond simple consideration of the biomass as a conceptual unit. The techniques available for the study of the microbial community are manifold, and one type of measurement may often supply information as to more than one parameter. The major focus of our considerations will be the size and activity of the microbial community, but we shall also briefly consider compositional measures, as these three parameters are inextricably linked.

2.1 SIZE

The total size of the viable biomass is a consistently good indicator of the status of the soil system taken as a whole, particularly when set in relation to the total organic matter pool. Many workers attempt to derive a biomass-carbon figure for their determination, even if carbon is not being determined directly, as in the case of adenosine triphosphate (ATP), in order to facilitate comparison with other studies. Exceptions to this occur when specific elemental pools are the subject of the investigation; this has been particularly so with respect to nitrogen in agricultural systems.

2.1.1. DIRECT OBSERVATION / MICROSCOPY

Microscopy is especially useful for determining the form and arrangement of microorganisms in soil. Thin section techniques permit observations of soil structure, root penetration and soil fauna. Fluorescent stains such as magnesium sulfonic acid and acridine orange allow the observation of microorganisms on soil particles.

2.1.1.1. Dilution Counting

This is where a known quantity of homogenized soil suspension is placed on a microscopic slide. The microorganisms are stained with a fluorescent dye, and numbers are counted under the microscope. The numbers in a small area or volume can then be used to calculate numbers in the total soil sample. It is very important that the soil is fully homogenized. Detergents and deflocculents have been used to prevent reaggregation of soil particles or microbial cells. Different stains can be used for different cell types/constituents.

- Fluorescein diacetate (FDA) is used for staining active fungal hyphae.
- Europium chelate stains DNA and RNA.
- Fluorescein isothiocyanate (FITC) stains proteins.

Microscopic methods have been criticized because of the difficulty of differentiating between live and dead microbial cells [8]. They may also miss the microbial cells that remain hidden within or between soil particles. In addition, weakly stained cells are difficult to count and some stains will fade with time. When counting microorganisms in very small samples homogenization is very important. If mixing is incomplete results may be misleading. Interpretation of results is also difficult with a wide range of conversion factors used to convert cell volume to microbial biomass [9]. Finally staining and counting methods are often considered too time consuming for routine analysis, although membrane filtration methods may be automated to a certain degree, and the advent of computer imaging techniques has reduced the variations occurring between different observers.

2.1.2. INDIRECT METHODS / CHEMICAL ANALYSIS

Any measurable component of microbial cells could be converted to a biomass figure. To give an accurate estimate of biomass this component should show little variation between cells of different microorganisms or between cells from the same microorganisms living under different conditions. It should also be relatively easy to measure under a wide variety of conditions. The measurement can be converted into a biomass carbon figure using a calibration factor determined by comparison with other methods.

2.1.2.1. Adenosine Triphosphate (ATP)

ATP has been considered as a parameter for measuring both microbial biomass and activity. It can be extracted from soil and estimated by the bioluminescence test system. Adenosine triphosphate (ATP) is common to all life. ATP in dead organisms and free ATP in soil are degraded rapidly. The luciferin+luciferase assay is very sensitive and ATP can be detected in concentrations as low as 10 pmol/L.

There are problems in the quantitative extraction of ATP from soil and with the inhibitory effects of substances found in soil extracts [10]. Often the extraction of ATP from cells is not complete and ATP is decomposed (by enzymic or chemical hydrolysis) during the extraction process. These problems can be largely overcome by the use of an internal standard. Another difficulty is that after its extraction ATP is strongly adsorbed by soil constituents. This problem is more difficult to overcome. Extraction efficiency can be evaluated by looking at the microbial biomass C: ATP ratio (where biomass C is determined by an alternative method). Efficient ATP estimation methods are characterized by a low value. Finally there are a relatively high number of ATP extraction methods. This makes it difficult to relate biomass or activity estimated by ATP to biomass and activity measured in other studies. Consequently there are a wide range of biomass C:ATP ratios [11]. It has been argued that ATP merely reflects activity of the community. This argument is not supported by the high degree of agreement between ATP measurements and biomass C determinations in a number of studies (e.g. [12]).

An extension of ATP methodology was attempted by Brookes and co-workers [13] in their development of an Adenylate Energy Charge (AEC) index. This is achieved by sequentially determining ATP, AMP and ADP in an extract. The amount of ATP present may be expressed as a percentage of the total adenylate pool. This will be around 80 % for a fully active biomass. This technique is very difficult to use as part of a large sampling program, but can be employed to resolve questions as to whether biomass or activity is being measured.

A high diversity of applied techniques for the extraction and measurement of ATP has led to biomass C:ATP ratios which vary between about 150 and 1000. Our own current investigations are expected to shed more light on the problems of ATP extraction. Preliminary results indicate a constant biomass C:ATP ratio of about 200.

2.1.2.2. Fumigation Incubation / Fumigation Flush

Störmer [14] found that plant growth was enhanced when soil was treated with a carbon disulfide fumigant. He postulated that this was caused by a liberation of additional nitrogen from the bodies of the organisms killed by the toxicant. He noted that after the treatment an increased proliferation of bacteria could be observed in the soil, and suggested that it was these bacteria which degraded the killed organisms liberating the nitrogen from their cells. Jenkinson [15] confirmed this explanation and concluded that the size of the CO_2 flush should provide a measure of the original biomass. The fumigation-incubation (FI) method was developed in a series of papers by Jenkinson and Powlson [16-20]. Briefly the initial soil sample is split in two and one half is incubated with the fumigant (usually $CHCl_3$). Both samples are then inoculated with a small portion of the original sample, and the CO_2 evolved is measured during a period of incubation. The difference in CO_2 evolved from the fumigated and nonfumigated samples is then converted to a biomass figure using a calibration or k_c factor.

With the FI method it is assumed that the amount of CO_2 generated from the decomposition of killed organisms must be much greater in the fumigated sample than that from dead organisms in the nonfumigated sample. So when using dried or frozen soil, in which there may be a large denatured biomass, a preincubation step is recommended. It is also assumed that the amount of CO_2 evolved from the degradation of nonliving soil organic matter is equal in the fumigated and nonfumigated samples, and that the proportion of carbon mineralized from the dead microbial biomass is constant over a given period in different soils, and under different conditions.

Some degree of uncertainty in the k_c factor must be expected if the composition of the soil microbial community differs. For example, dead fungi were found to degrade at a higher rate than dead bacteria over a 10-day incubation [21, 22]. Anderson and Domsch [22] found that variation in the k_c value was most sensitive to the rates of mineralization of fungi.

When a soil has recently received an organic amendment, the FI method can give negative biomass measurements. This is because the larger microbial community in the nonfumigated soil can mineralize the readily available C from the amendment more rapidly than the inocula in the fumigated sample [11]. In such cases, a preincubation step has been recommended. In soils with a low pH (below 5), FI has been found to underestimate biomass. This may be due to the inability of microbial recolonizers to mineralize nonmicrobial soil organic matter to the same extent as the native populations in the non-fumigated control [23]. It has been shown that this can be overcome at least to some extent by increasing the size of inoculum [24].

2.1.2.3. Fumigation Extraction

This is an adaptation of the FI method where the carbon from the lysed microbial cells is directly extracted in a salt solution (potassium sulfate) after fumigation [25]. In this instance the difference between the extracted carbon from the fumigated and nonfumigated samples is converted to a biomass figure using a calibration factor (k_{EC}). Here k_{EC} is the proportion of organic C extracted from the lysed microbial cells.

It is generally accepted that chloroform fumigation does not result in full sterilization of the soil. Figures for efficiency of sterilisation range from 82 % [26] to over 90 % [27]. Fungi appear to be more sensitive to fumigation than bacteria. Toyota et al. [27] suggested that this might be due to exopolysaccharides secreted by bacteria protecting them from the chloroform vapor. Soil moisture content also has considerable influence on the quantity of C extracted following chloroform fumigation [28]. In particular biomass is underestimated at low soil moisture contents [29].

The original k_{EC} value was calculated from a regression between E_C and biomass C estimated using the fumigation/incubation (FI) method [25]. Later studies found biomass measured by fumigation/extraction (FE) was significantly correlated with FI [30, 31], substrate-induced respiration (SIR) [32-34] and microscopic counting [35].

When the method was first developed total organic carbon (TOC) in the K_2SO_4 extracts was quantified by oxidation with potassium dichromate and a back titration of the remaining unreduced dichromate. More recently TOC has been determined using a TOC analyzer. This uses a UV-persulfate oxidation step before determining the resulting CO_2 using infrared spectroscopy [36]. The latter method gives a more complete oxidation of organic molecules resulting in higher k_{EC} factors.

In the FE method, the proportion of C extractable from lysed microbial cells will influence the range of k_{EC} values. Eberhardt et al. [37] found that the mean k_{EC} values for pure cultures ranged from 0.10 to 0.68, and that the range of values was particularly large for species of bacteria. They also found k_{EC} values to be higher, and less variable for fungal species. Martens [11] reviewed k_{EC} factors, and calculated an average value of 0.43 (when determining organic C with a TOC analyzer).

Fumigation extraction has the advantage that it can be used in soils that are not suitable for fumigation-incubation. These include soils with a low pH or soils containing readily degradable material. It can also be used for waterlogged soils when chloroform can be added as a liquid to a soil slurry [38]. In both fumigation methods, it is assumed that chloroform fumigation does not alter the decomposability of non-biomass soil organic matter [39]. This hypothesis has yet to be fully demonstrated [40].

The fumigation/extraction method has also been used to determine the quantities of extractable nitrogen [41], ninhydrin-reactive nitrogen (NRN) [42], phosphorus [43], sulfur [44] and carbohydrate [45] in the microbial biomass.

2.1.2.4. Alternatives to Fumigation

Islam and Weil [46] investigated the use of microwave irradiation followed by either incubation or extraction as an alternative to the chloroform fumigation methods. They found carbon flushes from irradiated soils to be strongly correlated to those determined using the fumigation/incubation and fumigation/extraction methods. This method is, however, less effective than chloroform fumigation, resulting in much smaller carbon flushes [47]. Freeze drying is another alternative to chloroform fumigation. Islam et al.[48] found freeze-dried extraction was very strongly correlated with biomass determined by fumigation incubation when studying 19 agricultural soils.

2.1.2.5. Substrate-Induced Respiration (SIR)

It has been observed that the addition of glucose to soil samples results in elevated respiration. This lasts for a few hours before cells start to proliferate. This has been termed the maximum initial response [49]. The respiration response was later measured against the fumigation/incubation method for a variety of soils, and found to give a linear response, indicating it was a suitable microbial biomass assay. The quantity of glucose required is the amount at which additional glucose does not result in an increased response. This will vary depending on the soil being studied.

Anderson and Domsch [50] calculated that 40 mg biomass C respires 1ml CO_2/h. This was amended to 30 mg biomass C per ml CO2/h by Kaiser et al. [32]. The conversion factor will depend on the efficiency in measuring the CO_2 evolved during the incubation. Martens [11]compared the fumigation/extraction, fumigation/incubation and substrate induced respiration methods in three soils which were amended with sewage sludge. The SIR method resulted in a much higher biomass value. It was suggested that the organic amendment stimulated microbial growth, increasing the number of 'active' microorganisms in the soil. These would respond more to the added substrate. Anderson and Domsch [50], studying cultured microorganisms, found that the SIR response of organisms in exponential growth was several times greater than that of organisms in stationary phase. Sparling [51] noted that the majority of regressions between SIR and biomass measured using fumigation methods do not pass through zero. It appears that proportion of the biomass measured by the fumigation methods that does not respond to glucose amendment (over a short incubation period). This supports the theory that SIR only measures the potentially active microbial biomass. An alternative explanation is that the chloroform fumigation solubilizes a portion of soil organic matter.

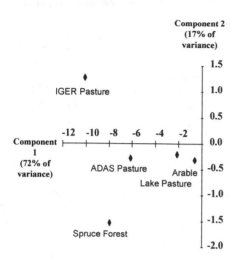

Figure 1. Positions of five contrasting soils in component space based SIR response profiles

2.2. ACTIVITY MEASUREMENTS

Microbial activity comprises all biochemical reactions catalyzed by microorganisms in soil. Some reactions involve the whole (active) soil microbial community – respiration and heat output, while others are confined to specific groups – nitrogen fixation, nitrification.

Estimation of microbial activity in the presence of an added substrate is often termed potential activity. If no substrate is added the term actual activity can be used. Of course if a soil has been removed from the field, sieved, moisture has been adjusted and has been incubated at a set temperature then the activity measured even without an added substrate is more a potential than an actual activity. Actual activity implies nondisturbed field conditions under which it is impossible to carry out many of these measurements.

2.2.1. RESPIRATION / SPECIFIC RESPIRATION

Active heterotrophic microbial cells derive energy from the transformation of organic material. The oxidation of organic matter by aerobic organisms results in the production of CO_2. Microbial metabolic activity can therefore be quantified by measuring CO_2 production (or O_2 uptake) [52]. Microbial activity has been frequently considered as an index of soil fertility [53]. Respiration rate has been used to determine biological activity in soil in relation to succession [54], the physical properties of soil [55, 56] and agricultural and silvicultural practices [57, 58].

Respiration may increase in response to an increase in microbial biomass or as the result of the increased activity of a stable biomass. Activity may be considered in relation to the microbial biomass [59]. The respiration to microbial biomass ratio has been termed specific activity or metabolic quotient (qCO_2). The metabolic quotient has been used as a bioindicator of disturbance and ecosystem succession [60, 61]. It has been suggested that the metabolic quotient has limited use as it confuses the effects of disturbance and stress [62]. Anderson [63] also noted that qCO_2 values could be influenced by changes in community structure and shifts in the ratio between active and dormant organisms.

Soil respiration can be determined using simple methods such as incubating soil in a jar and absorbing the CO_2 evolved in NaOH. The quantity of CO_2 trapped can then be determined by HCl titration. An alternative method is to use a continuous flow system measuring CO_2 by gas chromatography or infrared spectroscopy. Respiration can also be measured in the field. However it should be noted that this gives a measure of total biological activity including respiration from plant roots and the macro- and mesofauna.

2.2.2. NITROGEN TRANSFORMATIONS

2.2.2.1. Nitrogen Fixation
Nonsymbiotic N fixation in soils can be measured by the acetylene reduction method as acetylene is reduced to ethylene by the same enzyme system that reduces nitrogen gas to ammonia. Acetylene is added to a soil and, after a suitable incubation period, the ethylene produced is measured by gas chromatography. To get a figure for potential nonsymbiotic N fixation glucose can be added to the soil as a C source.

It is difficult to extrapolate from the amount of ethylene produced during a short incubation to long- term N_2 fixation. In theory the ratio of acetylene reduced to N_2 reduced is 3:1, in practice there are published observations from 1:1 to 7:1. In some systems acetylene inhibits nitrogenase activity – this would result in an underestimate of N_2 fixation. Acetylene is also known to suppress the oxidation of endogenously produced ethylene that is normally rapidly oxidized by aerobes, this resulting in an overestimate of N_2 fixation.

2.2.2.2. Nitrogen Mineralization + Nitrification
Net ammonification and nitrification can be measured in soils by splitting a soil sample into two, extracting one immediately (with a salt solution) and measuring ammonia, nitrite and nitrate. The second sample is incubated and a second measurement is taken. An organic source of N (commonly arginine) can be added to soil prior to incubation to

measure potential mineralization. In a similar way potential nitrification can be measured using an ammonia amendment prior to incubation. N-mineralization has often been measured by incubating soil cores in the field.

2.2.3. ENZYME ACTIVITIES

Many soil microorganisms depend on the activities of their extracellular enzymes to supply them with nutrients. Burns [64] separated 10 distinct categories of soil enzymes. These included enzymes within the cytoplasm of cells, enzymes attached to the outer surface of viable cells, enzymes secreted by living cells, enzymes released from lysed cells, and enzymes adsorbed to clay minerals or associated with humic material.

Care should be taken when interpreting the results of enzyme assays. These measurements are of potential enzyme activity when soil is incubated with excess substrate under optimum conditions in terms of temperature, pH and moisture content. Soil enzymes are often trapped in soil organic and inorganic colloids. Some soils will therefore have a large background of extracellular enzymes not directly associated with the microbial biomass.

2.2.3.1. Dehydrogenase Activity

Dehydrogenases are endogenous (within a cell) enzymes that catalyze the dehydogenation of organic compounds. In the dehydrogenase assay, triphenyl tetrazolium chloride (TTC) acts as an artificial electron acceptor (under anaerobic conditions). TTC is reduced to triphenyl formazan (TPF) which is colored and can be extracted in methanol and measured using a spectrophotometer.

Problems with this technique include the suggestion that TTC is an inefficient electron acceptor, only being reduced to TPF once other acceptors were exhausted. Ions such as NO_3^-, NO_2^- and Fe^{3+} have been shown to compete with TTC for electrons (Alef 1995). The method is also sensitive to environmental conditions such as pH, temperature, length of incubation, and addition of substrate. The dehydrogenase assay has the advantages that it is simple and fast, and it has yielded sensitive and consistent results in many studies [66].

2.3. COMMUNITY COMPOSITION

2.3.1. CULTURAL METHODS

The isolation and identification of microorganisms may provide useful information about the soil microbial community. Traditional plate counting methods involve shaking up a known weight of soil in sterile water or saline solution and progressively diluting the resulting solution. Aliquots of the resulting suspensions are added to a cooled, but still liquid, agar in sterile Petri dishes. Discrete colonies of microorganisms will appear at one (or more) of the dilutions. These may be counted and identified by traditional techniques of determinative bacteriology or mycology. Numbers can be multiplied by the dilution factor to obtain an estimate of the number of viable microorganisms in the original soil sample.

The main problem with this approach is that there are no artificial media that can simulate the complex soil environment. Consequently this method will only detect the small proportion of soil microorganisms that can be grown in culture [67]. Little is known about the remaining nonculturable portion of the microbial community.

2.3.2. BIOCHEMICAL METHODS – CELLULAR SUBSTANCES

Measurement of specific substances within cells has also been used to determine the biomass of different groups of microorganisms. Cell-wall components, such as muramic acid and chitin, have been used to determine bacterial and fungal biomass. However, both of these substances can originate from nonliving biomass [68]. Membrane-bound substances, such as ergosterol (specific to fungi) and phospholipid fatty acids (PLFAs), appear to be more rapidly mineralized in dead cells and therefore provide better markers for groups of microorganisms.

2.3.2.1. Muramic Acid
Muramic acid is only found in prokaryotes. It is a component of bacterial cell walls and can therefore be used as a measure of bacterial biomass. In Gram-positive bacteria it accounts for 10–50 % of the dry weight of the cell wall, in Gram-negative bacteria about 10 %. Muramic acid also occurs in humic acids after the incorporation of dead bacterial cells. It is therefore more suited to the study of soils with a low organic matter content. It has been shown to correlate well with other techniques in sub-surface soils [68].

2.3.2.2. Fungal Biomass – Ergosterol
Sterols represent 0.7%–1.0 % of fungal dry weight and are localized in the cell membrane. It is generally accepted that ergosterol is the main sterol of many fungi. Ergosterol is not produced by vascular plants, bacteria or soil fauna but may occur in protozoa and algae and may be stored after ingestion by arthropods [9]. As well as varying with species the ergosterol content of fungal cells varies with environmental conditions [69].

Ergosterol can be extracted from soil samples and determined by HPLC. The use of an internal standard allows calculation of extraction efficiency. The research using substrate-induced respiration and selective inhibitors produced a fungal biomass C/ergosterol ratio of 90 in temperate forest topsoil. A similar ratio of 79 was obtained from an average of all 37 conversion ratios collated by [9]. Ruzicka and co-workers [70] have developed a rapid method, involving an ultrasonication step, to estimate ergosterol in large numbers of samples.

2.3.2.3. Phospholipid Fatty Acids (PLFAs)
Fatty acids are ubiquitous in their distribution and occupy a wide range of structural and metabolic functions in both prokaryotes and eukaryotes [71]. They are the basic components of many lipids, including phospholipids, which are incorporated within the membranes of all living cells but do not occur in microbial storage products [72]. Different subsets of the microbial community have different PLFA patterns and the total amount of phospholipid in cells is considered to be well correlated to biomass [69]; methods for their use as biomass indicators have been widely developed [73].

PLFA profiles do not give information on species composition in soil but rather an overall picture, or fingerprint of the community structure. PLFA analysis has been used to study changes in the soil microbial community due to pollution [74-76], agricultural practices [77, 78], rhizosphere versus non-rhizosphere community dynamics [79] and with depth relative to the water table in peatlands [80].

2.3.3. FUNCTIONAL DIVERSITY

Catabolic profiling of microbial communities has seen a rapid increase in application. Degens and Harris [81] measured the response of microbial communities from a range of management types to the addition of a variety of carbon substrates. The responses to a total of 88 compounds over periods of time from 15 min to 4 h were determined, and the resulting activity profiles were analyzed by principal component and cluster analysis. Using this technique the authors were able to discriminate effectively between different management types (Fig. 1).

The use of Biolog™ plates to estimate catabolic diversity has also increased in recent years [82]. This system consists of 96 well plates, each containing a different substrate, ranging from glucose to complex polysaccharides. Each well also contains an indicator to show utilization of the substrate. Each well is inoculated with a dispersed soil suspension, and incubated for a set period. A note is made of each transformed/nontransformed substrate. It has been argued that this provides a catabolic profile of the community of interest but suffers from the following drawbacks:

- The organisms surviving the dispersal and carrying out the transformation are likely to come from the culturable fraction only.
- The incubation period can significantly affect the transformation profile.
- The degree of dilution of inoculum can alter the profile.

Consequently this technique must be used with caution, and preferably in conjunction with other techniques.

2.3.3.1. Selective Inhibition
Selective inhibition has been used in conjunction with substrate induced respiration (SIR) to determine the proportions of bacterial and fungal biomass in soil [49]. Antibiotics are added to the soil to exclude bacteria and fungicides to exclude fungal growth. With this method, it is difficult to detect the extent to which an antibiotic affects nontarget organisms [83]. Lin and Brookes [84] have suggested optimal concentrations of the antibacterial antibiotic streptomycin of 4 to 8 mg per g soil, and 8 to 12 mg per g for the antifungal antibiotic cycloheximide.

2.3.4. MOLECULAR METHODS

In recent years, major advances in our understanding of microbial community structure and dynamics have come from the application of molecular techniques to environmental samples. Rapid advances in equipment, automation, and decreases in prices of equipment have arisen as a result of the Human Genome project and the widespread commercial interest in genetic engineering. Consequently these technologies are now within the reach of most research laboratories, and they are being quickly taken up.

Most methods rely on the capture and amplification of sequences of interest. Phylogenetic relationships between bacteria within a sample may be determined, and communities from different habitats may be compared [85]. Gene probes may be used to track individual species, their dynamics and shifts in dominance. The techniques involved are summarized in Fig. 2.

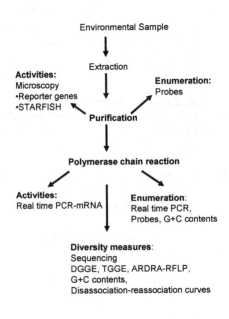

Figure 2. Molecular techniques used in microbial community analysis.

2.3.5. EXTRACTION

Two main approaches have been used for extraction of molecular material from soil samples:
- *Cell extraction* – isolation of cells followed by lysis and nucleic acid purification.
- *Direct extraction* – cells lysed in the sample followed by purification of the nucleic acid fraction.

The former suffers from the same drawbacks as all isolation techniques – only 1-5 % of cells are likely to be culturable, or even extractable from the soil matrix, a long recognized problem [86]. However, when the aim of the investigation is to track a well-characterized species, this methodology may be applied with some degree of confidence.

Direct extraction has the advantage that previously nondescribed microorganisms are included in the extraction procedure.

2.3.5.1. Cell Extraction

Cell extraction requires that as large a proportion of the viable microbial community as possible is isolated, in as pure a form as possible. Tradition of dilution-dispersal methods are notoriously poor at extracting either sufficient or representative community members. These disadvantages decrease in importance when more specific groups are targeted as the degree of confidence that a significant and representative number of the targeted cells has been extracted increases.

Centrifugation using polysaccharide or cesium chloride-ethidium bromide gradients have been used successfully to improve the concentration of microbial community members. Filtration of disperse samples can be most successfully in single species/genus studies, particularly when the final stage of the analysis is to be performed on the filter.

2.3.5.2. Direct Extraction

There are four component parts of the overall procedure [87]:
1. Cell lysis
2. Separation of DNA from other cellular components
3. Release of DNA from soil matrix material, and removal of humic substances
4. Precipitation of the DNA

These steps are often combined and may need to be repeated.

Both routes of extraction then allow further manipulation of nonamplified material, on filters, or other substrates, observable microscopically:

- **Reporter genes.** Sequences of genes are inserted into the species of interest, and detected by fluorescence. For example, antibiotic resistance can be incorporated into the species of interest, then the antibiotic included in the extraction procedure. There has also been extensive use of the *lux* gene sequence, allowing detection of microbial activity on filters and *in situ*.
- **STARFISH** combines total direct counting by multiple group-specific 16S rRNA probes with a simultaneously determined ^{3}H-substrate uptake for each group with microautoradiography [88, 89].

2.3.6. MEASUREMENTS OF DIVERSITY

2.3.6.1. Dissassociation-Reassociation Profiles

When nucleic acid material has been extracted, it is possible to obtain a gross estimate of community diversity by reassociation profiling [90]. Essentially, purified DNA is melted, then allowed to re-anneal over a controlled cooling period. The basic principle here is that highly diverse communities will produce re-association curves which take a longer time to re-associate than simple ones, as the complementary sequences take longer to "find" one another. Similarly, it is possible to compare the nucleic material from different communities by mixed annealing curves; similar communities re-associate more rapidly than dissimilar ones. This method has some advantages in terms of speed and reliability, but cannot indicate precisely which species are present or absent in different samples.

TABLE 1. Summary of techniques and the information they yield

General Approach	Technique	Size	Composition	Activity
Microscopy	Light	4	4	
	Electron	4	4	
	Epiflourescence	4	4	4
	Confocal	4	4	4
	Image analysis	4	4	4
	Fluorescent immunology	4	4	4
Chemical Analysis	Fumigation-(Extraction/Flush)	4		
	Lipids/Phospholipids	4	4	
	Ergosterol	4	4	
	ATP	4		4
Molecular techniques	Probes - phylogenetic		4	
	Probes - functional		4	
	Reporter genes		4	4
	ARDRA-RFLP		4	
	Real-time PCR	4	4	
	mRNA			4
Physiological analysis	Substrate induced respiration	4	4	4
	Respirometry			4
	Enzymology			4
Process studies	Uptake studies			4
	Nutrient transformations			4
	Isotopic tracers			4
	Organic matter fractions	4		
Culture and enrichment	Plate counts	4	4	
	MPN	4	4	

The total G+C content of soil DNA has been used both as a gross measure of diversity, and as an indicator of bacterial biomass [91]. This is particularly useful when either a large number of soils need to be compared rapidly, or a well characterised system requires routine monitoring to assay the effects of a particular factor, such as herbicide application [91].

2.3.6.2. Real-Time PCR

When there is a known sequence available, particular to a species or function, it is possible to use "real-time" PCR to determine how many copies of the original were present in the original sample. Real- time PCR works by attaching a "beacon" to a probe and hybridizing the sample over a number of PCR cycles with continuous measurement of light output at the wavelength of the beacon. As the light reaches a detectable level, it is possible to relate this to the number of copies in the original sample – more copies in the original will lead to earlier detection of a signal than samples starting with fewer copies. When mRNA is being targeted this will reflect the activity of the number of copies of the sequence which was active.

2.3.7. AMPLIFICATION OF NUCLEIC ACID MATERIAL

In most situations where genetic information is being sought, there is insufficient material in the native form for ready detection. Under these circumstances the polymerase chain reaction (PCR) is employed to multiply the number of copies of material available for analysis.

2.3.7.1. Diversity Measurements

The use of restriction fragment length polymorphism (RFLP) profiles has begun to be used widely to indicate diversity within and between systems, and to identify those organisms which are of major importance in differing systems.

Essentially, a conserved region from the organism of interest is isolated, amplified and cut by a restriction enzyme. Different species will produce different lengths of cut product. To increase the sensitivity of this approach this may be repeated with a different restriction enzyme to produce a second pattern, and so on until resolution of closely related strains is achieved. The data generated is subject to hierarchical clustering procedures to produce a similarity dendrogram. Dendrograms from different sites can also be compared.

The two further techniques available are denaturing gradient gel electrophoresis (DGGE) and amplified ribosomal DNA restriction analysis (ARDRA) [92].

3. Integrating Techniques

3.1. TECHNIQUES YIELDING MULTIPLE INFORMATION

It is clear that many of the techniques outlined above yield multiple information as to two or all three of our conceptual categories (Table 1). The reliability of these techniques for any given situation must, however, be used with some caution. Not all techniques are equally applicable in all circumstances, making comparison and interpretation of disparate studies problematic.

3.2. STATISTICAL APPROACHES

The volume and diversity of data arising from modern studies of the soil microbial community make inspection or simple gradient analysis extremely problematic in all but the simplest of controlled experimental systems. Certainly those studies of natural ecosystems involving both biological and physicochemical data require rigorous data reduction to enable interpretation. Fortunately powerful packages to facilitate multivariate data analysis and condensation are readily available, and often at reasonable cost [93]. The hierarchical dissimilarity tree is becoming increasingly common for comparing sites and organisms as to their similarity.

Using three conceptual categories (size, activity, and composition) it is possible to obtain a three- dimensional ordination of sites, allowing easy comparison of sites of a microbial community status [2]. This has potential for further development to provide a full thermodynamic model of the behavior of the systems under consideration, resulting in prescriptions for management.

4. Future Opportunities and Challenges

There are, as always, further lines of enquiry to be pursued, including the following:
- Yield and quenching.
- The problem of biomass acclimatization to pollutants.
- The need to employ multiple techniques to gain a full picture.

It is certainly of utmost importance to recognise the limitations of the techniques employed.

5. Conclusion

The burgeoning of techniques available and the ability to process much larger volumes of sample have led to a step jump in our understanding of the diversity and function of the soil microbial community, in the last ten years. As Andren and Balandreau [94] have pointed out, "*these recent advances have revealed a high diversity even at very small scales*", but make traditional Linnean approaches to classification and interpretation increasingly problematic in the genetically promiscuous world of the microorganism. We have many of the tools that we need to answer not only applied questions of the nature "what is the effect of compound x on population y, and how might I ameliorate it", but more fundamental questions of relationships between species diversity and functional redundancy, evolutionary adaptation and its mechanisms, patch dynamics, ecosystem architecture, and feedback between biotic and abiotic components. We may be close to providing definitive answers to many of these questions. The fact that we are studying microbial populations capable of producing successive generations in the course of a day gives us a distinct advantage over our colleagues working in animal and plant ecology.

References

1. Jenkinson, D.S. (1977) Studies on the decomposition of plant material in soil. *Journal of Soil Science* **28**, 424-434.
2. Bentham, H., Harris, J.A., Birch, P. and Short, K.C. (1993). Habitat classification and soil restoration assessment using analysis of soil microbiological and physico-chemical characteristics. *Journal of Applied Ecology* **29**, 711-718.
3. Rapport, D.J., Bohm, G., Buckingham, D., Cairns, J., Costanza, R., Karr, J.R., deKruijf, H.A.M., Levins, R., McMichael, A.J., Nielsen, N.O., and Whitford, W.G. (1999) Ecosystem health: the concept, the ISEH, and the important tasks ahead. *Ecosystem Health* **5**, 82-90.

4. Yachi, S. and Loreau, M. (1999) Biodiversity and ecosystem productivity in a fluctuating environment: the insurance hypothesis. *Proceedings of the National Academy of Science USA* **96**, 1463-1468.

5. Wardle, D.A. (1998) Controls of temporal variability of the soil microbial biomass: a global scale synthesis. *Soil Biology and Biochemistry* **30**, 1627-1637.

6. Teuben, A. and Verhoef, H.A. (1992) Relevance of micro- and mesocosm experiments for studying soil ecosystem processes. *Soil Biology and Biochemistry* **24**, 1179-1183.

7. Brookes, P.C. (1995) The use of microbial parameters in monitoring soil pollution by heavy metals. *Biology and Fertility of Soils* **19**, 269-279.

8. Bloem, J., Bolhuis, P.R., Veninga, M.R. and Wieringa, J. (1995) Microscopic methods for counting bacteria and fungi in soil. In: *Methods in Applied Soil Microbiology and Biochemistry*, Alef, K.; Nannipieri, P. (Eds.) , 162-173. Academic Press, London.

9. Newell, S.Y. (1992) Estimating fungal biomass and productivity in decomposing litter. In: *The Fungal Community: Its Organization and Role in the Ecosystem*. G.C. Caroll and D.T. Wicklow (Eds.), 521-561 Marcel Dekker, New York.

10. Alef, K. and Nannipieri, P. (1995) *Methods in Applied Soil Microbiology and Biochemistry*, pp 194. Academic Press, London.

11. Martens, R. (1995) Current methods for measuring microbial biomass C in soil: potentials and limitations. *Biology and Fertility of Soils* **19**, 87-99.

12. Barajas Aceves, M., Grace, C., Ansorena, J., Dendooven, L. And Brookes, P.C. (1999) Soil microbial biomass and organic C in a gradient of zinc concentrations in soils around a mine spoil tip. *Soil Biology and Biochemistry* **31**, 867-876.

13. Brookes, P.C., Tate, K.R., and Jenkinson, D.S. (1983) The adenylate energy charge of the soil microbial biomass. *Soil Biology and Biochemistry* **15**, 9-16.

14. Störmer, K. (1908) Über die Wirkungen des Schwefelkohlenstoffs und ähnlicher Stoffe auf den Boden (Effects of carbon disulphide and related compounds on soil). *Zentralbl Bakteriol Parasitenkd Infektionskr Hyg Abt 2* **20**, 282-288.

15. Jenkinson, D. S. (1966) Studies on the decomposition of plant material in soil II. Partial sterilization of soil and the soil biomass. *Journal of Soil Science* **17**, 280-302.

16. Jenkinson, D. S. (1976) The effects of biocidal treatments on metabolism in soil. IV. The decomposition of fumigated organisms in soil. *Soil Biology and Biochemistry* **8**, 203-208.

17. Jenkinson, D. S. and Powlson, D. S. (1976) The effects of biocidal treatments on metabolism in soil I. Fumigation with chloroform. *Soil Biology and Biochemistry* **8**, 167-177.

18. Jenkinson, D. S. and Powlson, D. S. (1976) The effects of biocidal treatments on metabolism in soil. V. A method for measuring soil biomass. *Soil Biology and Biochemistry* **8**, 209-213.

19. Jenkinson, D. S., Powlson, D. S. and Wedderburn, R. W. M. (1976) The effects of biocidal treatments on metabolism in soil III. The relationship between soil biovolume, measured by optical microscopy, and the flush of decomposition caused by fumigation. *Soil Biology and Biochemistry* **8**, 189-202.

20. Powlson, D. S. and Jenkinson, D. S. (1976) The effects of biocidal treatment on metabolism in soil. II. Gamma irradiation, autoclaving, air drying and fumigation. *Soil Biology and Biochemistry* **8**, 179-188.

21. Vance, E. D., Brookes, P. C. and Jenkinson, D. S. (1987) An extraction method for measuring soil microbial biomass-C. *Soil Biology and Biochemistry* **19**, 703-707.

22. Anderson, J. P. E. and Domsch, K. H. (1978) Mineralisation of bacteria and fungi in chloroform-fumigated soils. *Soil Biology and Biochemistry* **10**, 207-213.

23. Vance, E. D., Brookes, P. C. and Jenkinson, D. S. (1987) Microbial biomass measurements in forest soils: The use of the chloroform fumigation-incubation method in strongly acid soils. *Soil Biology and Biochemistry* **19**, 697-702.

24. Chapman, S. J. (1987) Inoculum in the fumigation method for soil biomass determination. *Soil Biology and Biochemistry* **19**, 83-87.

25. Vance, E. D., Brookes, P. C. and Jenkinson, D. S. (1987) Microbial biomass measurements in forest soils: determination of k_c values and tests of hypotheses to explain the failure of the chloroform fumigation incubation method in acid soils. *Soil Biology and Biochemistry* **19**, 689-696.

26. Lynch, J.M. and Panting, L.M. (1980) Cultivation and the soil biomass. *Soil Biology and Biochemistry* **12**, 29-33.

27. Toyota, K., Ritz, K. And Young, I.M. (1996) Survival of bacterial and fungal populations following chloroform- fumigation: effects of soil matric potential and bulk density. *Soil Biology and Biochemistry* **28**, 1545-1547.

28. Sparling, G. P. and West, A. W. (1989) Importance of soil water content when estimating soil microbial C, N and P by the fumigation-extraction methods. *Soil Biology and Biochemistry* **21**, 245-253.

29. Mueller, T., Lavahun, M.F.E., Joergensen, R.G. and Meyer, B. (1996) The problem of pretreatment and unintentional variations in the fumigation-extraction method for time-course measurements in the field.

46

Biology and Fertility of Soils **22**, 167-170.

30. Jordan, D. and Beare, M. H. (1991) A comparison of methods for estimating soil microbial biomass carbon. *Agriculture, Ecosystems and Environment* **34**, 35-41.

31. Tate, K. R., Ross, D. J. and Feltham, C. W. (1988) A direct extraction method to estimate soil microbial C: effects of experimental variables and some different calibration procedures. *Soil Biology and Biochemistry* **20**, 329-335.

32. Kaiser, E.-A., Mueller, T., Joergensen, R. G., Insam, H. and Heinemeyer, O. (1992) Evaluation of methods to estimate the soil microbial biomass and the relationship with soil texture and organic matter. *Soil Biology and Biochemistry* **24**, 675-683.

33. Sparling, G. P. and West, A. W. (1988) Modifications to the fumigation-extraction technique to permit simultaneous extraction and estimation of soil microbial C and N. *Communications in Soil Science and Plant Analysis* **19**, 327-344.

34. Sparling, G. P., Feltham, C. W., Reynolds, J., West, A. W. and Singleton, P. L. (1990) Estimation of soil microbial C by a fumigation-extraction method: use on soils of high organic matter content, and a reassessment of the k_{EC}-factor. *Soil Biology and Biochemistry* **22**, 301-307.

35. Martikainen, P. J. and Palojarvi, A. (1990) Evaluation of the fumigation-extraction method for the determination of microbial C and N in a range of forest soils. *Soil Biology and Biochemistry* **22**, 797-802.

36. Wu, J., Joergensen, R. G., Pommerening, B., Chaussod, R. and Brookes, P. C. (1990) Measurement of soil microbial biomass C by fumigation-extraction-an automated procedure. *Soil Biology and Biochemistry* **22**, 1167-1169.

37. Eberhardt, U., Apel, G., A. and Joergensen, R. G. (1996) Effects of direct chloroform fumigation on suspended cells of 14C and 32P labelled bacteria and fungi. *Soil Biology and Biochemistry* **28**, 677-679.

38. Inubushi, K., Brookes, P. C. and Jenkinson, D. S. (1991) Soil microbial biomass C, N and ninhydrin-N in aerobic and anaerobic soils measured by the fumigation-extraction method. *Soil Biology and Biochemistry* **23**, 737-741.

39. Jenkinson, D. S. (1988) Determination of microbial biomass carbon and nitrogen in soil. In: *Advances in Nitrogen Cycling in Agricultural Ecosystems*. J. R. Wilson (Ed.), pp. 368-386. C.A.B International, Wallingford.

40. Ross, D. J. (1990) Estimation of soil microbial C by a fumigation-extraction method: influence of seasons, soils and calibration with the fumigation-incubation procedure. *Soil Biology and Biochemistry* **22**, 295-300.

41. Brookes, P. C., Landman, A., Pruden, G. and Jenkinson, D. S. (1985) Chloroform fumigation and the release of soil nitrogen: a rapid direct extraction method to measure microbial biomass nitrogen in soil. *Soil Biology and Biochemistry* **17**, 837-842.

42. Joergensen, R. G. and Brookes, P. C. (1990) Ninhydrin-reactive nitrogen measurements of microbial biomass in 0.5 M K2SO4 soil extracts. *Soil Biology and Biochemistry* **22**, 1023-1027.

43. Brookes, P. C., Powlson, D. S. and Jenkinson, D. S. (1982) Measurement of microbial biomass phosphorous in soil. *Soil Biology and Biochemistry* **4**, 319-329.

44. Saggar, S., Bettany, J. R. and Stewart, J. W. (1981) Measurement of microbial sulphur in soil. *Soil Biology and Biochemistry* **13**, 493-498.

45. Joergensen, R. G., Mueller, T. and Wolters, V. (1996) Total carbohydrates of the soil microbial biomass in 0.5M K2SO4 soil extracts. *Soil Biology and Biochemistry* **28**, 1147-1153.

46. Islam, K.R. and Weil, R.R (1998) Microwave irradiation of soil for routine measurement of microbial biomass carbon. *Biology and Fertility of Soils* **27**, 408-416.

47. Monz, C.A. Reuss, D.E. and Elliott, E.T. (1991) Soil microbial biomass carbon and nitrogen estimates using 2450-MHZ microwave irradiation or chloroform fumigation followed by direct extraction. *Agriculture Ecosystems and Environment* **34**, 55-63.

48. Islam, K.R., Weil, R.R., Mulchi, C.L. and Glenn, S.D. (1997) Freeze-dried soil extraction method for the measurement of microbial biomass C. *Biology and Fertility of Soils* **24**, 205-210.

49. Anderson, J. P. E. and Domsch, K. H. (1973) Quantification of bacterial and fungal contributions to soil respiration. Archives of Microbiology 93, 113-127.

50. Anderson, J. P. E. and Domsch, K. H. (1978) A physiological method for the quantitative measurement of microbial biomass in soils. *Soil Biology and Biochemistry* **10**, 215-221.

51. Sparling, G. P. (1995) The substrate-induced respiration method. In: *Methods in Applied Soil Microbiology and Biochemistry*. K. Alef and P. Nannipieri (Eds.), pp. 397-404. Academic Press. London.

52. Nannipieri, P. (1990) Microbial biomass and activity measurements in soil: ecological significance. In: *Soil Biochemistry*. J -M Bollag and G. Stotzky (Eds.), pp. 293-355. Marcel Dekker, New York.

53. Parkinson, D. and Coleman, D. C. (1991) Microbial communities, activity and biomass. *Agriculture Ecosystems and Environment* **34**, 3-33.

54. Santruckova, H. (1992) Microbial biomass, activity and soil respiration in relation to secondary succession. *Pedobiologia* **36**, 341-350.

55. Sarathchandra, S. U., Perrott, K. W. and Littler, R. A. (1989) Soil microbial biomass: influence of simulated temperature changes on size, activity and nutrient-content. *Soil Biology and Biochemistry* **21**, 987-993.

56. Scheu, S., Maraun, M., Bonkowski, M. and Alphei, J. (1996) Microbial biomass and respiratory activity in soil aggregates of different sizes from three beechwood sites on a basalt hill. *Biology and Fertility of Soils* **21**, 69-76.

57. Wardle, D. A., Nicholson, K. S. and Rahman, A. (1994) Influence of herbicide applications on the decomposition, microbial biomass, and microbial activity of pasture shoot and root litter. *New Zealand Journal of Agricultural Research* **37**, 29-39.

58. Smolander, A., Kurka, A., Kitunen, V. and Malkonen, E. (1994) Microbial biomass C and N, and respiratory activity in soil of repeatedly limed and N- and P- fertilized Norway Spruce stands. *Soil Biology and Biochemistry* **26**, 957-962.

59. Anderson, T.-H. and Domsch, K. H. (1985) Determination of ecophysiological maintenance carbon requirements of soil microorganisms in a dormant state. *Biology and Fertility of Soils* **1**, 81-89.

60. Insam, H. and Domsch, K. H. (1988) Relationship between soil organic carbon and microbial biomass on chronosequences of reclamation sites. *Microbial Ecology* **15**, 177-188.

61. Ocio, J. A. and Brookes, P. C. (1990) An evaluation of methods for measuring microbial biomass in soils following recent additions of wheat straw and the characterisation of the biomass that develops. *Soil Biology and Biochemistry* **22**, 685-694.

62. Wardle, D. A. and Ghani, A. (1995) A critique of the microbial metabolic quotient (qCO2) as a bioindicator of disturbance and ecosystem development. *Soil Biology and Biochemistry* **27**, 1601-1610.

63. Anderson, T.-H. (1994) Physiological analysis of microbial communities in soil: Applications and Limitations. In: *Beyond the Biomass.* K. Ritz, J. Dighton and K.E. Giller (Eds.), pp. 67-75. Wiley-Sayce.

64. Burns R.G. (1982) Enzyme activity in soil: location and a possible role in microbial ecology. *Soil Biology and Biochemistry* **14**, 423-427.

65. Alef, K. (1995) Dehydrogenase activity. In: *Methods in Applied Soil Microbiology and Biochemistry.* K. Alef and P. Nannipieri (Eds.), pp. 228-231.Academic Press, London.

66. Harris J.A. and Birch P. (1989) Soil microbial activity in opencast coal mine restorations. *Soil Use and Management* **5**, 155-160.

67. Olsen, R. A. and Bakken, L. R. (1987) Viability of soil bacteria: optimisation of plate-counting technique and comparison between total counts and plate counts within different size groups. *Microbial Ecology* **13**, 59-74.

68. Grant, W.D. and West, A.W. (1986) Measurement of bound muramic and m-diaminopimelic acid and glucosamine in soil: evaluation as indicators of microbial biomass. *Journal of Microbial Methods* **6**, 47-53.

69. Tunlid, A. and White, D. (1992) Biochemical analysis of biomass, community structure, nutritional status and metabolic activity of microbial communities in soil. In: *Soil Biochemistry.* G. Stotzky and J.-M. Bollag (Eds.), pp. 229-262. Dekker, New York.

70. Ruzicka, S., Norman, M. and Harris J.A. (1995) Rapid ultrasonication method to determine ergosterol concentration in soil. *Soil Biology and Biochemistry,* **27**, 1215-1217.

71. Ratledge, C. and Wilkinson, S. G. (1988) An overview of microbial lipids. In: *Microbial Lipids.* C. Ratledge and S. G. Wilkinson (Eds.). Academic Press, London.

72. Kates, M. (1964) Bacterial lipids. *Advances in Lipid Research* **2**, 17-90.

73. Hill, T., McPherson, E.F., Harris, J.A. and Birch P. (1993) Microbial biomass estimated by phospholipid phosphate in soils with diverse microbial communities. *Soil Biology and Biochemistry* **25**, 1779-1786.

74. Bååth, E., Frostegård, Å. and Fritze, H. (1992) Soil bacterial biomass, activity, phospholipid fatty acid pattern, and pH tolerance in an area polluted with alkaline dust deposition. *Applied and Environmental Microbiology* **58**, 4026-4031.

75. Pennanen, T., Frostegård, Å., Fritze, H. and Bååth, E. (1996) Phospholipid fatty acid composition and heavy metal tolerance of soil microbial communities along two heavy metal -polluted gradients in coniferous forests. *Applied and Environmental Microbiology* **62**, 420-428.

76. Pennanen, T., Fritze, H., Vanhala, P., Kiikkilä, O., Neuvonen, S. and Bååth, E. (1998) Structure of a microbial community in soil after prolonged addition of low levels of simulated acid rain. *Applied and Environmental Microbiology* **64**, 2173-2180.

77.   Frostegård, Å., Bååth, E. and Tunlid, A. (1993) Shifts in the structure of soil microbial communities in limed forests as revealed by phospholipid fatty acid analysis. *Soil Biology and Biochemistry* **25**, 723-730.

78.   Zelles, L., Rackwitz, R., Bai, Q. Y., Beck, T. and Beese, F. (1995) Discrimination of microbial diversity by fatty acid profiles of phospholipids and lipopolysaccharides in differently cultivated soils. *Plant and Soil* **170**, 115-122.

79.   Steer, J. and Harris J.A. (2000) Shifts in the microbial community in rhizosphere and non-rhizosphere soils during the growth of *Agrostis stolonifera* . *Soil Biology and Biochemistry* **32**, 869-878.

80.   Sundh, I., Nilsson, M. and Borgå, P. (1997) Variation in microbial community structure in two boreal peatlands as determined by analysis of phospholipid fatty acid profiles. *Applied and Environmental Microbiology* **63**, 1476-1482.

81.   Degens, B. P. and Harris, J. A. (1997) Development of a physiological approach to measuring the catabolic diversity of soil microbial communities. *Soil Biology and Biochemistry* **29**, 1309-1320.

82.   Lee, C., Russell, N.J. and White, G.F. (1995) Rapid screening of bacterial phenotypes capable of biodegrading anionic surfactants: Development and validation of a microtitre plate method. *Microbiology* **141**, 2801-2810.

83.   Landi, L., Badalucco, L., Pomare, F. and Nannipieri, P. (1993) Effectiveness of antibiotics to distinguish the contributions of fungi and bacteria to net nitrogen mineralisation, nitrification and respiration. *Soil Biology and Biochemistry* **25**, 1771-1778.

84.   Lin, Q., and Brookes, P.C. (1999) An evaluation of the substrate-induced respiration method. *Soil Biology and Biochemistry* **31**, 1969-1983.

85.   Moffett, B.F., Walsh, K.A., Harris, J.A., and Hill, T.C.J. (2000) Analysis of Bacterial Community Structure using 16S rDNA analysis. *Anaerobe* **6**, 129-131.

86.   Brock, T.D. (1987) The study of microorganisms *in situ*: progress and problems. In: *Ecology of Microbial Communities* , Society for General Microbiology, Symposium No. 41. M. Fletcher, T.R.G. Gray, and J.G. Jones (Eds.), pp. 1-17. Cambridge University Press, Cambridge.

87.   Saano, A., and Lindstrom, K. (1995) Isolation and identification of DNA from soil. In: *Methods in Applied Soil Microbiology and Biochemistry*, Alef, K. and Nannipieri, P., (Eds.), p. 440. Academic Press, San Diego.

88.   Lee, N., Nielsen, P.H., Andreasen, K.H., Juretschko, S., Nielsen, J.L., Schleifer, K-H. and Wagner, M. (1999). Combination of fluorescent *in situ* hybridization and microautoradiography – a new tool for structure-function analyses in microbial ecology. *Applied and Environmental Microbiology* **65**, 1289-1297.

89.   Ouverney, C.C., and Fuhrman, J.A. (1999) Combined microautoradiography-16S rRNA Probe Technique for Determination of Radioisotope Uptake by Specific Microbial Cell Types *In Situ*. *Applied and Environmental Microbiology* **65**, 1746-1752.

90.   Torsvik, V., Goksoyr, J., Daae, F.L., Sorheim, Micjhalsen, and Salte, K. (1995) Use of DNA analysis to determine the diversity of microbial communities. In: *Beyond the Biomass*. Ritz, K., Dighton, J., and Giller (Eds.), K.E. Wiley-Sace, New York.

91.   Tiedje, J.M., AsumingBrempong, S., Nusslein, K., Marsh, T.L. and Flynn, S.J. (1999) Opening the black box of soil microbial diversity. *Applied Soil Ecology* **13**, 109-122.

92.   Muyzer, G. (1999) DGGE/TGGE a method for identifying genes from natural ecosystems. *Current Opinions in Microbiology* **2**, 317-322.

93.   Russek-Cohen, E. and Colwell, R.R. (1998) Numerical Classification Methods. In: *Techniques in Microbial Ecology* (Burlage, R.S., Atlas, R., Stahl, D., Geesey, G. and Sayler, G. Eds). Oxford University Press.

94.   Andren, O., and Balandreau, J. (1999) Biodiversity and soil functioning - from black box to can of worms? *Applied Soil Ecology* **13**, 105-108.

PART II

COMPARISON OF EFFECTIVE ORGANISMS IN BIOREMEDIATION PROCESS

RECENT ADVANCES IN THE BIODEGRADATION OF POLYCYCLIC AROMATIC HYDROCARBONS BY *Mycobacterium* SPECIES

C.E. CERNIGLIA

U.S. Food and Drug Administration, National Center for Toxicological Research, Division of Microbiology, 3900 NCTR Rd., Jefferson, AR 72079, USA

Abstract

Polycyclic aromatic hydrocarbons (PAHs) constitute a class of organic compounds containing two or more fused benzene rings in linear, angular, and cluster arrangements. The environmental fate of these ubiquitous contaminants is of concern because of the mutagenicity, ecotoxicity, and carcinogenic potential of high-molar-mass PAHs. A variety of bacterial species have been isolated that have the ability to degrade PAHs with two rings (naphthalene) and three rings (anthracene and phenanthrene). Most of the research on the degradative biochemical pathways, the genes involved in PAH metabolism, and genetic regulation has been on *Pseudomonas*, *Sphingomonas*, *Burkholderia*, and *Comamonas* strains. Recent reports have shown that various *Mycobacterium*, *Nocardia*, and *Rhodococcus* species have the ability to degrade PAHs containing more than three rings (such as pyrene, fluoranthene, and benzo[a]pyrene). *Mycobacterium* sp. PYR-1 (reclassified as *Mycobacterium vanbaalenii* strain PYR-1), which was originally isolated from oil-contaminated estuarine and marine sediments, is capable of mineralizing PAHs, such as naphthalene, pyrene, fluoranthene, phenanthrene, anthracene and benzo[a]pyrene. Biodegradation pathways have been elucidated which suggest that *Mycobacterium* sp. PYR-1 metabolizes PAHs through similar and unique catabolic pathways compared to that reported for Gram-negative bacteria. *Mycobacterium* sp. PYR-1 enhanced the degradation of four different aromatic ring classes of PAHs when inoculated into microcosms containing sediment and water from estuarine and freshwater environments. Molecular techniques to detect the *Mycobacterium* indicated that it survived and performed well in mixed-sediment microbial populations. Analysis of protein expression in *Mycobacterium* sp. PYR-1 using 2-dimensional polyacrylamide gel electrophoresis enabled detection of several major proteins whose activity was increased after induction by PAHs. The cloning and sequence analysis of some of the genes encoding PAH degradation have been conducted. These studies demonstrate the

V. Šašek et al. (eds.),
The Utilization of Bioremediation to Reduce Soil Contamination: Problems and Solutions, 51–73.
© 2003 *Kluwer Academic Publishers. Printed in the Netherlands.*

bioremediation potential of *Mycobacterium* species to degrade PAHs in aquatic and terrestrial environments.

1. Introduction

Polycyclic aromatic hydrocarbons (PAHs) constitute a class of organic compounds containing two or more fused benzene rings in linear, angular, and cluster arrangements. The environmental fate of these ubiquitous contaminants is of concern because of the mutagenicity, ecotoxicity, and carcinogenic potential of high-molar-mass PAHs. A variety of bacterial species have been isolated that have the ability to degrade PAHs with two rings (naphthalene) and three rings (anthracene and phenanthrene). Most of the research on the degradative biochemical pathways, the genes involved in PAH metabolism, and genetic regulation has been on *Pseudomonas, Sphingomonas, Burkholderia, and Comamonas* strains. Recent reports have shown that various *Mycobacterium, Nocardia,* and *Rhodococcus* species have the ability to degrade PAHs containing more than three rings (such as pyrene, fluoranthene, and benzo[a]pyrene). *Mycobacterium* sp. PYR-1 (reclassified as *Mycobacterium vanbaalenii* strain PYR-1), which was originally isolated from oil-contaminated estuarine and marine sediments, is capable of mineralizing PAHs, such as naphthalene, pyrene, fluoranthene, phenanthrene, anthracene and benzo[a]pyrene. Biodegradation pathways have been elucidated which suggest that *Mycobacterium* sp. PYR-1 metabolizes PAHs through similar and unique catabolic pathways compared to that reported for Gram-negative bacteria. *Mycobacterium* sp. PYR-1 enhanced the degradation of four different aromatic ring classes of PAHs when inoculated into microcosms containing sediment and water from estuarine and freshwater environments. Molecular techniques to detect the *Mycobacterium* indicated that it survived and performed well in mixed-sediment microbial populations. Analysis of protein expression in *Mycobacterium* sp. PYR-1 using 2-dimensional polyacrylamide gel electrophoresis enabled detection of several major proteins whose activity was increased after induction by PAHs. The cloning and sequence analysis of some of the genes encoding PAH degradation have been conducted. These studies demonstrate the bioremediation potential of *Mycobacterium* species to degrade PAHs in aquatic and terrestrial environments.

Polycyclic aromatic hydrocarbons (PAHs) are ubiquitous environmental pollutants produced from incomplete combustion of organic material [3]. PAHs are composed of fused aromatic rings in linear, angular, or cluster arrangements (Figure 1); their presence in contaminated soils and sediments poses a significant risk to the environment since they have ecotoxic, mutagenic, and in some cases carcinogenic effects on human tissues [77, 78]. Due to these properties, PAHs have been included in the U.S. *Environmental Protection Agency* and the *European Union* priority list of pollutants. PAHs are major constituents of crude oil, creosote and coal tar and contaminate the environment *via* many routes, including fossil fuel combustion, manufactured gas and coal tar production, wood treatment processes, automobile exhaust and waste incineration (Figure 2). PAHs enter the marine environment from a variety of sources, including petroleum pollution, fallout from air pollution, effluents from industries and sewage treatment plants and creosoted wharves [65]. Seafood is an interesting example of the various routes by which human foods may become contaminated by PAHs. Bivalve mollusks, such as clams, oysters, and mussels readily accumulate PAHs from contaminated waters.

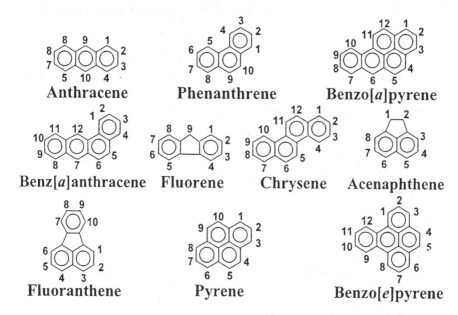

FIGURE 1.   Chemical structures of selected polycyclic aromatic hydrocarbons.

PAHs are hydrophobic compounds whose persistence within the environment is due chiefly to their low aqueous solubility. They adsorb onto soils and particulates, influencing their bioavailability and biodegradation [43] (Figure 3). Whereas low-molar-mass PAHs (naphthalene, phenanthrene, anthracene) are usually readily degraded, high-molar-mass PAHs (benzo[a]pyrene, benz[a]anthracene) resist extensive microbial degradation in soil and sediments [7, 31, 32]. The environmental fate of PAHs is important since they are the largest class of chemical carcinogens known today and their detection in air, water, soil and food has raised concern about the risk of exposure to public health [1]. Factors which affect the environmental loss of PAHs include atmospheric photolysis, sorption, water and lipid solubility, sedimentation, chemical oxidation, volatilization, and microbial degradation [7, 31] (Figure 4).

Removal of PAHs from soil is a necessary requirement of soil remediation to comply with relevant clean up standards based on individual PAH standards. To remediate contaminated sites, three methods are typically used: (1) physical removal, followed by concentration of the PAH, (2) biodegradation of the PAH and (3) chemical/thermal destruction of the PAH. The remediation strategy depends upon

54

the type, concentration and concentration distribution of the pollutants; physical state of the pollutant, type of soil, site specific aspects, such as size and history of the polluted site.

Natural oil seeps

Refinery and oil storage waste

Accidental spills from oil tankers and other ships

Municipal and urban wastewater discharge runoff

River-borne pollution

Atmospheric fallout of fly ash particulates

Petrochemical industrial effluents

Coal tar and other coal processing wastes

Automobile engine exhausts

Combustion of fossil fuels

Tobacco and cigarette smoke

Forest and prairie fires

Rural an urban sewage sludge

Refuse and waste incineration

Coal gasification and liquefaction processes

Creosote and other wood preservatives wastes

Chronic input associated with boating activities

FIGURE 2. Major sources of polycyclic aromatic hydrocarbons in the environment.

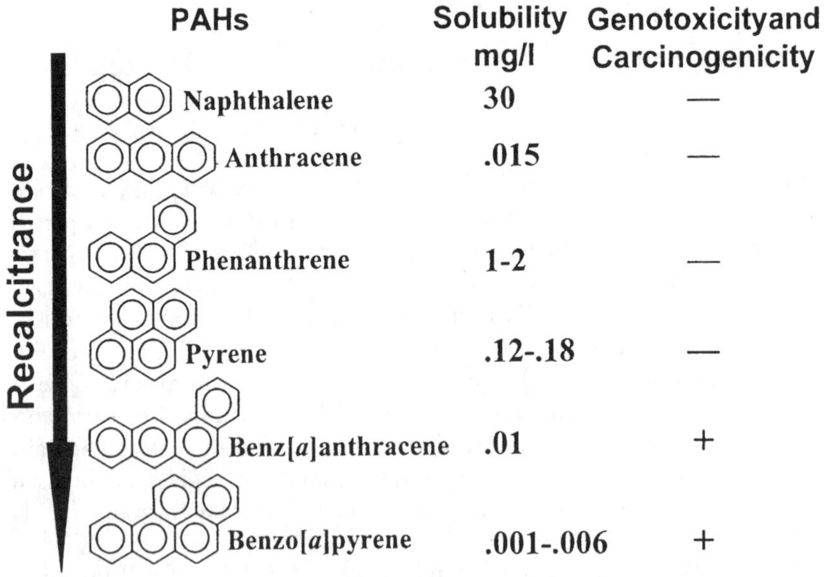

PAHs	Solubility mg/l	Genotoxicity and Carcinogenicity
Naphthalene	30	—
Anthracene	.015	—
Phenanthrene	1-2	—
Pyrene	.12-.18	—
Benz[a]anthracene	.01	+
Benzo[a]pyrene	.001-.006	+

FIGURE 3. Depiction of polycyclic aromatic hydrocarbons with solubility, genotoxicity, and biodegradation characteristics.

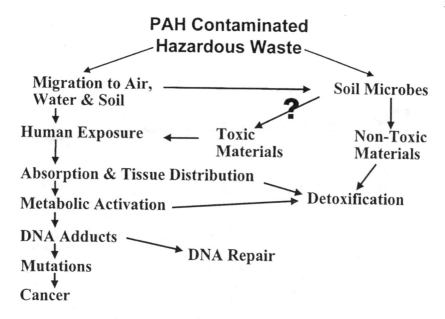

FIGURE 4.   Environmental fate and toxicological consequences of polycyclic aromatic hydrocarbons.

There is considerable interest in the use of biological systems to remediate PAH contaminated environments [77]. Research in PAH bioremediation has centered on four important aspects. First, the characterization of the biodegradation process useful in the treatment of PAHs. Second, the development of methods and procedures for enhancing the rate, extent and breadth of specificity of those processes. Third, the design and engineering of reaction systems and treatment strategies that successfully optimize the application of this biodegradation process and fourth, development of information on the ecological health risk associated with exposure to the PAHs [52].

2.   Toxicity of PAHs

A 1761 report by physician John Hill, recognizing the link between excessive use of tobacco snuff and nasal cancer, began over two centuries of research on PAH carcinogenesis [6]. In 1775, Percival Pott related chimney sweeps' exposure to soot with scrotal skin cancer. In 1915, Yamigiwa and Ichikawa reported that, after repeated applications of coal tar on the ears of rabbits, tumors were formed. From 1930 to 1955, Kennaway, Hieger, Cook and Hewett established that the carcinogenic fraction of coal tar contained PAHs. In the 1970s, James and Elizabeth Miller [49] showed that many chemicals require metabolic activation to express toxicity. It is now well established that PAHs must be metabolically activated by mammalian microsomal enzymes to elicit their latent mutagenic, genotoxic and carcinogenic properties.

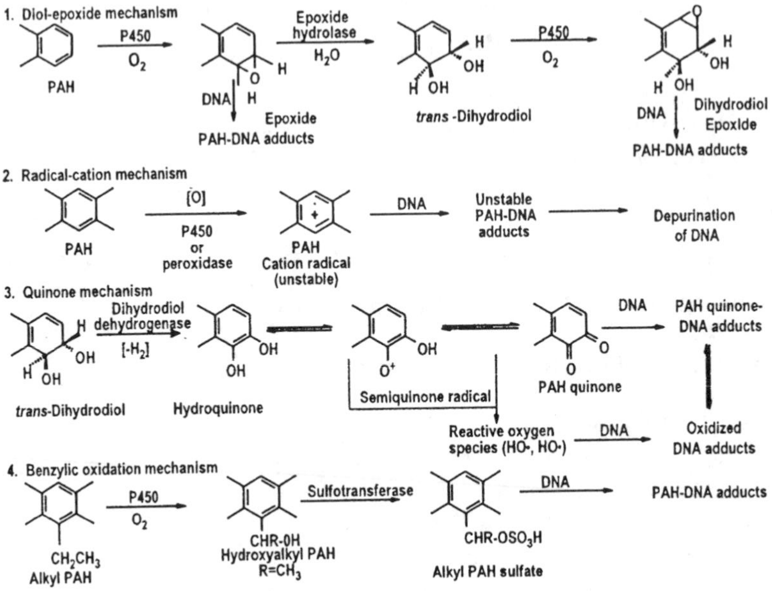

FIGURE 5. Mechanisms of PAH carcinogenesis (after [24]).

The mechanisms of PAH carcinogenesis (Figure 5) have recently been reviewed [24, 25] and indicated that there are at least four mechanisms: (1) the dihydrodiol epoxide mechanism, which involves metabolic activation of the PAH by microsomal cytochrome P450 enzymes to reactive epoxide and diol-epoxide intermediates that form covalent adducts with DNA, perhaps resulting in mutations that lead to tumorigenesis; (2) the radical-cation mechanism, which involves one-electron oxidation to generate radical-cation intermediates that may attack DNA, resulting in depurination; (3) the quinone mechanism, which involves enzymic dehydrogenation of dihydrodiol metabolites to yield quinone intermediates that may either combine directly with DNA or enter into a redox cycle with O_2 to generate reactive oxygen species capable of attacking DNA; and (4) the benzylic oxidation mechanism, which entails formation of benzylic alcohols that are converted by sulfotransferase enzymes to reactive sulfate esters that may attack DNA. The most significant mechanism of carcinogenesis by PAHs is the diol-epoxide pathway. Since microbial degradation of PAHs has the potential to form reactive intermediates, such as dihydrodiol epoxides or quinones, the metabolic profiles should be determined to see if potentially toxic intermediates are formed during the bioremediation of PAH-contaminated sites [9].

Some PAHs, but not all, are acutely toxic, mutagenic, or carcinogenic. For instance, the combination of anthracene and solar ultraviolet radiation is acutely toxic and immunosuppressive to fish [69]. Benzo[a]pyrene, benz[a]anthracene, chrysene, and several other PAHs, after metabolic activation by liver enzymes, induce mutations in bacteria [23]. Some PAH metabolites bind to DNA, RNA, and proteins; the resulting adducts may cause damage directly to cells and also have teratogenic or carcinogenic effects [23]. Exposure to high concentrations of PAHs in the workplace has been associated with lung and bladder cancers among industrial workers [44]. Although some PAHs show weak estrogenic or antiestrogenic activity [62], these effects are overshadowed by the carcinogenic properties of PAHs.

3. General Principles in the Microbial Degradation of PAHs

Bacteria and fungi metabolize a wide variety of PAHs, converting them either completely to CO_2 or else to various microbial metabolites that are considered dead-end products and do not result in the production of CO_2 [7, 8]. The elucidation of microbial degradation pathways is necessary to determine the extent of degradation and whether the metabolites that are formed are toxic or biologically inactive. Depending upon the enzymic repertoires of the microorganisms, different mechanisms are used to metabolize PAHs (Figure 6) [12, 18]. Bacterial degradation of PAHs generally proceeds *via* the action of multicomponent dioxygenases to form *cis*-dihydrodiols. These compounds are subsequently dehydrogenated to form dihydroxy-PAHs, which may be substrates for ring-fission enzymes [66]. Many bacteria are capable of complete mineralization of PAHs to form CO_2. Recent findings also indicate that PAHs can be metabolized by monooxygenases in bacteria to form *trans*-dihydrodiols, although this activity is generally lower than the dioxygenase activity in the same organism [28, 29].

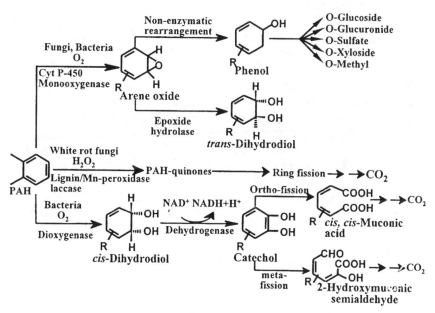

FIGURE 6.   Initial reactions in the degradation of PAHs by bacteria and fungi.

Non-ligninolytic fungi, such as *Cunninghamella elegans*, *Penicillium janthinellum*, and *Syncephalastrum* sp., metabolize PAHs *via* reactions catalyzed by cytochrome P450 monooxygenase and epoxide hydrolase to form *trans*-dihydrodiols. These reactions are highly regio- and stereoselective. Other polar metabolites formed include phenols, quinones, and conjugates [66,12]. The types of metabolites isolated are similar to those formed by mammals. Ligniolytic fungi, such as *Phanerochaete chrysosporium, Pleurotus ostreatus,* and *Trametes versicolor* degrade PAHs by nonspecific radical oxidation, catalyzed by extracellular ligninolytic enzymes, that leads primarily to PAH quinones. Some ligninolytic fungi can further metabolize PAH quinones by cleaving the aromatic rings, with subsequent mineralization of the PAH to CO_2 [22]. Since PAHs are relatively insoluble in water and bind strongly to organic matter in sediments and soils, they may not be accessible to microbial degradation so that they persist in anoxic environments [2]. There have been recent reports demonstrating the anaerobic degradation of PAHs when nitrate, sulfate or ferric iron serves as the terminal electron acceptor [15, 46, 48, 59, 80].

4. Bacterial Degradation of PAHs

The aerobic degradation of PAHs, especially the low-molar-mass ones, such as naphthalene, anthracene, and phenanthrene, has been extensively studied. Research on the purification and characterization and on the genes for the initial steps in the degradation of low molecular weight PAHs, such as naphthalene, have been elucidated from studies with *Pseudomonas, Sphingomonas, Burkholderia*, and *Comamonas* strains [47, 61, 79, 81, 82]. The catabolism of PAHs is initiated by oxidation *via* dioxygenases with the formation of the resultant *cis*-dihydrodiols. Dioxygenases catalyze the incorporation of both atoms of oxygen and two electrons from NADH into the resulting *cis*-dihydrodiol [7, 66]. This reaction is catalyzed by multi-component dioxygenases consisting of a reductase, a ferredoxin, and an iron−sulfur protein. The subsequent dehydrogenation by a NAD^+-dependent dehydrogenase yields dihydroxylated intermediates, which are further degraded through ring-cleavage pathways with the eventual formation of citric-acid cycle intermediates. Although most of what is known about the degradation of PAHs is based on studies with *Pseudomonas* and *Sphingomonas*, recently Gram-positive nocardioform organisms (*Mycobacterium, Nocardia*, and *Rhodococcus* species) have had the ability to degrade PAHs containing more than three rings [4, 5, 14, 16, 17, 19−21, 26−30, 33−37, 41, 42, 54, 55, 57, 58, 60, 63, 64, 67, 68, 71, 74, 75]. The aim of this article is to present an overview of current information on the metabolism of these environmentally hazardous chemicals, PAHs by members of the genus *Mycobacterium* and their potential for use in the bioremediation of PAH-contaminated aquatic and terrestrial ecosystems.

Mycobacterium is an old bacterial genus, which includes species from a broad range of ecological niches. Although most mycobacteria are free-living saprophytic organisms, much of the study of this genus has focused on those species that are pathogenic to humans. Pathogenicity ranges from strict specificity for humans and animals, such as *Mycobacterium tuberculosis, M. leprae,* and *M. paratuberculosis,* to specificity as saprophytes, such as *M. smegmatis* and *M. phlei.*

Recently members of the genus *Mycobacterium* have been reported to degrade a wide variety of environmentally hazardous chemicals, including high-molar-mass PAHs in soils and sediments [13, 45, 50, 56, 58, 70, 76]. Research in this laboratory has shown that *Mycobacterium* sp. PYR-1 can extensively degrade PAHs with up to five fused aromatic rings [28, 29, 34, 51].

The objectives of our present research program on the biodegradation of PAHs are: (*1*) to determine if *Mycobacterium* PYR-1 has the ability to metabolize PAHs, (*2*) to determine the metabolites produced by *Mycobacterium* sp. PYR-1 and elucidate the metabolic pathways, (*3*) to determine the biological and toxicological activity of the metabolites, (*4*) to characterize the enzymes involved in the metabolism of PAHs, (*5*) to elucidate the enzymic mechanisms and genes that mediate PAH metabolism, and (*6*) to determine the potential of *Mycobacterium* sp. PYR-1 for *in situ* biodegradation of PAHs.

5. Isolation of Polycyclic Aromatic Hydrocarbon-Degrading *Mycobacterium*

Mycobacterium sp. PYR-1 was isolated from sediments taken from a drainage pond in an estuarine area near an oil field, which was chemically exposed to petrogenic chemicals, in Port Aransas, Texas [26, 29]. The bacterium was isolated from a 500-mL microcosm containing 20 g of estuarine sediment, 180 mL of estuarine water, and 100 μg of pyrene [10]. After incubation of the microcosm for 25 d under aerobic conditions, the sediment samples were serially diluted and screened for the presence of pyrene-degrading microorganisms using the spray-plate technique [11, 40]. The surfaces of the agar plates were sprayed with a 2 % (*W/V*) solution of pyrene dissolved in acetone—hexane (1:1, *V/V*) and dried overnight at 35 °C to volatilize the solvent. Inocula (100 μL) from the 10^{-1}, 10^{-2}, 10^{-3} and 10^{-4} dilutions of microcosm sediments were gently spread with sterile glass rods onto the agar surface; the plates were inverted and incubated for 3 weeks at 24 °C in sealed plastic bags to conserve moisture. Colonies which can utilize the PAH will produce either zones of clearing or colored zones. After 2–3 weeks, colonies surrounded by clear zones due to pyrene uptake and utilization were observed (Figure 7). The colonies were picked and subcultured into fresh mineral salts medium containing 250 μg/L each of peptone, yeast extract, and soluble starch, and 0.5 μg/mL of pyrene dissolved in dimethylformamide. After three successive transfers, a bacterium was isolated that was able to degrade pyrene. The isolate was designated as strain PYR-1 and further identified as a *Mycobacterium* species based on the following morphological, biochemical, and 16S rRNA properties. It formed Gram-positive, acid-fast rods (1.4 μm in length and 0.7 μm in width). The colonies were smooth and saffron-yellow on Middlebrook medium. The 15 biochemical tests, mole percent G + C analysis of 66 %, characterization of mycolic acids with a carbon chain length of $C_{58}-C_{64}$, and fatty acid analysis were consistent with the assignment of this organism to the genus *Mycobacterium* [29, 53]. A comparison of the 16S rRNA sequence of PYR-1 with other known mycobacterial sequences available from Gen Bank found that strain PYR-1 is a member of the subgroup of rapidly-growing scotochromogenic mycobacteria [72] (Figure 8). Recent studies indicate that the 16S ribosomal RNA sequence of strain PYR-1 is similar to the sequence of the type strain of *M. austroafricanum*, except for one gap at the nucleotide 43 position. On the basis of phylogenetic analysis using 16S-RNA gene sequences, fatty acid analysis, pulse-field gel electrophoresis, physiological, and chemotaxomical characteristics, strain PYR-1 has been designed a new species and named *Mycobacterium vanbaalenii* [39].

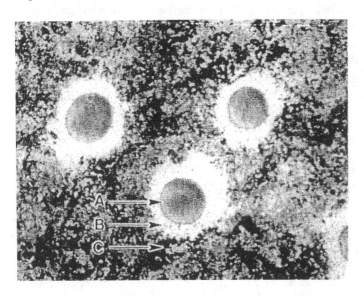

FIGURE 7.   Photograph of *Mycobacterium* sp. PYR-1 on an agar plate containing pyrene; **A** — *Mycobacterium* sp. PYR-1, **B** — the clear zone indicates pyrene utilization, **C** — pyrene.

6.   Metabolism of Polycyclic Aromatic Hydrocarbons by *Mycobacterium* sp. PYR-1

In mineral medium supplemented with yeast extract and starch, *Mycobacterium* sp. PYR-1 was able to mineralize a wide variety of PAHs, including naphthalene (60 % reduction), phenanthrene (51 % reduction), anthracene (48 % reduction), pyrene (63 % reduction), fluoranthene (90 % reduction), and benzo[a]pyrene (10 % reduction) to CO_2 within 2 d of incubation [26]. To test if the *Mycobacterium* sp. PYR-1 could degrade PAHs in a more complex mixture, 20 g of highly contaminated estuarine sediment from a creosote-contaminated site was extracted by Soxhlet extraction. The extract was further treated by applying it to a silica-alumina column and eluting it first with hexane and then with benzene. Flasks were incubated with *Mycobacterium* sp. PYR-1 and dosed with 512 µg of benzene-soluble PAH residue and incubated for 6 d. An average of 47 % of the total contaminants present was mineralized within 6 d [34]. In synthetic PAH mixture studies, we found that the order of degradation of individual PAHs on the mixture follows the solubility of these compounds in aqueous solution. Phenanthrene, fluoranthrene, pyrene, and anthracene were rapidly degraded by *Mycobacterium* sp. PYR-1. Significant degradation rates were observed among these four chemicals, the two tricyclic PAHs, anthracene and phenanthrene, appeared to be degraded to a greater extent with only 7.8 and 8.7 % remaining in the medium, respectively. Similarly, fluoranthene and pyrene, two tetracyclic compounds, were both degraded to 12.7 %. Benzo[*a*]pyrene appeared to be slowly degraded. Chrysene was minimally degraded [34]. These studies show the versatility and broad specificity for the enzyme systems involved in the initial oxidation of PAHs.

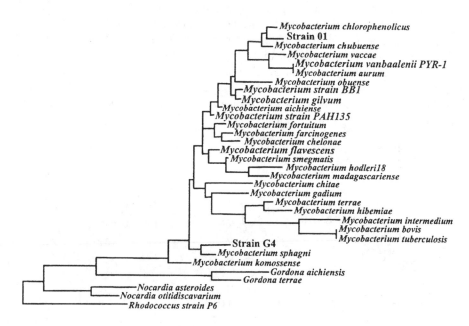

FIGURE 8.   Phylogenetic analysis of *Mycobacterium* species based on 16S RNA sequences.

7.   Mechanism of Oxidation of PAHs by *Mycobacterium* sp. PYR-1

Mycobacterium sp. PYR-1 initially oxidizes PAHs to form both *cis*- and *trans*-PAH dihydrodiols. A list of PAHs metabolized by *Mycobacterium* sp. PYR-1 and of the initial sites of oxidation is given in Figure 9. Oxygen-18 incorporation experiments showed that both atoms of the *cis*-dihydrodiol were derived from molecular oxygen,

62

but that only one atom of molecular oxygen was incorporated into the *trans*-dihydro-diol. The mechanism of oxidation is unique, since *Mycobacterium* sp. PYR-1 has both mono- and dioxygenases to catalyze the initial attack on the PAH [28, 35, 51]. The catabolic pathways for the degradation of anthracene, phenanthrene, pyrene, and benzo[a]pyrene by *Mycobacterium* species are shown in Figures 10–13. There are some common features in the oxidation of these compounds. For example, the initial attack on the pyrene nucleus occurs primarily in the K-region (4,5-positions) to form *cis*- and *trans*-pyrene dihydrodiols. Dehydrogenation to form 4,5-dihydroxy-pyrene followed by *ortho*-cleavage leads to the formation of 4,5-phenanthrene-dicarboxylic acid. Phenanthrene is also metabolized by *Mycobacterium* sp. PYR-1 with initial attack in the K-region to form *cis*- and *trans*-9,10-dihydrodiols. The formation of 2,2′-diphenic acid during phenanthrene degradation is analogous to the formation of 4,5-phenanthrene dicarboxylic acid during pyrene degradation. Therefore, it seems likely that the same mono- and dioxygenases and *ortho*-cleavage enzymes are involved in initial K-region attack and subsequent ring fission of the dihydroxylated intermediates. A naphthalene dioxygenase attack to form a *cis*-dihydrodiol in the 1,2-position of the naphthalene moiety is similar to the anthracene pathway. Pyrene is also metabolized at the 1,2-position on the aromatic ring to form the dead-end product 4-hydroxyperinaphthenone. *Mycobacterium* sp. PYR-1 also initially attacks phenanthrene in the 3,4-position to produce *cis*-3,4-phenanthrene dihydrodiol, with further metabolism to 1-hydroxy-2-naphthoic acid similar to pathways reported for Gram-negative organisms.

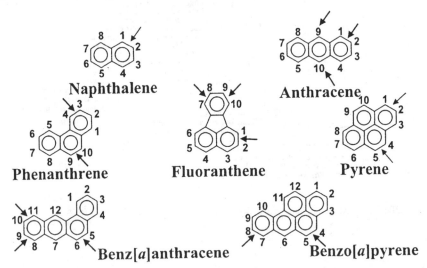

FIGURE 9. Sites of enzymic attack on the aromatic ring by *Mycobacterium* sp. PYR-1.

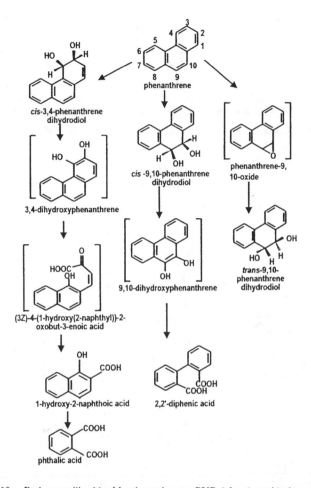

FIGURE 10. Pathway utilized by *Mycobacterium* sp. PYR-1 for the oxidation of phenanthrene.

8. Genetic Analysis of *Mycobacterium* sp. PYR-1

Very little is known about the enzymes and genes involved in the PAH degradation pathways in *Mycobacterium* species. Earlier studies have shown that pyrene metabolism is inducible in *Mycobacterium* sp. PYR-1. The ability of bacteria to degrade PAHs to less toxic compounds has been widely found to be encoded on plasmids. Induced and uninduced *Mycobacterium* sp. PYR-1 cells were analyzed for plasmids by pulsed-field gel electrophoresis. We did not detect any plasmids in *Mycobacterium* sp. PYR-1. These results suggest that the genes responsible for PAH degradation in *Mycobacterium* sp. PYR-1 reside on the chromosome. Previously, we showed that *Mycobacterium* sp. PYR-1 metabolizes PAHs at different sites of the aromatic ring, presumably *via* dioxygenase and monoxygenase attack on the PAH.

64

Initially, we screened the total genomic DNA with genes for known aromatic pathways by Southern hybridization. The genes utilized as probes included those for biphenyl degradation from *Pseudomonas* sp. LB400, those for naphthalene degradation using *P. putida* OU83, those for phenanthrene degradation using *Comamonas testosteroni* GZ37, those for toluene degradation using *P. putida* F1 and those for *m*-xylene, biphenyl, naphthalene, anthracene, and phenanthrene degradation using *Sphingomonas yanoikuyae* B1. There was no detectable genetic homology by Southern blotting experiments under low-stringency conditions capable of detecting 70 % identity between total DNA. These results suggest that genes for PAH degradation by *Mycobacterium* sp. PYR-1 are not closely related to other known genes with similar functions [38].

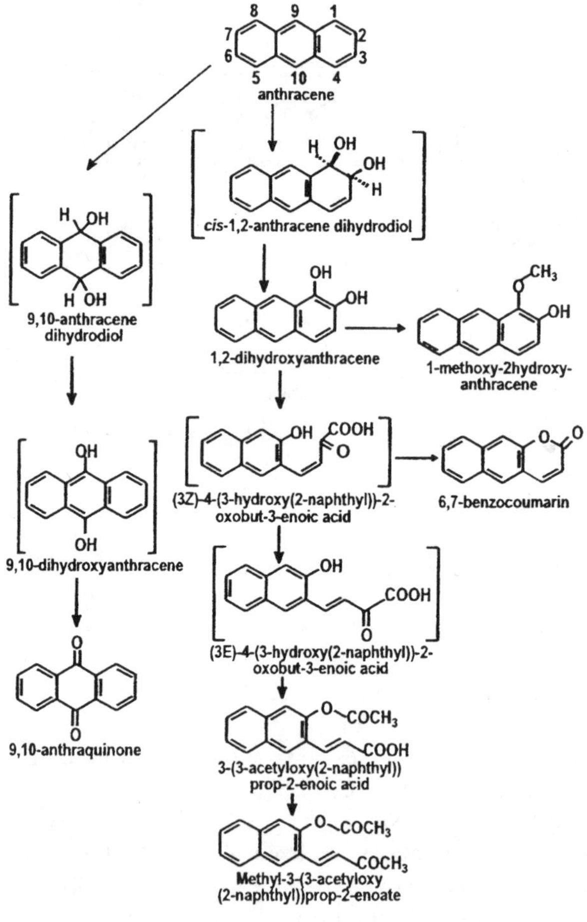

FIGURE 11.   Pathway utilized by *Mycobacterium* sp. PYR-1 for the oxidation of anthracene.

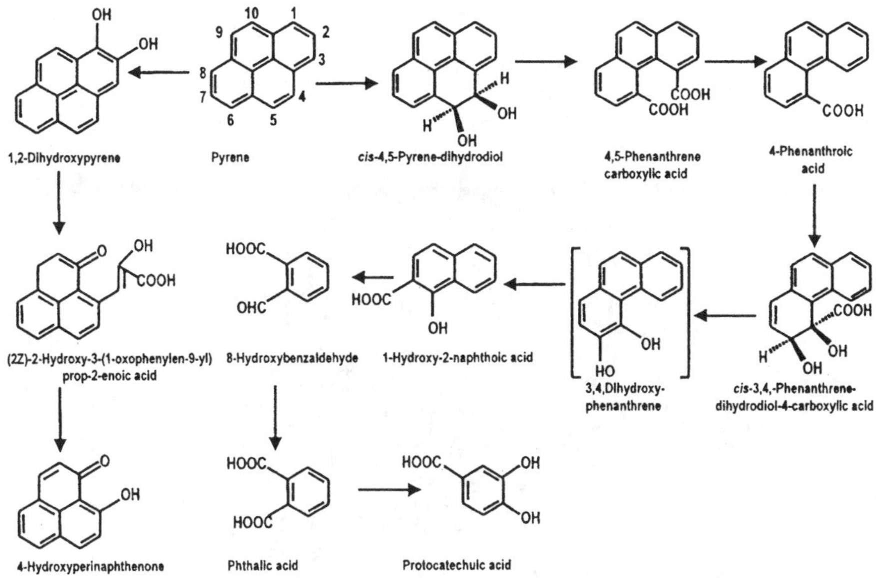

FIGURE 12.   Pathway utilized by *Mycobacterium* sp. PYR-1 for the oxidation of pyrene.

In order to clone genes involved in PAH degradation, two-dimensional (2D-SDS-PAGE) gel electrophoresis of pyrene-induced proteins from cultures of *Mycobacterium* sp. PYR-1 was conducted (Figure 14). Analysis of protein expression of *Mycobacterium* sp. PYR-1 using 2D-SDS-PAGE indicated that at least 5 major proteins were increased on induction of pyrene metabolism. An 81-kDa protein was induced in response to pyrene exposure. The N-terminal amino acid analysis showed that the 81-kDa protein is catalase−peroxidase, with significant homology to Kat GII of *Mycobacterium fortuitum*. The result raises the possibility that hydrogen peroxide participates in dioxygenase activity in *Mycobacterium* sp. PYR-1 [73]. Increased expression of a 50-kDa protein was also found in pyrene-induced cultures. The N-terminal amino acid sequences of the 50-kDa protein indicated that it is moderately similar to the large subunit of naphthalene-inducible dioxygenase. The amino acid sequences of the large, subunits of PAH dioxygenase from *Nocardioides* and *Rhodococcus* species showed about 50 to 42 % identical sequence with the PAH dioxygenase sequence of *Mycobacterium* sp. PYR-1 and quite distantly related (less than 40 %) to the genes encoding naphthalene dioxygenases of *Pseudomonas, Sphingomonas* and *Burkholderia* strain [38] (Figure 15). Studies are now being conducted to determine the gene organization for the dioxygenase system of *Mycobacterium* sp. PYR-1.

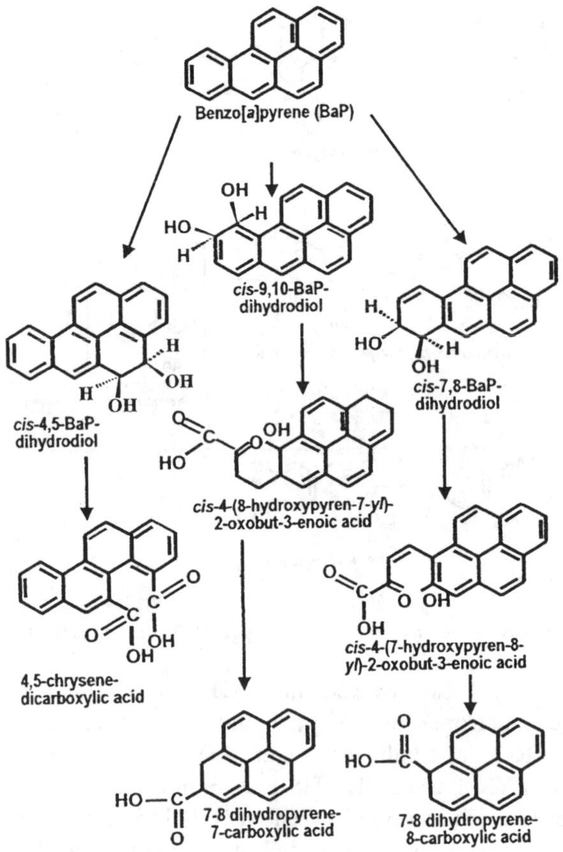

66

FIGURE 13. Pathway utilized by *Mycobacterium* spp. for the oxidation of benzo[a]pyrene.

9. Summary of Research on the Degradation of PAHs by *Mycobacterium* species

1. Many nocardioforms, including *Mycobacterium* sp. PYR-1, can mineralize high-molar-masst PAHs, such as anthracene, phenanthrene, fluoranthene, pyrene, and benzo[a]pyrene.
2. *Mycobacterium* sp. metabolizes PAHs *via* dioxygenase and monoxygenase attacks on the aromatic nucleus.
3. *Mycobacterium* sp. PYR-1 has an inducible system for PAH degradation.
4. Genes for PAH degradation are present in the chromosome for *Mycobacterium* sp. PYR-1, not plasmid encoded.

5. The genes for the degradation of PAHs have been cloned from *Mycobacterium* sp. PYR-1 and their nucleotide sequences determined.
6. *Mycobacterium* sp. PYR-1 has a novel dioxygenase and gene organization. No DNA hybridization was detected with any of the well-characterized toluene, naphthalene, phenanthrene, biphenyl and xylene dioxygenases.
7. Catalase–peroxidase plays an important role in PAH metabolism.
8. *Mycobacterium* sp. PYR-1 has potential application in the treatment of PAH-contaminated wastes.

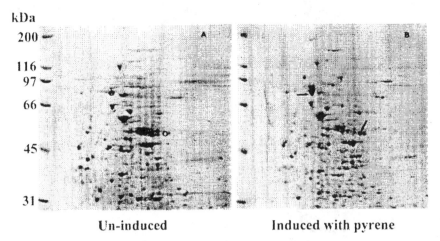

FIGURE 14.   2-D SDS PAGE of proteins obtained from *Mycobacterium* sp. PYR-1 uninduced and pyrene induced cells. The arrows indicate the 50- and 81-kDa proteins.

10. Conclusions

When one discusses the use of microorganisms in the remediation of PAH contaminated wastes, some research issues that should be addressed are the following:

1. Recent studies have shown that PAH-contaminated soil sites may contain a considerable number of PAH-degrading bacteria, although a minimum concentration of PAH must be present in the soil to select and stimulate the growth of the PAH-degrading microflora.
2. It is generally viewed that the recalcitrance of these compounds is primarily due to very low solubility and the hydrophobic nature of PAHs resulting in their partitioning into the soil matrix, thereby limiting their bioavailability for microbial degradation.
3. Cluster arrangements of PAHs, such as pyrene and benzo[a]pyrene, may be more difficult for microorganisms to metabolize than PAHs with angular arrangements (*e.g.*, chrysene and benz[a]anthracene).

4. The extent of PAH degradation does not always relate to the water solubility or $\log K_{ow}$. For example, solubility and $\log K_{ow}$ of chrysene are similar to that of benzo[*a*]pyrene, yet reports indicate that chrysene is metabolized to a greater extent than benzo[*a*]pyrene.

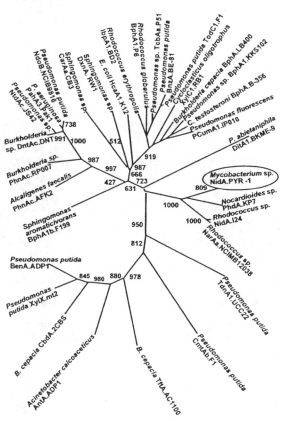

FIGURE 15.   Phylogenetic tree of PAH ring hydroxylating dioxygenases including the Nid A from *Mycobacterium* sp. PYR-1.

5. Screening studies suggest that the nocardioform actinomycetes (*e.g.*, *Rhodococcus*, *Nocardia*, *Mycobacterium*) may play a crucial role for the complete mineralization of PAH in soil.

6. If one compares the PAH-degradation pattern of all bacterial strains that are able to degrade high-molar-mass PAHs, it can be seen that the degradation of high-molar-mass PAH is still limited to very few substances. No microorganisms could be found that were able to degrade PAHs with five rings or more.

7. Microorganisms metabolizing high-molar-mass PAHs have probably developed mechanisms to increase their bioavailability by producing biosurfactant/bioemulsifier. They may be producing an extracellular lipopeptide or lipoprotein.

8. Partial oxidation of PAHs may be important in order to increase bioavailability of these compounds and thus enhance microbial degradation in contaminated environments. Studies are being conducted to explore the effects of chemical pretreatment, such as with Fenton's reagent (H_2O_2, Fe^{2+}) or ozone on biodegradation of PAHs.

9. Optimization of the biodegradation of PAHs in soil is required to achieve effective bioremediation of these compounds, especially those with more than three rings.

Acknowledgements

The author would like to thank Ms. Pat Fleischer for continuous clerical assistance and Ms. Sandra Malone for illustrations.

References

1. Ahn, Y., Sanseverino, J., and Sayler, G. (1999) Analyses of polycyclic aromatic hydrocarbon-degrading bacteria isolated from contaminated soils, *Biodegradation* 10, 149–157.
2. Atlas, R. and Cerniglia, C. (1995) Bioremediation of petroleum pollutants, *BioScience* 45, 332–339.
3. Blumer, M. (1976) Polycyclic aromatic hydrocarbons in nature, *Sci. Amer.* 234, 35–45.
4. Boldrin, B., Tiehm, A., and Fritzsche, C. (1993) Degradation of phenanthrene, fluorene, fluoranthene, and pyrene by a *Mycobacterium* sp., *Appl. Environ. Microbiol.* 59, 1927–1930.
5. Bouchez, M., Blanchet, D., and Vandecastelle, J. (1993) Degradation of polycyclic hydrocarbons by pure strains and by defined strain associations: inhibition phenomena and cometabolism, *Appl. Microbiol. Biotechnol.* 43, 156–164.
6. Cerniglia, C. (1984) Microbial metabolism of polycyclic aromatic hydrocarbons, *Adv. Appl. Microbiol.* 30, 31–71.
7. Cerniglia, C. (1992) Biodegradation of polycyclic aromatic hydrocarabons, *Biodegradation* 3, 351–368.
8. Cerniglia, C. (1993) Biodegradation of polycyclic aromatic hydrocarbons, *Curr. Opin. Biotechnolo.* 4, 331–338.
9. Cerniglia, C. (1997) Fungal metabolism of polycyclic aromatic hydrocarbons: past, present and future applications in bioremediation, *J. Industr. Microbiol. Biotechnol.* 19, 324–333.
10. Cerniglia, C. and Heitkamp, M. (1990) Polycyclic hydrocarbon degradation by a *Mycobacterium* sp., in M. Lindstrom (ed), *Methods in Enzymology*, Vol. 188, Academic Press, San Diego (CA) pp. 148–153.
11. Cerniglia, C. and Shuttleworth, K. (2002) Methods for isolation of polycyclic aromatic hydrocarbon (PAH)-degrading microorganisms and procedures for determination of biodegradation intermediates and environemntal monitoring of PAHs, in R. Crawford, M. McInerney, G. Knudsen, L. Stetzenback (eds), *Manual of Environmental Microbiology*, chap. 88, ASM Press, Washington (DC), pp. 972–986.
12. Cerniglia, C. and Sutherland, J. (2001) Bioremediation of polycyclic aromatic hydrocarbons by ligninolytic and non-ligninolytic fungi, in G. Gadd (ed), *Fungi in Bioremediation, British Mycological Society Symp. Ser. 23*, Cambridge University Press, Cambridge, pp. 136–187.
13. Cheung, P. and Kinkle, B. (2001) *Mycobacterium* diversity and pyrene mineralization in petroleum-contaminated soils, *Appl. Environ. Microbiol.* 67, 2222–2229.

70

14. Churchill, S. Harper, J., and Churchill, P. (1999) Isolation and characterization of a *Mycobacterium* species capable of degrading three- and four-ring aromatic and aliphatic hydrocarbons, *Appl. Environ. Microbiol.* 65, 549–552.

15. Coates, J., Anderson, R., and Lovley, D. (1996) Oxidation of polycyclic aromatic hydrocarbons under sulfate-reducing conditions, *Appl. Environ. Microbiol.* 62, 1099–1101.

16. Dean-Ross, D. and Cerniglia, C. (1996) Degradation of pyrene by *Mycobacterium flavescens*. *Appl. Microbiol. Biotechnol.* 46, 307–312.

17. Fritzsche, C. (1994) Degradation of pyrene at low defined oxygen concentrations by a *Mycobacterium* sp., *Appl. Environ. Microbiol.* 60, 1687–1689.

18. Gibson, D. (1982) Microbial degradation of hydrocarbons, *Toxicol. Environ. Chem.* 5, 237–250.

19. Grosser, R., Warshawsky, D., and Vestal, J. (1991) Endogenous and enhanced mineralization of pyrene, benzo[a]pyrene, and carbazole in soils, *Appl. Environ. Microbiol.* 57, 3462–3469.

20. Grosser, R., Warshawsky, D., and Vestal, J. (1995) Mineralization of polycyclic and N-heteropolycyclic aromatic compounds in hydrocarbon-contaminated soils, *Environ. Toxicol. Chem.* 14, 375–382.

21. Guerin, W. and Jones, G. (1988) Mineralization of phenanthrene by a *Mycobacterium* sp., *Appl. Environ. Microbiol.* 54, 937–944.

22. Hammel, K. (1995) Mechanisms for polycyclic aromatic hydrocarbon degradation by ligninolytic fungi, *Environ. Health Perspect.* 103, 41–43.

23. Harvey, R. (1991) *Polycyclic Aromatic Hydrocarbons: Chemistry and Carcinogenicity*, Cambridge University Press, Cambridge (UK).

24. Harvey, R. (1996) Mechanisms of carcinogenesis of polycyclic aromatic hydrocarbons, *Polycycl. Arom. Comp.* 9, 1–23.

25. Harvey, R., Penning, T., Jarabak, J., and Zhang, F. (1999) Role of quinone metabolites in PAH carcinogenesis, *Polycycl. Arom. Comp.* 16, 13–20.

26. Heitkamp, M. and Cerniglia, C. (1988) Mineralization of polycyclic aromatic hydrocarbons by a bacterium isolated from sediment below an oil field, *Appl. Environ. Microbiol.* 54, 1612–1614.

27. Heitkamp, M. and Cerniglia, C. (1989) Polycyclic aromatic hydrocarbon degradation by a *Mycobacterium* sp. in microcosms containing sediment and water from a pristine ecosystem, *Appl. Environ. Microbiol.* 55, 1968–1973.

28. Heitkamp, M., Freeman, J., Miller, D., and Cerniglia, C. (1988a) Pyrene degradation by a *Mycobacterium* sp.: identification of ring oxidation and ring fission products, *Appl. Environ. Microbiol.* 54, 2556–2565.

29. Heitkamp, M., Franklin, F., and Cerniglia, C. (1988b) Microbial metabolism of polycyclic aromatic hydrocarbons: isolation and characterization of a pyrene-degrading bacterium. *Appl. Environ. Microbiolol.* 54, 2549–2555.

30. Jimenez, I. and Bartha, R. (1996) Solvent-augmented mineralization of pyrene by a *Mycobacterium* sp., *Appl. Environ. Microbiol.* 62, 2311–2316.

31. Kanaly, R. and Bartha, R. (1999) Cometabolic mineralization of benzo[a]pyrene caused by hydrocarbon additions to soil, *Environ. Toxicol. Chem.* 18, 2186–2190.

32. Kästner, M., Breuer-Jammali, M., and Mahro, B. (1994) Enumeration and characterization of the soil microflora from hydrocarbon-contaminated soil sites able to mineralize polycyclic aromatic hydrocarbons (PAH), *Appl. Microbiol. Biotechnol.* 41, 267–273.

33. Kelley, I. and Cerniglia, C. (1991) The metabolism of fluoranthene by a species of *Mycobacterium*, *J. Industr. Microbiol.* 1, 19–26.

34. Kelley, I. and Cerniglia, C. (1995) Degradation of a mixture of high-molecular-weight polycyclic aromatic hydrocarbons by a *Mycobacterium* strain PYR-1, *J. Soil Contam.* 4, 44–91.

35. Kelley, I., Freeman, J., and Cerniglia, C. (1990) Identification of metabolites from degradation of naphthalene by a *Mycobacterium* sp., *Biodegradation* 1, 283–290.

36. Kelley, I., Freeman, J., Evans, F., and Cerniglia, C. (1991) Identification of a carboxylic acid metabolite from the catabolism of fluoranthene by a *Mycobacterium* sp., *Appl. Environ. Microbiol.* **57**, 636–641.

37. Kelley, I., Freeman, J., Evans, F., and Cerniglia, C. (1993) Identification of metabolites from the degradation of fluoranthene by *Mycobacterium* sp. PYR-1, *Appl. Environ. Microbiol.* **59**, 800–806.

38. Khan, A., Wang, R., Cao, W., Doerge, D., Wennerstrom, D., and Cerniglia, C. (2001) Molecular cloning, nucleotide sequence, and expression of genes encoding a polycyclic aromatic ring dioxygenase from *Mycobacterium* sp. strain PYR-1. *Appl. Environ. Microbiol.* **67**, 3577–3585.

39. Khan, A., Kim, S., Paine, D., and Cerniglia, C.E. (2002) Classification of a polycyclic aromatic hydrocarbon-metabolizing bacterium, *Mycobacterium* sp. strain PYR-1, as *Mycobacterium vanbaalenii* sp.nov., *Internat. J. System. Evolut. Microbiol.* (in press).

40. Kiyohara, H., K. Nagao, K., and Yana, K. (1982) Rapid screen for bacteria degrading water-insoluble, solid hydrocarbons on agar plates, *Appl. Environ. Microbiol.* **43**, 454–457.

41. Kleespies, M., Kroppenstedt, R., Rainey, F., Webb, L. and Stackebrandt, E. (1996) *Mycobacterium hodleri* sp.nov., a new member of the fast-growing mycobacteria capable of degrading polycyclic aromatic hydrocarbons, *Internat. J. System. Bacteriol.* **46**, 683–687.

42. Lloyd-Jones, G. and Hunter, D. (1997) Characterization of fluoranthene- and pyrene-degrading *Mycobacterium*-like strains by RAPD and SSU sequencing, *FEMS Microbiol. Lett.* **153**, 51–56.

43. Mackay, D. and Shiu, W. (1977) Aqueous solubility of PAH, *J. Chemi. Engin.* **22**, 399–402.

44. Mastrangelo, G., Fadda, E., and Marzia, V. (1996) Polycyclic aromatic hydrocarbons and cancer in man, *Environ. Health Persp.* **104**, 1166–1170.

45. McLellan, S., Warshawsky, D., and Shann, J. (2002) The effect of polycyclic aromatic hydrocarbons on the degradation of benzo[a]pyrene by *Mycobacterium* sp. strain RJGII-135. *Environ. Toxicol. Chem.* **21**, 253–259.

46. Meckenstock, R., Annweiler, E., Michaelis, W., Richnow, H., and Schink, B. (2000) Anaerobic naphthalene degradation by a sulfate-reducing enrichment culture. *Appl. Environ. Microbiol.* **66**, 2743–2747.

47. Menn, F., Applegate, B., and Sayler, G. (1993) NAH plasmid mediated catabolism of anthracene and phenanthrene by naphthoic acids, *Appl. Environ. Microbiol.* **59**, 1938–1942.

48. Milhelcic, J. and Luthy, R. (1988) Degradation of polycyclic aromatic hydrocarbon compounds under various redox conditions in soil–water systems, *Appl. Environ. Microbiolo.* **54**, 1182–1187.

49. Miller, E. and Miller J. (1985) Some historical perspectives on the metabolism of xenobiotic chemicals to reactive electrophiles, in M. Anders (ed), *Bioactivation of Foreign Compounds*, Academic Press, Orlando (FL), pp. 3–28.

50. Molina, M., Araujo, R., and Hodson, R. (1999) Cross-induction of pyrene and phenanthrene in a *Mycobacterium* sp. isolated from polycyclic aromatic hydrocarbon contaminated river sediments, *Can. J. Microbiol.* **45**, 520–529.

51. Moody, J., Freeman, J., Doerge, D., and Cerniglia, C. (2001) Degradation of phenanthrene and anthracene by cell suspensions of *Mycobacterium* sp. strain PYR-1, *Appl. Environ. Microbiol.* **67**, 1476–1483.

52. Mueller, J., Cerniglia, C., and Pritchard, P. (1996) Bioremediation of environments contaminated by polycylic aromatic hydrocarbons, in R. Crawford and D. Crawford (eds), *Bioremediation: Principles and Applications*, Cambridge University Press, Cambridge, pp. 125–194.

53. Rafii, F., Butler, W., and Cerniglia, C. (1992) Differentiation of a rapidly growing scotochromogenic polycyclic-aromatic-hydrocarbon-metabolizing strain of *Mycobacterium* sp. from other known *Mycobacterium* species, *Arch. Microbiol.* **157**, 512–520.

54. Rafii, F., Selby, A., Newton, R., and Cerniglia, C. (1994) Reduction and mutagenic activation of nitroaromatic compounds by a *Mycobacterium* sp. *Appl. Environ. Microbiol.* **60**, 4263–4267.

55. Rafii, F., Lunsford, P., Hehman, G., and Cerniglia, C. (1999) Detection and purification of a catalase–peroxidase from *Mycobacterium* sp. Pyr-1, *FEMS Microbiol. Lett.* **173**, 285–290.

72

56. Rehmann, K., Hertkorn, N., and Kettrup, A. (2001) Fluoranthene metabolism in *Mycobacterium* sp. strain KR20: identity of pathway intermediates during degradation and growth, *Microbiology*, 147, 2783–2794.

57. Rehmann, K., Noll, H., Steinberg, C., and Kettrup, A. (1996) Branched metabolic pathway for phenanthrene degradation in a pyrene-degrading bacterium, *PAC* 11, 125–130.

58. Rehmann, K., Noll, H., Steinberg, C., and Kettrup, A. (1998) Pyrene degradation by *Mycobacterium* sp. strain KR2, *Chemosphere* 36, 2977–2992.

59. Rockne, K., Chee-Sanford, J., Sanford, R., Hedlund, B., Staley, J., and Strand, S. (2000) Anaerobic naphthalene degradation by microbial pure cultures under nitrate-reducing conditions, *Appl. Environ. Microbiol.* 66, 1595–1601.

60. Saito, A., Iwabuchi, T., and Harayama, S. (1999) Characterization of genes for enzymes involved in the phenanthrene degradation in *Nocardioides* sp. KP7, *Chemosphere* 38, 1331–1337.

61. Sanseverino, J., Applegate, B., King, J., and Sayler, G. (1993) Plasmid-mediated mineralization of naphthalene, phenanthrene and anthracene, *Appl. Environ. Microbiol.* 59, 1931–1937.

62. Santodonato, J. (1997) Review of the estrogenic and antiestrogenic activity of polycyclic aromatic hydrocarbons: relationship to carcinogenicity, *Chemosphere* 34, 835–848.

63. Schneider, J., Grosser, R., Jayasimhulu, K., Xue, W., and Warshawsky. D. (1996) Degradation of pyrene, benz[a]anthracene, and benzo[a]pyrene by *Mycobacterium* sp. strain RJGII-135, isolated from a former coal gasification site, *Appl. Environ. Microbiol.* 62, 13–19.

64. Sepic, E., Bricelj, M., and Leskovsek, H. (1998) Degradation of fluoranthene by *Pasteurella* sp. IFA and *Mycobacterium* sp. PYR-1: isolation and identification of metabolites. *J. Appl. Microbiol.* 85, 746–754.

65. Storelli, M. and Marcotrigiano, G. (2001) Polycyclic aromatic hydrocarbons in mussels (*Mytilus galloprovincialis*) from the Ionian Sea, Italy, *J. Food Protect.* 64, 405–409.

66. Sutherland, J., Rafii, R., Khan, A., and Cerniglia, C. (1995) Mechanisms of polycyclic aromatic hydrocarbon degradation, in L. Young and C.E. Cerniglia (eds.), *Microbial Transformation and Degradation of Toxic Organic Chemicals*, Wiley-Liss, New York (USA), pp. 269–306.

67. Tiehm, A. and Fritzsche, C. (1995) Utilization of solubilized and crystalline mixtures of polycyclic aromatic hydrocarbons by a *Mycobacterium* sp., *Appl. Microbiol. Biotechnolol.* 42, 964–968.

68. Tongpim, S. and Pickard, M. (1999) Cometabolic oxidation of phenanthrene to phenanthrene *trans*-9,10-dihydrodiol by *Mycobacterium* strain S1 growing on anthracene in the presence of phenanthrene, *Can. J. Microbiol.* 45, 369–376.

69. Tuvikene, A. (1995) Responses of fish to polycyclic aromatic hydrocarbons (PAHs), *Ann. Zoolog. Fenn.* 32, 295–309.

70. Vila, J., Lopez, Z., Sabaté, J., Minguillón, Solanas, A., and Grifoll, M. (2001) Identification of a novel metabolite in the degradation of pyrene by *Mycobacterium* sp. strain AP1: actions of the isolate on two- and three-ring polycyclic aromatic hydrocarbons, *Appl. Environ. Microbiol.* 67, 5497–5505.

71. Walter, U., Beyer, M., Klein, J., and Rehm, H. (1991) Degradation of pyrene by *Rhodococcus* sp. UW1, *Appl. Microbiol. Biotechnol.* 34, 671–676.

72. Wang, R., Cao, W., and Cerniglia, C. (1995) Phylogenetic analysis of polycyclic aromatic hydrocarbon degrading mycobacteria by 16S rRNA sequencing, *FEMS Microbiol. Lett.* 130, 75–80.

73. Wang, R., Wennerstrom, D., Cao, W., Khan, A., and Cerniglia, C. (2000) Cloning, expression, and characterization of the *katG* gene, encoding catalase–peroxidase, from the polycyclic aromatic hydrocarbon-degrading bacterium *Mycobacterium* sp. strain PYR-1, *Appl. Environ. Microbiol.* 66, 4300–4304.

74. Weibenfels, W., Beyer, M., and Klein, J. (1990) Degradation of phenanthrene, fluorene and fluoranthene by pure bacterial cultures, *Appl. Microbiol. Biotechnol.* 32, 479–484.

75. Weißenfels, W., Beyer, M., Klein, J., and Rehm, H. (1991) Microbial metabolism of fluoranthene: isolation and identification of ring fission products, *Appl. Microbiol. Biotechnol.* 34, 528–535.

76. Willumsen, P., Karlson, U., Stackebrandt, E., and Kroppenstedt, R. (2001) *Mycobacterium frederiksbergense* sp.nov., a novel polycyclic aromatic hydrocarbon-degrading *Mycobacterium* species, *Internat. J. System. Evolut. Microbiol.* **51**, 1715–1722.

77. Wilson, S. and Jones, K. (1993) Bioremediation of soil contaminated with polynuclear aromatic hydrocarbons (PAHs): a review, *Environ. Pollut.* **81**, 229–249.

78. Wise, S., Benner, B., Burd, G., Chester, S., Rebbert, R., and Schantz, M. (1988) Determination of polycyclic aromatic hydrocarbons in a coal standard reference material, *Analyt. Chem.* **60**, 887–894.

79. Yang, Y., Chen, R., and Shiaris, M. (1994) Metabolism of naphthalene, fluorene, phenanthrene: preliminary characterization of a cloned gene cluster from *Pseudomonas putida* NCIB 9816, *J. Bacteriol.* **178**, 2158–2164.

80. Zhang, W. and Young, L. (1997) Carboxylation as an initial reaction in the anaerobic metabolism of naphthalene and phenanthrene by sulfidogenic consortia, *Appl. Environ. Microbiol.* **63**, 4759–4764.

81. Zylstra, G. and Kim, E. (1997) Aromatic hydrocarbon degradation by *Sphingomonas yanoikuyae* B1, *J. Industr. Microbiol. Biotechnol.* **19**, 408–414.

82. Zylstra, G., Wang, X., Kim, E., and Didolkar V. (1994) Cloning and analysis of the genes for polycyclic aromatic hydrocarbon degradation. *Ann. N. Y. Acad. Recomb. DNA Technol.* **721**, 386–398.

TOXIC METAL CONTAMINATION TREATMENT WITH MICROBES

G.M. GADD
Department of Biological Sciences, University of Dundee
Dundee DD1 4HN, Scotland, UK

1. Introduction

Environmental contamination by toxic metals is of considerable economic and environmental significance [1–5] although the influence of microbiological processes on such contamination and their potential for bioremediation are dependent on the organisms involved and on physical and chemical factors. Several microbial mechanisms can solubilize metals, increasing their bioavailability and potential toxicity, whereas others immobilize them and reduce bioavailability. The relative balance of mobilization and immobilization processes varies depending on the organism(s), their environment and physicochemical conditions. As well as being a natural component of biogeochemical cycles for metals and associated elements, these processes may be exploited for the treatment of contaminated solid and liquid wastes [2, 4, 6–8].

Metal mobilization can be achieved by autotrophic and heterotrophic leaching, chelation by microbial metabolites and siderophores, and methylation which can result in volatilization. Immobilization can result from sorption to cell components or exopolymers, transport and intracellular sequestration or precipitation as insoluble organic and inorganic compounds, *e.g.* oxalates [9–11], sulfides [8, 12], phosphates [13–15] and carbonates [16]. In addition, and depending on the metal species, reduction of higher-valency species may effect mobilization, *e.g.* Mn^{IV} to Mn^{II}, or immobilization, *e.g.* Cr^{VI} to Cr^{III} [17, 18]. For treatment of metal contamination, solubilization enables removal from solid matrices. Alternatively, immobilization may enable metals to be transformed *in situ* into insoluble and chemically inert forms and are particularly applicable to removing metals from mobile aqueous phases. This chapter will outline some microbiological processes of significance in determining metal mobility and which have actual or potential application in bioremediation of metal, metalloid and radionuclide pollution.

75

V. Šašek et al. (eds.),
The Utilization of Bioremediation to Reduce Soil Contamination: Problems and Solutions, 75–94.
© 2003 *Kluwer Academic Publishers. Printed in the Netherlands.*

2. Solubilization

2.1. AUTOTROPHIC LEACHING

Most autotrophic leaching is carried out by chemolithotrophic, acidophilic bacteria which fix carbon dioxide and obtain energy from the oxidation of ferrous iron or reduced sulfur compounds which causes the solubilization of metals because of the resulting production of Fe^{III} and H_2SO_4 [19, 20]. The microorganisms involved in autotrotrophic leaching include sulfur-oxidizing bacteria, e.g. *Thiobacillus thiooxidans*, iron- and sulfur-oxidizing bacteria, e.g. *Thiobacillus ferrooxidans* and iron-oxidizing bacteria, e.g *Leptospirillum ferrooxidans* [21, 22]. As a result of sulfur and iron oxidation by these bacteria, metal sulfides are solubilized, concomitant with the pH of their immediate environment being decreased and resulting in solubilization of other metal compounds. Such leaching of metal sulfides by *Thiobacillus* species and other acidophilic bacteria is well established for industrial-scale biomining [19, 21, 23]. In a bioremediation context, autotrophic production of sulfuric acid has also been used to solubilize metals from sewage sludge which can then be used as a fertilizer [24]. Leaching with *T. ferrooxidans* required acidification of the sludge to ≈pH 4.0 before adequate growth of the bacteria could occur although sulfur-oxidizing *Thiobacillus* sp. have been isolated which were able to grow at pH 7.0 [24]. The elemental sulfur which acts as an energy source can be of chemical or biological origin [25]. The use of solid sulfur may allow unused sulfur to be removed and therefore prevent further acidification of the sludge after disposal [26]. Simultaneous sewage sludge digestion and metal leaching under acidic conditions (pH 2.0−2.5) has an additional advantage over conventional aerobic and anaerobic digestion in that the acidity leads to a decrease in pathogenic microorganisms [27]

Autotrophic leaching has also been used to remediate metal-contaminated soils [28]. In a two-stage process, sulfur-oxidizing bacteria were used to acidify soil and solubilize toxic metals before treatment of the metal-contaminated liquid leachate using sulfate-reducing bacteria [8, 12]. Studies have also been conducted into the bioremediation of red mud, the main waste product of Al extraction from bauxite. Here, autotrophic leaching by indigenous *Thiobacillus* spp. was more efficient than heterotrophic leaching by a range of fungal strains, and decreased the toxicity of the waste product [29]. Bioleaching using elemental sulfur as substrate was better than sulfuric acid treatment for metal solubilization from contaminated aquatic sediments [30]. Iron- and sulfur-oxidizing bacteria were able to leach > 50 % of the metals present (As, Cd, Co, Cu, Ni, V, Zn, B, Be). Strains of *T. ferrooxidans* were able to remove all of the Cd, Co, Cu and Ni [31]. *T. ferrooxidans* has also been used to treat air pollution control residues (APCR: fly ash and used lime). Although viability of the thiobacilli was poor, removal of up to 95 % of Cd (from APCR of ≈270 mg Cd/kg) was achieved with ≈69 % Pb removal (from APCR of ≈5 g Pb/kg) and similar removals for Zn and Cu [32]. Note that most of the processes and laboratory studies involving autotrophic leaching have used mesophilic chemolithotrophs although moderately thermophilic species may prove important in the future.

2.2. HETEROTROPHIC LEACHING

Chemolithotrophic acidophilic bacteria have a high acidification capacity and are proven for *in situ* applications, meaning that most industrial bioleaching has been carried with these microorganisms [19, 23]. However, many species of fungi are able to leach metals (heterotrophic/chemoorganotrophic leaching) from industrial wastes and by-products [33], low-grade ores [34], and metal-bearing minerals [35, 36]. There are some advantages in using fungi in that they are more easily manipulated in bioreactors than thiobacilli and, by altering growth conditions, can be induced to produce high concentrations of organic acids [34, 37, 38]. Furthermore, many species can tolerate high concentrations of toxic metals [39] and can grow in both low and high pH environments [34]; the acidophilic nature of chemolithotrophic bacteria may make them unable to tolerate the higher pH values of many wastes [34].

Heterotrophic leaching by fungi occurs as a result of several processes, including the efflux of protons and the production of siderophores (in the case of iron), but the most important mechanism is the production of organic acids [9, 34, 40, 41]. Organic acids provide both a source of protons, which protonate the anionic component, and a metal-complexing anion [42]. Most metal citrates are highly mobile and are not readily degraded, degradation depending on the type of complex formed rather than the toxicity of the metal involved. Hence, the presence of citric acid in the terrestrial environment can leach potentially toxic metals [43]. Oxalic acid can act as a leaching agent for those metals that form soluble oxalate complexes, *e.g.* Al, Fe and Li [38].

Solubilization ability can be manipulated, *e.g.*, by changing the N and P balance or pH, to maximize organic acid production [44,45]. A deficiency of manganese (<10 nmol/L) in the growth medium leads to the production of large amounts of citric acid by *A. niger* [46], and typical concentrations of citric acid produced industrially can reach 600 mmol/L [37]. Oxalic acid production can be manipulated to yield up to 200 mmol/L on low-cost carbon sources with the optimum pH being around neutrality [38]. Effective leaching of a variety of wastes and low-grade minerals has been demonstrated, *e.g.* soils and muds, filter dust/oxides, lateritic ores, copper converter slag, fly ash and electronic waste materials [11, 33]. Laboratory-scale heterotrophic leaching of Ni and Co from low-grade laterite ore has been carried out, 55−60 % Ni being leached from the ore by *Aspergillus* and *Penicillium* strains [35, 47]. A strain of *Penicillium simplicissimum* has been used to leach Zn from insoluble ZnO contained in industrial filter dust. This fungus only developed the ability to produce citric acid (>100 mmol/L) in the presence of the filter dust [48−50]. Culture filtrates from *A. niger* have also been used to leach Cu, Ni and Co from copper converter slag [51]. Al has been leached from red mud with the most efficient fungal strain being *P. simplicissimum*, the fungus-derived acids (mainly citric) having a much greater ability to leach Al than pure citric acid [29]. Cd, Zn, Cu, Pb and Al were leached from municipal-waste fly ash in a two-stage process using *A. niger*. The environmental quality of the fly ash residue was deemed suitable for use in the construction industry [52].

A heterotrophic mixed culture has been employed for leaching manganiferous minerals. This was capable of reducing MnO_2, with the process being of potential value for the treatment of materials not treatable by conventional processes [53]. Another method for treatment of metal-contaminated sandy soil relies on siderophore-mediated metal solubilization by *Alcaligenes eutrophus*. Solubilized metals were adsorbed to the biomass and/or precipitated, with biomass separated from a soil slurry by a flocculation process. This process resulted in a sharp decrease in the bioavailability of Cd, Zn and Pb [54]. Related to heterotrophic solubilization is fungal translocation of, *e.g.* Cs, Zn and Cd, which can lead to concentration in specific regions of the mycelium and/or in fruiting bodies. Whether the concentration factors observed *in vitro* can be reproduced in the field and whether such amounts can contribute to soil bioremediation remains uncertain [55].

Heterotrophic solubilization can also have consequences for other remedial treatments for contaminated soils. Pyromorphite ($Pb_5(PO_4)_3Cl$) is a stable lead mineral which can form in urban and industrially-contaminated soils. Such insolubility reduces lead bioavailability and the formation of pyromorphite has been suggested as a remediation technique for lead-contaminated land, if necessary by means of phosphate addition. However, pyromorphite can be solubilized by phosphate-solubilizing fungi, *e.g. Aspergillus niger*, and plants grown with pyromorphite as a sole phosphorus source accumulated both P and Pb [56]. Further, during the fungal transformation of pyromorphite, the biogenic production of lead oxalate dihydrate was observed [56]. This emphasizes the importance of considering microbial processes in developing remediation techniques for metal-contaminated soils since mechanisms of metal transformation may have significant consequences for metal transfer between environmental compartments and organisms.

3. Immobilization

3.1. BIOSORPTION

Biosorption can be defined as the uptake of organic and inorganic metal species, both soluble and insoluble, by physicochemical mechanisms such as adsorption. In living cells, metabolic activity may also influence this process because of changes in the physicochemical characteristics of the cellular microenvironment. Almost all biological macromolecules have some affinity for metal species with cell walls and associated materials being of the greatest significance in biosorption. As well as this, cationic species can be accumulated by cells via transport systems of varying affinity and specificity. Once inside cells, metal species may be bound, precipitated, localized within intracellular structures or organelles, or translocated to specific structures, depending on the element concerned and the organism [4, 6, 8].

3.1.1.   Cell Walls and Associated Materials

Microbial exopolymers can be composed of polysaccharide, glycoproteins and lipopolysaccharide [57]. Many of these can act as polyanions and interact with cationic metal/radionuclide species while uncharged polymers are also capable of binding and entrapment of insoluble metal forms [58]. Peptidoglycan carboxyl groups are the main binding sites for metal cations in Gram-positive bacterial cell walls with phosphate groups contributing significantly in Gram-negative species [58]. Chitin is an important structural component of fungal cell walls and this is an effective biosorbent for metals and radionuclides as are chitosan and other chitin derivatives [59]. Fungal phenolic polymers and melanins possess many metal-binding sites with carboxyl, phenolic and alcoholic hydroxy, carbonyl and methoxy groups being particularly important [39].

3.1.2.   Free and Immobilized Biomass

Freely-suspended and immobilized biomass from bacterial, cyanobacterial, algal and fungal species has received attention with immobilized systems possessing several advantages including higher mechanical strength and easier biomass/liquid separation [60]. Living or dead biomass has been immobilized by encapsulation or cross-linking using, e.g. agar, cellulose, alginates, cross-linked ethyl acrylate−ethylene glycol dimethylacrylate, polyacrylamide, silica gel and glutaraldehyde [59−64]. Recent developments in reactor formats include hollow-fibre crossflow microfiltration using *Pseudomonas aeruginosa* for biosorption of Pb, Cu and Cd. Removal efficiencies in both single and ternary biosorption were Pb > Cu > Cd and at an influent (metal) of 200 μmol/L and flow rate of 350 mL/h, effluent concentrations of Cu and Pb satisfied discharge regulations. This study developed a rapid-equilibrium model and a mass-transfer model which accurately described single- and multi-metal biosorption, respectively [65]. *Arthrobacter* sp. was entrapped within a macro- and microporous matrix (poly)hydroxyethyl methacrylate cross-linked with trimethylolpropane trimethacrylate. The rate of biosorption was controlled by the rate of diffusion through the shell of saturated material and a maximum Cu uptake of ≈ 7 mg/g dry mass was observed for a resin-biomass complex containing 8 % (W/W) biomass [66, 67]. A model based on Fick's Law of Diffusion and the Langmuir adsorption model was able to predict experimentally determined kinetics of Ni^{2+} removal by fungal biosorbents [68]. Immobilized living biomass has mainly taken the form of bacterial biofilms on inert supports and is used in a variety of bioreactor configurations including rotating biological contactors, fixed-bed reactors, trickle filters, fluidized beds and air-lift bioreactors [3, 60, 69, 70]. Chemical modification of biomass may create derivatives with altered metal-binding abilities and affinities. *A. niger* mycelium was modified by introducing additional carboxy or ethyldiamino groups with the latter substitution increasing the maximal capacity for nickel from ≈ 70 to ≈ 1060 mmol/kg [71].

3.1.3. Metal-Binding Proteins, Polysaccharides and Other Biomolecules

A range of specific and nonspecific metal-binding compounds are produced by microorganisms. Nonspecific metal-binding compounds range from simple organic acids and alcohols to macromolecules such as polysaccharides, humic and fulvic acids [72,73]. Extracellular polymeric substances (EPS), a mixture of polysaccharides, mucopolysaccharides and proteins [74], are produced by bacteria, algae and fungi and also bind potentially toxic metals [75, 76]. Extracellular polysaccharides of microbial origin are able to both bind metals and also adsorb or entrap particulate matter, such as precipitated metal sulfides and oxides [77, 78]. One process has been described which uses floating cyanobacterial mats to remove metals from waters, the metal-binding process being due to large polysaccharides (>200 kDa) produced by the cyanobacteria [79].

Specific metal-binding compounds include siderophores which are low-molar-mass ligands (500−1000 Da) possessing a high affinity for Fe^{III} [80] and produced extracellularly in response to low iron availability. Siderophores can also bind other metals such as magnesium, manganese, chromium(III), gallium(III) and radionuclides, such as plutonium(IV) [72, 81]. Other metal-binding molecules have also been identified. Specific, low-molar-mass (6−10 kDa) metal-binding metallothioneins are produced by animals, plants and microorganisms in response to the presence of toxic metals [82]. Other metal binding proteins, phytochelatins and related peptides, have been identified in plants, algae and several microorganisms [83].

Metal-binding abilities of siderophores, metallothioneins, phytochelatins and other similar molecules have a potential for bioremediation. For example, a laboratory-scale process has been developed using one kind of metallothionein (ovotransferrin). Metal-contaminated water is passed through an affinity column containing ovotransferrin attached to CNBr-activated Sepharose 4B. The bound metal can be removed from the column using a low pH, weakly chelating buffer, such as HEPES, and the column reused [84]. Such processes may have potential for the remediation of large quantities of water which contain low metal concentrations. Eukaryotic metallothioneins and other metal-binding peptides have been expressed in *E. coli* as fusions to membranes or membrane-associated proteins, such as LamB, an outer membrane protein which functions as a coliphage surface receptor and is involved in maltose/maltodextrin transport. Such *in vivo* expression of metallothioneins provides a means of designing biomass with specific metal-binding properties [85−87]. Metal binding peptides of sequences

$$Gly-His-His-Pro-His-Gly \text{ (HP) and}$$
$$Gly-Cys-Gly-Cys-Pro-Cys-Gly-Cys-Gly \text{ (CP)}$$

were engineered into LamB protein and expressed in *E. coli*. Cd^{2+}:HP and Cd^{2+}:CP were 1:1 and 3:1, respectively. Surface display of CP increased the Cd^{2+} binding ability of *E. coli* four-fold. Some competition of Cu^{2+} with Cd^{2+} for HP resulted from strong Cu^{2+} binding to HP indicating that the relative metal-binding affinities of inserted peptides and the wall-to-metal ratio were important in the design of peptide sequences and their metal specificities [88]. Another gene encoding a *de novo* peptide sequence containing the metal-binding motif

$$Cys-Gly-Cys-Cys-Gly$$

was chemically synthesized and expressed in *E. coli* as a fusion with the maltose-binding protein. Such cells possessed enhanced binding of Cd^{2+} and Hg^{2+} [86].

Related to the application of metal-binding molecules is the identification of genes encoding phytochelatin synthases, phytochelatins playing major roles in metal detoxication in plants and fungi. Future research with microorganisms and plants may allow testing of the potential of *PCS* genes for bioremediation [89]. A variety of other metal-binding molecules may have future potential for metal recovery. Culture filtrates from the sulfate-reducing bacterium *Desulfococcus multivorans* exhibited copper binding with 12-day-old filtrates having a copper-binding capacity of $3.64 \pm 0.33\ \mu mol/mL$ [90]. A hollow-fiber reactor containing an engineered *E. coli* strain expressing Hg^{2+} transport and metallothionein accumulated Hg^{2+} effectively, reducing a concentration of $2\ mg/L$ to $\approx 5\ mg/L$ [91]. It can be noted that biosensors for the detection of metal bioavailability have been developed, based on the direct interaction between metal-binding proteins and metal ions. Here, capacitance changes of the proteins, *e.g.* a synechococcal metallothionein and a mercury-resistance regulatory protein, were detected in the presence of femto- to millimolar metal ion concentrations [92].

3.2. METAL PRECIPITATION

3.2.1. *Precipitation by Metal-Reducing Bacteria*

A diverse range of microorganisms can use oxidized species of metallic elements, *e.g.* Fe^{III}, Cr^{VI} or Mn^{IV} as terminal electron acceptors [93]. Fe^{III} and Mn^{IV} appear to be the most commonly utilized metals as terminal electron acceptors in the biosphere [94]. However, since the solubility of both Fe and Mn is increased by bacterial reduction, and neither is significantly toxic, other metals are targeted in waste-treatment. Molybdenum(VI) was reduced to molybdenum blue by a strain of *Enterobacter cloacae* which was isolated from a molybdate-polluted aquatic environment [95]. Another strain of *E. cloacae*, also isolated from a polluted habitat, was able to reduce Cr^{VI} to Cr^{III} under similar conditions [96, 97]. Dissimilatory Cr^{VI} reduction was also carried out by a strain of *Escherichia coli* under both anaerobic and aerobic conditions, albeit at a slower rate [98]. Another application of dissimilatory metal reduction is uranium precipitation which has potential for waste treatment and uranium recovery. While U^{VI} compounds are soluble, U^{IV} compounds such as the hydroxide or carbonate have low solubility and form precipitates at neutral pH. A strain of *Shewanella (Alteromonas) putrefaciens* which reduced Fe^{III} and Mn^{IV} also reduced U^{VI} to U^{IV} forming a black precipitate of U^{IV} carbonate [99]. U^{VI} was also reduced by *Desulfovibrio desulfuricans* producing a very pure precipitate of U^{IV} carbonate [100, 101]. It was also reported that *Desulfovibrio vulgaris* carried out a similar enzymic reduction of uranium(VI) [99]. Bacterial uranium reduction has also been combined with chemical extraction to produce a potential process for soil bioremediation [102]. Reductive precipitation of U^{VI} and Cr^{VI} in these organisms is mediated by multiheme cytochrome c proteins. The gene encoding cytochrome c_7 from *Desulfuromonas acetooxidans* has been cloned and expressed in *Desulfovibrio desulfuricans*, the recombinant organism exhibiting enhanced expression of metal reductase activities. Such overproduction of active cytochrome c_7 could be important in fixed-enzyme reactors or in organisms with enhanced metal reductase activities for bioremediation [103]. *D. desulfuricans* can couple the oxidation of a range of electron donors to reduction of Tc^{VII} which is precipitated as an insoluble oxide at cell peripheries. Resting cells, immobilized in a flowthrough membrane bioreactor accumulated substantial quantities of Tc when supplied with formate as electron donor, suggesting potential of this organism for treatment of Tc-contaminated wastewater [104, 105]. *D. desulphuricans* can also reduce Pd^{II} to cell-bound Pd^0 with hydrogen-dependent reduction being O_2-insensitive, thus providing a means of Pd recovery under oxic conditions [106]. The solubility of other radionuclides may be increased by reduction and this favor their removal from matrices such as soils. For example, iron-reducing bacterial strains solubilized 40 % of the Pu present in contaminated soils within 6−7 days [107] and both iron- and sulfate-reducing bacteria were able to solubilize Ra from uranium mine tailings, although solubilization occurred largely by disruption of reducible host minerals [108].

3.2.2. Precipitation by Sulfate-Reducing Bacteria

Sulfate-reducing bacteria (SRB) are strictly anaerobic heterotrophic bacteria commonly found in environments where oxygen is excluded and where carbon substrates and sulfate are available. SRB are almost entirely neutrophilic with maximum growth obtained in the range pH 6−8 [109]. However, some isolates can grow in moderately acid conditions such as mine and surface waters where the bulk phase pH is in the range 3−4. SRB utilize an energy metabolism in which the oxidation of organic compounds or hydrogen is coupled to the reduction of sulfate as the terminal electron acceptor, producing sulfide which may reach significant concentrations in sediments or chemostat cultures [14, 15]. With the exception of the alkali and alkaline-earth metals, the solubility products of most heavy metal sulfides are very low, in the range of 4.65×10^{-14} (Mn) to 6.44×10^{-53} (Hg) [110] so that even a moderate output of sulfide can remove metals to levels permitted in the environment [111, 112] SRB also create extremely reducing conditions which can chemically reduce metals such as uranium(VI), albeit at a slower rate than enzymic reduction. In addition, sulfate reduction partially eliminates acidity from the system as a result of the shift in equilibrium when sulfate (dissociated) is converted to sulfide (largely protonated) [14]. This can result in the further precipitation of metals such as copper or aluminum as hydroxides as well as increasing the efficiency of sulfide precipitation.

3.2.3. Processes Utilizing Metal Sulfide Precipitation

Acid mine drainage occurs through the activities of sulfur- and iron-oxidizing bacteria and, due to the quantities of sulfate available, sulfate reduction can be an important process controlling the efflux of metals and acidity [113–115]. Laboratory studies indicate that sulfate-reduction can provide both *in situ* [116] and *ex situ* metal removal from such wastes [117–119] and contribute to removal of metals and acidity in artificial and natural wetlands [120,121].

Large-scale bioreactors have also been developed using bacterial sulfate reduction for treating metal-contaminated waters. A pilot-scale study used either 3×200 L or single 4500 L fixed-bed vessels filled with spent mushroom compost at residence times between 9 and 17 days. Both systems raised the pH from 3.0–3.5 to 6–7 and removed almost all of the metals (Al, Zn, Cu Ni) in the inflow [122]. The most extensive use to date of SRB is in the treatment of contaminated ground water at the Budelco zinc-smelting works at Budel–Dorplein in the Netherlands. The pilot plant comprised a purpose-designed 9 m^3 stainless steel sludge-blanket reactor using SRB and was initially developed by Shell Research Ltd and Budelco B.V. This plant successfully removed toxic metals (primarily Zn) and sulfate from contaminated groundwater at the long-standing smelter site by precipitation as metal sulfides. The reactor used a selected but undefined consortium of SRB with ethanol as the growth substrate. It was capable of tolerating a wide range of inflow pH and operating temperatures, and yielded outflow metal concentrations below the ppb range [123, 124]. The process was expanded to commercial scale since 1992 by Paques B.V. (Netherlands) and is capable of treating over 7000 m^3/d. A process integrating bacterial sulfate reduction with bioleaching by sulfur-oxidizing bacteria has been developed to remove contaminating toxic metals from soils. In this process sulfur- and iron-oxidizing bacteria are employed to liberate metals from soils by the breakdown of sulfide minerals and production of sulfuric acid [8, 110]. Metals are liberated in the form of an acid sulfate solution which enables both a large proportion of the acidity and almost the entirety of the metals to be removed by bacterial sulfate reduction [14, 15, 125]. Metals were mainly precipitated as solid sulfides and removal of the target metals by the bioreactor achieved >98 % efficiency overall, with the exception of Mn and, to a lesser extent, Ni and Pb. The overall process effectively removed the contaminating metal load from the soil and converted it to sulfides which were concentrated 100- to 200-fold in the solid phase, while the concentrations of metals in the liquid effluent were low enough to meet environmental discharge criteria. SRB biofilm reactors may offer a means of process intensification and entrap or precipitate metals, *e.g.* Cu and Cd, at the biofilm surface [126, 127]. Mixed SRB cultures were more effective than pure cultures and metal removal was enhanced by the production of exopolymers [126, 127].

3.2.4. Phosphatase-Mediated Metal Precipitation

In this process, metal or radionuclide accumulation by bacterial biomass is mediated by a phosphatase, induced during metal-free growth, which liberates inorganic phosphate from a supplied organic phosphate donor molecule, *e.g.* glycerol 2-phosphate.

Metal/radionuclide cations are then precipitated as phosphates on the biomass [60, 62]. Most work has been carried out with a *Citrobacter* sp. and a range of bioreactor configurations, including those using immobilized biofilms, have been described [128, 129]. Zirconium was mineralized by a *Citrobacter* sp. to a mixture of $Zr(HPO_4)_2$ and hydrated zirconia (ZrO_2): biomineralization of uranium as HUO_2PO_4 was repressed by zirconium in the presence of excess P_i. Cell-bound HUO_2PO_4 facilitated Ni^{2+} removal by intercalative ion exchange into the polycrystalline lattice and also promoted Zr removal [130].

3.2.5. High-Gradient Magnetic Separation (HGMS)

This is a technique for metal ion removal from solution using bacteria rendered susceptible to magnetic fields. "Nonmagnetic" bacteria can be made magnetic by the precipitation of metal phosphates (aerobic) or sulfides (anaerobic) on their surfaces as described previously. For those organisms producing iron sulfide, it has been found that this compound is not only magnetic but also an effective adsorbent for metallic elements [131–133].

4. Transformations

4.1. METALLOID TRANSFORMATIONS

Two major transformation processes are reduction of metalloid oxyanions to elemental forms, *e.g.* SeO_4^{2-} and SeO_3^{2-} to Se^0, and methylation of metalloids, metalloid oxyanions or organometalloids to methyl derivatives, *e.g.* AsO_4^{3-}, AsO_2^- and methylarsonic acid to $(CH_3)_3As$ (trimethylarsine). These processes modify the mobility and toxicity of metalloids, have biogeochemical significance, and are also of remedial potential [17, 94, 134–138].

4.1.1. Microbial Reduction of Metalloid Oxyanions

Reduction of selenate (Se^{VI}) and selenite (Se^{IV}) to elemental selenium can be catalyzed by numerous microbial species [94, 139], and this often results in a red precipitate deposited around cells and colonies. Certain organisms can also use SeO_4^{2-} as an electron acceptor to support growth. The reduction can be coupled to a variety of organic substrates, *e.g.* lactate, acetate and aromatics, with the bacteria found in a range of habitats and not confined to any specific genus. These organisms, and perhaps the enzymes themselves, may have applications for bioremediation of selenium- and arsenic-contaminated environments [140]. *Sulfurospirillum barnesii* can simultaneously reduce both NO_3^- and Se^{VI}. Kinetic experiments with cell membranes of *S. barnesii* suggest the presence of constitutive selenate and nitrate reduction as well as an inducible, high-affinity nitrate reductase in nitrate-grown cells which also has a low affinity for selenate. Simultaneous reduction of micromolar Se^{VI} in the presence of millimolar nitrate suggests a role for these organisms in bioremediating nitrate-rich seleniferous wastewaters [141]. Reduction of TeO_3^{2-} to

Te^0 is also an apparent means of detoxication found in bacteria [142], the Te^0 being deposited in or around cells, particularly near the cytoplasmic membrane [143–146]. In contrast to bacteria, fungal reduction of metalloids has received less attention although numerous species are capable of SeO_3^{2-} reduction to Se^0, deposited intra- and extracellularly, resulting in a red color of colonies [139, 147, 148]. Fungal reduction of TeO_3^{2-} to Te^0 results in black or grey colonies [149] with elemental Te being deposited intra- and extracellularly [150].

4.1.2. Methylation of Metalloids

Microbial methylation of metalloids to yield volatile derivatives, e.g. dimethyl-selenide or trimethylarsine, can be mediated by a variety of bacteria, algae and fungi [134, 135]. Less work has been carried out on tellurium methylation by fungi although there is evidence of dimethyltelluride and dimethylditelluride production [134]. Several bacterial and fungal species can methylate arsenic compounds such as arsenate (As^V, AsO_4^{3-}), arsenite (As^{III}, AsO_2^-) and methylarsonic acid ($CH_3H_2AsO_3$) to volatile dimethylarsine ((CH_3)$_2$HAs) or trimethylarsine ((CH_3)$_3$As) [138].

4.1.3. Microbial Metalloid Transformations and Bioremediation

Oremland et al. [151, 152] described in situ removal of SeO_4^{2-}, by reduction to Se^0, by sediment bacteria in agricultural drainage regions of Nevada. Flooding of exposed sediments at Kesterson Reservoir with water (to create anoxic conditions) resulted in reduction and immobilization of large quantities of selenium that was present in the sediment [153]. Microbial methylation of selenium, resulting in volatilization, has also been used for in situ bioremediation of selenium-containing land and water at Kesterson Reservoir, California [154]. Selenium volatilization from soil was enhanced by optimizing soil moisture, particle size and mixing while in waters it was stimulated by the growth phase, salinity, pH and selenium concentration [155, 156]. The selenium-contaminated agricultural drainage water was evaporated to dryness until the sediment selenium concentration approached 100 mg Se/kg dry mass. Conditions such as carbon source, moisture, temperature and aeration were then optimized for selenium volatilization and the process continued until selenium levels in sediments declined to acceptable levels [154].

4.2. MERCURY

As the physicochemical interactions detailed previously, key microbial transformations of inorganic Hg^{2+} include reduction and methylation. The mechanism of bacterial Hg^{2+} resistance is enzymic reduction of Hg^{2+} to nontoxic volatile Hg^0 by mercuric reductase. Hg^{2+} may also arise from the action of organomercurial lyase on organomercurials [157]. Since Hg^0 is volatile, this could provide one means of mercury removal [158, 159]. Mercuric reductase, from a recombinant E. coli strain, has been immobilized on a chemically modifed diatomaceous earth support with immobilization enhancing stability and reusability: maximum activity was 1.2 nmol Hg per mg protein per s at an initial $[Hg^{2+}]$ of 50 μmol/L [160].

4.3. DEGRADATION OF ORGANOMETALS

As well as organomercurials, other organometals may be degraded by microorganisms. Organoarsenicals can be demethylated, while organotin degradation involves sequential removal of organic groups from the tin atom which results in a reduction in toxicity [135]. In theory, such mechanisms and interaction with the bioremediation possibilities described previously may provide a means of detoxication [135].

5. Concluding Remarks

Microorganisms play important roles in the environmental fate of toxic metals and metalloids with physicochemical and biological mechanisms effecting transformations between soluble and insoluble phases. Although the biotechnological potential of most of these processes has only been explored at laboratory scale, some mechanisms, notably bioleaching, biosorption and precipitation, have been employed at a commercial scale. Of these, autotrophic leaching is an established major process in mineral extraction but has also been applied to the treatment of contaminated soil. There have been several attempts to commercialize biosorption using microbial biomass but success has been limited, primarily due to competition with commercially produced ion-exchange media. As a process for immobilizing metals, precipitation of metals as sulfides has achieved large-scale application, and this holds out promise of further commercial development. Exploitation of other biological processes will undoubtedly depend on a number of scientific, economic and political factors. Finally, it should be emphasized that this area of research also provides understanding of the biogeochemistry of metal(loid) cycling in the environment and the central role of microorganisms in affecting metal mobility and transfer between different biotic and abiotic locations.

Acknowledgements

The author gratefully acknowledges financial support for his work described here from the *Biotechnology and Biological Sciences Research Council* and the *Royal Society*.

References

1. White, C., Wilkinson, S.C. and Gadd, G.M. (1995) The role of microorganisms in biosorption of toxic metals and radionuclides, *Int. Biodeter. Biodegrad.* 35, 17–40.

2. Gadd, G.M. (1992) Microbial control of heavy metal pollution, in J.C. Fry, G.M.Gadd, R.A. Herbert, C.W. Jones and I.A.Watson-Craik (eds.), *Microbial Control of Pollution. Society for General Microbiology Symposium Volume 48*, Cambridge University Press, Cambridge, pp. 59–88.
3. Gadd, G.M. (1992) Molecular biology and biotechnology of microbial interactions with organic and inorganic heavy metal compounds, in R.A. Herbert, and R.J. Sharp (eds.), *Molecular Biology and Biotechnology of Extremophiles,* Blackie, Glasgow, pp. 225–257.
4. Gadd, G.M. (1997) Roles of microorganisms in the environmental fate of radionuclides, in G.R. Bock and G. Cardew, (eds.), *CIBA Foundation 203: Health Impacts of Large Releases of Radionuclides,* John Wiley and Sons, Chichester, pp. 94–104.
5. Wainwright, M. and Gadd, G.M. (1997) Industrial Pollutants, in D.T. Wicklow and B. Soderstrom (eds.), *The Mycota, Volume V: Environmental and Microbial Relationships,* Springer-Verlag, Berlin, pp. 85–97.
6. Gadd, G.M. (1996) Influence of microorganisms on the environmental fate of radionuclides, *Endeavour* 20, 150–156.
7. Gadd, G.M. and White, C. (1993) Microbial treatment of metal pollution – a working biotechnology? *Trends Biotechnol.* 11, 353–359.
8. White, C., Sayer, J.A. and Gadd, G.M. (1997) Microbial solubilization and immobilization of toxic metals: key biogeochemical processes for treatment of contamination, *FEMS Microbiol. Rev.* 20, 503–516.
9. Sayer, J.A. and Gadd, G.M. (1997) Solubilization and transformation of insoluble inorganic metal compounds to insoluble metal oxalates by *Aspergillus niger, Mycol. Res.* 101, 653–661.
10. Gharieb, M.M., Sayer, J.A. and Gadd, G.M. (1998) Solubilization of natural gypsum (CaSO$_4$.2H$_2$O) and the formation of calcium oxalate by *Aspergillus niger* and *Serpula himantioides, Mycol. Res.* 120, 825–830.
11. Gadd, G.M. (1999) Fungal production of citric and oxalic acid: importance in metal speciation, physiology and biogeochemical processes, *Adv. Microb. Physiol.* 41, 47–92.
12. White, C., Sharman, A.K. and Gadd, G.M. (1998) An integrated microbial process for the bioremediation of soil contaminated with toxic metals, *Nature Biotechnol.* 16, 572–575.
13. Yong, P. and Macaskie, L.E. (1995) Enhancement of uranium bioaccumulation by a *Citrobacter* sp. via enzymically-mediated growth of polycrystalline NH$_4$UO$_2$PO$_4$, *J. Chem. Technol. Biotechnol.* 63, 101–108.
14. White, C. and Gadd, G.M. (1996a) Mixed sulphate-reducing bacterial cultures for bioprecipitation of toxic metals: factorial and response-surface analysis of the effects of dilution rate, sulphate and substrate concentration, *Microbiol.* 142, 2197–2205.
15. White, C. and Gadd, G.M. (1996b) A comparison of carbon/energy and complex nitrogen sources for bacterial sulphate-reduction: potential applications to bioprecipitation of toxic metals as sulphides, *J. Indust. Microbiol.* 17, 116–123.
16. McGenity, T.J. and Sellwood, B.W. (1999) New approaches to studying the microbial precipitation of carbonate minerals, *Sediment. Geol.* 126, 5–8.
17. Lovley, D.R. (1995) Bioremediation of organic and metal contaminants with dissimilatory metal reduction, *J. Indust. Microbiol.* 14, 85–93.
18. White, C. and Gadd, G.M. (1998) Reduction of metal cations and oxyanions by anaerobic and metal-resistant organisms: chemistry, physiology and potential for the control and bioremediation of toxic metal pollution, in W.D. Grant and T. Horikoshi (eds.), *Extremophiles: Physiology and Biotechnology,* John Wiley and Sons, New York, pp. 233–254.
19. Rawlings, D.E. (1997) Mesophilic, autotrophic bioleaching bacteria: description, physiology and role, in D.E. Rawlings (ed.), *Biomining: Theory, Micobes and Industrial Processes,* Springer-Verlag, Berlin, pp. 229–245.
20. Schippers, A. and Sand, W. (1999) Bacterial leaching of metal sulfides proceeds by two indirect mechanisms via thiosulfate or via polysulfides and sulfur, *Appl. Environ. Microbiol.* 65, 319–321.

21. Ewart, D.K. and Hughes, M.N. (1991) The extraction of metals from ores using bacteria, *Adv. Inorg. Chem.* **36**, 103–135.

22. Bosecker, K. (1997) Bioleaching: metal solubilization by microorganisms, *FEMS Microbiol. Rev.* **20**, 591–604.

23. Rawlings, D.E. and Silver, S. (1995) Mining with microbes, *Biotechnol.* **13**, 773–778.

24. Sreekrishnan, T.R. and Tyagi, R.D. (1994) Heavy metal leaching from sewage sludges: a technoeconomic evaluation of the process options, *Environ. Technol.* **15**, 531–543.

25. Tichy, R., Janssen, A., Grotenhuis, J.T.C., Lettinga, G. and Rulkens, W.H. (1994) Possibilities for using biologically-produced sulphur for cultivation of *Thiobacilli* with respect to bioleaching processes, *Biores. Technol.* **48**, 221–227.

26. Ravishankar, B.R., Blais, J.F., Benmoussa, H. and Tyagi, R.D. (1994) Bioleaching of metals from sewage sludge: elemental sulfur recovery, *J. Environ. Eng.* **120**, 462–470.

27. Blais, J.F., Meunier, N. and Tyagi, R.D. (1997) Simultaneous sewage sludge digestion and metal leaching at controlled pH, *Environ. Technol.* **18**, 499–508.

28. Zagury, G.J., Narasasiah, K.S. and Tyagi, R.D. (1994) Adaptation of indigenous iron-oxidizing bacteria for bioleaching of heavy metals in contaminated soils, *Environ. Technol.* **15**, 517–530.

29. Vachon, P., Tyagi, R.D., Auclair, J.C. and Wilkinson, K.J. (1994) Chemical and biological leaching of aluminium from red mud, *Environ. Sci. Technol.* **28**, 26–30.

30. Seidel, H., Ondruschka, J., Morgenstern, P. and Stottmeister, U. (1998) Bioleaching of heavy metals from contaminated aquatic sediments using indigenous sulfur-oxidizing bacteria: a feasibility study, *Wat. Sci. Technol.* **37**, 387–394.

31. Gomez, C. and Bosecker, K. (1999) Leaching heavy metals from contaminated soil by using *Thiobacillus ferrooxidans* or *Thiobacillus thiooxidans*, *Geomicrobiol. J.* **16**, 233–244.

32. Mercier, G., Chartier, M., Couillard, D. and Blais, J.F. (1999) Decontamination of fly ash and used lime from municipal waste, *Environ. Manag.* **24**, 517–528.

33. Brandl, H., Bosshard, R. and Wegmann, M. (1999) Computer-munching microbes: metal leaching from electronic scrap by bacteria and fungi, *Process Metall.* **9B**, 569–576.

34. Burgstaller, W. and Schinner, F. (1993) Leaching of metals with fungi, *J. Biotechnol.* **27**, 91–116.

35. Tzeferis, P.G., Agatzini, S. and Nerantzis, E.T. (1994) Mineral leaching of non-sulphide nickel ores using heterotrophic micro-organisms, *Lett. Appl. Microbiol.* **18**, 209–213.

36. Drever, J.I. and Stillings, L.L. (1997) The role of organic acids in mineral weathering, *Colloid. Surf.* **120**, 167–181.

37. Mattey, M. (1992) The production of organic acids, *Crit. Rev. Biotechnol.* **12**, 87–132.

38. Strasser, H., Burgstaller, W. and Schinner, F. (1994) High yield production of oxalic acid for metal leaching purposes by *Aspergillus niger*, *FEMS Microbiol. Lett.* **119**, 365–370.

39. Gadd, G.M. (1993) Interactions of fungi with toxic metals, *New Phytol.* **124**, 25–60.

40. Sayer, J.A., Raggett, S.L. and Gadd, G.M. (1995) Solubilization of insoluble metal compounds by soil fungi: development of a screening method for solubilizing ability and metal tolerance, *Mycol. Res.* **99**, 987–993.

41. Sayer, J.A., Kierans, M. and Gadd, G.M. (1997) Solubilization of some naturally-occurring metal-bearing minerals, limescale and lead phosphate by *Aspergillus niger*, *FEMS Microbiol. Lett.* **154**, 29–35.

42. Den vre, O., Garbaye, J. and Botton, B. (1996) Release of complexing organic acids by rhizosphere fungi as a factor in Norway Spruce yellowing in acidic soils, *Mycol. Res.* **100**, 1367–1374.

43. Francis, A.J., Dodge, C.J. and Gillow, J.B. (1992) Biodegradation of metal citrate complexes and implications for toxic metal mobility, *Nature* **356**, 140–142.

44. Dixon-Hardy, J.E., Karamushka, V.I., Gruzina, T.G., Nikovska, G.N., Sayer, J.A. and Gadd, G.M. (1998) Influence of the carbon, nitrogen and phosphorus source on the solubilization of insoluble metal compounds by *Aspergillus niger*, *Mycol. Res.* **102**, 1050–1054.

45. Gharieb, M.M. and Gadd, G.M. (1999) Influence of nitrogen source on the solubilization of natural gypsum (CaSO4.2H2O) and the formation of calcium oxalate by different oxalic and citric acid-producing fungi, *Mycol. Res.* 103, 473–481.

46. Meixner, O., Mischack, H., Kubicek, C.P. and Rohr, M. (1985) Effect of manganese deficiency on plasma-membrane lipid composition and glucose uptake in *Aspergillus niger*, *FEMS Microbiol. Lett.* 26, 271–274.

47. Tzeferis, P.G. (1994) Leaching of a low-grade hematitic laterite ore using fungi and biologically produced acid metabolites, *Int. J. Min. Proc.* 42, 267–283.

48. Schinner, F. and Burgstaller, W. (1989) Extraction of zinc from an industrial waste by a *Penicillium* sp., *Appl. Environ. Microbiol.* 55, 1153–1156.

49. Franz, A., Burgstaller, W. and Schinner, F. (1991) Leaching with *Penicillium simplicissimum*: influence of metals and buffers on proton extrusion and citric acid production, *Appl. Environ. Microbiol.* 57, 769–774.

50. Franz, A., Burgstaller, W., Muller, B. and Schinner, F. (1993) Influence of medium components and metabolic inhibitors on citric acid production by *Penicillium simplicissimum, J. Gen. Microbiol.* 139, 2101–2107.

51. Sukla, L.B., Kar, R.N. and Panchanadikar, V. (1992) Leaching of copper converter slag with *Aspergillus niger* culture filtrate, *Biometals* 5, 169–172.

52. Bosshard, P.P., Bachofen, R. and Brandl, H. (1996) Metal leaching of fly-ash from municipal waste incineration by *Aspergillus niger, Environ. Sci. Technol.* 30, 3066–3070.

53. Veglio, F. (1996) The optimisation of manganese-dioxide bioleaching media by fractional factorial experiments, *Process Biochem.* 31, 773–785.

54. Diels, L., DeSmet, M., Hooyberghs, L. and Corbisier, P. (1999) Heavy metals bioremediation of soil, *Molec. Biotechnol.* 12, 149–158.

55. Gray, S.N. (1998) Fungi as potential bioremediation agents in soil contaminated with heavy or radioactive metals, *Biochem. Soc. Trans.* 26, 666–670.

56. Sayer, J.A., Cotter-Howells, J.D., Watson, C., Hillier, S. and Gadd, G.M. (1999) Lead mineral transformation by fungi, *Curr. Biol.* 9, 691–694.

57. Geesey, G. and Jang, L. (1990) Extracellular polymers for metal binding, in H.L. Ehrlich and C.L. Brierley (eds.), *Microbial Mineral Recovery*, McGraw-Hill Publishing Company, pp. 223–275.

58. Beveridge, T.J. and Doyle, R.J. (1989) *Metal Ions and Bacteria*, New York: Wiley.

59. Tobin, J., White, C. and Gadd, G.M. (1994) Metal accumulation by fungi: applications in environmental biotechnology, *J. Indust. Microbiol.* 13, 126–130.

60. Macaskie, L.E. and Dean, A.C.R. (1989) Microbial metabolism, desolubilization and deposition of heavy metals: uptake by immobilized cells and application to the treatment of liquid wastes, in A. Mizrahi (ed.), *Biological Waste Treatment*, Alan R. Liss, New York, pp. 150–201.

61. Brierley, C.L. (1990) Bioremediation of metal-contaminated surface and groundwaters, *Geomicrobiol. J.* 8, 201–223.

62. Macaskie, L.E. (1991) The application of biotechnology to the treatment of wastes produced by the nuclear fuel cycle — biodegradation and bioaccumulation as a means of treating radionuclide-containing streams, *Crit. Rev. Biotechnol.* 11, 41–112.

63. Al Saraj, M., AbdelLatif, M.S., El Nahal, I. and Baraka, R. (1999) Bioaccumulation of some hazardous metals by sol-gel entrapped microorganisms, *J. Non-Crystalline Solids* 248, 137–140.

64. Aloysius, R., Karim, M.I.A. and Ariff, A.B. (1999) The mechanisms of cadmium removal from aqueous solution by nonmetabolizing free and immobilized live biomass of *Rhizopus oligosporus, World J. Microbiol. Biotechnol.* 15, 571–578.

65. Chang, J.-S. and Chen, C.-C. (1999) Biosorption of lead, copper and cadmium with continuous hollow-fiber microfiltration processes, *Sep. Sci. Technol.* 34, 1607–1627.

66. Veglio, F., Beolchini, F., Boaro, M., Lora, S., Corain, B. and Toro, L. (1999) Poly(hydroxyethyl methacrylate) resins as supports for copper (II) biosorption with *Arthrobacter* sp.: matrix nanomorphology and sorption performances, *Process Biochem.* 34, 367–373.

90

67. Veglio, F., Beolchini, F., Gasbarro, A. and Toro, L. (1999) *Arthrobacter* sp. as a copper biosorbing material: ionic characterisation of the biomass and its use entrapped in a poly-hema matrix, *Chem. Biochem. Eng. Quart.* **13**, 9–14.

68. Suhasini, I.P., Sriram, G., Asolekar, S.R. and Sureshkumar, G.K. (1999) Nickel biosorption from aqueous systems: studies on single and multimetal equilibria, kinetics, and recovery, *Sep. Sci. Technol.* **34**, 2761–2779.

69. Gadd, G.M. (1988) Accumulation of metals by microorganisms and algae, in H.-J. Rehm (ed.), *Biotechnology, Vol 6b*, VCN Verlagsgesellschaft, Weinheim, pp. 401–433.

70. Gadd, G.M. and White, C. (1990) Biosorption of radionuclides by yeast and fungal biomass, *J. Chem. Technol. Biotechnol.* **49**, 331–343.

71. Krämer, M. and Meisch, H.-U. (1999) New metal-binding ethyldiamino- and dicarboxy-products from *Aspergillus niger* industrial wastes, *BioMetals* **12**, 241–246.

72. Birch, L. and Bachofen, R. (1990) Complexing agents from microorganisms, *Experientia* **46**, 827–834.

73. Spark, K.M., Wells, J.D. and Johnson, B.B. (1997) The interaction of a humic acid with heavy metals, *Aust. J. Soil Res.* **35**, 89–101.

74. Zinkevich, V., Bogdarina, I., Kang, H., Hill, M.A.W., Tapper, R. and Beech, I.B. (1996) Characterization of exopolymers produced by different isolates of marine sulphate-reducing bacteria, *Int. Biodeter. Biodegrad.* **37**, 163–172.

75. Schreiber, D.R., Millero, F.J. and Gordon, A.S. (1990) Production of an extracellular copper-binding compound by the heterotrophic marine bacterium *Vibrio alginolyticus*, *Mar. Chem.* **28**, 275–284.

76. Beech, I.B. and Cheung, C.W.S. (1995) Interactions of exopolymers produced by sulphate-reducing bacteria with metal ions, *Int. Biodeter. Biodegrad.* **35**, 59–72.

77. Flemming, H.-K. (1995) Sorption sites in biofilms, *Wat. Sci. Technol.* **32**, 27–33.

78. Vieira, M.J. and Melo, L.F. (1995) Effect of clay particles on the behaviour of biofilms formed by *Pseudomomonas fluorescens*, *Wat. Sci. Technol.* **32**, 45–52.

79. Bender, J., Rodriguez-Eaton, S., Ekanemesang, U.M. and Phillips, P. (1994) Characterization of metal-binding bioflocculants produced by the cyanobacterial component of mixed microbial mats, *Appl. Environ. Microbiol.* **60**, 2311–2315.

80. Neilands, J.B. (1981) Microbial iron compounds, *Ann. Rev. Biochem.* **50**, 715–731.

81. Bulman, R.A. (1978) Chemistry of plutonium and the transuranics in the biosphere, *Struct. Bond.* **34**, 39–77.

82. Howe, R., Evans, R.L. and Ketteridge, S.W. (1997) Copper-binding proteins in ectomycorrhizal fungi, *New Phytol.* **135**, 123–131.

83. Rauser, W.E. (1995) Phytochelatins and related peptides, *Plant Physiol.* **109**, 1141–1149.

84. Spears, D.R. and Vincent, J.B. (1997) Copper binding and release by immobilized transferrin: a new approach to heavy metal removal and recovery, *Biotechnol. Bioeng.* **53**, 1–9.

85. Chen, W., Br hlmann, F., Richins, R.D. and Mulchandani, A. (1999) Engineering of improved microbes and enzymes for bioremediation, *Curr. Opin. Biotechnol.* **10**, 137–141.

86. Pazirandeh, M., Wells, B.M. and Ryan, R.L. (1998) Development of bacterium-based heavy metal biosorbents: enhanced uptake of cadmium and mercury by *Escherichia coli* expressing a metal binding motif, *Appl. Environ. Microbiol.* **64**, 4086–4072.

87. Valls, M., González-Duarte, R., Atrian, S. and De Lorenzo, V. (1998) Bioaccumulation of heavy metals with protein fusions of metallothionein to bacterial OMPs, *Biochimie* **80**, 855–861.

88. Kotrba, P., Dolečková, L., de Lorenzo, V. and Ruml, T. (1999) Enhanced bioaccumulation of heavy metal ions by bacterial cells due to surface display of short metal binding peptides, *Appl. Environ. Microbiol.* **65**, 1092–1098.

89. Clemens, S., Kim, E.J., Neumann, D. and Schroeder, J.I. (1999) Tolerance to toxic metals by a gene family of phytochelatin synthases from plants and yeast, *EMBO J.* **18**, 3325–3333.

90. Bridge, T.A.M., White, C. and Gadd, G.M. (1999) Extracellular metal-binding activity of the sulphate-reducing bacterium *Desulfococcus multivorans*, *Microbiol.* **145**, 2987–2995.

91. Chen, S.L., Kim, E.K., Shuler, M.L. and Wilson, D.B. (1998) Hg^{2+} removal by genetically engineered *Escherichia coli* in a hollow fiber bioreactor, *Biotechnol. Prog.* **14**, 667–671.

92. Corbisier, P., van der Leilie, D., Borremans, B., Provoost, A., de Lorenzo, V., Brown, N.L., Lloyd, J.R., Hobman, J.L., Csoregi, E., Johansson, G. and Mattiasson, B. (1999) Whole cell- and protein-based biosensors for the detection of bioavailable heavy metals in environmental samples, *Anal. Chimica Acta* **387**, 235–244.

93. Lovley, D.R. and Coates, J.D. (1997) Bioremediaition of metal contamination, *Curr. Opin. Biotechnol.* **8**, 285–289.

94. Lovley D.R. (1993) Dissimilatory metal reduction, *Ann. Rev. Microbiol.* **47**, 263–290.

95. Ghani, B., Takai, M., Hisham, N.Z., Kishimoto, N., Ismail, A.K., Tano, T. and Sugio, T. (1993) Isolation and characterisation of a Mo^{6+} reducing bacterium, *Appl. Environ. Microbiol.* **59**, 1176–1180.

96. Wang, P.-C., Mori, T., Komori, K., Sasatu, M., Toda, K. and Ohtake, H. (1989) Isolation and characterisation of an *Enterobacter cloacae* strain that reduces hexavalent chromium under anaerobic conditions, *Appl. Environ. Microbiol.* **55**, 1665–1669.

97. Fujie, K., Tsuchida T., Urano K. and Ohtake H. (1994) Development of a bioreactor system for the treatment of chromate wastewater using *Enterobacter cloacae* HO1, *Wat. Sci. Technol.* **30**, 235–243.

98. Shen, H. and Wang, Y.-T. (1993) Characterization of enzymatic reduction of hexavalent chromium by *Escherichia coli*, ATCC 33456, *Appl. Environ. Microbiol.* **59**, 3771–3777.

99. Lovley, D.R., Giovannoni, S.J., White, D.C., Champine, J.E., Phillips, E.J.P., Gorby, Y.A. and Goodwin, S. (1993) *Geobacter metallireducens* gen nov. sp.nov., a microorganism capable of coupling the complete oxidation of organic compounds to the reduction of iron and other metals, *Arch. Microbiol.* **159**, 336–344.

100. Lovley, D.R. and Phillips, E.J.P. (1992a) Bioremediation of uranium contamination with enzymatic uranium reduction, *Environ. Sci. Technol.* **26**, 2228–2234.

101. Lovley, D.R. and Phillips, E.J.P. (1992b) Reduction of uranium by *Desulfovibrio desulfuricans*, *Appl. Environ. Microbiol.* **58**, 850–856.

102. Phillips, E.J.P., Landa, E.R. and Lovley, D.R. (1995) Remediation of uranium contaminated soils with bicarbonate extraction and microbial U(VI) reduction, *J. Indust. Microbiol.* **14**, 203–207.

103. Aubert, C., Lojou, E., Bianco, P., Rousset, M., Durand, M.-C., Bruschi, M. and Dolla, A. (1998) The *Desulfuromonas acetoxidans* triheme cytochrome c7 produced in *Desulfovibrio desulfuricans* retains its metal reductase activity, *Appl. Environ. Microbiol.* **64**, 1308–1312.

104. Lloyd, J.R., Ridley, J., Khizniak, T., Lyalikova, N.N. and Macaskie, L.E. (1999) Reduction of technetium by *Desulfovibrio desulfuricans*: biocatalyst characterization and use in a flow-through bioreactor, *Appl. Environ. Microbiol.* **65**, 2691–2696.

105. Lloyd, J.R., Thomas, G.H., Finlay, J.A., Cole, J.A. and Macaskie, L.E. (1999) Microbial reduction of technetium by *Escherichia coli* and *Desulfovibrio desulfuricans*: enhancement via the use of high activity strains and effect of process parameters, *Biotechnol. Bioeng.* **66**, 122–130.

106. Lloyd, J.R. and Macaskie, L.E. (1998) Enzymatic recovery of elemental palladium using sulfate-reducing bacteria, *Appl. Environ. Microbiol.* **64**, 4607–4609.

107. Rusin, P.A., Sharp, J.E., Oden, K.L., Arnold, R.G. and Sinclair, N.A. (1993) Isolation and physiology of a manganese-reducing *Bacillus polymyxa* from an Oligocene silver-bearing ore and sediment with reference to Precambrian biogeochemistry, *Precambrian Res.* **61**, 231–240.

108. Landa, E.R. and Gray, J.R. (1995) US Geological Survey — results on the environmental fate of uranium mining and milling wastes, *J. Indust. Microbiol.* **26**, 19–31.

109. Postgate, J.R. (1984) *The Sulphate-Reducing Bacteria*, Cambridge University Press, Cambridge.

110. Chang, J.C. (1993) Solubility product constants, in D.R. Lide (ed.), *CRC Handbook of Chemistry and Physics*, CRC Press, Boca Raton, pp. 8–39.

111. Crathorne, B. and Dobbs, A.J. (1990) Chemical pollution of the aquatic environment by priority pollutants and its control, in R.M. Harrison, (ed.), *Pollution: Causes, Effects and Control*, The Royal Society of Chemistry, Cambridge, pp. 1–18.

112. Taylor, M.R.G. and McLean, R.A.N. (1992) Overview of clean-up methods for contaminated sites, *J. Inst. Wat. Environ. Man.* 6, 408–417.

113. Fortin, D., Davis, B., Southam, G. and Beveridge, T.J. (1995) Biogeochemical phenomena induced by bacteria within sulphidic mine tailings, *J. Indust. Microbiol.* 14, 178–185.

114. Ledin, M. and Pedersen, K. (1996) The environmental impact of mine wastes — roles of microorganisms and their significance in the treatment of mine wastes, *Earth Sci. Rev.* 41, 67–108.

115. Schippers, A., Von Rege, H. and Sand, W. (1996) Impact of microbial diversity and sulphur chemistry on safeguarding sulphidic mine waste, *Minerals Eng.* 9, 1069–1079.

116. Uhrie, J.L., Drever, J.I., Colberg, P.J.S. and Nesbitt, C.C. (1996) *In situ* immobilisation of heavy metals associated with uranium leach mines by bacterial sulphate reduction, *Hydrometall.* 43, 231–239.

117. Hammack, R.W. and Edenborn, H.M. (1992) The removal of nickel from mine waters using bacterial sulphate-reduction, *Appl. Microbiol. Biotechnol.* 37, 674–678.

118. Lyew, D., Knowles, R. and Sheppard, J. (1994) The biological treatment of acid-mine drainage under continuous-flow conditions in a reactor, *Process Safet. Environ. Protect.* 72, 42–47.

119. Christensen, B., Laake, M. and Lien, T. (1996) Treatment of acid-mine water by sulfate-reducing bacteria — results from a bench-scale experiment, *Wat. Res.* 30, 1617–1624.

120. Hedin, R.S. and Nairn, R.W. (1991) Contaminant removal capabilities of wetlands constructed to treat coal mine drainage, in G.A. Moshiri (ed.), *Proceedings of the International Symposium on Constructed Wetlands for Water-Quality Improvement*, Lewis Publishers, Chelsea MI, pp. 187–195.

121. Perry, K.A. (1995) Sulfate-reducing bacteria and immobilization of metals, *Mar. Geores. Geotechnol.* 13, 33–39.

122. Dvorak, D.H., Hedin, R.S., Edenborn, H.M. and McIntire, P.E. (1992) Treatment of metal-contaminated water using bacterial sulfate reduction: results from pilot-scale reactors, *Biotechnol. Bioeng.* 40, 609–616.

123. Barnes, L.J., Janssen, F.J., Sherren, J., Versteegh, J.H., Koch, R.O. and Scheeren, P.J.H. (1991) A new process for the microbial removal of sulfate and heavy metals from contaminated waters extracted by a geohydrological control system, *Trans. Inst. Chem. Eng.* 69, 184–186.

124. Barnes, L.J., Scheeren, P.J.M. and Buisman, C.J.N. (1994) Microbial removal of heavy metals and sulphate from contaminated groundwaters, in J.L. Means and R.E. Hinchee (eds.), *Emerging Technol. for Bioremediation of Metals*, Lewis Publishers, Boca Raton, pp. 38–49.

125. White, C. and Gadd, G.M. (1997) An internal sedimentation bioreactor for laboratory-scale removal of toxic metals from soil leachates using biogenic sulphide precipitation, *J. Indust. Microbiol.* 18, 414–421.

126. White, C. and Gadd, G.M. (1998) Accumulation and effects of cadmium on sulphate-reducing bacterial biofilms, *Microbiol.* 144, 1407–1415.

127. White, C. and Gadd, G.M. (2000) Copper accumulation by sulfate-reducing bacterial biofilms, *FEMS Microbiol. Lett.* 183, 313–318.

128. Macaskie, L.E., Jeong, B.C. and Tolley, M.R. (1994) Enzymically-accelerated biomineralization of heavy metals: application to the removal of americium and plutonium from aqueous flows, *FEMS Microbiol. Rev.* 14, 351–368.

129. Tolley, M.R., Strachan, L.F. and Macaskie, L.E. (1995) Lanthanum accumulation from acidic solutions using a *citrobacter* sp. immobilised in a flow-through bioreactor, *J. Indust. Microbiol.* 14, 271–280.

130. Brasnakova, G. and Macaskie, L.E. (1999) Accumulation of zirconium and nickel by *Citrobacter* sp., *J. Chem. Technol. Biotechnol.* 74, 509–514.

131. Watson, J.H.P. and Ellwood, D.C. (1994) Biomagnetic separation and extraction process fron heavy metals from solution, *Minerals Eng.* 7, 1017–1028.

132. Watson, J.H.P., Ellwood, D.C., Deng, Q.X., Mikhalovsky, S., Hayter, C.E. and Evans, J. (1995) Heavy metal adsorption on bacterially-produced FeS, *Minerals Eng.* 8, 1097–1108.

133. Watson, J.H.P., Ellwood, D.C. and Duggleby, C.J. (1996) A chemostat with magnetic feedback for the growth of sulphate-reducing bacteria and its application to the removal and recovery of heavy metals from solution, *Minerals Eng.* 9, 937–983.

134. Karlson, U. and Frankenberger, W.T. (1993) Biological alkylation of selenium and tellurium, in Metal Ions, in H. Sigel, and A. Sigel (eds.), *Biological Systems*, Marcel Dekker, New York, pp. 185–227.

135. Gadd, G.M. (1993) Microbial formation and transformation of organometallic and organometalloid compounds, *FEMS Microbiol. Rev.* 11, 297–316.

136. Zhang, Y.Q., Frankenberger, W.T. and Moore, J.N. (1999) Effect of soil moisture on dimethylselenide transport and transformation to nonvolatile selenium, *Environ. Sci. Technol.* 33, 3415–3420.

137. Brady, J.M., Tobin, J.T. and Gadd, G.M. (1996) Volatilization of selenite in aqueous medium by a *Penicillium* species, *Mycol. Res.* 100, 955–961.

138. Tamaki, S. and Frankenberger, W.T. (1992) Environmental biochemistry of arsenic, *Rev. Environ. Contam. Toxicol.* 124, 79–110.

139. Gharieb, M.M., Wilkinson, S.C. and Gadd, G.M. (1995) Reduction of selenium oxyanions by unicellular, polymorphic and filamentous fungi: cellular location of reduced selenium and implications for tolerance, *J. Indust. Microbiol.* 14, 300–311.

140. Stolz, J.F. and Oremland, R.S. (1999) Bacterial respiration of arsenic and selenium, *FEMS Microbiol. Rev.* 23, 615–627.

141. Oremland, R.S., Switzer Blum, J., Burns Bindi, A., Dowdle, P.R., Herbel, M. and Stolz, J.F. (1999) Simultaneous reduction of nitrate and selenate by cell suspensions of selenium-respiring bacteria, *Appl. Environ. Microbiol.* 65, 4385–4392.

142. Walter, E.G. and Taylor, D.E. (1992) Plasmid-mediated resistance to tellurite: expressed and cryptic, *Plasmid* 27, 52–64.

143. Taylor, D.E., Walter, E.g. Sherburne, R. and Bazett-Jones, D.P. (1988) Structure and location of tellurium deposited in *Escherichia coli* cells harboring resistance plasmids, *J. Ultrastruct. Mol. Struct. Res.* 99, 18–26.

144. Blake, R.C., Choate, D.M., Bardhan, S., Revis, N., Barton, L.L. and Zocco, T.G. (1993) Chemical transformation of toxic metals by a *Pseudomonas* strain from a toxic waste site, *Environ. Toxicol. Chem.* 12, 1365–1376.

145. Lloyd-Jones, G., Osborn, A.M., Ritchie, D.A., Strike, P., Hobman, J.L., Brown, N.L. and Rouch, D.A. (1994) Accumulation and intracellular fate of tellurite in tellurite-resistant *Escherichia coli*: A model for the mechanism of resistance, *FEMS Microbiol. Lett.* 118, 113–120.

146. Moore, M.D. and Kaplan, S. (1992) Identification of intrinsic high-level resistance to rare-earth oxides and oxyanions in members of the class *Proteobacteria*: characterization of tellurite, selenite and rhodium sequioxide reduction in *Rhodobacter sphaeroides*, *J. Bacteriol.* 174, 1505–1514.

147. Zieve, R., Ansell, P.J., Young, T.W.K. and Peterson, P.J. (1985) Selenium volatilization by *Mortierella* species, *Trans. Brit. Mycol. Soc.* 84, 177–179.

148. Morley, G.F., Sayer, J.A., Wilkinson, S.C., Gharieb, M.M. and Gadd, G.M. (1996) Fungal sequestration, mobilization and transformation of metals and metalloids, in Frankland, J.C., N. Magan and G.M. Gadd (eds.), *Fungi and Environmental Change*, Cambridge University Press, Cambridge, pp. 235–256.

149. Smith, D.G. (1974) Tellurite reduction in *Schizosaccharomyces pombe*, *J. Gen. Microbiol.* 83, 389–392.

150. Gharieb, M.M., Kierans, M. and Gadd, G.M. (1999) Transformation and tolerance of tellurite by filamentous fungi: accumulation, reduction and volatilization, *Mycol. Res.* **30**, 299–305.
151. Oremland, R.S., Steinberg, N.A., Maest, A.S., Miller, L.G. and Hollibaugh, J.T. (1990) Measurement of *in situ* rates of selenate removal by dissimilatory bacterial reduction in sediments, *Environ. Sci. Technol.* **24**, 1157–1164.
152. Oremland, R.S., Steinberg, N.A., Presser, T.S. and Miller, L.G. (1991) *In situ* bacterial selenate reduction in the agricultural drainage systems of Western Nevada, *Appl. Environ. Microbiol.* **57**, 615–617.
153. Long, R.H.B., Benson, S.M., Tokunaga, T.K. and Yee, A. (1990) Selenium immobilization in a pond sediment at Kesterton Reservoir, *J. Environ. Qual.* **19**, 302–311.
154. Thompson-Eagle, E.T. and Frankenberger, W.T. (1992) Bioremediation of soils contaminated with selenium, in R. Lal and B.A. Stewart (eds.), *Advances in Soil Science*, Springer-Verlag, New York, pp. 261–309.
155. Frankenberger, W.T. and Karlson, U. (1994) Soil-management factors affecting volatilization of selenium from dewatered sediments, *Geomicrobiol. J.* **12**, 265–278.
156. Rael, R.M. and Frankenberger, W.T. (1996) Influence of pH, salinity and selenium on the growth of *Aeromonas veronii* in evaporation agricultural drainage water, *Wat. Res.* **30**, 422–430.
157. Silver, S. (1996) Bacterial resistances to toxic metals — a review, *Gene* **179**, 9–19.
158. Barkay, T., Turner, R., Saouter, E. and Horn, J. (1992) Mercury biotransformations and their potential for remediation of mercury contamination, *Biodegradation* **3**, 147–159.
159. Brunke, M., Deckwer, W.-D., Frischmuth, A., Horn, J.M., Lunsdorf, H., Rhode, M., Rohricht, M., Timmis, K.N. and Weppen, P. (1993) Microbial retention of mercury from waste streams in a laboratory column containing *merA* gene bacteria, *FEMS Microbiol. Rev.* **11**, 145–152.
160. Chang, J.S., Hwang, Y.P., Fong, Y.M. and Lin, P.J. (1999) Detoxification of mercury by immobilized mercuric reductase, *J. Chem. Technol. Biotechnol.* **74**, 965–973.

AEROBIC BIODEGRADATION OF POLYCHLORINATED BIPHENYLS (PCBs)

The fate, distribution, kinetics and enhancement of PCB biodegradation efficacy in the bacterial cell suspension of Pseudomonas stutzeri

K. Dercová[1], Š. Baláž[2], B. Vrana[1], R. Tandlich[2]
[1]*Department of Biochemical Technology, Faculty of Chemical Technology, Slovak University of Technology, Radlinského 9, 812 37 Bratislava*
[2]*Department of Pharmaceutical Sciences, College of Pharmacy, North Dakota State University, Sudro Hall 108, Fargo, ND-58105*

1. Introduction

A large number of new organic chemicals are permanently being introduced in the environment. Among these, PCBs are persistent priority pollutants that have been used extensively in industrial applications. Based on the economical and environmental considerations, there is a growing interest in the use of the developing bioremediation technology to clean environmental sites that are contaminated with organic pollutants. Ideally, microorganisms convert organic substances into carbon dioxide and water with some salts as by-products containing chlorine or other halogens. However, the prerequisite of applying bioremediation technology in order to eliminate PCBs in contaminated soil or water is the development of microbial strains with enhanced biodegradation capabilities and/or adjustment to the environmental conditions.

Although certain biological methods have been already used to treat hazardous waste for more than a decade, there are still those who question the bioremediation technology's potential and those who overrate its use. The question is: where is the truth? What can we reasonably expect from a technology that some view as a panacea for all environmental problems, while others see it as an idea the time of which has not yet come [1]?

Some aspects of this problem will be also addressed in our work. Our contribution summarizes briefly the present knowledge concerning bacterial PCB oxidative catabolic pathways. Some of the factors that should be preferentially considered when looking for strategies to enhance aerobic PCB biodegradation efficacy will be discussed as well.

The present work deals with the fate of PCBs in the bacterial cell suspension and kinetics of PCB biodegradation in this system, and with the two following ap-

95

V. Šašek et al. (eds.),
The Utilization of Bioremediation to Reduce Soil Contamination: Problems and Solutions, 95–113.
© 2003 *Kluwer Academic Publishers. Printed in the Netherlands.*

proaches of PCB biodegradation improvement: chemical oxidation using Fenton reaction under abiotic conditions and the effect of terpenes on PCB biodegradation under biotic conditions. The results and discussion are largely based on the data obtained with *Pseudomonas stutzeri*, a bacterial strain isolated from a long-term PCB-contaminated soil [2].

PCBs are industrial compounds that have been detected as contaminants in almost every component of the global ecosystem including air, water, sediments, fish and wildlife, as well as human adipose tissue, milk, and serum. PCBs represent a serious ecological problem due to low degradability, high toxicity and strong bioaccumulation. These features are associated with their high hydrophobicity and low chemical reactivity. PCBs in commercial products and environmental extracts are complex mixtures of congeners that can be analyzed on a congener-specific basis using high-resolution gas chromatographic analysis. PCBs are metabolized primarily via mixed-function oxidases into a broad spectrum of metabolites. The metabolic activation is not required for PCB toxicity, and that the parent hydrocarbons are responsible for most of the biochemical and toxic responses elicited by these compounds. Structure–function relationships of PCB congeners have identified two major structural classes of PCBs that elicit "TCDD-like" responses, namely, the coplanar PCBs and their mono-*ortho* coplanar derivatives. These compounds competitively bind to the TCDD or aryl hydrocarbon (Ah) receptors. In addition, other structural classes of PCBs elicit biochemical and toxic responses that are not mediated through the Ah receptor.

1.1. BIODEGRADATION AND BIOREMEDIATION OF PCBs

The problems associated with the remediation of PCB-contaminated used mineral oils can be minimized using physicochemical processes such as molecular distillation [3]. For decontamination of environmental samples (soil, water and sediments), biological processes seem to be more cost effective and ecological than the physico-chemical ones.

Bioremediation is being evaluated by both industry and the US-EPA as the right technology for cleaning up hazardous waste sites. Bioremediation is formally defined as the controlled use of biodegradation to remove toxic chemicals from soil and groundwater. The process usually involves stimulation of indigenous subsurface microorganisms to degrade chemicals *in situ*, although there have been cases when selected microorganisms with some special metabolic capabilities have been added. The goal of the engineers is to find microorganisms, generally bacteria that will metabolize (or at least oxidize) the target contaminant under the conditions present at the contaminated site or in an above-ground reactor [4].

Bacteria may attack hazardous chlorinated organic waste in one of the three ways: (*a*) mineralize the compound directly, which means that the compound is converted to harmless inorganic molecules such as carbon dioxide and salts; (*b*) by mineralizing the compound only as a co-metabolite, which means that the bacteria require some other organic compounds for growth or to induce formation of the enzymes required for degradation of the target compounds; (c) by converting the compound to some other compound, which may be recalcitrant to further degradation. The main obstacles preventing mineralization of hydrophobic pollutants are: recalcitrance, toxicity of intermediates, and limited physical availability for microorganisms.

Because of many environmental and economic problems, there are efforts to develop technologies for bioremediation of the PCB-contaminated areas. Finding or development of PCB-degrading microorganisms with high degradation potential is the essential element of a successful bioremediation technology.

1.2. MICROORGANISMS CAPABLE OF AEROBIC PCB BIODEGRADATION

Some PCBs can undergo aerobic, as well as anaerobic biodegradation processes. Most of the bacterial isolates, capable of the aerobic PCB biodegradation, are Gram-negative rods and cocci. The most important bacteria belong to the following genera: *Acinetobacter* [5], *Achromobacter* [6], *Alcaligenes* [2, 7], *Comamonas* [8] and *Pseudomonas* [2, 6]. The Gram-positive PCB-degraders occur more rarely than the Gram-negative ones. The Gram-positive degraders include the following genera: *Bacillus* [9], *Micrococcus* [10], *Arthrobacter* [5] and *Corynebacterium* [11].

1.3. BACTERIAL PCB-CATABOLIC PATHWAYS

So far, there are no known aerobic bacterial strains that are able to degrade PCB congeners with a high number of chlorine atoms in the molecule (more than 6 chlorine atoms). Some aerobic bacteria are able to transform certain PCB congeners into the respective chlorobenzoate molecules. This pathway is called the upper catabolic pathway (further degradation of chlorobenzoate is known as the lower catabolic pathway) [9, 12, 13].

The first reaction of the upper catabolic pathway is catalyzed by biphenyl dioxygenase (BPO). This enzyme consists of four subunits. In the first step, two hydroxyl groups are introduced into the PCB congener molecule in positions 2 and 3. The first two subunits of the enzyme molecule are called the big and the small subunit, respectively. They are encoded by *bphA* and *bphB* genes and together they constitute a FeS protein molecule. This FeS protein contains sulfur and iron atoms and is called the terminal oxygenase. The terminal oxygenase interacts directly with the substrate and inserts two oxygen atoms into it. The last two subunits of the BPO molecule are ferredoxin, coded by the *bphF* gene, and the ferredoxin reductase, coded by the *bphG* gene. Two latter units enable the electron transport from the NADH molecule onto the oxidized form of the terminal oxygenase. A dihydrodiol is the product of this reaction. In the next step, the dihydrodiol is transformed to a catechol derivative. This reaction is catalyzed by dehydrogenase, a *bphB*-gene product. The catechol is further cleaved between positions 1 and 2 (*meta*-cleavage) by dioxygenase, a *bphC*-gene product. A 6-oxo-6-chlorophenylhexa-2,4-dienic acid molecule is the product of this reaction. In the next step, this molecule undergoes hydrolytic cleavage. The products of the latter reaction are a particular chlorobenzoate and a pentanoic acid derivative. All genes, encoding the PCB-conversion proteins, were cloned and sequenced for a wide range of bacterial species [14–18].

The PCB transformation into the respective chlorobenzoates by bacteria is a good example of the exploitation of the biphenyl catabolic pathway. This pathway is utilized in the metabolism of many biphenyl derivatives, including other halo-biphenyls and hydroxybiphenyls [19]. Some microorganisms have a wide range of isofunctional enzymes and are able to degrade different organic compounds of similar structure. Degradation of chlorobenzoates, the upper-catabolic-pathway end products, is achieved through the lower catabolic pathway. In this way, complete PCB removal can be observed [20].

1.4. FACTORS AFFECTING AEROBIC PCB-BIODEGRADATION BY BACTERIA

Many factors may influence the bacterial PCB-oxidation pathway. The most important are (*i*) the number of chlorine atoms in a PCB molecule, (*ii*) the substitution pattern of the PCB molecule (the substrate specificity of biphenyl dioxygenase), (*iii*) the congener toxicity, (*iv*) biphenyl as a co-substrate, (*v*) availability of a substrate, (*vi*) the production of metabolites with some inhibition effect on the catabolic pathway.

The substrate specificity of enzymes in the catabolic pathway is a very important factor of modulation of the bacterial strain reactivity towards substrate. The number of chlorine atoms and the substitution pattern of that particular molecule determine the biodegradability of a particular PCB congener. Some possibilities to increase the effectiveness of PCB degradation can be summarized as application of the following: (*a*) mixed cultures, (*b*) inducers — *e.g.* biphenyl, (*c*) structural analogs — terpenes, flavonoids, (*d*) chitin, (*e*) nutrients — *e.g.* phosphorus, (*f*) surfactants, (*g*) vitamin B_{12}, (*h*) genetically modified microorganisms, (*i*) chemical oxidation.

1.5. STRUCTURE–BIODEGRADABILITY RELATIONSHIPS OF PCBs

Bacteria have been divided into groups according to their PCB congener-substrate selectivity. This division is based on several systematic biodegradation studies. These have been conducted on different PCB congeners with different numbers of chlorine atoms in the molecule and different substitution patterns [6, 9, 11, 21]. All quantitative structure–biodegradability relationship (QSBR) studies on PCBs and their aerobic bacterial degradation show that *ortho*-chlorinated congeners are the most resistant against bacterial attack.

The majority of known aerobic bacterial PCB-degraders preferentially oxidize *meta*- and *para*-substituted PCB congeners. Bedard *et al.* [11] classified several bacterial isolates according to their PCB-congener-degradation spectrum. The *ortho*-chlorinated congeners were preferred to other biodegradation substrates by some of the isolated strains. Those included *Pseudomonas* sp. LB400 and *Alcaligenes euthrophus* H850. All these strains had a similar biodegradation spectrum of the PCB congeners. There have been several attempts to explain this interesting fact. The first explanation was that these bacteria used a different enzyme, 3,4-dioxygenase. However, the biochemical and genetic analyses showed the presence of the 2,3-dioxygenase [22]. Another explanation could be the influence of transport rate and mechanism on the PCB-degradation spectrum. This possibility was eliminated when no significant difference was found to exist between the intact bacterial cells and some enzymic extracts from these cells.

Genetic analysis is a significant source of information about the QSBR of individual PCB congeners. The above-mentioned studies on gene cloning and the sequence analysis showed the PCB/biphenyl degradation pathway to be conservative for all bacteria. All genes controlling the BPO subunits of the upper catabolic pathway of PCBs are of the same phylogenetic origin in all bacteria [15, 17, 23]. All these genes, together with the *bphB* and *bphC* genes, constitute the *bph*-operon which is similar in structure to the *tod*-operon of some strains of the *Pseudomonas* genus. The *tod*-operon encodes toluene dioxygenase in bacterial strains. The *bphG* gene, encoding the ferredoxin reductase molecule, is the only exception. It is located outside the *bph*-operon. However, it has been shown that this molecule affects the oxidation process [24]. The amino acid sequences of the BPO subunits have been compared for *Pseudomonas pseudoalcaligenes* KF707 and *Pseudomonas* sp. LB400 (an *ortho*-cleavage bacterium) [22]. A high degree of homology was observed. The homology was 100 % for the ferredoxin and ferredoxin reductase molecules, and ranged from 95 to 99 % for the subunits of the terminal oxygenase. It is interesting to see the results in the context of the oxidation spectrum of both strains. Induced point mutations have been conducted in the cloned *bphA* gene that encodes a big subunit of the BPO terminal oxygenase. It was found that the change of only three amino acids in positions 335, 338, and 341 was needed to give the mutated enzyme the ability to degrade *ortho*-, as well as *para*-substituted congeners [25]. This seems to be the way of influencing BPO reactivity and making cleavage of more resistant congeners possible.

1.6. CO-METABOLITE ENRICHMENT AND BACTERIAL PCB DEGRADATION

Since the early studies, it has been known that biphenyl is the biphenyl/PCB-catabolic pathway inducer [6]. The biphenyl degrading enzymes do not have a high degree of substrate specificity and can be used for the co-metabolite transformation of the biphenyl structural analogs. It also means that if the microbial cells are able to grow on biphenyl as the sole carbon source, they can transform the PCB congeners more effectively than other microbial suspensions. *Comamonas testosteroni* B-356 degrades AROCLOR 1242 quite slowly if there is no biphenyl present. The biodegradation can be accelerated by addition of the co-metabolites. This has been found for other bacterial strains [25]. The soil degradation of AROCLOR 1254 by *Comamonas testosteroni* B-356 also depends on biphenyl presence [26].

1.7. PRODUCTION OF METABOLITES INHIBITING THE PCB-METABOLIC PATHWAY

The reactivity spectrum of the BPO and other biphenyl/PCB-catabolic-pathway enzymes is not the only factor influencing the degradation capacity of the bacterial strains. To improve the properties of a strain, it is necessary to monitor the metabolites production.

1.7.1. The Upper-Catabolic-Pathway Metabolites

When growing on monochlorinated biphenyls, the biphenyl-degrading bacteria produce some defined metabolites, including the chlorobenzoates, as well as some other metabolites [12, 13]. The other metabolites result from the obstacles encountered in the direct-metabolite flow of the central metabolic pathway. Bacterial cells contain a large variety of enzymes with broad substrate specificity, which are able to transform the central-metabolic-pathway metabolites into some new compounds. The dehydrogenases are a good example. The *meta*-cleavage product of the biphenyl molecule can be transformed into a saturated side-chain compound as a result of the dehydrogenase activity. Further degradation of this compound is possible through the activity of some hydrolytic and decarboxylating enzymes.

In contrast to the induction of the aromatic-compound-catabolic pathways [27], only little is known about the genetic regulation of the biphenyl/PCB degradation. We know that biphenyl is an inducer of the enzymes of the upper catabolic pathway. However, the mechanism of this regulation is not clear. There is no available information on the regulation of the lower catabolic pathway, the degradation of benzoic acid and chlorobenzoates.

1.7.2. The Lower-Catabolic-Pathway Metabolites

The upper-catabolic-pathway enzymes convert a PCB-congener molecule into the respective chlorobenzoate that is further transformed by the lower-catabolic-pathway enzymes. If a microorganism possesses the enzymes of both catabolic pathways, a chlorobenzoate molecule can be transformed to inorganic substances such as water, carbon dioxide and chloride anions. However, this is rarely the case. Most of the PCB-degraders do not have both gene groups with the necessary level of enzyme activity. Consequently, chlorobenzoates are often accumulated as the dead-end products or are transformed only partially. These dead-end products are accumulated inside the bacterial cells and are later slowly eluted into the extracellular medium.

It is possible that some lower-catabolic-pathway metabolites influence the upper-catabolic-pathway enzyme activities. Most of the *Pseudomonas* sp. strains have a genome that includes the lower-catabolic-pathway enzyme genes. However, this information is not widely expressed. It seems that a biphenyl-induced degradation pathway of chlorobenzoate degradation is required for the ultimate PCB/biphenyl biodegradation. *Comamonas testosteroni* B-356 has two different catabolic pathways for the benzoic-acid degradation [28]. It was proved by the examination of that strain's ability to degrade chlorobenzoates in the presence of different growth substrates. If biphenyl is the growth substrate, the chlorobenzoate molecules are cleaved faster than in the case of benzoic acid. The metabolites are very different in the two cases. A logical explanation is the existence of two different benzoic-acid oxygenases. Biphenyl induces one of them while the other is induced in the presence of benzoic acid. These oxygenases differ from the BPO. In some bacterial strains, the biphenyl degrading genes are located inside the same operon as the benzoic acid ones [17]. The chlorobenzoate-degradation genes are mostly located on plasmids. In contrast, the benzoic acid ones is located on the chromosomes. The benzoic-acid-degradation genes have a narrower substrate spectrum. *Comamonas testosteroni* B-356 has one more unique feature which seems to be common to the PCB-degrading bacterial strains. If the cells of this strain are grown on biphenyl as the sole carbon source, they induce enzymes which transform chlorobenzoates into different metabolites. These metabolites are often full inhibitors of dioxygenases cleaving the aromatic ring. Very low amounts of these compounds inhibit the 2,3-dihydroxybiphenyl-1,2-dioxygenase activity. This leads to a complete stop of the substrate degradation [28]. These metabolites can also totally inhibit the growth and all metabolic activities of the degrading strain [29]. The biphenyl-induced benzoic acid oxygenase transforms 3-chlorobenzoic acids into the respective catechol and further to some toxic metabolites. This means that the presence of a *meta*-substituted congener in any PCB mixture has a negative influence on PCB degradation [30]. It was also found that AROCLOR 1242 bacterial degradation is inhibited by the presence of mono- and dichlorinated benzoates in the mixture.

1.8. GENETICALLY ENGINEERED BACTERIA

Molecular biology has provided highly useful techniques to modify the genetic composition of microorganisms and thus to allow for the potential construction of new organisms having the capacity to carry out catabolic sequences that are not available in existing organisms [31].

A variety of problems in the future may be solved by the use of genetically modified bacteria. The following advantageous contributions can be foreseen: (*a*) constructing bacteria able to grow on and mineralize pollutants that presently are only co-metabolized; (*b*) creation of new catabolic pathways to affect transformations that are not carried out efficiently or rapidly at present, by altering the range of substrates used by particular microorganisms; (*c*) increasing the amount or activity of specific enzymes in a microorganism; (*d*) construction of microorganisms — bacteria that not only can destroy target pollutants but also are resistant to inhibitors at the site that prevent degradation by indigenous microorganisms [32].

1.9. MIXED BACTERIAL CULTURES

As mentioned above, most of the isolated PCB-degrading bacteria do not have the complete genetic information for the ultimate PCB biodegradation. One of the ways to achieve the complete PCB biodegradation is to combine more bacterial or microbial strains into one stable mixed culture. One of the microorganisms could transform PCBs into chlorobenzoates while the other could remove those chlorobenzoates completely.

Another important factor is the competition of the inserted microorganism with the natural microflora in the given ecosystem, especially when the bioremediation is about to take place in nonsterile conditions. Once the degrader is inoculated in the bioremediation spot, a balance has to be achieved in the microbial composition after some time. That implies that a stable mixed culture has to be formed without oppressing the inoculated degrader. The growth rate of the degrader has to be higher than that of its predators and natural enemies, *i.e.* protozoa. At the same time, its growth rate should be comparable with the growth rate of other microorganisms. It could be achieved by the addition of a carbon substrate nonutilizable by other microorganisms (*e.g.* biphenyl). The PCB-degrader is able to utilize biphenyl as the sole carbon and energy source, and its growth would be preferred under *in situ* conditions. Biphenyl can be utilized more easily than PCBs, but constitutes a serious environmental threat itself.

1.9.1.   Application of Mixed Cultures for Ultimate PCB-Degradation

As mentioned above, one way how to achieve the full biodegradation of PCBs is to combine several different microorganisms into a stable mixed culture. Individual microorganisms catalyze different stages of the PCB-congener degradation. However, this solution has a major drawback. The PCB-degradation is an intracellular process and microorganisms transforming PCBs into chlorobenzoates tend to accumulate them inside the cells. They can subsequently inhibit the metabolism [33]. If a chlorobenzoate degrader is added, its growth rate will be rather low. The rate-limiting step will be a diffusion of chlorobenzoate molecule out of the PCB-degrader's cells. So far, it has only been possible to obtain some stable mixed cultures when the transfer substrate was thermodynamically stable and nontoxic. One example is the ultimate 4-chlorobiphenyl biodegradation into inorganic compounds by a mixed culture consisting of *Acinetobacter* sp. P6, a PCB-degrader, and some mono- and dichlorobenzoate-degrading pseudomonads [34]. Another example could be the complete removal of 4,4'-dichlorobiphenyl using a culture consisting of *Acinetobacter* sp. P6 (when grown on biphenyl) and *Acinetobacter* sp. 4CB1 (able to utilize 4-chlorobenzoate) [35]. 4-Chlorobenzoate is a nontoxic and stable transfer substrate in both cases and it is not metabolized by *Acinetobacter* sp. P6. If the transfer product (substrate) is reactive and its metabolites are toxic, it is almost impossible to isolate a mixed culture. This statement is supported by the fact that a mixed culture capable of the ultimate 3-chlorobiphenyl biodegradation has not yet been isolated.

Its degradation has two partial steps. The first is a rapid formation of 3-chloro-benzoate, a reactive substrate with toxic cleavage metabolites that inhibit the bacterial cell growth [36]. The above-mentioned problems could be overcome by regulation of the PCB to benzoate transformation rate in such a way that it would be equal to the rate of chlorobenzoate diffusion out from the cells. The metabolites inhibiting the bacterial metabolism would be prevented from the inside-cell accumulation.

Taking the last-mentioned facts into account, it is surprising that a mixed culture has been isolated that was capable of the complete removal of all monochlorinated PCBs, as well as 2,3- and 2,4-dichlorobiphenyls [37]. That culture consisted of *Pseudomonas* sp. BN10, a PCB-degrader, and of *Pseudomonas* sp. B13, able to degrade some chlorinated catechols. However, the PCB-removal mechanism differed from the above-mentioned one. Interstrain conjugation and gene exchange was found to take place between the PCB-degrader and the catechol degrader during cultivation. In that way, the biphenyl-degrading genes were transferred from the BN10 sp. into the B13 sp., and a new strain, called JHR2, was constituted. This example shows that the nature itself can help to avoid the genetic-engineering manipulations. Activity of ECO3 mixed culture could also be explained in the same way. This culture contains three bacterial strains and degrades some low chlorinated PCB congeners [38].

A mixed culture does not have to be formed even if the transfer product is not toxic. It could be the outcome of the mutual inhibition of the potential mixed-culture components. In such a case, the so-called "two-step" process can be applied [39].

1.9.2. Mixed Cultures Producing Surfactants

PCBs show low solubility in water and the biodegradation rate is determined by the rate of dissolution. The aqueous solubility is a function of the interface area of the solid or liquid PCB-phase in water. The addition of surfactants having emulsifying properties makes this surface larger and therefore it is expected to speed up the biodegradation process. Bacterial species and strains, which are able to excrete some emulsifying surfactants to the extracellular medium, could increase efficiency when added to a PCB-degrading mixed culture of bacteria [40, 41]. Other emulsifier producers include *Pseudomonas* sp. [40, 41], *Mycobacterium* sp. [42], *Acinetobacter* sp. [43] and a *Vibrionaceae* species [44]. The influence of biosurfactants on the PCB-degradation is not always positive. Some biosurfactants are toxic to the PCB-degrading bacteria and can decrease their substrate affinity [45].

1.10. PRETREATMENT TECHNIQUES

Many PCB mixtures have been used commercially. The congener composition and substitution patterns of these mixtures depend on the synthesis. Lower chlorinated congeners, with one or two chlorine atoms in the molecule, are easily degradable. However, the mixtures usually include large amounts of highly chlorinated congeners, which are more or less resistant to the aerobic biodegradation. If a PCB mixture is biodegraded, only the congeners that are accessible to microbial attack are removed. Pretreatment of the PCB mixture can be used prior to the biodegradation step. One of the pretreatment techniques is anaerobic biodegradation taking place before the aerobic one. From the environmental point of view this would be an ideal method. However, anaerobic biodegradation is a very slow process and it might take years before highly chlorinated congeners are removed. That makes it impractical if a rapid solution is needed. The pretreatment processes can be classified as physical and chemical techniques [46−48]. There are two chemical ways of improving the aerobic PCB biodegradation: (*i*) partial dechlorination−transformation of higher chlorinated congeners to lower chlorinated PCBs; (*ii*) partial congener oxidation in order to improve the aqueous solubility as well as to facilitate their microbial metabolism.

1.11. EXPERIMENTAL SYSTEMS FOR STUDYING PCB BIODEGRADATION

Model systems provide valuable information about the studied process. They serve scientific and application purposes. They can be used to determine the kinetic parameters of the process, including the biodegradation rate constants, different substrate utilizations, specific growth rates of microorganisms, as well as parameters of some physical processes. Thermodynamic parameters, such as the equilibrium constants of dissociation reactions and the PCB partition coefficients among different phases, can also be determined. Real remediation processes can be simulated in the following systems: batch, chemostat and immobilized experimental system [49−51].

1.12. SOIL MICROFLORA AND THE PCB-BIODEGRADATION PROCESS

Blasco *et al.* [52] found that addition of 4-chlorobiphenyl led to a rapid decrease in the cell number of every potential PCB-degrader. The toxic effects of 4-chlorobiphenyl, 4-chloropyrocatechol and 4-chlorobenzoate have been eliminated as the possible explanation. A product of the 4-chloropyrocatechol cleavage that was formed by the 3-oxoadipate-pathway enzymes caused the decrease. These enzymes are produced by the soil microflora. The substance was identified as protoanemonin, an antibiotic that has been previously isolated from plant material. If there is no elimination of chlorobenzoic acid in the mixed culture environment, this antibiotic is accumulated and the PCB-degrader's growth is stopped. Possible solutions to this problem include (*i*) co-inoculation of PCB- and chlorobenzoic acid degraders, and (*ii*) inoculation of a genetically manipulated bacteria which has the genetic information of the upper and lower catabolic pathway. The former possibility was found to work [53].

1.12.1. Degradation of PCBs in Soil

Soil is a very complex matrix. Bioavailability and degradation of PCBs is influenced by many physicochemical and environmental factors, as well as mobility of indigenous or supplemented bacteria. Paya-Perez et al. [54] found linear correlation between the distribution constants and soil organic carbon content. Sorption was favored by a high degree of chlorination and the absence of ortho-substitution in the biphenyl. Sorption kinetics seemed to be first-order controlled (two steps) with rate constants of the order of many hours. Authors presented correlations between the sorption data and published physical data, which will be useful for modeling the transport and the fate of PCBs and chlorobenzenes in the environment.

The degradation of PCBs (DELOR 103) in soils by the biphenyl-utilizing strain *Alcaligenes xylosoxidans* was studied. In addition to the congener specificity, significant differences in the degradation of PCBs by the strain in the different soil types were observed. Efficiency of degradation was generally better in sterilized soils, but the differences were not as significant as the differences observed between the different soil types. These results indicate that the degradation of PCBs is probably related not only to the capabilities of the strain employed and quality and amount of competitive species inhabiting the soils, but also to the soil sorption of the PCB congeners: the degradation is faster in soils containing an intermediate amount of organic carbon with a high portion of total and aromatic carbon in humic acids [55].

1.12.2. In situ Application of PCB-Degrading Bacteria

The whole biodegradation research aims at using bacteria for bioremediation of contaminated areas, predominantly soils. However, the PCB-degrading bacteria can also play an important role in wastewater treatment. The ability to transform a pollutant is very important, but it is not the only prerequisite for the application of bacteria in situ. There are several possible explanations for the failure of a laboratory-tested strain, when applied in situ [56]. They include: (i) presence of the substances in the environment that inhibit metabolism of the degrading microorganism; (ii) the inoculated microorganism having lower specific growth rate than its predators, that means it is not able to compete for an essential growth factor; (iii) the inoculated microorganism having higher affinity for another substrate than to the pollutant; (iv) the inoculated microorganism being not able to reach the contaminated site or spot. Unfortunately, only little is presently known about the oxidative PCB-biodegradation in natural ecosystems, such as soils, river sediments, etc. The only bioremediation application based on PCB aerobic biodegradation and applied in a contaminated site was a field experiment with the Hudson River sediments carried out in 1991 [57]. Some river sediments with a very low redox potential contain an unidentified culture of anaerobic bacteria. These are known to catalyze a very slow but highly effective biodegradation of highly chlorinated PCBs.

2. Results

This part will briefly concentrate on the goals and major results obtained in our research of the fate, distribution and kinetics of PCB degradation in the bacterial cell suspension, as well as on the study of improvement of PCB degradation efficiency by selected chemical and biological methods. Knowledge of the biodegradation kinetics is essential for the evaluation of the persistence of organic pollutants and for the assessment of the potential risk associated with exposure of susceptible individuals and animals to certain chemicals. The biodegradation rate constant is an important parameter for the prediction of the time required to eliminate PCB congener from the contaminated site. However, there are only few kinetic data on the aerobic PCB biodegradation under the defined conditions available to date. The goals of our work can be summarized as follows: (1) to design a PCB distribution model describing the concurrent processes of evaporation, biosorption, and biodegradation in a suspension of PCB-degrading bacteria, (2) to use the distribution model for the determination of the primary biodegradation rate constants of individual PCB congeners present in the commercial PCB formulation DELOR 103, in a *Pseudomonas stutzeri* suspension, (3) to conduct an empirical structure-degradability analysis of the obtained data, (4) to improve aerobic biodegradation by chemical oxidation — Fenton reaction, (5) to enhance the efficacy of PCB degradation using inducers — investigation of the effect of terpenes.

2.1. BIODEGRADATION OF PCBs

Kinetics of the PCB distribution in an active bacterial suspension of *Pseudomonas stutzeri* (an isolate from long-term PCB-contaminated soil) [2] was studied by monitoring the evaporated amounts and the concentration of PCBs remaining in the aqueous medium with the bacterial biomass. A simple apparatus for effective monitoring of the PCB evaporation kinetics in batch biodegradation experiments has been described [58, 59].

A model that takes into account biosorption, evaporation, and primary biodegradation of the individual PCB congeners was constructed [60, 61]. Four processes take place simultaneously: evaporation, reversible adsorption to biomass and desorption from biomass, and biodegradation of PCBs. Each congener in the mixture is assumed to behave independently. At the beginning of the experiment, the PCB mixture is spiked into the aqueous medium. After bacterial biomass addition each congener can be: (i) partitioned into biomass, (ii) released from the biomass, (iii) metabolized inside the cells and (iv) evaporated. After evaporation, each congener is immediately and irreversibly bound to the sorbent (Silipor C18). The uptake into and release from the cells are treated as fully reversible processes. The presented model can be described by a set of the first-order linear differential equations with constant coefficients. The resulting model was fitted to the experimental data, with the parameters characterizing biosorption and evaporation determined in separate experiments [62].

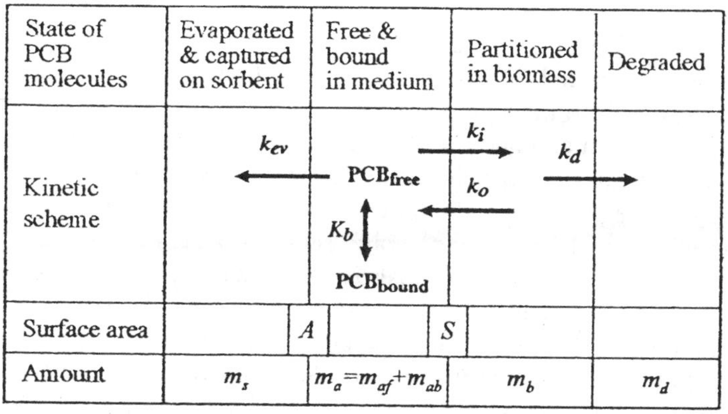

State of PCB molecules	Evaporated & captured on sorbent	Free & bound in medium	Partitioned in biomass	Degraded
Kinetic scheme	$\xleftarrow{\;k_{ev}\;}$ PCB$_{free}$	$K_b \updownarrow$ PCB$_{bound}$	$\xrightarrow{k_i}$ $\xleftarrow{k_o}$ $\xrightarrow{k_d}$	
Surface area	A	S		
Amount	m_s	$m_a = m_{af} + m_{ab}$	m_b	m_d

Figure 1. A kinetic scheme of the fate of a PCB congener in the degrading bacterial biomass suspension. The rate constants k characterize four concurrent processes: evaporation (subscript ev) through the air/water interface A, uptake into the biomass (i), release from the biomass (o), both through the membrane/water interface S, and intracellular degradation (d). Binding to the extracellular colloidal particles is much faster than the above processes and is described by the association constant K_b. Distribution is expressed in terms of amounts m. The evaporated molecules bind immediately and irreversibly to the sorbent (subscript s). In the extracellular aqueous phase (a), the congener molecules are present as free (af) and bound (ab). After partitioning in the biomass (m), the congener molecules can be degraded (d).

In this way, the biodegradation rate constants were determined for the individual congeners of the PCB commercial mixture DELOR 103. For a simple empirical determination of the influence of the chlorine substitution pattern on biodegradability, rate constants of biodegradation of di- and trichlorobiphenyl derivatives were analyzed using multiple linear regression analysis. The main results obtained can be summarized as follows: *1)* The PCB distribution kinetics in an active bacterial suspension of *Pseudomonas stutzeri* was described by a detailed model that took biosorption, evaporation and primary biodegradation of PCBs into account. *2)* The rate constants of biodegradation are structure-dependent and decrease with increasing number of chlorine atoms. *3)* The effect of the chlorine substitution pattern on biodegradability was empirically assessed. The biodegradability of PCB congeners decreases with increasing number of chlorine atoms. In particular, *ortho*-substitution seems to cause a decrease of biodegradability. The increasing number of free *ortho-* and *meta*-positions in the biphenyl molecule leads to the increased biodegradability of a congener.

2.2. ENHANCEMENT OF PCB BIODEGRADATION EFFICACY

2.2.1. *Pretreatment Techniques — Chemical Oxidation*
The focus of the first approach was on oxidation using hydroxyl radical-based processes generated from hydrogen peroxide by ferrous ion catalysis (Fenton type re-

action). Fenton's reaction (FR) was applied and its effect on the rate and extent of the abiotic degradation of individual mono-, di-, tri- and tetrachlorobiphenyls in DELOR 103 was estimated. Partial chemical oxidation modifies molecular structures that are resistant to biodegradation, enhances water solubility of the organic compounds, facilitates microbial action, and increases the biodegradation rate.

Based on our results we can make the following conclusions about mutual relationships of PCB degradation and the Fenton type reaction: *1*. The oxidation effect of FR is amplified with increasing molar ratio $Fe^{2+} : H_2O_2$. Raising the concentration of peroxide solution also enhances the oxidation. The most efficient oxidation of PCB solution ($10\ \mu g/mL$) was achieved by using FR containing 1 mol/L H_2O_2 and 1 mmol/L Fe^{2+}; further increase of the concentration of peroxide did not enhance oxidation effect significantly. *2*. The FR elimination rate constants of PCB congeners decrease with increasing number of chlorine substituents in the biphenyl molecule and show good correlation with the values of molar mass of the PCB congeners and their 1-octanol/water partition coefficients. *3*. Comparison of the rate constants of elimination for abiotic and biotic degradations implies that abiotic chemical degradation by FR proceeds much faster and more effectively than the biotic one, in the bacterial cell suspension. *4*. Half-times of abiotic elimination of individual PCB congeners by FR are prolonged with increasing number of chlorine substituents. *5*. Chemical elimination by FR seems to be a structure-independent process, whereas the rate of biodegradation depends on the chlorine substitution pattern of PCBs [63].

Chemical pretreatment (pre-oxidation) using FR can be evaluated as a realistic and advantageous in comparison with the remediative technologies, since it does not burden the environment and because only catalytic amounts of Fe^{2+} are used in the reaction together with hydrogen peroxide which is irreversibly degraded.

The pretreatment techniques can increase the number of biodegradable compounds in the biodegraded PCB mixture. However, some nonspecific products can also be formed during these processes. The products can be toxic to the microorganisms, as well as to higher animals. This should be taken into account when applying any pretreatment technique in a remediation technology.

2.2.2. Effect of Terpenes

Use of PCB-degrading bacteria for *in situ* bioremediation is hindered by the impossibility to use biphenyl in soil amendments, due to its adverse health effects, high cost, and low water solubility. Biphenyl serves as a primary substrate and as an inducer of the first reaction of metabolic pathway catalyzed by BDO of inoculated or indigenous microorganisms with PCB degradation ability.

Recently, it was speculated that certain plant compounds, including flavonoids, lignin and terpenes, may serve as natural substrates for induction of *bph* genes. Several plant materials that contain terpenes (eucalyptus and ivy leaves, orange peels, pine needles) caused disappearance of PCBs in the soil samples under laboratory conditions [64–66].

In the second approach of our work, the effect of two terpenes (potential inducers of biphenyl 2,3-dioxygenase), carvone and limonene, in the presence of glucose, biphenyl, glycerol, and xylose as sole carbon and energy sources on biodegradation of DELOR 103 has been evaluated. Biodegradation experiments were carried out in the presence of the PCB-degrading bacterium *Pseudomonas stutzeri* that could not utilize carvone and limonene as the sole carbon source. The main results of our work can be summarized as: *1*. Addition of terpenes, carvone and limonene exerted an enhancing effect on PCB biodegradation when glycerol and xylose were used as carbon sources whereas no such effect could be demonstrated with biphenyl and glucose as substrates. *2*. Promising biodegradation values were determined with xylose as carbon source and carvone as terpene inducer. In this system, 30−70 % of congeners were degraded in the presence of 10 and 20 mg/L carvone, irrespective of the used concentration, whereas only 7−37 % of individual PCB congeners were eliminated from the system without terpene addition. Xylose belongs to hardly utilizable substrates under aerobic conditions. In contrast to biphenyl known for its toxic effect, xylose is a nontoxic compound.

The fact, that *Pseudomonas stutzeri* was able to utilize xylose (available for selective pressure and nontoxic properties) as a sole carbon source while simultaneously maintaining its degradation potential of PCBs, make it attractive and perspective for application as a microorganism suitable for bioaugmentation of biotopes contaminated with PCBs. The aqueous solubility, low cost and effectivenes of carvone make it a potentially practical inducer for cleaning up DELOR-contaminated soil in biphenyl-free bioremediation technology.

3. Conclusions

The aerobic biodegradation is one of the possibilities to remove PCBs from the environment. Effectivity of this process can be increased in two ways: by application of the molecular biology techniques or by an adjustment of the environmental conditions.

The main goal of the remediation technologies is the ultimate PCB removal from the environment. The only way how to achieve this is probably a combination of physical, chemical and biological approaches.

Acknowledgements

Financial support from the *Slovak Grant Agency* (grants no. 1/4201/97 and 1/6252/99), *NATO International Scientific Exchange Programs* (Linkage Grant Envir.LG.940637) and the Environmental Technologies RTD Programme (DG XII/D-1) of the *Commission of the European Communities* under contract number EV5V-CT92-0211 is gratefully acknowledged.

References

1. Sweet, G.H. (1992): Bioremediation: Myth *vs.* realities. *Environ. Protect., May,* 1–4.
2. Dercová, K., Baláž, Š., Haluška, Ľ., Horňák, V., and Holecová, V. (1995) Degradation of PCB by bacteria isolated from long-time contaminated soil, *Int. J. Environ. Anal. Chem.* 58, 337–348.
3. Cvengroš, J., Filištein, V. (1999) Separation in a PCB-contaminated mineral oil system, *Environ. Eng. Sci.* 16, 15–20.
4. Bennett, G.F., Olmstead, K.P. (1992) Microorganisms get to work, *Chem. Brit.* 2, 133–137.
5. Kohler, H.P.E., Kohler-Staub, D., and Focht, D.D. (1988) Cometabolism of polychlorinated biphenyls: enhanced transformation of Aroclor 1254 by growing bacterial cells, *Appl. Environ. Microbiol.* 54, 1940–1945.
6. Furukawa, K., Matsumura, F., and Tonomura, K. (1978) *Alcaligenes* and *Acinetobacter* strains capable of degrading polychlorinated biphenyls, *Agric. Biol. Chem.* 42, 543–548.
7. Bedard, D.L., Wagner, R.E., Brennan, M.J., Haberl, M.L., and Brown, J.F. Jr. (1987) Extensive degradation of Aroclors and environmental transformed polychlorinated biphenyls by *Alcaligenes eutrophus* H850, *Appl. Environ. Microbiol.* 53, 1094–1102.
8. Sylvestre, M. and Fateux, J. (1982) New facultative anaerobe capable of growth on chlorobiphenyls, *J. Gen. Appl. Microbiol.* 28, 61–72.
9. Massé, R., Messier, F., Péloquin, L., Layotte, C., and Sylvestre, M. (1984) Microbial degradation of 4-chlorobiphenyl, a model compound of chlorinated biphenyls, *Appl. Environ. Microbiol.* 47, 947–951.
10. Bevinakatti, B.G. and Ninnekar, H.Z. (1992) Degradation of biphenyl by a *Micrococcus* species, *Appl. Microbiol. Biotechnol.* 38, 273–275.
11. Bedard, D.L., Unterman, R., Bopp, L.H., Brennan, M.J., Haberl, M.L., and Johnson, C. (1986) Rapid assay for screening and characterizing microorganisms for the ability to degrade polychlorinated biphenyls, *Appl. Environ. Microbiol.* 51, 761–765.
12. Ahmad, D., Sylvestre, M., Sondossi, M., and Massé, R. (1991*a*) Bioconversion of 2-hydroxy-6-oxo-6-(4′-chlorophenyl)hexa-2,4-dienoic acid, the *meta*-cleavage product of 4-chlorobiphenyl, *J. Gen. Microbiol.* 137, 1375–1385.
13. Ahmad, D., Sylvestre, M., and Sondossi, M. (1991*b*) Subcloning of *bph* from *Pseudomonas testosteroni* B-356 in *Pseudomonas putida* and *Escherichia coli*: evidence for dehalogenation during initial attack on chlorobiphenyls, *Appl. Environ. Microbiol.* 57, 2880–2887.
14. Hayase, N., Taira, K., and Furukawa, K. (1990) *Pseudomonas putida* KF715 *bphabcd* operon encoding biphenyl and polychlorinated biphenyl degradation-cloning, analysis, and expression in soil bacteria, *J. Bacteriol.* 172, 1160–1164.
15. Taira, K., Hirose, J., Hayashida, S., and Furukawa, K. (1992) Analysis of *bph* operon from the polychlorinated biphenyl-degrading strain of *Pseudomonas pseudoalcaligenes* KF707, *J. Biol. Chem.* 267, 4844–4853.
16. Hofer, B., Eltis, L.D., Dowling, D.N., and Timmis, K.N. (1993) Engineering of alkyl-responsive and haloaromatic-responsive gene-expression with mini-transposons containing regulated promoters of biodegradative pathways of *Pseudomonas*, *Gene* 130, 47–55.
17. Kikuchi, Y., Nagata, Y., Hinata, M., Kimbara, K., Fukuda, M., Yano, K., and Takagi, M. (1994) Identification of *bphA4* gene encoding ferredoxin reductase involved in biphenyl and polychlorinated biphenyl degradation in *Pseudomonas* sp. strain KKS102, *J. Bacteriol.* 176, 1689–1694.
18. Sylvestre, M., Hurtubise, Y., Barriault, J., Bergeron, J., and Ahmad, D. (1996) Characterization of active recombinant 2,3-dihydro-2,3-dihydroxybiphenyl dehydrogenase from *Comamonas testosteroni* B-356 and sequence of the encoding gene (*bphB*), *Appl. Environ. Microbiol.* 62, 2710–2715.

19. Sondossi, M., Sylvestre, M., Ahmad, D., and Massé, R. (1991) Metabolism of hydroxybiphenyl by biphenyl/chlorobiphenyl degrading *Pseudomonas testosteroni*, strain B-356, *J. Ind. Microbiol.* 7, 77–88.

20. Asturias, J A. and Timmis, K.N. (1993) Three different 2,3-dihydroxybiphenyl-1,2-dioxygenase genes in the gram-positive polychlorobiphenyl-degrading bacterium *Rhodococcus globerulus* P6, *J. Bacteriol.* 175, 4631–4640.

21. Bedard, D.L., Haberl, M.L. (1991) Influence of chlorine substitution pattern on the degradation of polychlorinated biphenyls by eight bacterial strains, *Microb. Ecol.* 20, 87–102.

22. Gibson, D.T., Cruden, D.L., Haddock, J.D., Zylstra, G.J., and Brand, J.M. (1994) Oxidation of polychlorinated biphenyls by *Pseudomonas* sp. strain LB400 and *Pseudomonas pseudoalcaligenes* KF707, *J. Bacteriol.* 175, 4561–4564.

23. Bergeron, J., Ahmad, D., Barriault, D., Larose, A., Sylvestre, M., and Powlowski, J. (1994) Identification and mapping of the gene translation products involved in the first steps of the *Comamonas testosteroni* B-356 biphenyl and chlorobiphenyl biodegradation pathway, *Can. J. Microbiol.* 35, 329–336.

24. Hirose, J., Suyama, A., Hayshida, S., and Furukawa, K. (1994) Construction of hybrid biphenyl (*bph*) and toluene (*tod*) genes for functional analysis of aromatic ring dioxygenases, *Gene* 183, 27–33.

25. Seeger, M., Timmis, K.N., and Hofer, B. (1995) Conversion of chlorobiphenyls into phenylhexadienoates and benzoates by the enzymes of the upper pathway for polychlorobiphenyl degradation encoded by the *bph* locus of *Pseudomonas* sp. strain LB400, *Appl. Environ. Microbiol.* 61, 2654–2658.

26. Barriault, D. and Sylvestre, M. (1993) Factors affecting PCB degradation by an implanted bacterial strain in soil microcosms, *Can. J. Microbiol.* 39, 594–602.

27. Furukawa, K., Simon, J.R., and Chakrabarty, A.M. (1983) Common induction and regulation of biphenyl, xylene toluene, and salicylate catabolism in *Pseudomonas paucimobilis, J. Bacteriol.* 154, 1356–1362.

28. Sondossi, M., Sylvestre, M.M, and Ahmad, D. (1992) Effects of chlorobenzoate transformation on the *Pseudomonas testosteroni* biphenyl and chlorobiphenyl degradation pathway, *Appl. Environ. Microbiol.* 58, 485–495.

29. Arensdorf, J.J. and Focht, D.D. (1994) Formation of chlorocatechol *meta*-cleavage products by a *Pseudomonad* during metabolism of monochlorobiphenyls, *Appl. Environ. Microbiol.* 60, 2884–2889.

30. Guilbeault, B., Sondossi, M., Ahmad, D., and Sylvestre, M. (1994) Factors affecting the enhancement of PCB degradative ability of soil microbial populations, *Int. Biodeter. Biodegr.* 33, 73–91.

31. Neilson, A.H. (1996) An environmental perspective on the biodegradation of organochlorine xenobiotics, *Int. Biodeter. Biodegr.* 3–21.

32. Alexander, M. (1994) *Biodegradation and Biodeterioration*, Academic Press Ltd., San Diego.

33. Vrana, B., Dercová, K., Baláž, Š., and Ševčíková, A. (1996a) Effect of chlorobenzoates on the degradation of polychlorinated biphenyls (PCB) by *Pseudomonas stutzeri*, *World J. Microb. Biotech.* 12, 323–326.

34. Furukawa, K. and Chakrabarty, A.M. (1982) Involvement of plasmids in total degradation of chlorinated biphenyls, *Appl. Environ. Microbiol.* 44, 619–626.

35. Adriaens, P., Kohler, H.P.E., Kohler-Staub, D., and Focht, D.D. (1989) Bacterial dehalogenation of chlorobenzoates and coculture biodegradation of 4,4'-dichlorobiphenyl, *Appl. Environ. Microbiol.* 55, 887–892.

36. Fava, F. and Marchetti, L. (1991) Degradation and mineralization of 3-chlorobiphenyl by mixed aerobic bacterial culture, *Appl. Microbiol. Biotechnol.* 36, 240–245.

37. Havel, J., Reineke, W. (1991) Total degradation of various chlorobiphenyls by cocultures and *in vivo* constructed hybrid pseudomonads, *FEMS Microbiol. Lett.* 78, 163–170.

112

38. Fava, F., Di Gioia, D., Cinti, S., Marchetti, L., and Quattroni, G. (1994) Degradation and dechlorination of low chlorinated biphenyls by a three-membered bacterial co-culture, *Appl. Microbiol. Biotechnol.* **41**, 117–123.

39. Hiramoto, M., Ohtake, H., and Toda, K. (1989) A kinetic study on total degradation of 4-chlorobiphenyl by a 2-step culture of *Arthrobacter* and *Pseudomonas* strains, *J. Ferment. Bioeng.* **68**, 68–70.

40. Berg, G., Seech, A.G., Lee, H., and Trevors, J.T. (1990) Identification and characterization of a soil bacterium with extracellular emulsifying activity, *J. Environ. Sci. Health* **25**, 753–764.

41. McElwee, C.G., Lee, H., and Trevors, J.T. (1990) Production of extracellular emulsifying agent by *Pseudomonas aeruginosa* UG1, *J. Ind. Microbiol.* **5**, 25–31.

42. Cooper, D.G., Liss, S.N., Longgay, R., and Zajic, J.E. (1981) Surface activity of *Mycobacterium* and *Pseudomonas*, *J. Ferment. Technol.* **59**, 97–101.

43. Goldman, S., Shabrtai, Y., Rubinosits, C., and Rosenberg, E. (1987) Emulsan in *Acinetobacter calcoaceticus* RAG1 – distribution of cell-free and cell-associated cross-reacting material, *Appl. Environ. Microbiol.* **44**, 165–170.

44. Persson, A., and Molin, G. (1987) Capacity for biosurfactant production of environmental *Pseudomonas* and *Vibrionaceae* growing on carbohydrates, *Appl. Microbiol. Biotechnol.* **26**, 439–442.

45. Viney, I. and Bewley, R.J.F. (1990) Preliminary studies on the development of a microbiological treatment for polychlorinated biphenyls, *Arch. Environ. Contam. Toxicol.* **19**, 789–796.

46. Wright, M.A., Knowles, Ch.J., Stratford, J., Jackman, S.A., and Robinson, G.K. (1996) The dechlorination and degradation of Aroclor 1242, *Int. Biodeter. Biodeg.* **10**, 61–67.

47. Aronstein, B.N., Paterek, J.R., Kelley, R.L., and Rice, L.E. (1995) The effect of chemical pretreatment on the aerobic microbial degradation of PCB congeners in aqueous system, *J. Ind. Microbiol.* **15**, 55–59.

48. Baxter, R.M., Sutherland, D.A. (1984) Biochemical and photochemical processes in the degradation of chlorinated biphenyls, *Environ. Sci. Technol.* **18**, 608–610.

49. Fava, F. Baldoni, F., Marchetti, L., and Quattroni, G. (1996*a*) A bioreactor system for mineralization of low-chlorinated biphenyls, *Process Biochem.* **31**, 659–667.

50. Fava, F., Di Gioia, D., Marchetti, L., and Quattroni, G. (1996*b*) Aerobic dechlorination of low-chlorinated biphenyls by bacterial biofilms in packed bed batch bioreactors, *Appl. Microbiol. Biotechnol.* **45**, 562–568.

51. Vrana, B., Tandlich, R., Baláž, Š., and Dercová, K. (1998) The aerobic biodegradation of polychlorinated biphenyls by bacteria, *Biológia* **53**, 251–256.

52. Blasco, R., Mallavarapu, M., Wittich, R.M., Timmis, K.N., and Pieper, D.H. (1997) Evidence that formation of protoanemonin from metabolites of 4-chlorobiphenyl degradation negatively affects the survival of 4-chlorobiphenyl-cometabolizing microorganisms, *Appl. Environ. Microbiol.* **63**, 427–434.

53. Hickey, W.J., Searles, D.B., and Focht, D.D. (1993) Enhanced mineralization of polychlorinated biphenyls in soil inoculated with chlorobenzoate-degrading bacteria, *Appl. Environ. Microbiol.* **59**, 1194–1200.

54. Paya-Perez, A.B., Riaz, M., and Larsen, B.R. (1991) Soil sorption of 20 PCB congeners and six chlorobenzenes, *Ecotox. Environ. Safe* **21**, 1–17.

55. Haluška, Ľ., Barančíková, G., Baláž, Š., Dercová, K., Vrana, B., Furčiová, E., Paz-Weisshaar, M., and Bielek, P. (1995) Degradation of PCB in different soils by inoculated *Alcaligenes xylosoxidans*, *Sci. Total Environ.* **175**, 275–285.

56. Goldstein, R.M., Mallory, L.M., and Alexander, M. (1985) Reasons for possible failure of inoculation to enhance biodegradation, *Appl. Environ. Microbiol.* **50**, 977–983.

57. Harkness, M.R., McDermott, J.B., Abramowicz, D.A., Salvo, J.J., Flanagan, W.P., Stephens, M.L., Mondello, F.J, May, R.J., Lobos, J.H., Caroll, K.M., Brennan, M.J., Bracco, A.A., Fish, K.M., Warner, G.L., Wilson, P.R., Dietrich, D.K., Lin, D.T., Morgan, C.B., and Gately, W.L. (1993)

In situ stimulation of aerobic PCB biodegradation in Hudson River sediments, *Science* **259**, 503–507.

58. Vrana, B., Dercová, K., and Baláž, Š. (1995) Monitoring evaporation polychlorinated biphenyls (PCB) in long-term degradation experiments, *Biotechnol. Tech.* **9**, 333–338.

59. Vrana, B., Dercová, K., and Baláž, Š. (1996*b*) A kinetic distribution model of evaporation, biosorption and biodegradation of polychlorinated biphenyls (PCBs) in the suspension of *Pseudomonas stutzeri*, *Biotechnol. Tech.* **10**, 37–40.

60. Dercová, K., Vrana, B., Baláž, Š., and Šándorová, A. (1996) Biodegradation and evaporation of polychlorinated biphenyls (PCBs) in liquid media, *J. Ind. Microbiol.* **16**, 325–329.

61. Dercová, K., Vrana, B., and Baláž, Š. (1999*a*) A kinetic distribution model of evaporation, biosorption and biodegradation of polychlorinated biphenyls (PCBs) in the suspension of *Pseudomonas stutzeri*, *Chemosphere* **38**, 1391–1400.

62. Vrana, B., Baláž, Š., Tandlich, R., and Dercová, K. *Environ. Toxicol. Chem.* (in press).

63. Dercová, K., Vrana, B., Tandlich, R., and Šubová, Ľ. (1999*b*) Fenton's type reaction and chemical pretreatment of PCBs, *Chemosphere* **39**, 2621–2628.

64. Hernandez, B.S., Koh, S.C., Chial, M., and Focht, D.D. (1997) Terpene-utilizing isolates and their relevance to enhanced biotransformation of polychlorinated biphenyls in soil, *Biodegradation* **8**, 153–158.

65. Gilbert, E.S., Crowley, D.E. (1997) Plant compounds that induce polychlorinated biphenyl biodegradation by *Arthrobacter* sp. strain B1B, *Appl. Environ. Microbiol.* **63**, 1933–1938.

66. Gilbert, E.S. and Crowley, D.E. (1998) Repeated application of carvone-induced bacteria to enhance biodegradation of polychlorinated biphenyls in soil, *Appl. Microbiol. Biotechnol.* **50**, 489–494.

ADSORPTION OF HEAVY METALS TO MICROBIAL BIOMASS

Use of Biosorption for Removal of Metals from Solutions

P. BALDRIAN and J. GABRIEL
Laboratory of Biochemistry of Wood-Rotting Fungi
Institute of Microbiology, Academy of Sciences of the Czech Republic
Vídeňská 1083, 142 20 Prague, Czechia

The finding that several microorganisms are able to bind and accumulate heavy metals suggested the possibility to use them for removal of metals from solutions by the process called biosorption. Although biosorption of heavy metals was studied also with plants (mosses, leaves of trees) and marine macroalgae, microorganisms seem to be the most promising for practical use. The use of different microorganisms for biosorption was reviewed several times [1−7]. Here we want to concentrate on the mode of metal binding, the possibilities to increase the binding capacity of microbial biomass, and the usefulness of microbes for biosorption in large-scale industrial processes.

1. Binding Sites for Metals in Biomolecules

The ability of microorganisms to bind metal ions is due to the presence of cationic and anionic functional groups present in the cell, above all on the cell surface. Heavy metals can be classified according to their "hardness" — the ability to bind F^- and I^-. The metallic ions forming strong bonds with F^- are called hard. In biological systems, hard ions form stable bonds with OH^-, HPO_4^{2-}, CO_3^{2+}, $R-COO^-$, and $=C=O$. All of these groups include oxygen atoms. The soft ions form very strong bonds with CN^-, $R-S^-$, $-SH^-$, $-NH_2^-$, and imidazol, *i.e.*, groups containing N and S atoms. Several studies have been undertaken with the aim to elucidate the involvement of different functional groups in metal binding. Their results — based mainly on selective blocking of individual types of functional groups — revealed that mostly carboxyl groups, phosphate and amines were involved in metal binding. Cell envelopes of *Pseudomonas fluorescens* bound Ni, Cu and Zn (but not Cd) preferentially to carboxyl groups [8]. Alkali-extracted mycelium of *Aspergillus niger* lost 90 % of its metal-binding capacity after chemical modification of carboxyl groups [9]. *Chlorella vulgaris* cells with modified carboxyl groups also showed a major decrease in biosorption of Zn and Cd, some decrease was found also in cells with modified amine groups [10]. Also uranium was adsorbed mainly to amino and carboxyl groups

V. Šašek et al. (eds.),
The Utilization of Bioremediation to Reduce Soil Contamination: Problems and Solutions, 115–125.
© 2003 *Kluwer Academic Publishers. Printed in the Netherlands.*

of chitosan and glutamate glucan [11]. Chemical modification of *Aspergillus niger* functional group revealed the major importance of carboxyl and amino groups in metal biosorption and insignificance of phosphate and lipidic components of mycelia [12]; Zn and Pb were bound by carboxyl and phosphoryl groups in *Penicillium chrysogenum* [13]. Although the main role of cell-wall polysaccharides in metal binding is generally accepted, the contribution of other cell-wall components (protein, polyphenols *etc.*) must also be taken into account. In *Saccharomyces cerevisiae*, degradation of cell-wall protein led to 30 % reduction of metal binding [14]. Cell-wall polymers, isolated from microbial material were usually more efficient in metal sorption than whole cells: glucan, mannan and chitin isolated from *Saccharomyces cerevisiae* bound more Cd, Co and Cu than the cell walls as a whole. Cu was bound to mannan with a strong preference [14]. Dark-colored fungi bind metals efficiently to melanin molecules, due to the presence of multiple nonequivalent binding sites [15, 16]. The cell wall is the most important metal-binding component of the cell (isolated cell walls can bind up to 70 % of the total metal bound by the cell), however, the metals are also bound to intracellular compounds (mostly protein) and the role of extracellular components also should not be underestimated [17].

2. Microorganisms Used for Biosorption

All major groups of microorganisms have been tested for their biosorption applicability. Dissimilar metal-binding performances of different taxonomic groups are mostly due to the composition of the cell wall, ability of metal accumulation by living cells, production of extracellular matrix and inducibility of metal-binding systems (melanins, phytochelatins *etc.*). From the biotechnological viewpoint, the costs of propagation and modification of biosorbents must also be taken into account. In this regard, waste biomass, produced as a by-product of fermentation biotechnologies, is the cheapest source of biosorbent production (Table 1).

TABLE 1. Biosorbents from biotechnologically important microorganisms produced as wastes in fermentation processes

Bacteria	*Streptomyces erythraeus, Streptomyces noursei, Streptomyces pimprina, Streptoverticillium cinnamoneum*
Cyanobacteria	*Spirulina platensis*
Fungi	*Aspergillus niger, Aspergillus terreus, Claviceps paspali, Mucor miehei, Penicillium chrysogenum, Rhizopus arrhizus, Saccharomyces cerevisiae*
Complex biosorbents	brewery waste (mostly yeasts), distillery waste (yeasts and bacteria)

Bacterial biosorbents differ with respect to the structure of the cell wall. The cell wall of G⁻ bacteria contains in addition to peptidoglycan also lipids and protein,

G$^+$-bacterial cell wall consists almost solely of peptidoglycan, but it binds significant amount of bivalent cations, and the sorption can also involve ion exchange. Bacteria can be easily propagated and several species are used in biotechnology. They also offer the best possibilities for genetic engineering.

Fungal cell wall is composed mostly of polysaccharides: chitin, chitosan, cellulose and various glucans are present in filamentous fungi, whereas yeast cell walls contain glucans and mannans, and a significant amount of protein. Some fungi can produce melanins or other pigments, the formation of which is sometimes metal-regulated [18, 19]. Wood-rotting fungi were found to accumulate high levels of Ca^{2+}, which is replaced by heavy metal ions during biosorption [20]. Since many species of filamentous fungi and yeasts are commonly used in biotechnological processes, they represent a large source of cheap biosorbent.

Biosorption properties of some species of cyanobacteria and microalgae have also been tested for use as biosorbent. The cell wall of cyanobacteria and algae consists mostly of polysaccharides (cellulose), in some cases extracellular matrix is formed. The disadvantage of their use is a relatively slow growth. However, some cyanobacteria can be obtained as a fermentation industry by-product. Sorption of uranium to the lichen species *Peltigera membranacea* has been tested by Haas *et al.* [21]. Although lichen biosorbents could combine the advantages of fungal and algal sorbents, the practical impossibility of cultivation renders them useless for large-scale processes.

In addition to individual species, mixed cultures of microorganisms (including complex wastes from industrial fermentation processes) can be used for metal binding. Although the contribution of individual components of mixed cultures is different, the advantage of cheap production is important for biosorption application [22].

3. Improvement of Biosorption Properties of Biomass

The biosorption capacity of biological material can be improved in several ways:

- — selection of active strains and screening among strains of individual species
- — modification of cultivation conditions
- — post-cultivation modifications of the biosorbent

3.1. ORIGIN OF THE BIOSORBENT

Strains with high metal-binding capacity can sometimes be obtained from metal-contaminated sites or after adaptation of the strains to higher metal contents. The resistance of these strains can be due to overproduction of metal-binding components, both intracellularly and as a part of the cell wall. Several metal-resistant microorganisms tested showed excellent biosorption properties. A culture of microorganisms, isolated from an acid mine drainage impacted site produced biosorbent material with about three-fold metal-binding capacity compared with other biosorbents examined in the study of Xie *et al.* [23]. A strain of *Neurospora crassa* resistant to Co was more effective than the wild type in removing Co^{2+} [24]. The same observations were made with Cd-resistant *Pseudomonas aeruginosa* [25]. Biosorbent prepared by alkali treatment of Cd-tolerant *Curvularia lunata* was able to accumulate 6 % (W/W) Cd [26]. Current results show that fungal strains of the same species can differ significantly in metal tolerance (Baldrian and Gabriel, *unpublished data*). In addition to isolation of microbial strains from metal-polluted sites, screening of culture collections for metal-resistant strains can help to find new prospective sorbents.

3.2. CULTIVATION CONDITIONS

Several studies confirmed that the composition of cultivation medium affects the composition of cell components and thus influences the metal-binding capacity of the biomass. The cells of *Saccharomyces cerevisiae* grown in the presence of cysteine accumulated more Cu, Zn, Ag and Cd. Addition of phosphate to cultivation media led to increased Zn uptake [27, 28]. Also pretreatment of *Saccharomyces cerevisiae* with glucose increased biosorption efficiency [29]. The chemical form of nutrients may also influence metal sorption: *Neurospora crassa* cultivated on nitrate-N medium accumulated more Co than after cultivation on ammonia-N medium. The difference between cell-wall-binding capacities was 3−5 fold in this case [24].

The presence of metals in cultivation medium can also improve the biosorption properties of microorganisms. Isolated cell walls of *Fusarium oxysporum* cultivated in the presence of copper adsorbed more Co, Cu, Fe, Ni, and Zn compared to non-treated biomass [30]. The metal treatment led to an increase of protein, sugar and chitin content in cell walls. In another case, pretreatment of *Thiobacillus ferrooxidans* with Cd led to increased Cd uptake by its biomass [31].

Biosorption properties of microorganisms often vary with age (or growth phase) of the cultures, although literature offers rather opposing data. Older *Saccharomyces cerevisiae* cells had only a half of Ag-binding capacity of young cells but the binding capacity of isolated walls did not differ significantly [28]. Older biosorbent prepared from *Fusarium flocciferum* was less effective in Cu, Cd and Ni removal [32]. Resting cells of *Pseudomonas aeruginosa* exhibited higher Pb-binding capacity, Cd was best taken up by exponential cells, and no effect of age was found on Cu sorption [33]. Age of cultures did not affect Pb, Cd and Zn binding to *Citrobacter* sp. [34]. Huang and Chiu [35] found that the optimum size of pellets for Cd^{2+} removal by *Rhizopus oryzae* was 3.8 mm, probably also due to differences between the metal-binding properties of young and old mycelium.

3.3. MODIFICATION OF BIOMASS FOR BIOSORPTION

3.3.1. Chemical and Physical Treatment

Exposure of biomass to different chemicals as well as further physical treatment affected in several cases the metal-binding capacities of the biosorbent. For example, chemical modifications of *Saccharomyces cerevisiae* biomass including boiling in alkali, treatment with methionine or sodium dodecyl sulfate had a negative effect on kali, treatment with methionine or sodium dodecyl sulfate had a negative effect on Ag biosorption [36] but the same treatment enhanced Co uptake [37]. Treatment with NaOH, formaldehyde, dimethyl sulfoxide and detergent improved Pb, Cd, Cu, Ag and Ni sorption of *Aspergillus niger* biomass [38, 39]. NaOH treatment improved Cd sorption capacity of *Phanerochaete chrysosporium* pellets, whereas HCl decreased it to 25 % of the control [40]. On the other hand, acid-treated mycelia of *Rhizopus oryzae* and *Aspergillus oryzae* bound more Cd then untreated biomass [35, 41]. Washing of *Saccharomyces cerevisiae* biomass (leading to partial removal of cations present in the cell wall) significantly reduced its U-binding capacity [42]. Treatments of biosorbent with sodium chloroacetic acid, carbon disulfide, phosphorus oxychloride, and ammonium thiosulfate resulted in 33, 74, 133 and 155 % improvements in metal-binding capacity, respectively [23]. Attention has also been paid to the so-called "cationic activation" of biomass: it has been reported that calcium saturation of the biomass of *Rhizopus arrhizus*, *Mucor miehei* and *Penicillium chrysogenum* prior to Pb, Cd, Cu and Zn biosorption increased metal accumulation. Also the neutralization of pH during the biosorption process led to improvement of accumulation [43].

In addition to chemical treatment, some physical procedures were found to affect metal biosorption. Lee and Lee [44] reported different sorption capacities in dead mycelia of *Phanerochaete chrysosporium* dried at different temperatures: the sorption was a linear function of mycelial surface area. Boiling of dried granulated biomass of *Streptoverticillium cinnamoneum* increased Pb and Zn binding by 40−50 % [45]. The relatively unusual treatment with electric pulses improved uranium biosorption of a mixed culture of yeast and bacteria obtained as whiskey distillery waste [46].

3.3.2. Use of Separate Components

In addition to whole cells or cultures, biosorbent can be prepared by fractionation of the biomass. However, the use of separate components (mostly of cell-wall origin) need not always provide better results than total biomass. Binding of Cr, Co, Ni and Cu to isolated cell walls from G^- and G^+ bacteria was compared by Churchill *et al.* [47]. Isolated cell walls of *Escherichia coli* and *Pseudomonas aeruginosa* showed >8-fold and >2-fold higher metal-binding capacity than whole cells. Loaec *et al.* [48] successfully used exopolysaccharide from the hydrothermal bacterium *Alteromonas macleodii* for Pb, Cd, and Zn biosorption. Compared to whole cells, isolated mother cell walls of the green alga *Chlorella fusca* had a lower biosorption capacity [49]. Because the preparation of cell components is usually expensive and time-consuming, the use of isolated cell-wall components for biosorption is important rather from the theoretical viewpoint.

3.3.3. Genetic Engineering

First attempts have been undertaken to construct genetically engineered microorganisms with improved metal-binding abilities. Genetically engineered *Escherichia coli* overexpressing Hg^{2+} transport system and metallothionein was used for Hg^{2+}

Cu, Ni and Ag was cleaned using *Escherichia coli* expressing *Neurospora crassa* met-allothionein gene in the periplasm [51]. The use of genetically modified microorgan-isms is limited by the viability of strains, cost of their production and cultivation and current considerations about the safety of their use in practice.

3.3.4. *Immobilization*

The properties of immobilized cells in reactor operation (mechanical integrity, easy separation) can be advantageous under specific circumstances. So far, different car-riers have been tested for immobilization of bacteria, algae and fungi used as biosorbents. In addition to the widely used calcium alginate beads [52, 53], also reticulated foam [54], polysulfonate resins [23], granular activated carbon or acti-vated carbon-carrying biofilm [53, 55], polyurethane [56], hydroxymethyl methacry-late-based macroporous resins [57], magnetite [58], aquacel and ceramic beads [59] and polyacrylamide gel [60] were used for immobilization of biomass, or the biosor-bent was cross-linked using aldehyde [61]. The sorption of metal to materials used for immobilization also substantially contributes to metal binding by the whole sys-tem [62], but not in all cases did the immobilization lead to better performance. Im-mobilized *Rhizobia* in ceramic beads and in aquacel (porous cellulose polymer) were more effective in Cr sorption than free cells. Immobilization on ceramics was best [59]. Contrary to this, immobilization of biomass had only a little effect on Cr bio-sorption of *Rhizopus arrhizus* [63]. The entrapment of cells in alginate beads can sometimes significantly reduce their intrinsic biosorption capacity due to loss of available binding sites. For cases where biosorption is the primary objective, free-cell systems should be therefore preferred [64].

4. Industrial Solutions as Targets for Biosorption

Although most biosorption experiments were performed with the aim to develop a practical method for metal removal from industrial sources on a large scale, infor-mation about applications of biosorption to practical samples of polluted or metal-containing water is relatively scarce. Use of biosorption in large-scale industrial pro-cesses has been reviewed by McKay *et al.* [65] who concentrated on copper biosorp-tion and, in the case of radionuclides, by McEldowney [66] and Ashley and Roach [67]. Some recent attempts to treat industrial effluents and solutions with the use of a microbial biosorbent are summarized in Table 2.

The practical use of biosorption is often limited by metal-ion competition in so-lutions containing a mixture of metals [54], competition with protons (pH) for bind-ing sites [54, 77]. In the case of polluted sea water salinity can substantially influence the biosorption efficiency, due to competition of Na^+ and other cations with the metal of interest [78].

TABLE 2. Use of microorganisms as biosorbents in industrial processes

Biosorbent	Industrial facility / process	Metals present	Reference
Aspergillus niger	bioleaching of clays	Al	[68]
Azolla filiculoides	electroplating	Zn	[70]
Cyanobacteria	electroplating	Ni	[69]
Escherichia coli[1]	plating factory	Cd, Cu, Ni, Ag	[51]
Ganoderma lucidum	monazite-processing industry	rare-earth elements	[73]
Mucor miehei	tannery	Cr	[71]
Penicillium chrysogenum	uranium waste water clean-up	Ra	[72]
Rhizopus arrhizus	separation of uranium	U	[74]
Saccharomyces cerevisiae	electroplating	Cu, Cr, Cd, Ni, Zn	[29]
Saccharomyces cerevisiae[2]	electroplating	Zn, Cr, Cu	[14]
	tannery	Cr	[14]
Thiobacillus ferrooxidans	extraction from arsenopyrite	Au, Ag	[68]
Activated sludge	uranium waste-water clean-up	Ra	[72]
	electroplating	Zn, Cr, Cu	[75]
Anaerobic sludge	wine industry waste	Cu	[76]
	municipal waste water	Cu, Ni, Cr, Zn	[76]

[1]Genetically engineered strain overproducing *Neurospora crassa* metallothionein.   [2]Alkali-treated.

5. Advantages of Biosorption Application

Compared to classical technologies of waste-water treatment, biosorption offers the following advantages:

— *The system offers low capital investment and low operation costs*: the costs of biosorbents are either very low and contain only the costs for transportation and/or drying (when it is obtained as waste biomass) or they are similar to the costs of activated carbon (in the case that the biosorbent must be cultivated). Industrial ion-exchange resins are 3–10 times more expensive than the same amount of biosorbent (according to [6]). The costs of propagation can further be decreased by the use of cheap substrates, *e.g.*, starch wastewater [79].

— *The system is effective over a broad temperature and pH range and can be regenerated.*

— *Metals can be selectively removed*: Biosorption performances of closely related species can significantly differ — among four wood-rotting fungi tested, the uptake of Cd, Pb, Al and Ca differed by two orders of magnitude [80, 81]. The fungi also exhibit a high level of selectivity regarding metal accumulation [80]. Biosorbents, selectively accumulating one metal from a mixture can be found [82].

Above all the advantages of cheap production and metal selectivity are the most promising properties of microbial biomass for the development of novel industrial applications based on biosorption. Although further improvement can be achieved by modification of the biomass with the use of modified cultivation protocols, increased costs of such treatments must be taken into account.

References

1. Volesky, B. (1990) *Biosorption of heavy metals*, CRC Press, Boca Raton.
2. Volesky, B. (1994) Advances in biosorption of metals — selection of biomass types, *FEMS Microbiol. Rev.* **14**, 291–302.
3. Volesky, B. and Holan, Z.R. (1995) Biosorption of heavy metals, *Biotechnol. Prog.* **11**, 235–250.
4. Kratochvil, D. and Volesky, B. (1998) Advances in the biosorption of heavy metals. *Trends Biotechnol.* **16**, 291–300.
5. White, C., Wilkinson, S.C., and Gadd, G.M. (1995) The role of microorganisms in biosorption of toxic metals and radionuclides, *Internat. Biodeterior. Biodegrad.* **35**, 17–40.
6. Kapoor, A. and Viraraghavan, T. (1995) Fungal biosorption — an alternative treatment option for heavy metal bearing wastewaters: a review, *Biores. Technol.* **53**, 195–206.
7. Siegel, S.M., Galun, M., and Siegel, B.Z. (1990) Filamentous fungi as metal biosorbents: a review, *Water Air Soil Pollut.* **53**, 335–343.
8. Falla, J. and Block, J.C. (1993) Binding of Cd^{2+}, Ni^{2+}, Cu^{2+}, Zn^{2+} by isolated envelopes of *Pseudomonas fluorescens*, *FEMS Lett.* **1993**, 347–352.
9. Akthar, M.N., Sastry, K.S., and Mohan, P.M. (1996) Mechanism of metal-ion biosorption by fungal biomass, *Biometals* **9**, 21–28.
10. Cho, D.Y., Lee, S.T., Park, S.W., and Chung, A.S. (1994) Studies on the biosorption of heavy metals onto *Chlorella vulgaris*, *J. Environ. Sci. Health A* **29**, 389–409.
11. Janssoncharrier, M., Saucedo, I., Guibal, E., and Lecloirec, P. (1995) Approach of uranium sorption mechanisms on chitosan and glutamate glucan by Ir and C-13-NMR analysis, *React. Funct. Polymers* **27**, 209–221.
12. Kapoor, A. and Viraraghavan, T. (1997) Heavy metal biosorption sites in *Aspergillus niger*, *Biores. Technol.* **61**, 221–227.
13. Sarret, G., Manceau, A., Spadini, L., Roux, J.C., Hazemann, J.L., Soldo, Y., Eybertberard, L., and Menthonnex, J.J. (1998) Structural determination of Zn and Pb binding sites in *Penicillium chrysogenum* cell walls by EXAFS spectroscopy, *Environ. Sci. Technol.* **32**, 1648–1655.
14. Brady, D., Stoll, A.D., Starke, L., and Duncan, J.R. (1994) Chemical and enzymatic extraction of heavy metal binding polymers from isolated cell walls of *Saccharomyces cerevisiae*, *Biotechnol. Bioeng.* **44**, 297–302.
15. Gadd, M.H. and deRome, L. (1988) Biosorption of copper by fungal melanin. *Appl. Microbiol. Biotechnol.* **29**, 610–617.
16. Fogarty, R.V. and Tobin, J.M. Fungal melanins and their interactions with metals. *Enzyme Microb. Technol.* **19**, 311–317.
17. Pradhan, A.A. and Levine, A.D. (1992b) Role of extracellular components in microbial biosorption of copper and lead, *Water Sci. Technol.* **26**, 2153–2156.
18. Butler, M.J. and Day, A.W. (1998) Fungal melanins: a review, *Can. J. Microbiol.* **44**, 1115–1136.
19. Baldrian, P. and Gabriel, J. (1997) Effect of heavy metals on the growth of selected wood-rotting basidiomycetes, *Folia Microbiol.* **42**, 521–523.
20. Muraleedharan, T.R. and Venkobachar, L.I. (1994) Further insight into the mechanism of biosorption of heavy metals by *Ganoderma lucidum*, *Environ. Technol.* **15**, 1015–1027.
21. Haas, J.R., Bailey, E.H., and Purvis, O.W. (1997) Bioaccumulation of metals by lichens — uptake of aqueous uranium by *Peltigera membranacea* as a function of time and pH, *Amer. Mineral.* **83**, 1494–1502.

22. Pradhan, A.A. and Levine, A.D. (1992a) Experimental evaluation of microbial metal uptake by individual components of a microbial biosorption system, *Water Sci. Technol.* **26**, 2145–2148.

23. Xie, J.Z., Chang, H.L., and Kilbane, J.J. (1996) Removal and recovery of metal ions from waste water using biosorbents and chemically modified biosorbents, *Biores. Technol.* **57**, 127–136.

24. Karna, R.R., Sajani, L.S., and Mohan, P.M. (1996) Bioaccumulation and biosorption of Co^{2+} by *Neurospora crassa*, *Biotechnol. Lett.* **18**, 1205–1208.

25. Wang, C.L., Michels, P.C., Dawson, S.C., Kitisakkul, S., Baross, J.A., Keasling, J.D., and Clark, D.S. (1997) Cadmium removal by a new strain of *Pseudomonas aeruginosa* in aerobic culture, *Appl. Environ. Microbiol.* **63**, 4075–4078.

26. Vepachedu, S.K.V.R.R., Akthar, N., and Mohan, P.M. (1997) Isolation of a cadmium tolerant *Curvularia* sp. from polluted effluents, *Curr. Sci.* **73**, 453–455.

27. Engl, A. and Kunz, B. (1995) Biosorption of heavy metals by *Saccharomyces cerevisiae* — effects of nutrient conditions, *J. Chem. Technol. Biotechnol.* **63**, 257–261.

28. Simmons, P. and Singleton, I. (1996) A method to increase silver biosorption by an industrial strain of *Saccharomyces cerevisiae*, *Appl. Microbiol. Biotechnol.* **45**, 278–285.

29. Stoll, A. and Duncan, J.R. Enhanced heavy metal removal from waste water by viable, glucose pretreated *Saccharomyces cerevisiae* cells, *Biotechnol. Lett.* **18**, 1209–1212.

30. Hefnawy, M.A. and Razak, A.A. (1998) Alteration of cell wall composition of *Fusarium oxysporum* by copper stress, *Folia Microbiol.* **43**, 453–458.

31. Baillet, F., Magnin, J.P., Cheruy, A., and Ozil, P. (1997) Cadmium tolerance and uptake by a *Thiobacillus ferrooxidans* biomass, *Environ. Technol.* **18**, 631–637.

32. Delgado, A., Anselmo, A.M., and Novais, J.M. (1998) Heavy metal biosorption by dried powdered mycelium of *Fusarium flocciferum*, *Water Environ. Res.* **70**, 370–375.

33. Chang, J.S., Law, R., and Chang, C.C. (1997) Biosorption of lead, copper and cadmium by biomass of *Pseudomonas aeruginosa* Pu21, *Water Res.* **31**, 1651–1658.

34. Puranik, P.R. and Paknikar, K.M. (1999) Biosorption of lead, cadmium, and zinc by *Citrobacter* strain Mcm B-181 — characterization studies, *Biotechnol. Prog.* **15**, 228–237.

35. Huang, C.P. and Chiu, H.H. (1994) Removal of trace Cd(II) from aqueous solutions by fungal adsorbents — an evaluation of self-immobilization of *Rhizopus oryzae*, *Water Sci. Technol.* **30**, 245–253.

36. Singleton, I. and Simmons, P. (1996) Factors affecting silver biosorption by an industrial strain of *Saccharomyces cerevisiae*, *J. Chem. Technol. Biotechnol.* **65**, 21–28.

37. Ting, Y.P. and Teo, W.K. (1994) Uptake of cadmium and zinc by yeast — effects of Co-metal ion and physical-chemical treatments, *Biores. Technol.* **50**, 113–117.

38. Akthar, M.N., Sastry, K.S., and Mohan, P.M. (1995) Biosorption of silver ions by processed *Aspergillus niger* biomass, *Biotechnol. Lett.* **17**, 551–556.

39. Kapoor, A. and Viraraghavan, T. (1998) Biosorption of heavy metals on *Aspergillus niger* — effect of pretreatment, *Biores. Technol.* **63**, 109–113.

40. Gabriel, J. and Baldrian, P. (1996) Applicability of cultures of higher fungi to biosorption of cadmium, *Mineralia Slovaca* **28**, 343–344.

41. Huang, C.P. and Huang, C.P. (1996) Application of *Aspergillus oryzae* and *Rhizopus oryzae* for Cu(II) removal, *Water Res.* **30**, 1985–1990.

42. Riordan, C., Bustard, M., Putt, R., and McHale, A.P. (1997) Removal of uranium from solution using residual brewery yeast — combined biosorption and precipitation, *Biotechnol. Lett.* **19**, 385–387.

43. Fourest, E., Canal, C., and Roux, J.C. (1994) Improvement of heavy metal biosorption by mycelial dead biomasses (*Rhizopus arrhizus*, *Mucor miehei* and *Penicillium chrysogenum*): pH control and cationic activation, *FEMS Microbiol. Rev.* **14**, 325–332.

44. Lee, J.Y. and Lee, E.K. (1998) Drying temperature can change the specific surface area of *Phanerochaete chrysosporium* pellets for copper adsorption, *Biotechnol. Lett.* **20**, 531–533.

124

45. Puranik, P.R. and Paknikar, K.M. (1997) Biosorption of lead and zinc from solutions using *Streptoverticillium cinnamoneum* waste biomass, *J. Biotechnol.* **55**, 113–124.

46. Bustard, M., Rollan, A., and McHale, A.P. (1998) The effect of pulse voltage and capacitance on biosorption of uranium by biomass derived from whiskey distillery spent wash, *Bioprocess Eng.* **18**, 59–62.

47. Churchill, S.A., Walters, J.V., and Churchill, P.F. (1995) Sorption of heavy metals by prepared bacterial cell surfaces, *J. Environ. Eng. ASCE* **121**, 706–711.

48. Loaec, M., Olier, R., and Guezennec J. (1997) Uptake of lead, cadmium and zinc by a novel bacterial exopolysaccharide, *Water Res.* **31**, 1171–1179.

49. 49.Wehrheim, B. and Wettern, M. (1994) Biosorption of cadmium, copper and lead by isolated mother cell walls and whole cells of *Chlorella fusca*, *Appl. Microbiol. Biotechnol.* **41**, 725–728.

50. Chen, S.L., Kim, E.K., Shuler, M.L., and Wilson, D.B. (1998) Hg^{2+} removal by genetically engineered *Escherichia coli* in a hollow fiber bioreactor, *Biotechnol. Prog.* **14**, 667–671.

51. Brower, J.B., Ryan, R.L., Pazirandeh, M. (1997) Comparison of ion exchange resins and biosorbents for the removal of heavy metals from plating factory waste-water, *Environ. Sci. Technol.* **31**, 2910–2914.

52. DaCosta, A.C.A. and Leite, S.G.F. (1991) Metals biosorption by sodium alginate immobilized *Chlorella homosphaera* cells, *Biotechnol. Lett.* **13**, 559–562.

53. Wilkins, E. and Yang, Q.L. (1996) Comparison of the heavy metal removal efficiency of biosorbents and granular activated carbons, *J. Environ. Sci. Health A* **31**, 112–118.

54. Zhou, J.L. and Kiff, R.J. (1991) The uptake of copper from aqueous solution by immobilized fungal biomass, *J. Chem. Technol. Biotechnol.* **52**, 317–330.

55. Scott, J.A. and Karanjkar, A.M. (1992) Repeated cadmium biosorption by regenerated *Enterobacter aerogenes* biofilm attached to activated carbon, *Biotechnol. Lett.* **14**, 737–740.

56. Hu, M.Z.C. and Reeves, M. (1997) Biosorption of uranium by *Pseudomonas aeruginosa* strain Csu immobilized in a novel matrix, *Biotechnol. Prog.* **13**, 60–70.

57. Veglio, F., Beolchini, F., Gasbarro, A., Lora, S., Corain, B., and Toro, L. (1997) Polyhydroxyethylmethacrylate (Polyhema)–Trimethylolpropanetrimethacrylate (Tmptm) as a support for metal biosorption with *Arthrobacter* sp., *Hydrometallurgy* **44**, 317–320.

58. Chua, H., Wong, P.K., Yu, P.H.F., and Li, X.Z. (1998) The removal and recovery of copper(II) ions from waste water by magnetite immobilized cells of *Pseudomonas putida* 5-X, *Water Sci. Technol.* **38**, 315–322.

59. Mamaril, J.C., Paner, E.T., and Alpante, B.M. (1997) Biosorption and desorption studies of chromium(III) by free and immobilized *Rhizobium* (Bjv-R-12) cell biomass, *Biodegradation* **8**, 275–285.

60. Yong, P. and Macaskie, L.E. (1998) Bioaccumulation of lanthanum, uranium and thorium, and use of a model system to develop a method for the biologically-mediated removal of plutonium from solution, *J. Chem. Technol. Biotechnol.* **71**, 15–26.

61. Ashkenazy, R., Yannai, S., Rahman, R., Rabinovitz, E., and Gottlieb, L. (1999) Fixation of spent *Saccharomyces cerevisiae* biomass for lead sorption, *Appl. Microbiol. Biotechnol.* **52**, 608–611.

62. Sag, Y., Nourbakhsh, M., Aksu, Z., and Kutsal, T. (1995) Comparison of Ca-alginate and immobilized *Z. ramigera* as sorbents for copper(II) removal, *Process Biochem.* **30**, 175–181.

63. Prakasham, R.S., Merrie, J.S., Sheela, R., Saswathi, N., and Ramakrishna, S.V. (1999) Biosorption of chromium-VI by free and immobilized *Rhizopus arrhizus*, *Environ. Pollut.* **104**, 421–427.

64. Gourdon, R., Rus, E., Bhende, S., and Sofer, S.S. (1990) A comparative study of cadmium uptake by free and immobilized cells from activated sludge, *J. Environ. Sci. Health A* **25**, 1019–1036.

65. McKay, G., Ho, Y.S., and Ng J.C.Y. (1999) Biosorption of copper from waste waters — a review, *Separ. Purif. Meth.* **28**, 87–125.

66. McEldowney, S. (1990) Microbial biosorption of radionuclides in liquid effluent treatment, *Appl. Biochem. Biotechnol.* **26**, 159–179.

67. Ashley, N.V. and Roach, D.J.W. (1990) Review of biotechnology application to nuclear waste treatment, *J. Chem. Technol. Biotechnol.* **49**, 381–394.

68. Torma, A.E. (1988) Use of biotechnology in minig and metalurgy, *Biotechnol. Adv.* **6**, 1–8.

69. Corder, S.L. and Reeves, M. (1994) Biosorption of nickel in complex aqueous waste streams by cyanobacteria, *Appl. Biochem. Biotechnol.* **45–46**, 847–859.

70. Zhao, M., Duncan, J.R., and Vanhille, R.P. (1999) Removal and recovery of zinc from solution and electroplating effluent using *Azolla filiculoides*, *Water Res.* **33**, 1516–1522.

71. Tobin, J.M. and Roux, J.C. (1998) *Mucor* biosorbent for chromium removal from tanning effluent, *Water Res.* **32**, 1407–1416.

72. Tsezos, M., Baird, M.H.I., and Shemilt, L.W. (1986) Adsorptive treatment with microbial biomass of ^{226}Ra-containing waste-waters, *Chem. Eng. J.* **32**, B29–B38.

73. Muraleedharan, T.R., Philip, L., Iyengar, L., and Venkobachar, C. (1994) Application studies of biosorption for monazite processing industry effluents, *Biores. Technol.* **49**, 179–186.

74. Tsezos, M., McCready, R.G.L., and Bell, J.P. (1989) The continuous recovery of uranium from biologically leached solutions using immobilized biomass, *Biotechnol. Bioeng.* **34**, 10–17.

75. Atkinson, B.W., Bux, F., and Kasan, H.C. (1998) Waste activated-sludge remediation of metal-plating effluents, *Water SA* **24**, 355–359.

76. Morper, M. (1985) Anaerobic biosorption: removal of heavy metals from waste-water, *Chem. Anlagen-Verfahren* **18**, 103–106.

77. Huang, C.P. and Morehart, A.L. (1991) Proton competition in Cu(II) adsorption by fungal mycelia, *Water Res.* **25**, 1365–1375.

78. Garnham, G.W., Codd, G.A., and Gadd, G.M. (1991) Effect of salinity and pH on cobalt biosorption by the estuarine microalga *Chlorella salina*, *Biol. Metals* **4**, 151–157.

79. Yin, P.H., Yu, Q.M., Jin, B., and Ling, Z. (1999) Biosorption removal of cadmium from aqueous solution by using pretreated fungal biomass cultured from starch wastewater, *Water Res.* **33**, 1960–1963.

80. Gabriel, J., Mokrejš, M., Bílý, J., and Rychlovský, P. (1994) Accumulation of heavy metals by some wood-rotting fungi, *Folia Microbiol.* **39**, 115–118.

81. Gabriel, J., Vosáhlo, J., and Baldrian, P. (1996) Biosorption of cadmium to mycelial pellets of wood-rotting fungi, *Biotechnol. Tech.* **10**, 345–348.

82. Andres, Y., MacCordick, H.J., and Hubert, J.C. (1995) Selective biosorption of thorium ions by an immobilized mycobacterial biomass, *Appl. Microbiol. Biotechnol.* **44**, 271–276.

NONENZYMIC DEGRADATION AND DECOLORIZATION OF RECALCITRANT COMPOUNDS

F. NERUD, P. BALDRIAN, J. GABRIEL and D. OGBEIFUN
Laboratory of Biochemistry of Wood-Rotting Fungi
Institute of Microbiology, Academy of Sciences of the Czech Republic,
Vídeňská 1083, 142 20 Prague, Czechia

The ability of a nonenzymic system containing Cu^{II}/pyridine/peroxide to decolorize structurally different synthetic dyes and to degrade selected PAHs was followed. An intense and rapid (after 1 h) decolorization has been obtained with phenol red (89 %), Evans blue (95 %), eosin yellowish (84 %) and Poly B-411 (92 %). The use of radical scavengers, thiourea and superoxide dismutase, showed that hydroxyl radicals rather than superoxide anions are involved in the decolorization. The intensive degradation of PAHs has been obtained after 24 h. Benzo[a]pyrene has also been degraded by the Cu^{II}/H_2O_2 peroxide system.

1. Introduction

Industrial wastes and effluents are undesirable by-products of economic development and technical advancement. Widespread contamination of soil as well as groundwater and surface water by organopollutants like PAHs and synthetic dyes represents severe ecological problems. Most of these compounds are toxic, mutagenic or cause esthetic problems in the receiving water, are highly persistent and therefore difficult to remove from the environment. A variety of physical, chemical and biological techniques for environmental remediation have been recently summarized [1]. It is evident that every method has its disadvantages. Most of physical-chemical cleaning technologies are costly and rather inefficient. On the other hand, the complexity of microbial mechanisms for degradation of organopollutants as well as the time period before microbial degradation starts, ranging from weeks to many months, makes the technology slow to emerge as a viable method of remediation. It becomes apparent that the study of principles of degradation and the development of efficient methods of decontamination is needed to solve the hazardous waste problem. Microbial degradation and some of the oxidation methods, especially those using hydrogen peroxide, are efficient and cost-competitive in comparison with other cleaning methods.

V. Šašek et al. (eds.),
The Utilization of Bioremediation to Reduce Soil Contamination: Problems and Solutions, 127–133.
© 2003 *Kluwer Academic Publishers. Printed in the Netherlands.*

2. Microbial Degradation

Numerous bacterial and fungal species capable to degrade PAHs or to decolorize synthetic dyes have been reported [2−4]. Generally, bacterial systems require pre-conditioning to a particular pollutant. Bacteria must be pre-exposed to a pollutant to allow the enzymes that degrade the pollutant to be induced. The pollutant must also be present in a significant concentration, otherwise induction of enzyme synthesis will not occur. Therefore, there is a finite level to which pollutant can be degraded by bacteria. In the case of dyes, most decolorization proceeds through simple adsorption to cell-mass. Besides bacteria, greatest interest was focused on the group of *Basidiomycetes* that cause white rot of wood. The xenobiotics metabolism of white-rot fungi is at least in part a consequence of the mechanisms that these organisms use for lignin degradation. This capacity is assumed to result from the activities of non-specific free-radical-based mechanisms of ligninolytic enzymes, lignin peroxidases, manganese peroxidases and laccases. Each of these enzyme classes has been implicated in the degradation of pollutants [5]. However, not all fungi produce all these enzymes simultaneously [6−7]. The variability is attributed to differences in the enzyme production by various fungi. On the other hand, it is difficult to find a distinct correlation between the pollutant degradation and production of enzymes, even if the enzymes were detected [8]. Similar results have been obtained in the case of degradation of PCB and PAH by isolates of *Pleurotus ostreatus* overproducing ligninolytic enzymes, manganese peroxidase and laccase [9, 10]. It is evident that the mechanisms by which white-rot fungi degrade pollutants are still not well understood and further research is needed to select the most appropriate fungus for bioremediation program. Most of this work has been done in the laboratory, where the conditions can be easily controlled. However, in the field, environmental factors, like temperature, water content and physicochemical properties of the soil, pH, competitions with the autochthonous microflora and bioavailability of pollutants make the remediation by white-rot fungi more complicated.

3. Degradation by Nonenzymic Radical Reactions

An alternative approach to bioremediation is the use of free-radical-generated reactions. The participation of activated oxygen species, like hydroxyl radical, hydrogen peroxide and superoxide anion radical during degradation processes have been discussed in the literature many times [11−14]. Due to their great oxidizing ability hydroxyl radicals have received great attention. Among the systems known to produce these species, the Fenton reagent plays an important role. The Fenton reagent has also been used for the degradation of chlorophenols [15]. The methods of decolorization based on oxidative reaction of hydroxyl radicals generated by various methods such as O_3/UV, H_2O_2/UV, TiO_2 photocatalysis and photo-assisted Fe^{III}/H_2O_2 processes have been described [16, 17]. The use of hydrogen peroxide as an oxidizing agent for environmental applications has also been described [1]. Recently Watanabe *et al.* [18] published a system containing Cu^{II}, hydrogen peroxide and fungal metabolite pyridine, which is able to depolymerize lignin under physiological conditions. This system we used for decolorization of structurally different synthetic dyes and degradation of PAHs.

4. Decolorization of Synthetic Dyes by the Cu^{II}/Pyridine/Peroxide System

The effectiveness to decolorize the representatives of azo (Evans blue), triphenylmethane (phenol red), heterocyclic (eosin yellowish) and polymeric (Poly B 411) dyes has been followed. The use of this system resulted in very efficient decolorization of all dyes tested (Table 1). The decolorization was very fast and usually completed within a few minutes after mixing the reaction components. No significant differences were observed when hydrogen peroxide was replaced with cumene hydroperoxide.

TABLE 1.   Decolorization (rate, %) of synthetic dyes by Cu^{II}/pyridine/peroxide system using H_2O_2 and cumene hydroperoxide

Dye	H_2O_2		Cumene hydroperoxide	
	1 h	24 h	1 h	24 h
Poly B-411	89	89	91	94
Phenol red	58	58	50	78
Evans blue	84	94	95	95
Eosin yellowish	92	92	97	97

As shown in Figure 1, the dyes were also partially decolorized with Cu^{II} and hydrogen peroxide but the addition of pyridine markedly accelerated the decolorization. No decolorization was observed when only pyridine and hydrogen peroxide were used, indicating that pyridinium oxide is not involved in the reaction. The system also failed to decolorize the dyes when Mn^{II} or Fe^{II} were used instead of Cu^{II}.

Thiourea and superoxide dismutase, known scavengers of hydroxyl radical and superoxide anion radical were used. Addition of 1 mmol/L thiourea to Cu^{II}/hydrogen peroxide and Cu^{II}/pyridine/hydrogen peroxide systems led to complete inhibition of decolorization. On the other hand, the addition of superoxide dismutase (200 U/mL) did not affect the decolorization reactions. This result is in contrast with the finding of Watanabe *et al.* (1998) who demonstrated the involvement of superoxide anion in the degradation of lignin by the copper/ /pyridine/hydrogen peroxide system. Our results indicate that in the decolorization

130

reactions rather hydroxyl radicals are involved. The decolorization activity was not affected by the changes of pH in the range 3–9, but at pH 10 the decolorization was inhibited by 20–30 %. Increased temperature did not affect the rate of decolorization but the reaction was significantly faster at higher temperature.

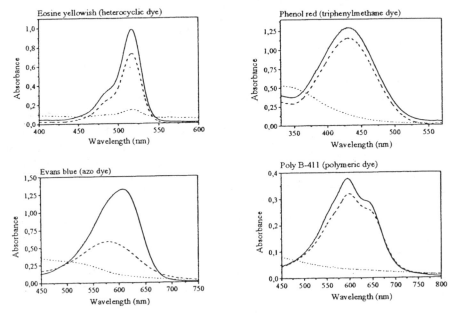

FIGURE 1.　Visible spectrum of dyes before (*solid lines*) and after 1-day decolorization with $H_2O_2/CuSO_4$ (*dashed lines*) and $H_2O_2/CuSO_4$/pyridine (*dotted lines*).

5. Degradation of PAH by Cu^{II}/pyridine/H_2O_2 system

The degradation of different PAH by the complete system after 24 h is shown in Figure 2. It is seen that the degradation is very effective.

The degradation of benzo[*a*]pyrene in acetonitrile and water solutions has been followed in a greater detail. In this experiment the complete system and system without pyridine have been used. The results of degradation after 24 h are shown in Figure 3.

In acetonitrile 70 % degradation has been obtained only by the complete system, in water solution only about 50 %. However, in water solution the most effective degradation has been obtained with Cu^{II}/H_2O_2 system. The time course of benzo[*a*]pyrene degradation by this system is shown in Figure 4.

To our knowledge this is the first study to demonstrate the ability of Cu^{II}/pyridine/H_2O_2 system to decolorize synthetic dyes and to degrade PAHs. Attempts to apply this system in soil contaminated by the above pollutants are in progress. The results show that nonenzymic degradation is very powerful and fast and provides new perspectives for the use of these or modified systems in environmental biotechnology.

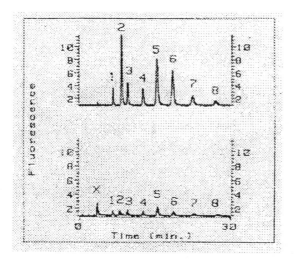

FIGURE 2. HPLC analysis of PAH degradation by Cu^{II}/pyridine/H_2O_2 system; 1 — pyrene, 2 — benzo[a]anthracene, 3 — chrysene, 4 — benzo[b]fluoranthene, 5 — benzo[k]fluoranthene, 6 — benzo[a]pyrene, 7 — dibenzo[a,h]anthracene, 8 — benzo[g,h,i]perylene, × — unidentified products.

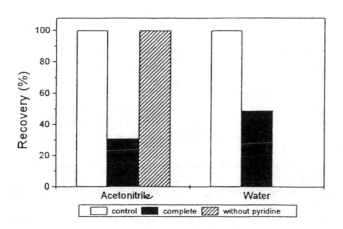

FIGURE 3. Recovery of benzo[a]pyrene after 24-h incubation with Cu^{II}/pyridine/H_2O_2 and with the omission of pyridine in acetonitrile and in water.

132

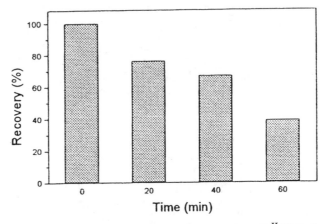

FIGURE 4. Time course of benzo[*a*]pyrene degradation by Cu^{II}/H$_2$O$_2$ in water.

References

1. Hamby, D.M. (1996) Site remediation techniques supporting environmental restoration activities — a review, *Sci. Total Environ.* **191**, 203–224.

2. Cerniglia, C.E. (1992) Biodegradation of polycyclic aromatic hydrocarbons, *Biodegradation* **3**, 351–368.

3. Paszynski, A. and Crawford, R.L. (1995) Potential for bioremediation of xenobiotic compounds by the white rot fungus *Phanerochaete chrysosporium*, *Biotechnol. Prog.* **11**, 368–379.

4. Banat, I.M., Nigam, P., Singh, D., and Marchant, R. (1996), Microbial decolorization of textile dye containing effluents: A review, *Bires. Technol.* **58**, 217–227.

5. Field, J.A., de Jong, E., Feijoo-Costa, G., and de Bont, J.A.M. (1993) Screening for ligninolytic fungi aplicable to the biodegradation of xenobiotics, *Trends Biotechnol.* **11**, 44–49.

6. Nerud, F., Zouchová, Z., and Mišurcová, Z. (1991) Ligninolytic properties of different white rot fungi, *Biotechnol. Lett.* **13**, 657–660.

7. Hatakka, A. (1994) Lignin-modifying enzymes from selected white rot fungi: production and role in lignin degradation, *FEMS Microbiol Rev.* **13**, 125–135.

8. Nerud, F. and Mišurcová, Z. (1996) Distribution of ligninolytic enzymes in selected white rot fungi, *Folia Microbiol.* **41**, 264–266.

9. Bezalel, L., Hadar, Y., and Cerniglia, C.E. (1996) Mineralization of polycyclic aromatic hydrocarbons by the white rot fungus *Pleurotus ostreatus*, *Appl. Environ. Microbiol.* **62**, 292–295.

10. Nerud, F., Homolka, L., Eichlerová, I., and Gabriel, J. (1999) Biodegradative ability of *Pleurotus ostreatus* isolates overproducing ligninolytic enzymes. The 5th International Symposium *In situ* on site Bioremediation, Abstract Book (posters) B6, April 19–22, San Diego, California.

11. Koenigs, J.W. (1974) Hydrogen peroxide and iron: A proposed system for decomposition of wood by brown rot basidiomycetes, *Wood Fiber Sci.* **6**, 66–80.

12. Guillén, F., Martínez, M.J., Munoz, C., and Martínez, A.T. (1997) Quinone redox cycling in the ligninolytic fungus *Pleurotus eryngii* leading to extracellular production of superoxide aninon radical, *Arch. Biochem. Biphys.* **339**, 190–197.

13. Kotterman, M.J., Wasseveld, R.A., and Field, J.A. (1996) Hydrogen peroxide as limiting factor in xenobiotic oxidation by nitrogen-sufficient cultures of *Bjerkandera* sp. strain BOS55 overproducing peroxidases, *Appl. Environ. Microbiol.* **62**, 880–885.

14. Wood, P.M. (1994) Pathways for production of Fenton's reagent by wood rotting fungi, *FEMS Microbiol Rev.* **13**, 313–320.

15. Barbeni, M., Minero, C., and Pellizetti (1987): Chemical degradation of chlorophenols with Fenton's reagent. *Chemosphere* **16**, 2225–2237.

16. Ollis, D.F. and Al-Akabi, H. (1993): *Photocatalytic Purification and Treatment of Water and Air*, Elsevier, Amsterdam.

17. Herrera, F., Kiwi, A., Lopez, A., and Nadtochenko, V. (1999): Photochemical decoloration of remazol brilliant blue and uniblue A in the presence of Fe^{3+} and H_2O_2. *Environ. Sci. Technol.* **33**, 3145–3151.

18. Watanabe, T., Koller, K., and Messner, K. (1998) Copper-dependent depolymerization of lignin in the presence of fungal metabolite pyridine. *J. Biotechnol.* **62**, 221–230.

The Impact of Sulfonation Pattern on Indigo Degradation by *Phanerochaete chrysosporium* Ligninolytic Enzymes

H. PODGORNIK and A. PERDIH

University of Ljubljana, Faculty of Chemistry and Chemical Technology, Aškerčeva 5, 1000 Ljubljana, Slovenia

Abstract

Indigo dyes can be successfully decolorized by both groups of *Phanerochaete chrysosporium* extracellular ligninolytic enzymes, *i.e.* by manganese (MnP) and lignin (LiP) peroxidases. The decolorization rate of indigo di-, tri-, and tetrasulfonate was measured using isolated isoenzymes LiPH2 and MnP2. Although the decolorization was successful by MnP and LiP on all three tested indigo sulfonates, the degree of indigo ring sulfonation had an important impact on the decolorization rate. In the case of MnP, the specific decolorization rate dropped by approximately one order of magnitude with each additional sulfonic acid in the dye structure. This diminution followed the increase of the dyes standard potential which accompanies a higher degree of sulfonation. Using the LiP isoenzyme the process was reversed; indigo tetrasulfonate was decolorized most rapidly, while the other two dyes were decolorized considerably more slowly. An analogy with the HSAB Principle it is proposed to explain this unusual behavior observed in the decolorization of indigo sulfonates by LiP.

1. Introduction

The white-rot fungus *Phanerochaete chrysosporium* produces under nitrogen limitation lignin peroxidase (LiP) and manganese peroxidase (MnP), which are widely examined not only due to their strong ligninolytic potential but also due to their ability to degrade different xenobiotics [1, 2]. Among the substrates metabolized by this fungus there are also several synthetic dyes.

Extracellular peroxidases degrade xenobiotics *via* a free-radical mechanism [2] with the help of very reactive species — the veratryl alcohol (VA) cation radical $(VA^{•+})$, and Mn^{III} at LiP and MnP, respectively. Mn^{II} is the best reducing substrate for the MnPII compound in the MnP catalytic cycle. The resulting Mn^{III}, stabilized by chelation with an ω-organic acid, acts as an obligatory redox couple, oxidizing various organic substrates [3]. In contrast to the indisputable role of Mn^{III} in MnP

V. Šašek et al. (eds.),
The Utilization of Bioremediation to Reduce Soil Contamination: Problems and Solutions, 135–142.
© 2003 *Kluwer Academic Publishers. Printed in the Netherlands.*

catalysis, the necessity of mediation by VA$^{\bullet+}$ is highly dependent on the initial redox potential of the substance to be degraded by LiP [4]. Regarding their redox potential, three groups of substances can be formed. The first group of substances cannot be degraded by LiP due to their high redox potential. In reactions between LiP and the second group of substrates, which have a redox potential between that of LiPI and LiPII compounds, VA indirectly acts as a mediator by completing the LiP catalytic cycle [5]. Direct mediation of VA is important in LiP reactions of the third group of substrates having their redox potentials lower than LiPI and LiPII [6].

Extracellular ligninolytic enzymes are involved in complex reactions which are not entirely understood [4]. Therefore, the general relationship between the substitution pattern and dye decolorization by this group of enzymes is far from being simple and clear [7]. Indigo dyes are rare representatives of dyes which can be successfully decolorized by both groups of peroxidases, by MnP and LiP [8, 9]. Differently sulfonated indigo dyes have also been used as redox indicators, therefore some of their properties are known. This enabled us to compare their degradation by *P. chrysosporium* LiP and MnP isoenzymes.

2. Materials and Methods

LiP and MnP Production. *P. chrysosporium* MZKI B-223 (ATCC 24725) was grown in a nitrogen-limited medium [7]. The growth medium contained 40 µmol/L Mn^{II} for LiP production and 1 mmol/L Mn^{II} for MnP production [10].

Isoenzymes Isolation. LiP isoenzyme H2 was isolated as previously reported [11]. For MnP purification, separation on DEAE-Sepharose was applied [12]. A 10-fold concentrated growth medium was dialyzed against 10 mmol/L sodium acetate (pH 6). After that, 25 mL of the medium was loaded to the DEAE-Sepharose column previously equilibrated by potassium phosphate (20 mmol/L, pH 6.5). The column was thoroughly washed with potassium phosphate (20 mmol/L, pH 6) as well as by a sodium succinate (pH 5) solution. The bound MnP isoenzymes were washed with 1 mmol/L sodium acetate (pH 6) and desalted on a PD-10 column (*Pharmacia*, Uppsala, Sweden). MnP2 was isolated by HPLC separation using a similar concentration gradient as described above for the separation of LiP isoenzymes.

The main data about the concentration and activity of both used peroxidases are collected in Table 1.

Enzyme Activities and Protein Concentration. *LiP activity* was determined by a method of Tien and Kirk [1]. *MnP activity* was determined by the oxidation of 2,6-dimethoxyphenol [13]. *Protein concentration* was determined by the method of Lowry *et al.* [14] using bovine serum albumin as standard.

Decolorization Experiments. Decolorization rate was measured at the absorption maximum wavelength of 609, 600, and 594 nm for indigo disulfonate, trisulfonate and tetrasulfonate, respectively. Experiments were done at pH 4.5 (50 mmol/L malonate) or at pH 3 (50 mmol/L sodium tartrate), to study the effect of MnP or LiP, respectively. Experiments employing LiP were performed at 2 mmol/L VA, whereas 1 mmol/L $MnSO_4$ was added to the reaction mixtures at decolorization experiments by MnP. The dye concentration was 15 or 30 mg/L. The enzyme activity was usually above 80 mU/mL of the reaction mixture. The reaction was initiated by addition of 0.3 mmol/L H_2O_2. The relative decolorization rate [7] was calculated as the maximum decolorization rate normalized to the enzyme activity in 1 mL of the reaction mixture. It is given as the $(dA/min)/U$, where U is the unit of MnP or LiP activity. The extent of decolorization was calculated from the absorbance of the solution after the decolorization was completed.

All experiments were performed in at least three replicates.

TABLE 1.   Characteristics of peroxidases used in decolorization of indigo sulfonates

Peroxidase	pI	Protein concentration mg/L	Enzyme activity U/L	Specific activity U/mg
LiPH2	4.3	96.5	1760	18.2
MnP2	4.6	142	17300	121.9

3.   Results and Discussion

The experiments performed with the growth medium with high LiP or MnP activity showed that although decolorization was successful with all three tested indigo sulfonates by MnP and LiP, the degree of indigo ring sulfonation had an important impact on the decolorization process. The cultivated growth medium is a complex solution, containing also VA, Mn^{II}, Tween 80, and chelated Mn^{III}, which can all considerably influence MnP or LiP catalytic cycles or nonenzymic degradation reactions [15]. In order to exclude the influence of the above-mentioned substances, the isolated LiPH2 and MnP2 were used in experiments (Table 1), which are the most abundant isoforms obtained under optimal conditions applied for LiP and MnP production by *P. chrysosporium*, respectively.

The first main difference between the decolorization of indigo dyes by MnP and LiP is easily observed comparing the initial spectra of the reaction solution and the final spectra after decolorization stopped (*data not shown*). Decolorization of all three tested dyes by LiP resulted in a bright yellow solution, although the process in the case of disulfonate and trisulfonate was very slow. According to our previous studies [15], 5-isatinsulfonic acid with the absorption maximum at 372 nm is mainly formed through these reactions as a consequence of indigo bridge breakdown. In contrast, during the decolorization of indigo disulfonate by MnP, besides the yellow product also the red colored dimer of the dye with the absorption maximum at 550 nm is formed [16].

The second great difference, which was particularly surprising for us, was that the decolorization of indigo disulfonate by MnP and of indigo tetrasulfonate by LiP was considerably faster than that of all other reactions. The differences were evaluated by relative reaction rates for all three indigo dyes and both groups of peroxidases (Table 2). Since the used indigo dyes have also been used as redox potential indicators, we expected that at least the general tendency of decolorization rates would be in accordance with the standard potential increase which accompanies the sulfonation degree. The standard potential increases almost linearly at both pH values available in the literature (Table 3). From these data it can be assumed that at pH 3 and pH 4.5, which are used in decolorization experiments with LiP and MnP, respectively, the tendency is the same. In the case of MnP the specific decolorization rate dropped by approximately one order of magnitude with each additional sulfonic group in the dye structure (Table 2). This diminution followed the increase of the redox potential which accompanies a higher degree of sulfonation as can be expected considering the general correlation between the redox potential of the substrates and their degradability by ligninolytic enzymes. Using the LiP isoenzyme H2 (or the growth medium) the process was reversed; indigo tetrasulfonate was decolorized most rapidly, while the other two dyes were decolorised considerably more slowly.

TABLE 2. Decolorization of indigo sulfonates by MnP2 and LiPH2[1]

Indigo	MnP2[2]		LiPH2[3]	
	RDR, (dA/min)/U	DE, %	RDR, (dA/min)/U	DE, %
Disulfonate	−1.26	99.4	−0.205	100*
Trisulfonate	−0.156	68.6	−0.604	100*
Tetrasulfonate	−0.0075	4	−9.47	100

[1]Dye concentration was 30 mg/L; RDR — relative decolorization rate, DE — decolorization extent.
[2]Activity 0.1–0.3 U/mL (50 mmol/L sodium malonate, pH 4.5).
[3]Activity 0.1–0.12 U/mL (50 mmol/L sodium tartrate, pH 3).
*Decolorization was completed only at LiP activity >0.11 U/L.

Different chemical parameters, which can lead to such unexpected decolorization patterns were tested. As shown in Table 3, the molar absorption coefficients of the three dyes are similar, excluding the possibility that the big differences in the measured reaction rates could arise from different absorptivities of the dyes. The next possible factor can be the different composition and pH of the buffer medium, tartrate (pH 3) or malonate (pH 4.5) at decolorization experiments by LiP and MnP, respectively. As can be expected from the literature data that malonate is a much better chelator of Mn^{III} than tartrate [3] the change of buffer decreased the MnP activity much more than a considerable decrease of pH value (Table 4). Decolorization experiments, performed with MnP2 in malonate buffer of different pH values in

TABLE 3. Standard potentials [17] and estimated molar absorption coefficients of indigo sulfonates

Indigo	Standard potential V (30 °C)		Estimated ε L mol^{-1} cm^{-1}
	pH 7	pH 0	
Disulfonate	−0.125	0.291	16 000
Trisulfonate	−0.081	0.332	16 600
Tetrasulfonate	−0.046	0.365	17 100

the range 3−4.5 showed diminished decolorization rates as well as MnP activity, not changing, however, the prominent drop in decolorization rates from disulfonate to tetrasulfonate at each employed pH value (*data not shown*). Also the addition of 1 mmol/L VA to the reaction mixture had no influence on decolorization by MnP.

TABLE 4. The influence of reactive mixture composition on MnP and LiP activity (%)

pH	LiP activity			MnP activity	
	tartrate[1]	malonate	malonate + 1 mmol/L MnSO$_4$	tartrate	malonate[2]
3.0	100	73.0	71.3	17.6	45.6
3.5	99.5	57.2	59.5	46.7	79.3

[1]100 % was arbitrarily set in tartrate buffer of pH 3.0.
[2]100 % was arbitrarily set in malonate buffer of pH 4.5.

On the other hand, decolorization of indigo sulfonates by LiPH2 can be completed only up to pH 3.8 probably as a consequence of considerable inhibitory effect of increased pH value on LiP activity (Table 4). On the basis of the described experiments we decided to use malonate buffer at pH 3.5 for direct comparison of indigo sulfonate decolorization by LiP and MnP (Table 5). Under these conditions, LiP activity was only slightly above one-half of the optimum (Table 4), while the decolorization rate of indigo tetrasulfonate was ≈75 % of the optimal one (Tables 2, 5). Whereas the addition of 1 mmol/L MnSO$_4$ decreased the overall decolorization rate of indigo tetrasulfonate, it did not influence the general trend. Since addition of

$MnSO_4$ to the reaction mixture had no inhibitory influence on LiP activity (Table 4) we suspect that Mn^{II} influences the indigo tetrasulfonate, affecting the overall reaction rate. Although the differences between decolorization with LiP and MnP were at pH 3.5 not as prominent as under conditions optimal for each particular enzyme, under identical composition of the reaction mixture the tendency of decolorization rates was still reversed considering the sulfonation degree of indigo dyes.

TABLE 5. The relative decolorization rate $((dA/min)/U)$ of indigo sulfonates by MnP2 and LiPH2[1]

Indigo	LiPH2		MnP2
		+ 1 mmol/L $MnSO_4$	+ 1 mmol/L $MnSO_4$
Disulfonate	−0.35	−0.37	−0.60
Trisulfonate	−2.48	−0.61	−0.09
Tetrasulfonate	−7.10	−1.75	−0.02

[1]Malonate buffer of pH 3.5; 1 mmol/L veratryl alcohol was added into the reaction mixture.

Therefore, we believe that the degradation mechanism should differ significantly between various specific substances, such as at the tested indigo dyes. During decolorization of indigo sulfonate by MnP, where Mn^{III} plays the central role, the decrease of reaction rate with higher sulfonation rate is thus in agreement with expectation. On the other hand, it is very difficult to explain the opposite tendency observed in the degradation of indigo sulfonates by LiP, where the $VA^{•+}$ radical plays a central role. In order to find an explanation for such unusual behavior, we first tried to determine the exact role of VA in the degradation of indigo sulfonates by LiP. It was observed that in the absence of VA in the reaction mixture the decolorization by LiP did not start at all. Two sets of experiments confirmed a direct mediation of VA in indigo tetrasulfonate decolorization. There was no such inhibition by higher VA concentrations in the reaction mixture (*data not shown*) as was observed in reactions where VA is necessary only for completing the LiP catalytic cycle [5]. Concentrations up to 10 mmol/L were tested, which exceeded by far the K_m of VA (135 μmol/L). Second, an increasing dye concentration in the reaction mixture resulted in an increasing lag in VA oxidation to veratraldehyde (310 nm). The length of the lag phase was proportional to the dye concentration, as was previously observed in the case of chlorpromazine [6]. These results seem to support the hypothesis that VA really acts as a mediator in the decolorization of indigo dyes by LiP. Similarly, the complexed Mn^{III} had been previously shown to act as an oxidation intermediate in the degradation of various dyes, among which was also indigo disulfonate [15].

One possible explanation for the observed phenomenon may be an analogy with Pearson's HSAB (Hard and Soft Acid and Base) principle [18]. The main statement of this principle is that hard acids prefer to react with hard bases, and soft acids with soft bases. Due to their strong inductive effect, sulfonic groups intensively decrease the electronic potential of the indigo dye molecule, withdrawing electrons from their aromatic rings. Indigo disulfonate should be the most and indigo tetrasulfonate the least prone to give away its electron. This is the case with indigo decolorization by MnPs but not by LiPs. By analogy with the HSAB theory, $VA^{\cdot+}$, which is a bigger and more polarizable molecule, would in this case be a "soft oxidant". The smaller, unpolarizable Mn^{III} with a high charge density, on the other hand, can be considered as a "hard oxidant". As could be expected, the "hard oxidant" Mn^{III} reacts most rapidly with the compound having the lowest redox potential (indigo disulfonate). The "softer" $VA^{\cdot+}$ reacts most rapidly with the substance which is most polarizable although it has the highest redox potential (indigo tetrasulfonate).

In the present study the reverse influence of the sulfonation pattern caused an unusual effect of LiP, which we believe is interesting for the global understanding of ligninolytic enzymes action. It cannot be simply explained on the basis of redox potential, which is usually used for the description of the degradation mechanism by ligninolytic enzymes. As was shown in the case of sulfonated indigo dyes, an analogy with the HSAB Principle can give a completely new insight into the complicated and not entirely understood mechanism of action of ligninolytic enzymes.

Acknowledgement

This work was supported by grant no. L4-3068 from the *Slovenian Ministry of Science and Technology*.

References

1. Tien, M. and Kirk, T.K. (1984) Lignin-degrading enzyme from *Phanerochaete chrysosporium*: purification, characterization, and catalytic properties of a unique H_2O_2-requiring oxygenase. *Proc. Nat. Acad. Sci. USA* 81, 2280–2284.
2. Reddy, C.A. (1995) The potential for white-rot fungi in the treatment of pollutants. *Curr. Opin. Biotechnol.* 6, 320–328.
3. Wariishi, H., Dunford, H.B., Mac-Donald, I.D., and Gold, M.H. (1989) Manganese peroxidase from the lignin-degrading basidiomycete *Phanerochaete chrysosporium*. *J. Biol. Chem.* 264, 3335–3340.
4. Mester, T. and Tien, M. (2000). Oxidation mechanism of ligninolytic enzymes involved in the degradation of environmental pollutants. *Internat. Biodeter. Biodegr.* 46, 51–59.
5. Koduri, R. and Tien, M. (1994) Kinetic analysis of lignin peroxidase: explanation for the mediation phenomenon by veratryl alcohol. *Biochemistry* 33, 4225–4230.
6. Goodwin, D.C., Aust, S.D., and Grover, T.A. (1995) Evidence for veratryl alcohol as a redox mediator in lignin peroxidase-catalyzed oxidation. *Biochemistry* 34, 5060–5065.

7. Podgornik, H., Grgič, I., and Perdih, A. (1999) Decolorization rate of dyes using lignin peroxidases of *Phanerochaete chrysosporium*. *Chemosphere* **38**, 1353–1359.

8. Knapp, J.S., Newby, P.S., and Reece, L.P. (1995) Decolorization of dyes by wood-rotting basidiomycete fungi. *Enzyme Microb. Technol.* **17**, 664–668.

9. Glenn, K.J. and Gold, M.H. (1985) Purification and characterization of an extracellular Mn(II)-dependent peroxidase from the lignin-degrading basidiomycete, *Phanerochaete chrysosporium*. *Arch. Biochem. Biophys.* **242**, 329–341.

10. Podgornik, H., Stegu, M., Žibert, E., and Perdih, A. (2001) Laccase production by *Phanerochaete chrysosporium* — an artefact caused by Mn(III)? *Lett. Appl. Microbiol.* **32**, 407–411.

11. Podgornik, H., Podgornik, A., and Perdih, A. (1999) A method of fast separation of lignin peroxidases using CIM (convective interaction media) disks. *Anal. Biochem.* **272**, 43–47.

12. Youngs, H.L., Sundaramoorthy, M., and Gold, M.H. (2000) Effects of cadmium on manganese peroxidase — competitive inhibition of Mn-II oxidation and thermal stabilization of the enzyme. *Eur. J. Biochem.* **267**, 1761–1769.

13. Field, J.A., Vledder, R.H., van Zelst, J.G., and Rulkens, W.H. (1996) The tolerance of lignin peroxidase and manganese-dependent peroxidase to miscible solvents and the *in vitro* oxidation of anthracene in solvent: water mixtures. *Enzyme Microb. Technol.* **18**, 300–308.

14. Lowry, O.H., Rosenbrough, N.J., Farr, A.L., and Randall, R.J. (1951) Protein measurement with the Folin phenol reagent. *J. Biol. Chem.* **193**, 265–275.

15. Podgornik, H., Stegu, M., Podgornik, A., and Perdih, A. (2001) Isolation and characterization of Mn(III) tartrate from *Phanerochaete chrysosporium* culture broth. *FEMS Microbiol. Lett.* **201**, 265–269.

16. Podgornik, H., Poljanšek, I., and Perdih, A. (2001) Transformation of Indigo carmine by *Phanerochaete chrysosporium* ligninolytic enzymes. *Enzyme Microb. Technol.* **29**, 166–172.

17. *Langes Handbook of Chemistry* (1985) (J.A. Dean, Ed.), Mc Graw Hill, New York, pp. 6–20.

18. Pearson, R.G. (1963) Hard and soft acids and bases. *J. Am. Chem. Soc.* **85**, 3533–3539.

SCREENING OF FUNGAL STRAINS FOR REMEDIATION OF WATER AND SOIL CONTAMINATED WITH SYNTHETIC DYES

Č. NOVOTNÝ[1], B. RAWAL[2], M. BHATT[1], M. PATEL[3], V. ŠAŠEK[1] and H. P. MOLITORIS[2]

[1] Institute of Microbiology, Academy of Sciences of the Czech Republic, Vídeňská 1083, 14220 Prague 4, Czech Republic; [2] Botanical Institute, University of Regensburg, Universitätsstrasse 31, D-93040 Regensburg, Germany; [3] Department of Biosciences, S.P. University, Vidyanagar-388120, India

Abstract

Using Remazol Brilliant Blue R (RBBR), strains of terrestrial white rot (WRF) and marine fungi (MF) were screened for efficient decolorization. Dye degradation potential of selected strains was studied with chemically different dyes (azo, anthraquinone, heterocyclic, triphenylmethane). *Irpex lacteus* and *Pleurotus ostreatus* (WRF) efficiently degraded dyes from all groups whereas less efficient and selective degradations were observed with *Dactylospora haliotrepha* and *Aspergillus ustus* (MF). Seawater salinity often reduced decolorization efficiency of WRF but increased decolorization ability of MF. In soil *I. lacteus* removed 77 % of RBBR used at 150 μg/g within 6 weeks. The work presents fungi as suitable candidates to be applied to re-mediation of dye-contaminated water and soil.

1. Introduction

More than 10 000 commercially available dyes representing a total of 700 000 tons of dyestuff are produced worldwide annually, 10 % of which enters industrial effluents. Many dyes are resistant to conventional wastewater treatment (Shaul et al. 1991) and, therefore, the need exists for development of new processes for treating those effluents. Bioremediation by fungi offers an alternative since some ligninolytic species are efficient dye degraders (Paszczynski and Crawford 1995; Banat et al. 1996; etc.).

The biodegradation potential of the majority of ligninolytic fungal species existing in nature as well as of marine fungal species is poorly known (Paszczynski and Crawford 1995; Raghukumar et al. 1996; Swamy and Ramsay 1999). The purpose of our study was to select strains of terrestrial white rot and marine fungi that would effectively decolorize recalcitrant dyes and would be applicable to bioremediation of dye-contaminated water and soil. Since in some countries the pollution with dyes affects large areas of marine coast environment, the effect of seawater salinity on decolor-ization capacities of the selected fungal strains was documented.

143

V. Šašek et al. (eds.),
The Utilization of Bioremediation to Reduce Soil Contamination: Problems and Solutions, 143–148.
© 2003 *Kluwer Academic Publishers. Printed in the Netherlands.*

2. Material and Methods

2.1. MICROORGANISMS

The following strains of terrestrial fungi were used: *Bjerkandera adusta* 606/93; *Ceriporia metamorphosa* 193/93; *Daedaleopsis confragosa* 491/93; *Ganoderma lucidum* 530/93; *Irpex lacteus* 617/93; *Mycoacia* sp. 446/93; *Pachykytospora tuberculosa* 505/93; *Phellinus pseudopunctatus* 538/93; *Phellinus punctatus* 421/93; *Pleurotus ostreatus* 670/93; *Stereum rugosum* 210/93; *Trametes versicolor* 167/93 and *Tyromyces chioneus* 616/93.

The following marine micromycetes were used: *Anguillospora longissima* M234; *Aniptodera mangrovei* M235; *Aspergillus ustus* M224; *Banhegia setispora* M236; *Bathyascus* sp. M237; *Corollospora pulchella* M217; *Dactylospora haliotrepha* M240; *Flagellospora curvula* M241; *Hypoxylon oceanicum* M244; *Lulworthia grandispora* M245; *Paraliomyces lentiferus* M248; and *Zalerion* sp. M249.

All strains were maintained on MEG agar slants at 4 °C.

2.2. SCREENING ON AGAR PLATES AND BIODEGRADATION OF DYES

The following agar (20 g/L; Oxoid, U.K.) media were used, containing (per liter): MEG pH 4.5, 5 g malt extract and 10 g glucose; YEPG pH 4.5, 2 g yeast extract, 5 g peptone, 10 g glucose, 1 g KH_2PO_4 and 0.5 g $MgSO_4.7H_2O$; low nitrogen mineral medium (Tien and Kirk 1988) pH 4.5 (LNMM); and GPY pH 6.0, 1 g glucose, 0.5 g peptone and 0.1 yeast extract. Artificial seawater (3 % salinity, W/V) was according to Lorenz and Molitoris (1997). Sterilization was done by autoclaving (120 °C, 20 min).

Decolorization of dyes was evaluated after inoculation of the dye-containing (200 µg/mL) agar plate with a mycelium-covered disc removed from a fresh MEG culture and subsequent incubation at the appropriate temperature. Decolorization and growth were read in time. Control plates without fungal inoculation were always included in the experiments. All cultures were run in duplicate.

The following dyes were used: Methyl Red (monoazo; Sigma, Germany), Methyl Orange (monoazo; Lachema, Czech Republic), Reactive Orange 16 (monoazo; Aldrich, USA), Congo Red (disazo; Merck, Germany), Reactive Black 5 (disazo; Aldrich, Germany), Naphthol Blue Black (disazo; Aldrich, USA), Remazol Brilliant Blue R (RBBR) (anthraquinone; Sigma, USA), Disperse Blue 3 (anthraquinone; Aldrich, USA), Fluorescein (heterocyclic; Lachema, Czech Republic), Methylene Blue (heterocyclic thiazine; Merck, Germany), Bromophenol Blue (triphenylmethane; Lachema, Czech Republic).

2.3. SCREENING WITH RBBR-CELLULOSE

Two-layer agar medium based on artificial seawater (Lorenz and Molitoris 1997) containing (per liter) 0.5 g peptone, 0.1 g yeast extract and 16 g agar was used, the upper layer contained 5 g RBBR-cellulose per liter. The medium was inoculated on the top with a mycelium-covered disc removed from a fresh fungal culture and incubated at for 7 weeks to detect cellulase activity and decolorization of RBBR.

2.4. DYE DEGRADATION IN SOIL

An amount of 8 g of dry tyndalized brown soil (particles <2 mm; native humidity 5.4 %, pH 6.5, organic carbon 4.7 % glucose equivalent) was used in 100-mL Erlenmeyer flasks. The soil was homogenously spiked with RBBR (final concentration 150 μg/g; moisture content 20 %, W/W). Freshly prepared, straw-grown *I. lacteus* (8.0 g) was mixed with the RBBR-spiked soil. Controls contained either heat-killed inoculum or sterilized straw without the fungus. All flasks were incubated at 28°C.The contaminated soil was extracted and analyzed on days 0, 14, 28, 35, 42 and 49 after inoculation. All live samples and controls were extracted and analyzed in triplicate.

RBBR was extracted from the spiked soil using a multisolvent system made up of chloroform, methanol and distilled water (1:1:1, V/V). Each solvent was added separately in the above order, followed by mixing and vigorous shaking of the soil sample. The soil sample was sonicated (15 min) and filtered, the filtrate was collected and chloroform removed using a separation funnel. After centrifugation the filtrate was placed in an open glass Petri dish and the solvents evaporated at 100 °C for 6-8 h. The residues were redissolved in distilled water and centrifuged. After dilution, absorbance was read at 578 nm using a Perkin-Elmer λ 11 UV-VIS spectrophotometer.

3. Results and Discussion

3.1. SCREENING OF TERRESTRIAL AND MARINE FUNGI

A broader preselection (not shown) using RBBR of terrestrial, wood-rotting fungal strains capable of efficient decolorization of dyes involved more than 100 strains collected in forests of central Europe or obtained from the Culture Collection of Basidiomycetes, Institute of Microbiology, Prague, to cover the following genera: *Abortiporus, Agaricus, Anthrodia, Aspergillus, Aurantioporus, Aureobasidium, Bjerkandera, Ceriporia, Cerrena, Coriolopsis, Daedalea, Daedaleopsis, Dictyostelium, Flammulina, Fomes, Fomitopsis, Ganoderma, Gleophyllum, Grifola, Hapalopilus, Hymenochaete, Inonotus, Irpex, Ischnoderma, Laricifomes, Lentinus, Leucoagaricus, Mycoacia, Pachykytospora, Panus, Phaeolus, Phallus, Phanerochaete, Phellinus, Phlebia, Phlebiopsis, Phlyctochytrium, Pycnoporus, Pilatoporus, Pleurotus, Podospora, Schizophyllum, Sphaerobolus, Spongipellis, Stereum, Trametes, Tyromyces.*

Table 1 documents that the most efficient of the preselected RBBR degraders were able to grow well in the presence of the dye and decolorized the plate completely within 10 days. The decolorization efficiency was tested in mineral and complex media conditions. The content of nitrogen source, known to control the production of ligninolytic peroxidases in white rot fungi (Kaal et al. 1995), was from 0.002-0.003 % (LNMM, MEG) to 0.07 % (YEPG). Some strains decolorized RBBR rapidly under broad nutritive conditions (*I. lacteus, T. versicolor*) others preferred specific conditions to a different degree (*C. metamorphosa, B. adusta, P. ostreatus*) (Table 1).

I. lacteus and *P. ostreatus* efficiently colonized sterile soil by exploratory mycelium whereas *T. versicolor* was capable of only a partial colonization and *B. adusta* did not grow into the soil at all (not shown). Therefore, *I. lacteus* and *P. ostreatus* were chosen for the comparison of their decolorizing capacities with those of marine fungi.

Screening of dye-degrading strains of marine fungi originating from the Botanical Institute, University of Regensburg with RBBR-cellulose was carried out under the conditions of both freshwater and seawater salinity (not shown). Only two of the fungi, *A. ustus* and *D. haliotrepha*, were able to decolorize the plates at both salinities and were therefore used in the comparative study.

TABLE 1. Screening of wood-rotting fungal strains on RBBR-containing agar media.

| Fungus | Medium [1,2] | | | | | |
| | YEPG | | MEG | | LNMM | |
	A [3]	B [4]	A [3]	B [4]	A [3]	B [4]
Bjerkandera adusta	7	9	10	9	P	17
Cerioporia metamorphosa	6	-	7	7	7	7
Daedaleopsis confragosa	P	21	P	-	P	-
Ganoderma lucidum	P	21	12	14	12	17
Irpex lacteus	5	9	7	7	9	7
Mycoacia sp.	14	14	14	14	14	14
Pachykytospora tuberculosa	P	-	P	-	P	P
Phellinus pseudopunctatus	19	19	17	14	14	14
Phellinus punctatus	17	17	17	16	P	17
Pleurotus ostreatus	10	10	10	9	14	20
Stereum rugosum	17	-	17	16	P	20
Trametes versicolor	9	10	7	7	P	10
Tyromyces chioneus	P	27	10	10	14	14

[1] Culture medium contained 200 mg RBBR per litre; [2] cultures were incubated at 28°C for 27 days;
[3] A - growth expressed as a number of days necessary for colonization of the whole plate; P , only a part of the plate surface was colonized within 27 days; [4] B - decolorization expressed as a number of days necessary for complete decolorization of the plate; P, partial decolorization after 27 weeks; - , no decolorization.

3.2. DECOLORIZATION OF VARIOUS DYES

In order to compare dye degradation capacities of the selected representatives of terrestrial ligninolytic basidiomycetes and marine micromycetes, a set of compounds belonging to four major groups of recalcitrant dyes was used (Table 2). Terrestrial fungi were superior to marine fungi in having the ability to decolorize all types of dyes. *I. lacteus* and *P. ostreatus* thus belong to those few white rot fungal species, such as *T. versicolor*, *P. chrysosporium* and *Bjerkandera* sp., that are capable of efficient decolorization of a broad spectrum of chemically different dyes (Knapp *et al.* 1995; Swamy and Ramsay 1999). The results obtained with *P. ostreatus* (Table 2) confirmed those of Knapp et al. (1995) who demonstrated the ability of their natural isolate of this fungal species to decolorize high concentrations of various azo-, anthraquinone-, Cu-phthalocyanine- and triphenylmethane dyes. *D. haliotrepha* was able to attack all dyes except for Reactive Orange 16 and Disperse Blue 3, but many dyes were decolorized only partially. *A. ustus* was able to attack only less than a half of the dyes tested and the decolorization was mostly incomplete (Table 2). We cannot compare our data obtained with the two latter fungi with those of other studies since the ability of marine fungi to degrade synthetic dyes is mostly unknown (Raghukumar et al. 1996).

Decolorization by terrestrial fungi of some dyes was slowed down or even completely blocked by seawater salinity (Table 2). In some cases, however, the decolorization was not affected by the salinity or even a positive effect could be observed (Congo Red). Different effects of the salinity in *I. lacteus* and *P. ostreatus* were observed in the case of degradation of Disperse Blue 3, Fluorescein and Bromophenol Blue (Table 2). Reasons for such diverse salinity effects on decolorization of various dyes by white-rot fungi are unknown.

As far as the marine fungi were concerned, seawater salinity enabled them to degrade disazo-, anthraquinone-, heterocyclic- and triphenylmethane dyes (Table 2). On the other hand, Methyl Red and, in the case of *D. haliotrepha*, also Congo Red, could be decolorized only if deionized water was used to prepare the medium.

In general, the results confirm the hypothesis of a lower degree of salinity stress (cf. Capone and Bauer 1992) in the marine fungi compared to the terrestrial ones. However, the decolorization by *I. lacteus* and *P. ostreatus* under seawater salinity conditions shows that even under those stressing conditions these fungi can use their biochemical machinery for dye degradation.

3.3. DYE DEGRADATION IN SOIL

The ability of *I. lacteus* to degrade dyes in soil was tested using RBBR at 150 µg/g

TABLE 2. Decolorization of synthetic dyes by terrestrial and marine fungi on agar medium containing deionized water or articifial seawater.

Dye	Decolorization efficiency [1]							
	Irpex lacteus		*Pleurotus ostreatus*		*Dactylospora hal.*		*Aspergillus ustus*	
	DW [2]	SW [3]	DW [2]	SW [3]	DW [2]	SW [3]	DW [2]	SW [3]
Methyl Red	9	29	15	P	10	-	17	-
Methyl Orange	9	11	10	25	12	P	-	-
Reactive Orange16	10	14	15	29	-	-	-	-
Congo Red	20	14	22	14	P	-	-	-
Reactive Black 5	11	16	13	14	-	25	-	P
Naphtol Blue Black	27	25	17	16	-	38	-	P
RBBR	10	10	13	14	-	P	-	P
Disperse Blue 3	17	16	12	38	-	-	-	P
Fluorescein	15	11	13	21	-	P	-	-
Methylene Blue	22	25	20	-	-	P	-	-
Bromophenol Blue	12	12	27	14	-	P	-	-

[1] Cultures were incubated at 22°C for 6 weeks. Decolorization is expressed as a number of days necessary for complete decolorization of the plate; P, partial decolorization after 6 weeks; - , no decolorization; [2] DW, GPY medium based on deionized water; [3] SW, GPY medium based on artificial seawater.

as the model dye. The tyndalized, RBBR-spiked soil was inoculated by mixing with freshly prepared, straw-grown fungal inoculum and the dye removal was detected in time. After 6 weeks, 77 % of the original amount of RBBR was removed, the disappearance being linear at a constant rate of 2.8 µg RBBR/g soil per day (not shown). The results suggest that recalcitrant dyes, like other organopollutants, can be removed from the soil environment by the action of ligninolytic fungi (cf. Andersson and Henrysson 1996; Novotný et al. 1999; etc.).

The study demonstrates that effective dye degraders applicable to remediation of freshwater, seawater and soil environments contaminated with recalcitrant dyes can be screened from among unknown strains of terrestrial and marine fungal organisms.

Acknowledgements

The work was supported by the Czech-German project WTZ-TSR-040-97 and by grants of the Grant Agency of the Czech Republic No. 526/99/0519 and 526/00/1303. The authors thank P. Vampola for providing the strains of basidiomycete fungi. Part of the results was presented at the Symposium on Biotechnology in the Textile Industry, May 3-7, 2000, Portugal.

References

1. Andersson, B.E. and Henrysson, T. (1996) Accumulation and degradation of dead-end metabolites during treatment of soil contaminated with polycyclic aromatic hydrocarbons with five strains of white-rot fungi, *Appl Microbiol. Biotechnol.* **46**, 647-652.
2. Banat, I.M., Nigam, P., Singh, D. and Marchant, R. (1996) Microbial decolorization of textile-dye-containing effluents: a review, *Bioresource Technol.* **58**, 217-227.
3. Capone, D.G. and Bauer, J.E. (1992) Microbial processes in coastal pollution, in R. Mitchell (ed.), *Environmental Microbiology*, Wiley-Liss, New York, pp. 191-237.
4. Kaal, E.J., Field, J.A., Joyce, T.W. (1995) Increasing ligninolytic enzyme activities in several white rot basidiomycetes by nitrogen-sufficient media, *Bioresource Technol.* **53**, 133-139.
5. Knapp, J.S., Newby, P.S. and Reece, L.P. (1995) Decolorization of dyes by wood-rotting basidiomycete fungi, *Enzyme Microb. Technol.* **17**, 664-668.
6. Lorenz, R. and Molitoris, H.P. (1997) Cultivation of fungi under simulated deep sea conditions, *Mycol. Res.* **101**, 1355-1365.
7. Novotný, Č., Erbanová, P., Šašek, V., Kubátová, A., Cajthaml, T., Lang, E., Krahl, J. and Zadražil, F. (1999) Extracellular oxidative enzyme production and PAH removal in soil by exploratory mycelium of white rot fungi, *Biodegradation* **10**, 159-168.
8. Paszczynski, A. and Crawford, R.L. (1995) Potential for bioremediation of xenobiotic compounds by the white rot fungus *Phanerochaete chrysosporium*, *Biotechnol. Progr.* **11**, 368-379.
9. Raghukumar, C., Chandramohan, D., Michel, F.C. and Reddy, C.A. (1996) Degradation of lignin and decolorization of paper mill bleach plant effluent (BPE) by marine fungi, *Biotechnol. Lett.* **18**, 105-106.
10. Shaul, G.M., Holdsworth, T.J., Dempsey, C.R. and Dostall, K.A. (1991) Fate of water soluble azo dyes in the activated sludge process, *Chemosphere* **22**, 107-119.
11. Swamy, J. and Ramsay, J.A. (1999) The evaluation of white rot fungi in the decoloration of textile dyes, *Enzyme Microbial. Technol.* **24**, 130-137.
12. Tien, M. and Kirk, T.K. (1988) Lignin peroxidase of *Phanerochaete chrysosporium*, *Methods Enzymol.* **161**, 238-249.

EFFECT OF COUNTERMEASURES OF RADIONUCLIDE UPTAKE BY AGRICULTURAL PLANTS

N. GONCHAROVA [1] and P. KISLUSHKO [2]

[1] *International Sakharov Environmental University, Dolgobrodskaya str.23, Minsk, 220009,BELARUS;*

[2] *Institute of Plant Protection, Lesnaya str.,9, Priluki, 223011,BELARUS*

1. Introduction

Mineral fertilizers have been used extensively in Belarus to reduce radionuclide uptake to plants from soil and many experimental studies have been carried out to evaluate the optimum ratios of nitrates, phosphates and potassium to give the best reduction in root uptake. Mineral fertilizers are applied under conditions of radioactive contamination of a territory by changing the ratio of main elements of nutrition. Optimum N: P: K ratio is 1:1.5:2.0. The reduction of transfer of ^{137}Cs in a grass-stand under the action of fertilizers is usually by a factor of 2.0. Radical improvement of meadows is one of the most effective countermeasures that have been implemented. Reductions of 2-4 times in root uptake are more realistic under field conditions. Radical improvement is more efficient on organic than on mineral soils. In maximal cases the cesium uptake in plants may be decreased by a factor of 10.

In order to study radionuclide transfer to plants and work out measures aimed at its reduction the scientific research program "Agroradiology" has been worked out in Belarus. Experiments in ECP-2 confirmed the efficiency of traditional countermeasure which were used in Ukraine, Russia and Belarus: mechanical treatment, such as polishing and mulching and the use of fertilisers [1-2]. At present the economic basis of the protective measures has a priority, primarily directed at the decrease of radionuclide penetration into the food products, reduction of the net cost and improvement of the quality of agricultural products.

2. Materials and Methods

In 1988-1999 oats, lupine, winter rye, barley and potato were grown at Vetka, Vetkovky state farm, Gomel Region, Belarus, in protected and control treatments, each replicated four times in 40 m^2 plots. The level of radiation background in experimental areas varied from 1 µGy to 0.3 mGy per h which corresponded to total soil activity of 555-1480 kBq/ m^2. The radioactivity content in the different plant fractions (Bq/kg dry mass) was related to the soil contamination level expressed on a surface basis (Bq/ m^2), referred to as the transfer factor (TF).

V. Šašek et al. (eds.),
The Utilization of Bioremediation to Reduce Soil Contamination: Problems and Solutions, 149–152.
© 2003 *Kluwer Academic Publishers. Printed in the Netherlands.*

The ^{137}Cs contents were measured by direct gamma-spectrometry using a high-purity Ge(Li) detector with a volume of 190 cm^3 and the multichannel analyzer NUC with 8192 channels [3].

Fertilized and limed and synthetic and natural biologically active substances (BAS) were used: azofos - a pesticide, campozan - a herbicide and plant growth regulator, hydrohumates and oxyhumates - peat extracts, potassium polysulfides (PPS) and magnesium sulfate ($MgSO_4$) - microfertilizers and Cu-, Zn-, Mg-biologically active microelement contents compounds produced by Grodno PA "Azot" from caprolactam production wastes. Campozan was applied to the soil before and three times during the growing season of winter rye.

Oats and lupine seeds were cultivated in aqueous solution of azofos, potassium polysulfides (PPS) and magnesium sulfate ($MgSO_4$) before planting. Potato haulm was sprinkled three times with aqueous solutions of hydrohumates, oxyhumates and biologically active microelement (at 10-day intervals) in budding and flowering phases. The ^{137}Cs concentration in plant straw and grain was compared in the protected and control plots.

Our methods are based on optimization of the vital activity processes in plants, using biologically active compounds [4]. Use of biologically active compounds gives the principal possibility of regulating radionuclide penetration into agricultural plants and of obtention of a crop with permissible levels of radionuclides.

3. Results and Discussion

3.1. THE EFFECTIVENESS OF PROTECTIVE MEASURES

The protective measures in the Belarus were carried out in two stages:
The first stage (1986-1991):

- Highly contaminated lands where it was impossible to obtain agricultural products with the permissible radionuclides content were excluded from use;
- Cultures accumulating a lot of radionuclides were excluded from the sowing cycle;
- Acidic soils were limed;
- Increased doses of phosphorus and potassium fertilizers were applied;
- On swamp sites drainage and deep turf plowing was carried out.

The protective measures reduced the ^{137}Cs penetration into agricultural production by 75 %.

The ^{90}Sr penetration into food products has decreased twice during the post-accident period. However, the availability of ^{90}Sr for plants still remains high, with a tendency to increase.

The second stage (since 1992):

Detailed oriented countermeasures in agriculture; methods of reduction of plant product contamination at the expense of mineral feed regulation; application of biologically active substances application and new forms of fertilizers.

It was found that BAS extended substantially the time of intensive life activity of plant leaves and enhanced photosynthetic activity. Increasing the harvest of plants was a result of higher activity of these processes (Table 1).

It was also found that during plant ontogeny ^{137}Cs was redistributed in the plant organs. In particular, ^{137}Cs concentrated in the green material in the flowering phase was higher after BAS application than in the control.

TABLE 1. Effect of BAS on harvest (kg/ha) of principal agricultural plants

Plant	With BAS	Without BAS	With BAS/Without AS
Oats (Grain)	3650	3500	150
Lupine (Seeds)	2140	1900	240
Winter rye (Grain)	3280	2900	380
Potato (Straw)	19750	15900	3850

Before the harvest, ^{137}Cs (Table 2) and ^{90}Sr (Table 3) concentrations in the grain and straw were lower with BAS case than in the control.

TABLE 2. Effect of BAS on ^{137}Cs accumulation (Bq/kg) in principal agricultural crops.

Plant	With BAS	Without BAS	(Without AS/ With BAS)
Oats (Grain)	96.2	110.0	1.14
Lupine (Seeds)	270.0	288.0	1.10
Winter rye (Grain)	35.0	52.3	1.50
Potato (Straw)	74.6	91.4	1.23

Thus, treatment of plants with BAS and plant growth regulator results in a higher harvest, increasing the radionuclide accumulation in green material and reducing radionuclide contents in the generative organs.

TABLE 3. Effect of biologically active substances (BAS) on ^{90}Sr accumulation (Bq/kg), in the harvest of principal agricultural crops

Plant	With BAS	Without BAS	Without AS/ With BAS
Oats (Grain)	51.0	55.0	1.08
Lupine (Seeds)	29.3	38.4	1.31
Winter rye (Grain)	20.0	23.0	1.15
Potato (Straw)	7.4	11.4	1.54

4. Conclusions

A great number of biologically active compounds have been tested, which increased the disease resistance of plants and simultaneously activated the physiological and biochemical processes that control the transport of micro- and microelements (radionuclides included) and their "soil-root-stem-leaf"distribution. It will be interesting to study the mechanism of this action. But today we can suggest that the BAS block the ^{137}Cs outflow from leaves to generative organs.

Use of biologically active compounds gives a principal possibility for regulating and suggest a new method of lowering radionuclide penetration into agricultural plants to obtain a crop with permissible levels of radionuclide content [5].

These data can make a valuable contribution to agricultural remediation strategies

References

1. Grebenshichikova, N.,. Firsakova, S., and NovikA.. (1992) Investigation of regularities in radiocaesium behaviour, in: Soil - Plant Cover of Byelorussian Polesye after the accident at the Chernobyl NPP, *Agrokhimia* **1,** 9 –99 (in Russian).
2. Fesenko, S., Alexakhin R., Spiridonov, S., and Sanzharova, N. (1995) Dynamics of ^{137}Cs concentration in agricultural production in areas of Russia subjected to contamination after the accident at the Chernobyl Nuclear Power Plant, *Rad. Prot. Dos.,* **60,** 155 -166.
3. Knatko, V., Skomorokhov, A., and Asimova, V. (1996) Characteristics of ^{90}Sr, ^{37}Cs and ^{239,240}Pu migration in undisturbed soil of Southen Belarus after the Chernobyl accident, *J. Enviro. Radioactiv.* **30,** 185-196.
4. Gaponenko, V., Goncharova., N., and Kislushko P. (1992) The influence of biologically active compounds and growth regulators on accumulation of radionuclides, photosynthetic and productive activity of winter rye, *Izv. AN .Resp. Belarus.* **11–12,** 1018 –1921 (in Russian).
5. Goncharova , N., Kislushko, P., Zhebrakova, I., an . Matsko, V. (1994). New ways of enhancing the vital activity of plants to increase crop yields and suppress radionuclide accumulation. Proceeding Soil-Plant Relationships, Forschungzentrum, Seibesdorf. Austria 1994, pp. 258 - 266.

PART III

ECOTOXICOLOGY AND TOXICITY MONITORING OF BIOREMEDIATION MEASURES

LOW-COST MICROBIOTESTS FOR TOXICITY MONITORING DURING BIOREMEDIATION OF CONTAMINATED SOILS

G. PERSOONE [1] and B. CHIAL [1,2]

[1] *Laboratory of Environmental Toxicology and Aquatic Ecology, Ghent University, J. Plateaustraat 22, 9000 Ghent, Belgium;*

[2] *Centro de Ciencias del Mar y Limnologia, Universidad de Panama, Panama City, Panama.*

Abstract

This paper addresses in the first place the need to apply toxicity tests with batteries of species from different trophic levels for an (ecologically meaningful) determination of the toxic hazard of contaminated soils. Furthermore, "direct contact" tests must be performed as well as assays on pore waters, leachates or percolates.

Most toxicity tests available to date are dependent on the continuous maintenance of live stocks and their application is restricted to a limited number of highly specialized laboratories. Repeated toxicity testing in bioremediation programs is very expensive and there is an urgent need for cost-effective alternatives. Microbiotests with micro-algae, protozoans and various invertebrate test species have been developed over the last few years, which depart from dormant or immobilized stages of the test biota and are hence independent of the burden of continuous stock cultur-ing/maintenance. The sensitivity of these new assays (named Toxkits) has been compared to that of "conventional" aquatic and direct-contact tests in a variety of studies on pure chemicals as well as on natural samples, and found to be equivalent, if not better. Recently a direct contact microbiotest, the Ostracodtoxkit, was developed for sediments, which now also seems to be applicable to contaminated soils. besides numerous applications in aquatic toxicology, the low-cost and user-friendly Toxkit microbiotests now also appear to be an attractive tool for routine toxicity monitoring and/or follow-up of detoxication in bioremediation programs of contaminated soils.

1. Toxicity testing of contaminated soils

The approach mostly used to date to determine the degree of pollution of aquatic and terrestrial environments is the quantification of the contaminants by chemical analysis. In the field of soil contamination, for example, the large majority of the papers published in the Proceedings of the International "Consoil" FZK/TNO Conferences on contaminated soils, only report chemical facts and figures.

It is, however, now generally accepted that a meaningful evaluation of the toxic hazard of pollutants cannot be performed exclusively by chemical analyses. First of all,

155

it is indeed not feasible in practice to analyze all the contaminants (each of which may have a toxic impact); secondly, the chemical approach does not take into account the bioavailability of the contaminants, which is the real driver of the magnitude of the toxic effects.

The "health" of ecosystems is totally dependent on the biological communities with their specific biota. When toxic thresholds are reached, the biological chain of producers (plants), consumers (animals) and decomposers (bacteria and fungi) is affected in one or more of its links and the functioning of the system is jeopardised. In order to find out whether and to what degree aquatic or terrestrial systems are threatened, toxicity tests are performed with selected test species exposed to samples of polluted waters or soils. For a correct approach, a battery of tests with representatives of the different trophic levels is needed, since toxicity is "species-" as well as "chemical"-specific and one cannot predict at which link(s) the biological chain will be broken. For soil toxicity assessment, the need for a test battery has clearly been demonstrated [1,2].

Table 1 gives an overview of some frequently used test species in terrestrial toxicology. The types of assays mentioned are all "direct contact" tests whereby the biota are in continuous contact with the soil samples. These tests, however, do not reveal the indirect hazard of soil contamination resulting from horizontal runoff or rainwater percolation into groundwaters. Contaminants indeed partition between the solid phase (soil particles) and the liquid phase (pore water) in a ratio that depends on the physico/chemical characteristics of the soil. Contrary to the chemicals which remain adsorbed to the soil particles, the dissolved fraction is directly available for uptake by biota and its impact must thus also be taken in consideration in soil ecotoxicology. The hazard of the pollutants which partition to the liquid phase necessitates bioassays on pore waters, percolates or leachates, with aquatic species, such as e.g. micro-algae (growth-inhibition test), crustaceans (e.g. the *Daphnia* immobilization assay) and fish (mortality test).

2. Toxicity testing in the framework of bioremediation

The change of the toxic hazard of contaminated soils during bioremediation operations has to be assessed by both types of tests mentioned above which implies repeated and concurrent application of two different test batteries. This repeated testing is similar to the TRE (Toxicity Reduction Evaluation) approach for industrial effluents, during which the benefits (in terms of toxicity reduction) of the changes made in the production or treatment processes are evaluated.

Faced with the necessity of repeated testing during bioremediation of contaminated soils, one is *de facto* confronted with one of the major problems in ecotoxicology, namely the (high) costs of bioassays which are due in the first place to the need of culturing stocks of test species. Persoone and Van de Vel [3] calculated that culturing makes up a substantial fraction (often more than 50 %) of the total cost of a toxicity test. Because of the technological, biological problems inherent in year-round culturing/maintenance of live stocks and the resulting financial implications, toxicity testing is to date restricted worldwide to a small number of highly specialized institutes or laboratories. For the same reasons testing is also limited qualitatively (in terms of the size of the test battery) as well as quantitatively (application in routine).

TABLE 1. Overview of some frequently used test species with the period of exposure and the endpoints of the respective bioassays

Test species	Common name	Exposure time	Test criterion
Tests on bacteria			
Indigenous populations	Bacteria	Test dependent	CO_2 respiration enzymic activity
Tests on plants			
Lolium perenne [a]	Rye grass	14 days	seed germination/growth
Raphanus sativus [a]	Radish	14 days	seed germination/growth
Tests on invertebrates			
Eisenia foetida	Compost worm	2-8 weeks	survival [b]/reproduction [c]
Folsomia candida [d]		2-4 weeks	survival/reproduction
Enchytraeus albidus [e]	Springtail (Collembola)	2-6 weeks	survival/reproduction
	Pot worm		

[a] International protocol available [4]; [b] International protocol available [5]; [c] Draft International Protocol available [6]; [d] Draft International Protocol available [7]; [e] Draft Protocol available [8]

3. Microbiotests as alternatives to conventional toxicity tests

The awareness of the bottlenecks associated with conventional bioassays and the increasing demand for routine toxicity testing and biomonitoring triggered the development of alternative assays. Attempts have been made to miniaturize bioassay technologies by using mainly unicellular or small multicellular organisms, hence the name small-scale tests or "microbiotests".

According to Blaise [9], one of the pioneers of this new approach, a microbiotest should have the following characteristics :

158

inexpensive or cost-effective
generally not labor-intensive
have a high sample throughput potential
cultures that are easily maintained or are maintenance-free
modest laboratory and incubation space requirement
low costs of consumables (e.g. disposable test containers)
low sample-volume requirements

These prerequisites should, however, not lead to loss of precision, repeatability or sensitivity in comparison with conventional toxicity tests.

An excellent overview of bioassays with smaller life forms, enzyme systems, tissue cultures and young life forms was recently compiled by Wells *et al.* [10] in the book "Microscale Testing in Aquatic Toxicology". From this review, it clearly appears that by far the largest majority of the small-scale assays described, however, still rely on (continuous) culturing or maintenance of live stocks of test species. Actually until a few years ago only two internationally accepted microbiotests could claim for full independence from stock culturing, namely the luminescence-inhibition assay with lyophilized marine bacteria [11], commercially available under the name Microtox® and Lumistox®, and the seed germination/growth test with plant seeds [4].

Extensive research performed over the last 15 years in the Laboratory for Biological Research in Aquatic Pollution (LABRAP) at the Ghent University in Belgium, presently renamed Laboratory of Environmental Toxicology and Aquatic Ecology (LETAE), has gradually led to the development of additional maintenance-free microbiotests with various test species (micro-algae, protozoans, rotifers, crustaceans). These small-scale bioassays which were given the generic name "Toxkits" [12] can be performed anywhere anytime since they depart from biological material in dormant (cryptobiotic) or immobilized form, with a shelf life of several months up to several years. Adhering to the miniaturization prerequisites of small-scale assays described by Blaise [9], the Toxkits contain, besides the test biota, all test materials (concentrated media, small test containers, micropipettes, etc.), which makes their application simple, practical and cost-effective. The precision and the repeatability of these assays is also better than that of conventional ecotoxicological tests and they can be qualified as highly standardized since the same standard materials are used by everyone.

To date about a dozen acute and short-chronic Toxkit microbiotests have been completed and are now also available commercially. Table 2 gives an overview of the major characteristics of these new small-scale assays.

During the development of the Toxkit assays, many of which make use of the same test species as in conventional toxicity tests, efforts have been made to adhere to the experimental procedures prescribed by international organizations (e.g. OECD and ISO). The only difference from the corresponding conventional bioassays is that the Toxkits contain dormant or immobilized test biota, which can be hatched or set free on demand at the time of performance of the assays, thus bypassing all the problems and costs associated with stock culturing.

As demonstrated in many papers [13-19] the sensitivity of the Toxkits (for pure chemicals as well as for natural samples) is very similar to that of the corresponding conventional bioassays.

TABLE 2. Overview of the Toxkit microbiotests with different test species

Toxkit	Type of biota	Test species	Inert stocks	Type of test	Test duration	Test criterion
Algaltoxkit F[TM]	Microalgae	*Raphidocelis subcapitata*	algal beads	chronic	72 h	growth inhibition
Protoxkit F[TM]	Ciliated protozoa	*Tetrahymena thermophila*	stationary cultures	chronic	24 h	growth inhibition
Rotoxkit F[TM] acute	Rotifers	*Brachionus calyciflorus*	cysts	acute	24 h	mortality
Rotoxkit F[TM] short chronic	Rotifers	*Brachionus calyciflorus*	cysts	chronic	48 h	reproduction
Thamnotoxkit F[TM]	Anostracan crustaceans	*Thamnocephalus platyurus*	cysts	acute	24 h	mortality
Daphtoxkit F[TM] magna	Cladoceran crustaceans	*Daphnia magna*	ephippia	acute	48 h	immobility/ mortality
Daphtoxkit F[TM] pulex	Cladoceran crustaceans	*Daphnia pulex*	ephippia	acute	48 h	immobility/ mortality
Ceriodaphtoxkit F[TM] acute	Cladoceran crustaceans	*Ceriodaphnia dubia*	ephippia	acute	24 h	mortality
Ceriodaphtoxkit F[TM] chronic	Cladoceran crustaceans	*Ceriodaphnia dubia*	ephippia	chronic	7 d	reproduction

4. Toxkit microbiotests for routine toxicity monitoring

The development of additional cost-effective microbiotests with biota from different trophic levels has rapidly triggered their incorporation in research and in toxicity-monitoring programs in several countries. Toxkit assays are now used in many countries worldwide for investigations and routine applications in a variety of domains, often in combination with the bacterial microbiotest. The state of the art of the development and application of microbiotests was recently highlighted in the International Symposium on New Microbiotests for Routine Toxicity Screening and Biomonitoring, the Proceedings of which contain about 40 papers dealing with Toxkits [20].

5. A new Toxkit microbiotests for "direct-contact" applications

As indicated above, the hazard of contaminated soils is not only related to water soluble pollutants and should hence not be assessed exclusively by tests on the water fraction (which is the case with all the Toxkit microbiotest assays developed so far). Bispo *et al.* [2] performed toxicity tests on water leachates of contaminated soils and on soils extracted with organic solvents after leaching and found that the toxicity of the leachates was only part of the total hazard of the soils. As shown in Table 1, direct contact tests are now available and used in soil ecotoxicology. For contaminated river sediments such bioassays have also gradually been developed and included in the test battery and complement the assays formerly performed only on pore waters [21-23].

Yet, like all other aquatic tests, sediment contact assays also suffer from the burden of stock culturing. Endeavors were therefore made in LETAE at the Ghent University to develop a culture/maintenance-free Toxkit microbiotest for whole sediment assays [24,25]. The new test is a short-chronic (6 day) bioassay which proceeds from the dormant eggs (cysts) of a small benthic crustacean, the ostracod *Heterocypris incongruens*. The "Ostracodtoxkit" [26] has already been applied in different studies on contaminated river sediments from Flanders, Belgium, and from Canada, to determine its potential and its limitations and to compare its sensitivity with that of "conventional" whole-sediment assays [27-29]. These studies showed that the ostracod microbiotest had a very good precision and was sensitive as the direct-contact sediment assays with the invertebrate test species mostly used (i.e. the crustacean amphipod *Hyalella azteca* and the midge larva *Chironomus riparius*). Taking into account its many advantages and not in the least its low costs, the Ostracodtoxkit is to date used more and more for toxicity monitoring of contaminated sediments.

6. Application of the Ostracodtoxkit to contaminated soils

For soils all the direct contact bioassays with invertebrates listed in Table 1 require a substantial exposure time (from 2 up to 8 weeks); these tests, however, also suffer from the burden of culturing/maintenance of live stocks. The question was hence posed whether the new ostracod microbiotest would not be a possible alternative to the lengthy and costly conventional whole-soil assays. The reasoning was that sediments are in fact "underwater soils", so why not confront the ostracod crustacean with "submerged" polluted soils to find out whether they constitute a hazard for invertebrate biota ? A first study was hence undertaken on 16 soil samples taken at former zinc smelters in Flanders, Belgium, and from sites on which contaminated river sludges had been spread. The sensitivity of the *H. incongruens* 6-day microbiotest was compared to that of the 28-day reproduction-inhibition assay with the springtail *Folsomia candida* (the latter data taken from a study by Lock and Janssen [30]).

The outcome of the investigations revealed that for most samples the ostracod assays were substantially more sensitive than the springtail tests [31]. Leachates were subsequently prepared with the same soils to determine in how far the detected toxicity was due to the "soluble" compounds that partitioned to the leachates or to the chemicals which remained adsorbed to the soil particles after leaching. The ostracod tests performed on both the solid and the liquid fractions indicated that the mortality of the organisms was still very high in the direct-contact tests with the soils after leaching, whereas it was low in most leachates. The latter findings clearly confirm the necessity of direct-contact tests on soils, as amply demonstrated for sediments [21,23].

A small interlaboratory intercalibration exercise with Ostracodtoxkit has also been performed in the meantime between LETAE in Ghent, Belgium, and the Institute for Agrobiotechnology in Tulln, Austria, on PAH-contaminated soils (data not published). The correlation coefficient between the data obtained by both laboratories was as high as 0.99 which confirmed the high precision of the new microbiotest and the low variation coefficient (6-7 %) between repeated tests performed on the reference chemical potassium dichromate.

7. Conclusions

Determination of the hazard of contaminated soils by toxicity tests is to date a complex, time-consuming and expensive exercise, which inherently precludes large-scale application and/or toxicity follow-up in bioremediation programs. Routine toxicity testing of contaminated soils will hence only become possible when low-cost alternatives are fully validated, in comparison to the traditional assays for both the solid and the liquid fractions. As reported in this paper, stock culture/maintenance-independent small-scale assays now appear on the horizon which, on the basis of the results obtained so far, seem to show that they could become dependable and attractive low-cost tools for routine toxicity assessment of terrestrial environments.

References

1. Bierkens, J., Klein, G., Corbisier, P., Van den Heuvel, R., Verschaeve L., Weltens, R., and Schoeters, G. (1998). Assessment of soil quality using a multitiered test battery of bioassays. *In : Contaminated Soil 98*. Thomas Telford, London, 1109-1110.

162

2. Bispo, A., Jourdain, M.J., and Jauzein, M. (1998). A procedure to assess contaminated soil ecotoxicity. *In :* *Contaminated Soil 98.* Thomas Telford, London, 355-364.

3. Persoone, G., and Van de Vel, A. (1987). Cost-analysis of 5 current aquatic ecotoxicological tests. *Report EUR 1134 EN.* Commission of the European Communities.

4. OECD (1984a). Terrestrial plants, growth test. OECD Guidelines for the Testing of Chemicals. N° 208. *Organization for Economic Co-operation and Development,* Paris.

5. OECD (1984b). Earthworm acute toxicity test. OECD Guidelines for the Testing of Chemicals N° 207. *Organization for Economic Co-operation and Development,* Paris.

6. ISO (1995). Soil Quality – Effects of pollutants on earthworms (*Eisenia fetida*). Part 2. Determination of effects on reproduction. Draft *International Standard. International Organization for Standardization* ISO/DIS 11268-2.2.

7. ISO (1994). Soil Quality – Effects of soil pollutants on Collembola (Folsomia candida) : Method for the determination of effects on reproduction. Draft International Standard. *International Organization for Standardization* ISO/DIS 11267.

8. Römbke, J., Moses, T., and Pessina, G. (1998). Protocol of the final assessment workshop of the International Ringtest : The Enchytraeid Reproduction Test (ERT). March 30/31, Ispra, Italy.

9. Blaise, C. (1991). Microbiotests in aquatic ecotoxicology : characteristics, utility and prospects. *Environmental Toxicology and Water Quality :* an international Journal. 6, 145-155.

10. Wells, P.G., Lee, K., and Blaise, C. (Eds). (1998). Microscale Testing in Aquatic Toxicology. Advances, Techniques and Practice. CRC Publishers. 679 pages.

11. Bulich, A.A. (1979). Use of luminescent bacteria for determining toxicity in aquatic environments. *In :* *Aquatic Toxicology :* Second Conference. L.L. Marking and R.A. Kimerle (Eds). ASTM STP 667. American Society for Testing and Materials, Philadelphia PA., 98-106.

12. Persoone, G. (1991). Cyst-based toxicity tests. I. A promising new tool for rapid and cost-effective toxicity screening of chemicals and effluents. *Zeitschr. für Angew. Zoologie* 78, 235-241.

13. Persoone, G. (1998a). Development and first validation of a "stock-culture free" algal microbiotest : the Algaltoxkit. *In : Microscale Testing in Aquatic Toxicology. Advances, Techniques and Practice,* Wells, P.G., Lee, K., and Blaise, C. (Eds), C.R.C. Publishers. Chapter 20, 311-320.

14. Persoone, G. (1998b). Development and first validation of Toxkit microbiotests with invertebrates, in particular crustaceans. *In : Microscale Testing in Aquatic Toxicology. Advances, Techniques and Practice,* Wells, P.G., Lee, K., and Blaise, C. (Eds), C.R.C. Publishers. Chapter 30, 437-449.

15. Fochtman, P. (2000) Acute toxicity of nine pesticides as determined with conventional assays and alternative microbiotests. *In : New Microbiotests for Routine Toxicity Screening and Biomonitoring,* Persoone, G., Janssen, C., and De Coen, W. (Eds). Kluwer Academic/Plenum Publishers, 233-241.

16. Latif, M., and Zach, A. (2000). Toxicity studies of treated residual wastes in Austria using different types of conventional assays and cost-effective microbiotests. *In : New Microbiotests for Routine Toxicity Screening and Biomonitoring,* Persoone, G., Janssen, C., and De Coen, W. (Eds). Kluwer Academic/Plenum Publishers, 367-383.

17. Ulm, L., Vrzina, J., Schiesl, V., Puntaric, D., and Smit, Z. (2000). Sensitivity comparison of the conventional acute *Daphnia magna* immobilization test with the Daphtoxkit F^{TM} microbiotest for household products. *In : New Microbiotests for Routine Toxicity Screening and Biomonitoring,* Persoone, G., Janssen, C., and De Coen, W. (Eds). Kluwer Academic/Plenum Publishers, 247-252.

18. Vandenbroele, M.C., Heijerick, D.G., Vangheluwe, M.L., and Janssen, C.R. (2000). Comparison of the conventional algal assay and the Algaltoxkit F^{TM} microbiotest for toxicity evaluation of sediment pore waters. *In : New Microbiotests for Routine Toxicity Screening and Biomonitoring,* Persoone, G., Janssen, C., and De Coen, W. (Eds). Kluwer Academic/Plenum Publishers, 261-268.

19. Van der Wielen, C. and Halleux, I. (2000). Shifting from the conventional ISO 8692 algal growth inhibition test to the Algaltoxkit F^{TM} microbiotest ? *In : New Microbiotests for Routine Toxicity Screening and Biomonitoring,* Persoone, G., Janssen, C., and De Coen, W. (Eds). Kluwer Academic/Plenum Publishers, 269-272.

20. Persoone, G., Janssen, C., and De Coen, W. (Eds). 2000. *New Microbiotests for Routine Toxicity Screening and Biomonitoring* Kluwer Academic/Plenum Publishers, 565 pages.

21. Burton G.A., Ingersoll C.G., Burnett L.C., Henry M., Hinmann M.L., Klaine S.J., Landrum P.F., Ross P., Tuchman M. (1998). A comparison of sediment toxicity test methods at three Great Lake areas of concern. *J Great Lakes Res.* 22 (3) 495-511.

22. ASTM (1998). Standard test methods for measuring the toxicity of sediment-associated contaminants with fresh water invertebrates. E 1706-95b. *American Society for Testing and Materials,* 1141-1223.

23. Vangheluwe, M.L., Janssen, C.R., and Van Sprang, P.A. (2000). Selection of bioassays for sediment toxicity screening. *In : New Microbiotests for Routine Toxicity Screening and Biomonitoring,* Persoone, G., Janssen, C. and De Coen, W. (Eds). Kluwer Academic/Plenum Publishers, 449-458.

24. Chial B. and Persoone G. (2002a). Cyst-based toxicity tests XIII. Development of a short chronic sediment toxicity test with the ostracod crustacean *Heterocypris incongruens*. Selection of test parameters. *Environmental Toxicology* (in press).

25. Chial B. and Persoone G. (2002b). Cyst-based toxicity tests XIV. Development of a short chronic sediment toxicity test with the ostracod crustacean *Heterocypris incongruens*. Methodology and precision. *Environmental Toxicology* (in press).

26. Ostracodtoxkit F™. (2001). Chronic « direct contact » Toxicity Test for Freshwater Sediments. Standard Operational Procedure. Creasel, Deinze, Belgium. 18 pages.

27. Chial B. and Persoone G. (2002c). Cyst-based toxicity tests XIV. Application of the ostracod solid phase microbiotest for toxicity monitoring of river sediments in Flanders (Belgium). *Environmental Toxicology* (in press).

28. Chial B. and Persoone G. (2002d). Cyst-based toxicity tests XVI. Sensitivity comparison of the solid phase *Heterocypris incongruens* microbiotest with the *Hyalella azteca* and *Chironomus riparius* contact assays on freshwater sediments from Peninsula Harbour (Ontario, Canada). *Chemosphere* (in press).

29. Blaise C., Gagné F., Chèvre N., Harwood M., Lee K., Lappalainen J., Persoone G., Doe K., Chial B. (2002). Toxicity assessment of oil-contaminated freshwater sediments. *J.Bioremediation* (in press).

30. Lock, K. and Janssen, C.R. (2001). Ecotoxicity of zinc in spiked artifical soils versus contaminated field soils. *Environ.Sci.Technol.* 35, 4295-4300.

31. Chial and Persoone. (2002e). Cyst-based toxicity tests XV – Application of the ostracod solid phase microbiotest for toxicity monitoring of contaminated soils. *Ecotoxicology* (in press).

APPLICATION OF BIOASSAYS FOR SITE SPECIFIC RISK ASSESSMENT
Evaluation of Toxic Effects Exhibited by Organic Pollutants Present in Soil

A. P. LOIBNER, R. BRAUN, V. BOLLER and O. SZOLAR
IFA-Tulln, Department of Environmental Biotechnology
Konrad Lorenz Str. 20, A-3430 Tulln, AUSTRIA

1. Introduction

Bioassays are widely applied to measure the toxic effects of aquatic samples such as surface water and groundwater, leachates and elutriates, as well as sediments, sludges, and soils. Contrary to (eco)toxicity tests, which translate single chemical concentrations into toxic effects (dose-response relationship giving EC50, NOEC, etc.), bioassays are applied to samples from contaminated sites without detailed knowledge of qualitative and quantitative extent of pollution [1]. Depending on the type of test, bioassays may deliver fast and relevant information on adverse effects that contaminated samples pose to certain test organisms. Currently, bioassays are used to prioritize contaminated sites and to establish site specific remediation goals.

1.1. RISK ASSESSMENT STRATEGIES

Over the last decades, several assessment methodologies have been developed to characterize the risks that chemical substances pose to humans and the environment. Early techniques have been focusing on human risk assessment which were then followed by an increasing effort to evaluate risks for ecosystems. Common to both is a principle framework involving the three phases: problem formulation, risk analysis, and risk characterization. The first phase includes information gathering, the definition of assessment endpoints (for ecological risk assessment only), and the development of a conceptual model (defines linkages between source, stressor, and response) and analysis plan. The second phase contains exposure and effect assessment which finally leads to risk estimation and description in the third phase [2,3,4]. Figure 1 shows the inter-relationship between the different components for ecological risk assessment and, moreover, the embedding of the risk assessment process in a contaminated soil management framework.

Most risk assessments are predictive and use models to describe potential exposure scenarios for (new) chemicals such as pesticides that will be released to the environment some time in the future. Retrospective assessments, on the other hand, may measure effects directly (epidemiological approach) or use data from chemical analysis of historically polluted environments to estimate risk [3]. The latter, however, in many cases faces the problem that a wide range of contaminants may be present at a site with only selected compounds detected by chemical analysis. Thus, a possibly significant toxic potential will not be considered in risk characterization. Moreover, the possibility

V. Šašek et al. (eds.),
The Utilization of Bioremediation to Reduce Soil Contamination: Problems and Solutions, 165–175.
© 2003 *Kluwer Academic Publishers. Printed in the Netherlands.*

of combined toxicity in terms of synergistic, antagonistic, or additive chemical interactions is another source of uncertainty.

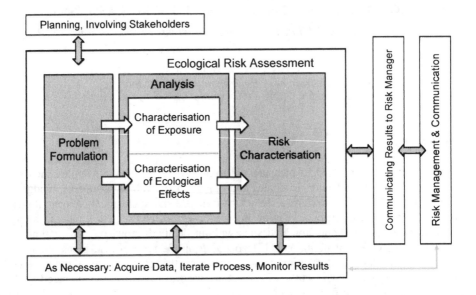

Figure 1: Framework for ecological risk assessment (simplified from [3])

Risk assessment can be conducted in a generic or a site-specific mode. The first predicts the risk posed from a hypothetical concentration of a chemical in soil based on data obtained from laboratory toxicity testing and/or models that provide generic standards (guideline values, screening values, soil quality criteria, etc.) that should be applicable to an extended range of contaminated sites [6, 7]. In general, this approach is rather conservative since it does not integrate site specific information. However, the integration of location relevant information in site specific risk assessment allows standard quality objectives to be adjusted to values that are less restrictive. [5].

Recent activities in the Netherlands focus on the development of soil function dependent ecological risk assessment procedures [8]. Therefore, a classification of different soil uses has been established: industrial areas/infrastructure, urban and rural residential areas, agricultural areas and recreational/natural areas. After describing the desired soil use for a contaminated site, distinct ecological aspects, which relate soil use and ecological soil functioning, are determined. A final evaluation step suggests the application of site-specific measurement instruments to determine the ecological damage present at the contaminated location. This step implies the use of a set of different site specific indicators, including bioassays and ecological investigations besides chemical analysis.

1.2. TOXICITY MEASUREMENT

A large number of toxicity tests have been developed and applied to measure the toxic effects of existing chemicals that are ubiquitously present in the environment. These tests are also applied to newly synthesized substances that are planned to be released to the environment and, therefore, require a toxicological evaluation. Moreover, appropriate toxicity tests have been adopted for ecotoxicological evaluations of various contaminated media and results have been used for ecological risk assessment.

Toxicity may be measured at different biological levels. Suborganism-level measures exhibit the highest toxicological relevance since they are diagnostic and sensitive to pollutants. However, their response is only vaguely linked to population- and community level assessment endpoints, which themselves are often resistant to effects due to their compensatory an adaptive mechanisms. [2] Thus, toxicity testing on organism level is a good compromise between sensitivity and ecological relevance. [9]

1.3. ECOTOXICITY TESTS

Ecotoxicity tests are applied to measure a dose-response relationship of a known chemical substance for a certain test organism. Depending on the type of organism, the substance of concern can be supplied as dosage (for higher organisms under consideration of route of entry) or in a medium (water, sediment, soil) that is relevant for the endangered ecosystem. Commonly, test results are used to calculate toxicity levels, such as EC50, LC50, or NOEC values. The application of safety factors to such toxicity levels accounts for deviations such as intra- and interspecies differences, acute to chronic exposure, and the quality and quantity of data available for calculations [2]. Such a procedure enables the derivation of screening levels which are used for comparative risk analysis. For establishing screening levels for contaminated soil, standardized and internationally accepted ecotoxicity test methods have been employed. Numerous test procedures are available for testing aquatic samples, which are relevant for surface water (contaminant run off) and groundwater (leaching of pollutants) contamination. However, only a reduced number of methods are available for testing soil samples. Moreover, when using ecotoxicity tests in ecological risk assessment an appropriate number of relevant test species including microorganisms, plants, and animals (invertebrates) should be covered in order to detect possible influences of pollutants on the ecosystem present at a contaminated site.

For microorganisms, single-species tests as well as test systems on community level are suggested for application [10]. For practical reasons, single species (selected test strains) are very often exposed to soil elutriates whereas community level tests investigate the microorganisms already present in soil. In addition, the relevance of microorganisms as an appropriate assessment endpoint is discussed in scientific literature rather frequently [11]. Protection of selected microbial species in soil is still questioned. However, the importance of the soil microflora for soil functioning (e.g. nitrogen cycling) is recognized [12, 13].

A large number of phytotoxicity tests are available for the investigation of toxic effects of chemicals like pesticides. Some of them are internationally recognized standards [14, 15], others are suggested by scientists and national or regional working groups. Several measurement endpoints like root elongation, germination or growth (biomass production) are used for toxicity evaluation.

Soil fauna includes a broad range of soil dwelling organisms with different types of exposure to contaminated soil. Several standardized test methods are available using different test species such as earthworms or collembola [16, 17].

When deriving soil screening values by using ecotoxicity tests, several limitations have to be considered, among others the bioavailability of pollutants in soil, the evaluation of pollutant mixtures or multi-component pollutants (e.g. crude oil), the selection of appropriate endpoints and exposure scenarios, and multiple sources of stress [1].

1.4. BIOASSAYS

In general, comparable protocols are used for the application of ecotoxicity test and bioassays. However, for the measurement of toxic effects of a certain chemical substance in soil, the compound of concern is usually added to the soil (spiking) and ecotoxicity tests are applied to estimate toxicity levels, such as EC50, LC50, or NOEC values. Unlike ecotoxicity tests, bioassays are used to measure toxic effects of field-contaminated soils without having in-depth information on the number and concentration of contaminants present in soil [1]. Although limited data from chemical analysis in general may be available for most samples, it is not a prerequisite for the application of bioassays to have knowledge about the pollutants present at a site. In this case, contaminated soil as a whole instead of a single or a set of pollutants is investigated for exhibiting adverse effects.

By considering site specific conditions, bioassays circumvent some problems inherent to ecotoxicity tests such as the bioavailability of contaminants in soil, the presence of not analyzed toxic substances at a site, and the occurrence of toxicity interactions of pollutant mixtures. When adding organic pollutants to solid phases like soil, sequestration of contaminants caused by pollutant/matrix interactions may occur resulting in a reduced bioavailability of the respective pollutants. The phenomenon of bioavailability is influenced by factors, such as the physico-chemical characteristics of the pollutant, the soil composition (mainly amount and type of soil organic matter), the presence of co-pollutants (e.g., USEPA-PAH in oily matrix), and the residence time of contaminants in soil (ageing). Differences in bioavailability influence the dose-response relationship for a certain amount of a toxic substance in soil. In particular, for hydrophobic contaminants like high-molar-mass PAH, and for soils rich in organic matter, lower toxicity can be expected. Site specific variations in pollutant availability will be considered when applying bioassays. Moreover, an increase in toxicity may be detected if no analyzed pollutants are present in a sample indicating that extended efforts in chemical analysis are necessary. Bioassays should be accomplished together with conventional chemical analysis which will result in additional information to be used for site characterization. For the case that several contaminants are present at a site, the response of a biological system may be synergistic, antagonistic, or additive. These interactions may be detected when performing a bioassay directly on the contaminated sample.

There exist also limitations for the application of bioassays in risk assessment approaches. For the reason of pollutant interactions, the calculation of toxicity levels (e.g., NOEC) for single chemicals may not be possible if more than one substance is present in the sample. The derivation of toxicity values might also be confounded by toxic effects exhibited by unknown contaminants such as pollutant metabolites, or by toxicants not analyzed. In addition, calculations can only be accomplished if significant effects are detected, and if different concentrations for the same set of contaminants in the same matrix are available which, in fact, will be difficult to proof. Therefore, data are often compared to a noncontaminated reference soil and results are given in percentage inhibition. However, it might be difficult to find an appropriate reference soil with characteristics comparable to the contaminated sample. Moreover, when considering the application of different bioassays including several test species (test batteries), it may be a large amount of work and thus, rather costly to accomplish these tests for all soil samples taken at a site. In addition, results face the problem of uncertainty, as, in contrast to toxicity values derived from ecotoxicity tests, only a limited number of replications are made for each sample.

2. Bioassays: Applicability and Limitations

Various bioassays including microorganisms, plants and invertebrates have been applied to PAH-contaminated soils. Results of toxicity testing and information on the applicability of bioassays are given in the following section.

2.1. INFLUENCE OF SOIL COMPOSITION

Two model soils different in soil texture and organic matter content have been used to evaluate the influence of soil composition on the toxicity of a non-aged soil contamination. Soil 1 (silty loam soil, organic matter content of 8 %), and soil 2 (loamy sand soil, organic matter content of 0.8 %) have been spiked to a total PAH concentration of 500 mg/kg dry soil using anthracene oil which is a fraction of the tar-oil distillation. In an oily matrix, it contains large amounts of low- and middle-molar-mass USEPA-PAH (including four-ring substances). Heavy USEPA-PAH are also present in anthracene oil, but only in traces.

Fig. 2 shows the acute toxicity for the earthworm *Eisenia fetida* of soil 1 and soil 2. Experiments have been done according to OECD Guideline No. 207 [16]. PAH-spiked soil has been added to non-spiked soil at portions of 5, 15, 25, and 35 %. For soil 1 a 100 % mortality has been observed at 35 % of spiked soil, whereas for soil 2 (low in organic matter) already 15 % contaminated soil have caused total mortality.

Similar differences between the two model soils have been found for the phytotoxicity test (applied according to [14]) using cress, *Lepidium sativum*. Results are shown in Fig. 3.

The LUMISTOX test (equivalent to the MICROTOX test) has been used to evaluate elutriates of spiked soils by measuring the inhibition of the bioluminescence of *Vibrio fischeri* after an exposure time of 30 min [18]. Again, the elutriates obtained for soil 2 exhibited a 19 % higher toxicity than soil 1 (data not shown). In both model soils, the same contaminants were present at equal concentrations. In conclusion, it appears that the composition of the soil (in particular the organic matter content) influences the extent of toxic effects. Hydrophobic organic pollutants may show lower toxic effects in soils higher in organic matter.

2.2. SELECTION OF A REFERENCE SOIL

When applying bioassays to terrestrial samples, the choice of suitable reference soils is of high importance, in particular, for the application of plant growth inhibition tests. Influence of soil characteristics of noncontaminated soil on plant growth are shown in Fig. 4. Although nutrients have been adlusted, differences in biomass

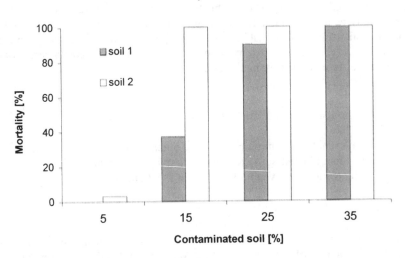

Acute earthworm toxicity in PAH-spiked soils

Figure 2. Comparison of the acute earthworm toxicity (*Eisenia fetida*) in two PAH-spiked soils. Reduced mortality has been observed for the soil higher in organic matter (soil 1). Portions of 5, 15, 25, and 35 % of spiked soil were added to non-spiked model soil.

production have been observed between soil 1 and soil 2, which are small for cress (*Lepidium sativum*) but high for millet (*Sorghum bicolor*) and rape (*Brassica napus*). Moreover, growth was less for all plants on both model soils relative to standard soil, a reference soil for compost evaluations (Austrian standard procedure; [19]). This indicates that the choice of an inappropriate reference soil may strongly influence toxicity results.

For experiments using spiked soil samples (e.g. in research work), the non-contaminated soil material should be used as a reference soil. In case organic solvents are used to dissolve the contaminants prior to their addition to soil, these solvents should also be applied to the reference soil in order to keep possible alterations of the soil matrix on a comparable level. However, it has to be assured that solvent residues are removed before toxicity testing.

When investigating field samples from a contaminated site, no reference soil is available in most cases. If possible, reference soil samples should be taken from a non-contaminated location on the site. However, to find such a location is not very likely in particular for industrial sites, where the heterogeneity of the surface and subsurface is assumed to be rather high. Other possibilities include the physico-chemical characterization of the contaminated soil and the use of an almost equivalent noncontaminated

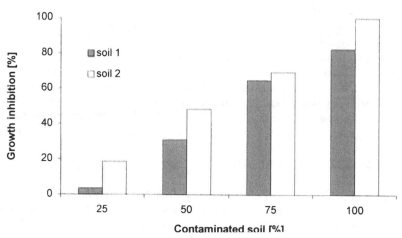

Figure 3. Comparison of the growth inhibition of cress (*Lepidium sativum*) in two PAH-spiked soils. Reduced toxicity has been observed for the soil higher in organic matter (soil 1). Portions of 25, 50 and 75 % of spiked soil were added to non-spiked model soil and tested besides 100 % spiked soil.

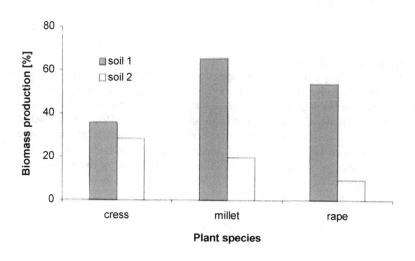

Figure 4. The biomass production of three plant species on two non-contaminated model soils is compared to growth on a standard soil normally used for compost evaluations. Biomass production was lowest on soil 2 and highest (100 %) on standard soil.

soil as reference. This reference soil may be selected from a soil collection (if available), or may be blended to the required composition by using soil compounds like sands of different mesh sizes, organic matter (e.g., peat), and clay.

Moreover, pollutants could be extracted from soil using solvents. However, to remove all pollutants possibly present, different solvents would have to be applied several times. This is a time- and solvent-consuming procedure and it cannot be taken for granted that all contaminants and total solvent have been removed prior to the application of bioassays. From the different methods tested in our laboratories for selection of an appropriate reference soil, the blending of an artificial soil seemed to be most promising and, therefore, was used for phytotoxicity testing of industrial soil.

2.3. TESTING OF INDUSTRIAL SOILS

The toxicity of several industrial soils from different European regions has been evaluated by the use of bioassays. Soils have been sampled from industrial sites such as former gas plants or steel works, with PAH as main contamination.

In Fig. 5, data are shown for a soil sample from an Austrian industrial site containing all 16 USEPA-PAH (total PAH-concentration 4160 mg/kg dry soil). Bioassays applied to the soil included test organisms, such as cress, millet, rape, earthworms, autochthonous nitrifying bacteria [18] and *Bacillus cereus* [20]. The rotifer *Brachionus calyciflorus* and crustaceans like *Thamnocephalus platyurus* and *Artemia salina* (TOXKITS [21]), as well as luminescent bacteria (*Vibrio fischeri*) were exposed to elutriates of the respective soil. Moreover, the SOS-Chromotest [18] was applied to elutriates in order to detect potential genotoxic effects of leachable contaminants. No genotoxicity was detected and only low toxicity has been observed for most of the bioassays applied, except for nitrifying bacteria. However, in contrast with the other bioassays, the ammonia-oxidizing capacity was measured for autochthonous microorganisms (no nitrifying bacteria added). Almost no nitrification activity was detected, indicating that either no nitrifying bacteria were present in soil or, if present, these bacteria were inhibited by pollution.

When measuring elutriates using spectroscopic methods, colored samples should be evaluated critically. Light absorption, fluorescence quenching (algal-growth-inhibition test), or bioluminescence quenching (luminescent-bacteria-inhibition test) often caused by dissolved humic matter may simulate toxic effects. If available, color correction methods should be applied or alternative detection techniques have to be chosen (e.g., counting of cells instead of fluorescence measurement to assess growth of algae). Fig. 6 shows bioluminescence inhibition data observed for two soil samples. Due to visible coloration of both elutriates color correction has been applied with both samples which resulted in a significant difference (change from 22 to 14 %) for soil B, indicating that only low toxicity was exhibited by this elutriate and higher readings prior to color correction were due to absorption of emitted light.

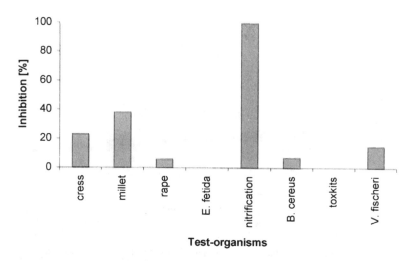

Figure 5. Application of bioassays to evaluate PAH-contaminated soil (industrial soil A). In general, low inhibitions was measured for all tested organisms except nitrifying bacteria.

3. Conclusions

In general, bioassays deliver important information on toxic effects exhibited by contaminated soil. They consider site specific conditions, such as the influence of the matrix on bioavailability, and chemical interactions if multiple contamination is present. However, as results cannot be linked to concentrations of single pollutants, the calculation of toxicity levels (LC50, NOEC, etc.) for particular contaminants is not possible but needed for risk characterization (linking exposure to effects). It might be appropriate to calculate, for instance, an LC50 for an effluent of an industrial plant which emits into a river. Based on flow rates, the exposure to organisms present in the river can be modelled and the risk may be calculated. However, further research is necessary to develop similar applications for contaminated soil. Moreover, at most sites, pollutants are heterogeneously distributed in soil. Therefore, greater number of samples has to be taken to meet statistical needs. This may increase the cost for site characterization dramatically, in particular, when applying test batteries for each sample. Nevertheless, when selecting appropriate test organisms, bioassays provide useful data to screen for hazards present and to establish a priority ranking for contaminated sites.

174

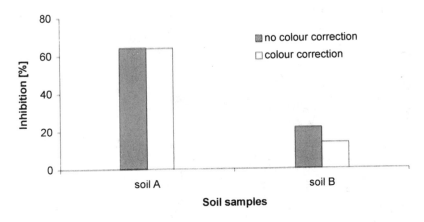

Figure 6. Application of a color correction method for the bioluminescence measurement.

References

1. Ferguson, C., Darmendrail, D., Freier, K., Jensen, B.K., Jensen, J., Kasamas, H., Urzelai, A. and Vegter J. (1988) *Risk Assessment for Contaminated Sites in Europe.* Volume 1. *Scientific Basis,* LQM Press, Nottingham.
2. Gaudet, C. (1994) A Framework fo Ecological Risk Assessment at Contaminated Sites in Canada: Review and Recommendations, Scientific Series No. 199, Ecosystem Conservation Directorate, Environment Canada, Ottawa, Ontario.
3. Risk Assessment Forum (1998) *Guidelines for Ecological Risk Assessment,* EPA/630/R-95/002F, US-EPA, Washington, DC.
4. European Commission, DGXI (1994) *Risk Assessment of Existing Substances, Technical Guidance Document,* DGXI, Brussels.
5. Vegter, J. (1998) Development of Risk Assessment and Risk Management, in *Contaminated Soil '98,* Thomas Thelford Publishing, London, 291-299.
6. Faber, J.H. (1998) *Ecological Risks of Soil Pollution,* Technical Soil Protection Committee, P.O. Box 30947, 25000 GX The Hague, The Netherlands.
7. Office of Solid Waste and Emergency Response (1996) *Soil Screening Guidance, Fact Sheet,* EPA/540/F-95/041, US-EPA, Washington, DC.
8. Rutgers, M., Faber, J.H., Postma, J.F. and Eijsackers, H. (2000) *Site-specific ecological risks: A basic approach to the function-specific assessment of soil pollution,* Rapporten Programma Geintegreerd Bodemonderzoek, Vol. 28, Wageningen, The Netherlands
9. Sutter II, G.W. (1993) *Ecological Risk Assessment,* Lewis Publishers, Boca Raton.
10. Rutgers, M. and Breure, A. M. (1999) Risk assessment, microbial communities, and pollution-induced community tolerance, *Human and Ecological Risk Assessessment* **5,** No.4, 661-670.
11. Chapman, P.M. (1999) Invited Debate/Commentary: The Role of Soil Microbial Tests in Ecological Risk Assessment, *Human and Ecological Risk Assessessment* **5,** No.4, 657-660.
12. Efroyson R.A. and Suter II, G.W. (1999) Finding a niche for soil microbial toxicity tests in ecological risk assessment, *Human and Ecological Risk Assessessment* **5,** No.4, 715-727.
13 Sheppard, S.C. (1999) Soil microbial bioassays: quick and relevant but are they useful? *Human and Ecolological Risk Assessessment* **5,** No.4, 697-705.

14. OECD (1984) *OECD Guidelines for Testing of Chemicals,* Terrestrial Plant, Growth Test, OECD Guideline No. 208.

15. International Standards Organisation - ISO (1995) Soil Quality – Determination of the Effects of Soil Pollutants on Soil Flora: Effects of Chemicals on the Emergence and Growth of Higher Plants, ISO 11269-2.

16. OECD (1984) *OECD Guidelines for Testing of Chemicals,* Earthworm, Acute Toxicity Test, OECD Guideline No. 207.

17. International Standards Organisation - ISO (1991) Soil Quality – Effects of Soil Pollutants on Collembola: Determination of the Inhibition of Reproduction, ISO TC 190/SC4/W6 2N 34.

18. Dott, W. (1995) *Bioassays for Soils,* Inhibition of the Luminescence of *Vibrio fischeri,* DECHEMA, Frankfurt am Main.

19. ÖNORM 2023 (1986) *Untersuchungsmethoden und Güteüberwachung von Komposten,* Vienna

20. Rönnpagel, K., Liß, W. and Ahlf, W. (1995) Microbial Bioassays to Assess the Toxicity of Solid-Associated Contaminants, *Ecotoxicology and Environmental Safety* **31**, 99-103.

21. Persoone, G. (1991) Cyst-based Toxicity Tests: I. A Promising New Tool for Rapid and Cost Effective Toxicity Screening of Chemicals and Effluents, *Zeitschrift für Angewandte Zoologie* **78**, 235-241.

EFFECT OF PAH-CONTAMINATED SOIL ON SOME PLANTS AND MICROORGANISMS

B. MALISZEWSKA-KORDYBACH and B. SMRECZAK

Institute of Soil Science and Plant Cultivation,
24-100 Pulawy, Poland

Abstract

A very difficult task in bioremediation processes is verification of the start-up and the end points. Establishing SQC values requires not only human risk characterisation but also the availability of ecotoxicological information. The aim of the studies was to evaluate the effect of soil artificially contaminated with a mixture of PAHs on plants and soil microorganisms. Phytotoxicity studies included two kinds of tests: acute toxicity tests (PAH concentration range 1-100 mg/kg) and chronic tests (initial PAH content in soil 100 mg/kg). The effect of PAHs on soil microbiological properties was assessed on the basis of pot experiments carried out for 90 days at the initial level of soil contamination with PAHs 10 and 100 mg/kg. Different parameters of soil microbial activity were determined periodically and related to PAH content changes in soils. The effects (NOEC, LOEC, EC_x) were evaluated when possible.

1. Introduction

One of the groups of persistent organic pollutants are polycyclic aromatic hydrocarbons (PAHs) which are probably the largest and structurally most diverse class of organic compounds known [15]. The major source of PAHs are crude oils, coal and oil shale [4,7,15]. PAHs present in the atmosphere derive principally from combustion of fossil fuels in heat and power generation, refuse burning and coke ovens [7,15,18]. In the urban areas of industrialized countries vehicle emissions are also an important PAH source. The stationary fuel combustion sources (mainly domestic heating) are responsible for about 70 % of total PAH emission in Europe while industrial processes contribute by about 20 % of PAH emission [7,18].

The main sink for most of the hydrophobic organic contaminants including PAHs is soil [7,15,18]. It is estimated [18] that more than 90 % of the total burden of PAHs resides in the surface soil layers although soil profile data show that these compounds may extend into much deeper layers [6,18]. The main cause of contamination of terrestrial environment with PAHs is atmospheric deposition [4,6,7,15,18]. The other

V. Šašek et al. (eds.),
The Utilization of Bioremediation to Reduce Soil Contamination: Problems and Solutions, 177–185.
© 2003 *Kluwer Academic Publishers. Printed in the Netherlands.*

sources of PAHs to soil are disposal of waste materials, creosote use, road runoffs and car tire shredding, accidental fuel spills and leakages as well as industrial wastewaters, sewage sludges and compost applied to agricultral land [4-7, 15,18]. The highest PAH contents correspond to industrially contaminated soils at sites involved with such activities as gasification/liquefaction of fossil fuels (gas works), coke production, asphalt production, coal tar production, wood treatment and preservation processes (creosote), fuel processing, *etc.* [3,7,11,13,18]. Very high levels of PAHs are noted also at military base areas (fuel contamination) [11,13,18]. Comprehensive information on contaminated sites in European countries was given recently in reports prepared by the Austrian Federal Environment Agency [13] and the Danish Environment Protection Agency [11]. In industrial countries, like the UK, contaminated sites contribute by about 13 % of total environmental PAH burden [18]. Most of these sites have to be reclaimed.

A very difficult task in bioremediation is verification of the start-up and the end points. The decisions are most often based on the existing soil quality criteria (SQC) which should ensure human health protection and proper functioning of the soil ecosystem within the expectation of given land use (soil functions) [5,17]. In some countries [6,17,18] recommended specific land-use "cleanup values" define levels of contamination, which constitute a concrete hazard to humans, animals and plants – if the cleanup values are exceeded, then measures to eliminate the hazard (e.g. soil bioremediation) must be taken. At lower levels of soil contamination, pathway specific "trigger values" - indicating maximum levels of contaminant at which hazard for risk receptors (human, animal, microorganism, plants) is not anticipated - are often applied. Establishing SQC values requires not only human risk characterization but also the availability of ecotoxicological information [5]. Unfortunately, the data on ecotoxicological activity of organic pollutants, including PAHs, in soil environment are very limited.

The aim of the work was to investigate the effect of soil pollution with PAHs on the growth of some plants and on soil microbiological properties. The studies were performed as laboratory and pot experiments with soils artificially contaminated with PAHs.

2. Materials and Methods

Five soils with different properties were used in the studies – Table 1. Four soils (no. I–IV) were collected from the surface (0–20 cm) of a typical agricultural horizon from an area distant from possible PAH sources. The fifth soil (no. V) was obtained by amendment of soil no. II with horticulture compost at 1:1 mass ratio. Before the test the soil samples were air-dried and sieved (2-mm diameter). All soils were artificially contaminated with PAHs. To simulate soil-environment conditions a mixture of four hydrocarbons of different properties (fluorene, anthracene, pyrene and chrysene), identified in soils from polluted regions [3,7,18], was applied in the study (as a CH_2Cl_2 solution).

TABLE 1. Basic properties of the soils

Property	Soils				
	no. I	no. II	no. III	no. IV	no. V
Texture	Loamy sand	Loamy sand	Loamy sand	Silty clay loam	Loamy sand
C_{org} (%)	0.72	0.75	1.04	3.21	5.20
Ph_{KCl}	5.5	6.6	6.9	7.0	6.5

Studies of the effect of soil contamination with PAHs on plants were performed for the plants at the initial period of their development (laboratory tests) and for the mature plants (pot experiments). Laboratory phytotoxicity tetsts included four different plant species (wheat - *Triticum vulgare*, maize – *Zea mays,* tomato – *Lycopersicon esculentum*, bean – *Phaseolus vulgaris*). In pot experiments two plants (maize – *Zea mays* and carrot – *Daucus carota*) of different root system and different root system and different utilization were applied. The conditions of the phytotoxicity experiments are given in Table 2 and the conditions of the studies on the effect of soil contamination with PAHs on soil microorganisms are presented in Table 3.

TABLE 2. Conditions of measuring the effect of soil contamination with PAHs on plants.

Conditions	Acute toxicity test (initial period of plant growth)	Chronic test (mature plant)
Type of experiment:	laboratory	pot
Time:	10 – 21 days	90 days
PAHs:	mixture of 4 compounds (fluorene+anthracene+pyrene+chrysene)	
Factors (levels) Plants	4 (wheat, maize, bean, tomato)	2 (maize, carrot)
Soils	3 (no. I, no. III, no. IV)	3 (no. I, no. II, no. III, no. IV, no. V)
ΣPAH	3 (1, 10, 100 mg/kg)	1 (PAH initial concentration 100 mg/kg)
Parameters:	growth (2) (root length, stem length)	yield (2) (fresh weight, dry weight)
Effect data:	NOEC, EC_x,	NOEC

TABLE 3. Conditions of the experiments; effect of soils contamination with PAHs on microorganisms.

Conditions	1st series	2nd series
Type of experiment; time	pot experiment ; 90 days	
PAHs	equivalent weight mixture of 4 compounds (fluorene+anthracene+pyrene+chrysene)	
Initial $\sum$PAH level	10 mg/kg	100 mg/kg
Factors (levels):		
Soils	2 (no. II, no. V)	3 (no. I, no. III, no. IV)
Time	5 (7, 15, 30, 60, 90 days)	
Parameters	dehydrogenases activity (DH), acidic phosphatases activity (PhAc), alkaline phosphatases activity (PhAl.), intensity of respiration (IR) total bacteria number (TBN)	
Effect data	NOEC, EC_x	

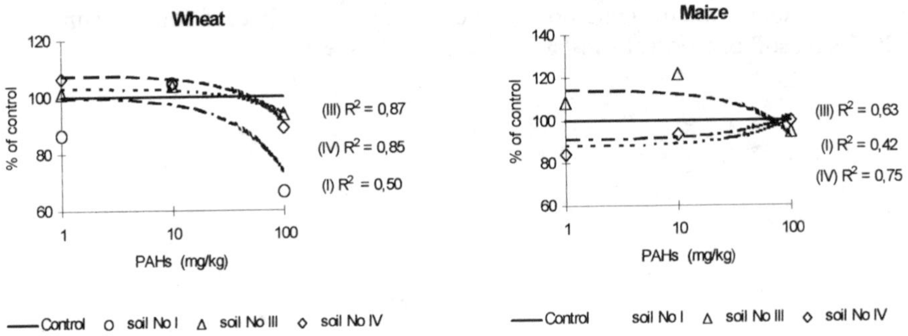

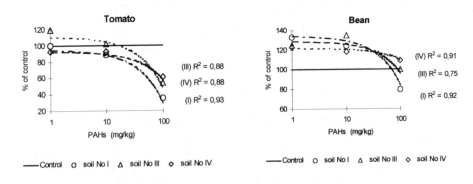

Figure 1. The effect of soil contamination with PAHs on monocotyledonous plant growth (stem length) in the initial period of their development.

Figure 2. The effect of soil contamination with PAHs on dicotyledonous plant growth (stem length) in the initial period of their development.

3. Results

The results of the acute phytotoxicity tests with soils contaminated with a mixture of PAHs are presented in Fig. 1 (monocotyledonous plants) and in Fig. 2 (dicotyledonous plants). The presented data (stem-length measurements) indicate that pollution of the soils with PAHs at lower concentrations ($\sum$PAH $\leq$ 10 mg/kg) rather stimulate than inhibit the growth of plants at the initial period of their development. For the less sensitive plants – bean and maize – characterized by larger seeds with hard cover, the inhibitory effect was not observed at all within the range of the applied PAH concentrations (up to 100 mg/kg). The most sensitive plant appeared to be tomato (smallest seeds). The response of dicotyledonous plants (tomato, bean – Fig. 4) was better related to the changes of PAH content in soil (higher determination coefficient values) than that of monocotyledonous plants (wheat, maize – Fig. 3). The phytotoxic effect of PAHs was not only plant related but also depended on soil properties – mainly organic matter (OM) content. Significant negative correlation between the percentage of the root length inhibition and OM content in soil was observed at PAH concentrations of 100 mg/kg for maize ($r^2 = 0.98$), wheat ($r^2 = 0.99$) and bean ($r^2 = 0.99$). PAHs, as most hydrophobic organic pollutants, are sorbed mainly on soil organic substances and the quantity and quality of these substances control the range of sorption [4–8], thus influencing bioavailibility of these pollutants often defined as PAH concentration in soil water [2]. It has to be pointed out that although all four PAH compounds were introduced into soil in equal amounts (25 % of total content) their percentage in soil water phase, evaluated on the basis of the equilibrium partition model [16] was significantly different and corresponded to about 71–85 % for fluorene, 5–20 % for anthracene, 7–9 % of pyrene and 0.2–1.0 % for chrysene, depending on soil properties and the level of PAH amendment [8].

The results obtained enable the calculation of EC_{20} values within the limit of the applied PAH concentrations only for tomato (22 - 48 mg $\sum$4PAH/kg) in all three soils and for wheat in soil no. I (75 mg $\sum$4PAH/kg). The extrapolation of the data gives values of 110–350 mg $\sum$4PAH/kg for bean and 140 mg $\sum$4PAH/kg for maize (in soil no. III). The NOEC values were 100 mg $\sum$4PAH/kg for bean and maize, and 10 mg $\sum$4PAH/kg for wheat and tomato. More detailed discussion of the studies on phytotoxic activity of soil contaminated with PAHs in relation to the existing literature information is given elsewhere [5,8].

There was no effect of soil contamination with PAHs on the yield of the plants observed in chronic experiments (90 days), with the exception of the carrot cultivated in soil No V – Table 4. Although the initial level of PAHs in soils used in these experiments was 100 mg/kg the content of these hydrocarbons decreased significantly in the course of the study to about 5–15 % of the initial content [10]. Hence, the NOEC values calculated for each soil were estimatrd on the basis of the mean PAH content during the time of experiment. They were within the limit of 25–35 mg $\sum$4PAH/kg, depending on the plant and soil properties.

TABLE 4. Effect of soil contamination with PAHs (initial content of Σ4PAH = 100 mg/kg) on the yield of plants after 90 days of the experiment .

Plant	Soil no. I	Soil no. II	Soil no. III	Soil no. IV	Soil no. V
			% of control		
Maize	97	-	103	106	-
Carrot	-	107	-	-	70*

*/ s - Statistically different from control at p=0.05

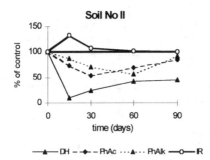
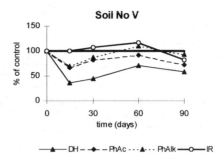

Figure 3. Effect of soil contamination with PAHs on some parameters of soil microbial activity (explanations – Table 3). The initial content of Σ4PAH = 10 mg/kg.

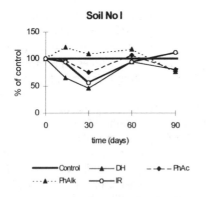
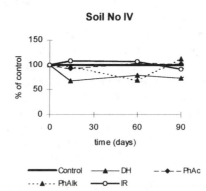

Figure 4. Effect of soil contamination with PAHs on some parameters of soil microbial activity (explanations – Table 3). The initial content of Σ4PAH = 100 mg/kg.

The examples of the results of the investigation of the effect of soil contamination with PAHs on soil microbial activity are presented in Fig. 3 (Σ4PAH initial content 10 mg/kg) and in Fig. 4 (Σ4PAH initial concentration 100 mg/kg). The presented data indicate that pollution of the soil with the mixture of PAH compounds at the level of 10 mg/kg strongly affected soil enzymic activity (DH, PhAc, PhAl), dehydrogen-ases activity (DH) being the most sensitive parameter. The strongest DH inhibition (90 % of control) was observed in light, low-OM content soil no. II, 15 days after soil pollution with PAHs. The same level of soil pollution with PAHs had no effect or stimulated the intensity of soil respiration (IR) and caused strong increase of total bacterial number (up to nine fold of the control value in soil no. II – data not shown). In the second series, where the higher level (100 mg/kg) of soil pollution with PAHs was applied – Fig. 4, the trends were the same (enzymic activity inhibition, total bacterial number stimulation) although reaction of soil microorganisms was less distinct than in the first series of experiments. In this case the lowest observed effect was 64 % of DH inhibition in sandy soil no. I (low organic matter content) after 30 days of study. Although there was no significant correlation between the effect of PAHs on soil microbial parameters and soil OM content, nevertheless the data conf-irm the earlier findings [4,5,9,12] that toxicity of PAHs toward soil microorganisms is lower in soils richer in organic substances.

Some of the effect data evaluated on the basis of the results obtained are given in Table 5. The presented values does not indicate a high ecotoxic activity of soils freshly polluted with PAHs – the majority of the obtained NOEC and LOEC values correspond to the levels of PAHs noted in very highly contaminated soil environment (about 100 mg/kg) [2,7]. However, some of the results suggest that the risk of a harmful effect of PAHs on soil biota, particularly on soil micoroflora, exists even at a rather low level of soil contamination with PAHs such as observed in agricultural lands [4,7,18].

Further research is necessary that will enable us to set up guidlines that determine "safe" concentrations of PAHs in soils with respect to human health and environemnt protection criteria as well as contaminated-site management. Special attention should be paid to elaboration of the methods of determination of the potentially bioavailable fraction of PAHs in relation to ecotoxicological activity of these pollutants.

TABLE 5. Effect of soil pollution with PAHs on some plants and microorganisms – toxicity parameters.

Organism/parameters	Effect data (mg Σ4PAH /kg)	Comments	
NOEC			
Plants - early period of development / growth			
Tomato	10	21 days	(soil no. I)
Wheat	10	14 days	(soil no. I)
Wheat	100*	14 days	(soils no. III and IV)
Maize	100*	16 days	(soils no. I, III, IV)
Bean	100*	12 days	(soils no. I, III, IV)
Plants – mature / yield			
Maize	100*	90 days	(soils no. I, III, IV)
Carrot	100*	90 days	(soil no. II)
Microorganisms / IR, TBN	10*	15 days	(soils no. II and V)
/ IR, TBN	100*	15 days	(soils no. I and IV)
LOEC			
Plants - early period of development / growth			
Tomato	$EC_{20} = 22$	21 days	(soil no. I)
Wheat	$EC_{20} = 76$	14 days	(soil no. I)
Plants – mature / yield			
Carrot	$EC_{30} = 100$	90 days	(soil no. V)
Microorganisms / DH	$EC_{90} = 10$	15 days	(soil no. II)
/DH	$EC_{64} = 10$	15 days	(soil no. V)
/DH	$EC_{36} = 100$	15 days	(soil No. I)
/DH	$EC_{33} = 100$	15 days	(soil No. IV)

*) Highest applied concentration; IR - intensity of respiration; TBN – total bacterial number, DH – dehydrogenases activity.

References

1. Boesten, J.J.T.I. (1993) Bioavailability of organic chemicals in soil related to their concentration in the liquid phase: a review. *Sci. Total Environ., Supplement*, 397-407.
2. Bradley, L.J.N., Magee, B.H., and Allen, S.L. (1994) Background levels of polycyclic aromatic hydrocarbons (PAH) and selected metals in New England urban soils, *J. Soil Contam.* 3 (4), 349-361.
3. Holoubek, Y., Kocan, A., Holoubkova, I., Kohoutek,J.,Falandysz, J., Roots, O. (2000) Persistent, boaccumulative and toxic chemicals in Central and Eastern European Countries - State-of-the-art. Report. *TOCOEN Report No. 150A*, Brno, CR, May 2000 (http://recetox.chemi.muni.cz.)
4. Jensen, J., Folker-Hansen, P., (1995) Soil quality criteria for selected organic compounds. Danish Environmental Protection Agency. Working Report No.47.
5. Jones, K.C., Alcock, R.E., Johnson, D.L., Northcott, G.L., Semple, K.T., and Woolgar, P.J. (1996) Organic chemicals in contaminated land: analysis, significance and research priorities, *Land Contam.& Reclam.* 4, 189-197.
6. Maliszewska-Kordybach, B. (1999) Persistent organic contaminants in the environment; PAHs as a case study. In: J.C. Block, V.V. Goncharuk and P.Baveye (eds.), *Bioavalibility of organic xenobiotics in the environment.* pp.3-37, NATO ASI Series, Kluwer Academic Publishers, Dordrecht/Boston/ London.
7. Maliszewska-Kordybach, B. Smreczak, B. (2000) Ecotoxicological activity of soils polluted with polycyclic aromatic hydrocarbons (PAHs) – effect on plants. *Environ. Technol.* 21, 1090-1110.

8. Maliszewska-Kordybach, B. Smreczak, B. Martyniuk, S. (2000) The effect of polycyclic aromatic hydrocarbons on microbiological properties of soils of different acidity and organic matter content *Roczniki Glebozn.* – in press   (in Polish).

9. Maliszewska-Kordybach, B.. Martyniuk, S. Smreczak, B., Stuczyński T. (2000) Degradation and ecotoxicological activity of polycyclic aromatic hydrocarbons (PAH) in soil contaminated with heavy metals. Report - Project No 6PO4G 031 12 of Polish Committee for Scientific Research, 28 pp., Pulawy, May 2000   (in Polish)

10. *Management of Contaminated Sites and Land. The Baltic States, CEE, Balkan and CIS Countries* (2000). Ad Hoc International Working Group on Contaminated Land. Meeting in Copenhagen, June, 1999, Ed. J.N. Andersen. 145 pp.

11. Pankhurst, C.E., Rogers, S.L., Gupta V.S.R., (1998) Microbial parameters for monitoring soil pollution. In: *Environmental Biomonitoring: The Biotechnology Ecotoxicology Interface*, 46-68, Lynch J.M. and Wiseman A. (eds), Cambridge University Press, Cambridge.

12. Procop, G. Edelgaard, I. Schamann, M. Bonilla A. (1998) *Contaminated Sites.*, European Topic Centre Soil under contract to the European Environment Agency. Vienna, 129 pp.

13. Rossel, D., Tarradellas, J., Bitton, G., Morel, J.L., (1997) Use of enzymes in soil ecotoxicology: a case of dehydrogenase and hydrolitic ensymes.In : *Soil Ecotoxicology*, 179-206., Tarradellas J., Bitton G. And Rossel D. (eds.), Lewis Publishers, Boca Raton, New York, London, Tokyo.

14. Sims, R.C. and Overcash, M.R. (1983) Fate of polynuclear aromatic compounds (PNAs) in soil-plant systems, *Residue Reviews* **88**, 1-68.

15. Swartz R., Schults D.W., Ozretich R., Lamberson J.O., Cole F.A., DeWitt T.H., Redmond M.S. and Ferraro S.P. (1995) ΣPAH: A model to predict the toxicity of polynuclear aromatic hydrocarbon mixtures in field-collected sediments. *Env. Toxicol. Chem.*,14, 1977-1987.

16. Vollmer, M.K., Gupta, S.K. (1995) Risk assessment plan for contaminated soils in Switzerland. General procedure and case study for cadmium, Swiss Federal Research Station for Agricultural Chemistry and Hygiene of Environment , 39 pp. Liebefeld-Bern, Switzerland,

17. Wild, S.R. and Jones, K.C. (1995) Polynuclear aromatic hydrocarbons in the United Kingdom environment: a preliminary source inventory and budget, *Environ. Poll.* 88, 91-108.

DEVELOPMENT OF IMMUNOMICROSCOPIC METHODS FOR BIOREMEDIATION

K.C. RUEL and J.-P. JOSELEAU

Centre de Recherches sur les Macromolécules Végétales, CNRS, BP 53, 38041 Grenoble, cedex 9, France

1. Introduction

In nature, wood can be completely degraded by lignolytic fungi. Basidiomycetous white-rot fungi are particularly efficient wood degraders because of their ability to completely mineralize lignin [1,2]. The enzymes produced by white-rot fungi are directed against the two types of macromolecules that make up the walls of wood cells, polysaccharides and polyphenols.. These enzymes consist of polysaccharide hydrolases that break down cellulose and hemicelluloses, and of a series of oxidizing enzymes that are able to attack and depolymerize lignins. The latter group consists of a variety of extra-cellular enzymes among which peroxidases – principally lignin peroxidase (LiP) and manganese-dependent peroxidase (MnP) – together with phenol oxidases laccase (Lac), aryl-alcohol oxidase (AAO) and other nonphenol oxidases, contribute to the highly complex mechanisms leading to lignin depolymerization, aromatic rings opening and demethylation [3-5]. An essential step in the enzymic attack of lignin is the formation of aryl cation radicals originating from both phenolic and nonphenolic aromatic rings [1,4,6]. This is the typical mode of action of LiP which uses hydrogen peroxide to generate aryl cation radicals. The other major peroxidase, particularly efficient in lignin oxidation, is manganese peroxidase which oxidizes manganese ions from $Mn+2$ to $Mn+3$. These ions that have a short life-time, are stabilized by organic acids, mostly dicarboxylic acids, and are thought to diffuse in the tight lignocellulosic network of the wood cell walls [7]. Another component of the enzyme complex participating in lignin oxidation is laccase, a phenol oxidase which seems now to be rather widespread amont the most efficient wood-degrading white-rot fungi [8]. This latter enzyme can oxidize lignin without the need of hydrogen peroxide. Laccase utilizes molecular oxygen which it reduces into two molecules of water, in a way similar to that of cytochrome *c* oxidase. The enzyme catalyzes the one electron oxidation of a large array of methoxylated phenols, diphenols and polyphenols. The specificity of action of laccase on the phenolic subunits of lignin is not absolute and it has been shown that a large number of nonphenolic substrates could also be oxidized [9]. The oxidation mechanism involves redox mediators. The role of such mediators, artificial or natural, is of particular importance for the oxidation of nonphenolic substrates [10]. Other enzymes, such as cellobiose:quinone 1-oxido-reductase (CBQ1),

187

V. Šašek et al. (eds.),
The Utilization of Bioremediation to Reduce Soil Contamination: Problems and Solutions, 187–197.
© 2003 *Kluwer Academic Publishers. Printed in the Netherlands.*

which oxidize cellobiose and reduce quinone, as well as cellobiose dehydrogenase (CDH) have been reported as often associated with lignin biodegradation [11].

White-rot fungi are the only microorganisms capable of carrying out complete degradation of lignin to carbon dioxide and water [4]. Because of this unique efficacy against aromatic compounds, white-rot fungi have been considered for their potential application in other aromatic oxidation in the field of detoxication of environmental pollutants [12-14]. This was the case for bioremediation of soils contaminated with compounds such as polycyclic aromatic hydrocarbons (PAHs) (15). Degradation of chlorophenols to complete mineralization by indigenous microorganisms may take several years [16].

The growth of white-rot fungi was shown to increase greatly in sterile soil when a carbohydrate carbon source, nitrogen source or a complex nutrient, such as lignocellulosic material, was added to the soil [16]. Wheat straw and straw composts have been studied as a means to accelerate bioremediation of chlorophenol-contaminated soil [17,18]. However, most studies of the secretion of ligninolytic enzymes of white rot fungi have been carried out in liquid media but there have been only few studies of the production of these enzymes in solid-state fermentation [18-20].

The aim of this study was to better understand the mode of development of white rot fungi in wheat straw before considering to use this lignocellulosic material as a nutrient source and supporting material for microbial development in soil. The ultrastructural study of wheat straw as a lignocellulosic substrate and the influence of lignin topochemistry on the mode of degradation by white rot fungi were studied by electron microscopy with immunocytochemical approaches. The transcription of ligninolytic enzyme genes has been shown to be different in solid state fermentation of wheat straw from that in liquid culture [20]. In view of identifying *in situ* the induction of transcripts of laccase when *Phlebia radiata* was grown in liquid culture, a method of *in situ* hybridization was adapted in transmission electron microscopy.

2. Results

2.1. ULTRASTRUCTURAL TOPOCHEMISTRY OF WHEAT STRAW

Wheat straw contains a variety of tissues and like all gramineous plants, its lignin is made up of three units, p-hydroxyphenyl propane (H), guaiacyl (G) and syringyl (S). To get a precise description of the relative distribution of the different types of lignins in the cell walls, we recently developed immunological probes [21] directed toward the major types of lignin structures on the basis of their monomeric composition, and of the inter-unit linkages, differentiated as condensed (essentially C-C bonds) and non-condensed (essentially ether bonds) [22]. Immunolabeling was done on ultrathin sections (50 nm) by incubating them in solutions of the antisera at optimal dilution. The primary antibodies were then labeled with a secondary marker coupled to gold particles. The results of lignin epitope distribution provided evidence that the heterogeneity identified by chemical analysis [23] was actually due to microheterogeneity of lignin-type distribution within tissues and within an individual cell wall. Staining with potassium permanganate delineated four distinct layers of different intensity in the secondary wall of vessel of metaxylem (Fig. 1A). When the mixed GS immunoprobes were applied to metaxylem walls the four layers did not react identically. Interestingly, the antiserum against noncondensed units labeled essentially layers 2 and 4 (Fig. 1B) which were the most reactive to $KMnO_4$, whereas the antiserum against condensed units labeled layers 1 and 3 (Fig. 1C). The preferential labeling of layers 2 and 4 agree with the high reactivity of noncondensed syringyl lignin with $KMnO_4$ [24]. This complementary labeling reveals that, although GS lignin is present throughout the metaxylem walls, a difference exists in the distribution of the condensed and noncondensed forms of these lignins.

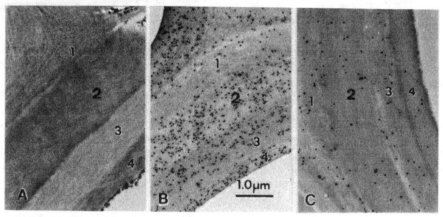

Figure 1. Lignin distribution within the walls of a vessel of metaxylem. A, KMnO4 staining delineates four zones of inequal intensities within the vessel wall; B, anti-GS (zt) gives a strong positive labeling on layers 2 and 4; C, anti-GS (zl) labels layers 1 and 3 which were only weakly reactive to the "zt" antiserum. 1,2,3,4: concentric layers of the secondary wall of metaxylem from the middle lamella (ml) to the lumen. ProteinA (5 nm) and silver enhancement procedure for the two immunolabeling.

The results of labeling with antibodies directed against homoguaiacyl (G) lignins and homo p-hydroxyphenyl (H) lignins showed that these epitopes were less abundant than the GS epitopes (25).

2.2. LIGNIN TOPOCHEMISTRY INFLUENCES BIODEGRADATION BY LIGNINOLYTIC FUNGI.

The anti-lignin antisera were used in this study to investigate the influence of the nature of lignin upon the morphology of lignocellulosic cell wall during degradation by two typical white-rot fungi, *Phanerochaete chrysosporium* and *Phlebia radiata*. The former attacks fibers from the lumen, first forming a narrow fringe of degradation which progresses outward in S2 to achieve a thinning of the secondary wall. However, in this process, it appears that all the layers do not offer the same susceptibility to the attack. This is clearly exemplified by the particular resistance of S1 and cell corners which are degraded only in the final stages of the attack (26). Immunocytological analysis of the lignins present in these different zones shows that noncondensed guaiacyl-syringyl lignin is largely predominant throughout S2, and therefore corresponds to an easily degradable lignin (Fig. 2A). This form of lignin is almost totally absent from S1 which on the other hand, contains *p*-hydroxyphenyl-propane lignin, as shown by the localized positive reaction provided by the anti-H antiserum (25). Considering that polymerization of p-coumaryl alcoohol favors the formation of condensed interunit linkage in H lignins, and that the guaiacyl lignin which is present in cell corners is also known to correspond to a condensed form of lignin, it follows that the parts of the cell wall most resistant to biodegradation are those in which condensed lignin forms are present. Another example of that was observed in the degradation of wheat straw by *P. radiata*. In this case, fibers and vessels do not exhibit the same susceptibility to degradation. Whereas at an early stage of attack lignin appears uniformly removed throughout the secondary walls of fibers, only a limited area of delignification is formed in the internal part of the vessel wall (Fig. 2B). The localization of the mixed guaiacyl-syringyl lignin in the respective cell walls showed that the noncondensed copolymer is uniformly distributed in the fiber secondary wall, whereas it is deposited down in layers in the secondary wall of the vessel (Fig. 2A). In the latter, the narrow internal layer rich in GS lignin corresponds to the layer which undergoes early delignification. The next layer, delineated by a weak response to anti-GS antibody, corresponds to a lower content of guaiacyl-syringyl lignin (Fig. 1C). This suggests that the nature of the lignin that prevails in this zone is not as easily degradable as the noncondensed GS first encountered from the lumen.

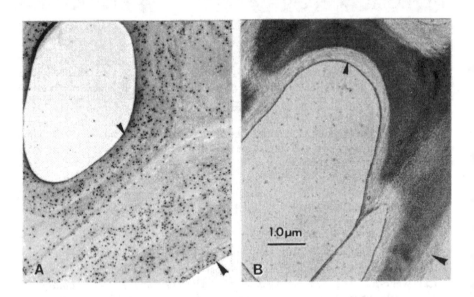

Figure 2. A, Topochemical distribution of noncondensed mixed guaiacyl-syringyl lignin units in fiber and vessel of wheat straw. Immunolabeling with the anti-GS zt antibody labeled with Protein A-gold 5nm followed by silver enhancement. B, Same tissues after four weeks of degradation by *P. radiata*. The clear areas correspond to the decayed parts of the walls. PATAg staining.

The above results are consistent with biochemical analyses of lignin biodegradation products indicating that syringyl-type substructures are more susceptible to oxidative cleavage of Cα-Cβ bonds than guaiacyl substructures . This also agrees with *in situ* demonstration by solid state ^{13}C-NMR that white-rot-decayed birch lignin contains fewer syringyl units.

2.3. IMPORTANCE OF THE EXTRACELLULAR HYPHAL SHEATH IN THE ENZYMIC ATTACK

Phanerochaete chrysosporium possesses β-1,3-1,6 linked glucans in its cell walls and as an extracellular sheath but can also excrete glucan in the culture medium. The extracellular forms of the polysaccharides are not produced at all stages of growth by the fungus and therefore are not always present. The fungus is generally grown in low-form culture flasks [27] and in these conditions several factors can affect the production of the glucan. The exopolysaccharide was not produced under high nitrogen conditions but was excreted when the glucose concentration in the medium fell below a certain level.

Visualization of the extracellular sheath was carried out with various staining and contrasting approaches, including chemical staining with potassium permanganate,

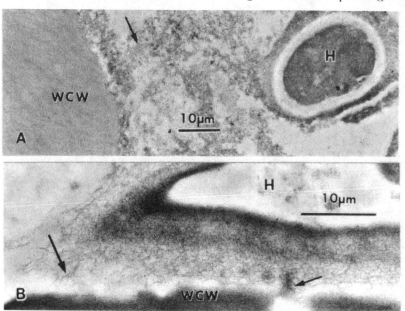

Figure 3. The sheath of white rot fungi. A, *P. chrysosporium* growing in poplar wood: KMnO₄ staining of the fungal sheath. B, PATAg staining.(H: hypha; WCW: wood cell wall; arrows: sheath)

periodate-silver staining (PATAg) and immunolabeling with a β-1,3-glucan-specific antibody [28]. The sheath constitutes an extension of the hyphal wall and seems to have, among other functions, a role in establishing contact with the wood cell wall. Hyphal cells, agglutinated in the same mucilage material were kept in close proximity with the wood by the interaction between the sheath and the inner surface of the wood cell wall (Fig. 3). The nonspecific KMnO$_4$ staining showed that the hyphal sheath could establish a material junction between the fungus and the wood cell wall, thus acting as an adhesive agent between the microorganism and the wood. The more selective PATAg staining revealed that the β-glucan network effectively bound to the degraded wood cell wall.

Such an attachment to cellulosic material was recently demonstrated in the case of *Trichoderma reesei*. In more advanced stages of degradation, fragments of the wood cell walls could be seen bound to the mucilagenous sheath.

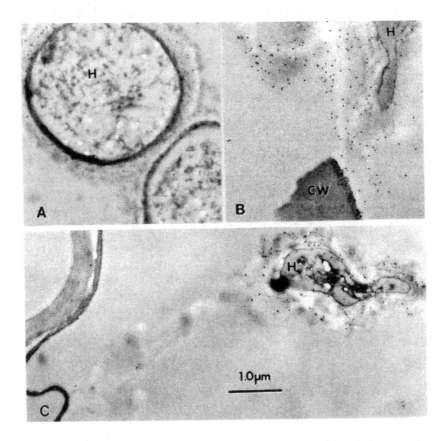

Figure 4. Immunolabeling of the enzymes excreted during degradation of wheat straw. A, Lignin peroxidase excreted by *P. chrysosporium* growing in wood. B and C: *C. subvermispora* in wheat straw B, Manganese peroxidases C, laccases

Immuno-gold cytochemical labeling of the principal types of P. chrysosporium enzymes implicated in the degradation of the wood cell wall polymers showed that cellulases and hemicellulases, as well as LiP, could have different localizations in the hyphae [7, 29, 30]. They may be intracellular or extracellular, depending on the physiological state of the fungus. Microscopy studies have shown that extracellularly released peroxidases can be found at a certain distance from the hyphae in advanced stages of wood decay. However, no explanation was provided for the mechanisms of transport of the enzymes from the hypha to the degrading cell wall region.

In the present study, the possible participation of the glucan sheath in the transport of extracellular glycanases and peroxidases was studied. The combination of two modes of staining, i.e., by chemical reaction for the underlying structure of the mucilage and by immuno-gold labeling for visualizing the excreted enzymes, demonstrated that the proteins could be adsorbed in the β-glucan mesh. When an anti-LiP was applied in combination with uranyl acetate, the nonspecific uranyl acetate staining showed the mass of mucilage making a junction between the hyphae and wood cell wall, and the anti-LiP immuno-gold marker demonstrated both the hyphal wall localization of the LiP and its interaction with the slime (Fig. 4). When the periodate-silver marker of the glucan was used with the anti-LiP primary antibody, gold particles appeared scattered onto the thin silver grains, thereby delineating the mesh of the glucan sheath. Direct interaction of the enzyme with the β-glucan was thus evidenced by this double-staining approach. The same observations were made when MnPs were excreted by *Ceriporiopsis subvermispora* growing on wheat straw (Fig. 4B) or glycohydrolases in the case of *P. chrysosporium* [31]. Because of this association, it seems that the diffusion of the enzymes was restricted in distance to the extent of the glucan network.

2.4. EXPRESSION OF LIGNINOLYTIC ENZYMES

To assess the production of Lac and MnP during the secondary growth of *P. radiata* in liquid culture, enzyme activities were measured and the proteins were localized by immunocytochemistry in transmission electron microscopy (TEM). The precocious activity of Lac gene was visualized by *in situ* hybridization specifically adapted to TEM.The results showed that just before transfer into LN+VA (low nitrogen + veratryl alcohol medium), Lac activity was present. This result, not reported in the literature, suggests that, in primary growth phase and in the absence of inducer, Lac is a constitutive enzyme in *P. radiata*. In LN+VA, we observed a strong decrease in activity on days 1 to 4 and then an increase from day 5. The presence of Lac activity, both in HN and LN medium, suggests that re-induction of Lac synthesis was effective in the LN+VA medium. MnP was also present but with a much lower activity and reached a maximum on day 5. Thus, extracellular Lac activity precedes MnP.

Additional information was obtained by Western blotting analysis using polyclonal anti-Lac and anti-MnP antibodies. Before transfer into LN+VA medium, only one band corresponding to Lac at 67 kDa was observed. This band disappeared on days 3 and 4 and reappeared from day 6, accompanied by low molar masses. The lower-mass bonds could correspond to Lac fragments resulting from proteolytic activity, likely to be present in the medium.

In order to gain insight in the intracellular synthesis of Lac and MnP, immuno-cytolocalization of the proteins was carried out in TEM on ultrathin sections. A net labeling of Lac was seen on some hyphal walls one day before transfer. On day 1, the Lac labeling became very weak with a few cytoplasmic vesicles which were positive indicating that Lac synthesis started. On day 3 a clear enhancement of the labeling intensity in the hyphal was obvious (Fig. 5 A,B,C). This entrapment of Lac inside the hyphal walls correlates with the absence of activity in the extracellular liquid culture medium.

To determine at what stage the *lac* transcripts were most abundant, dots and Northern blots were done on total RNA. On dots as well as Northern blots, a specific signal was obtained with the Dig-labeled antisense probe, whereas no signal was obtained with the sense probe. On day one in LN+VA, a single band (1.6 kb) corresponding to the size of the *lac* gene reported by Saloheimo *et al.* [32] was seen, whereas two bands at 1.6 and 0.8 kb were visible on days 2 and 3. The general

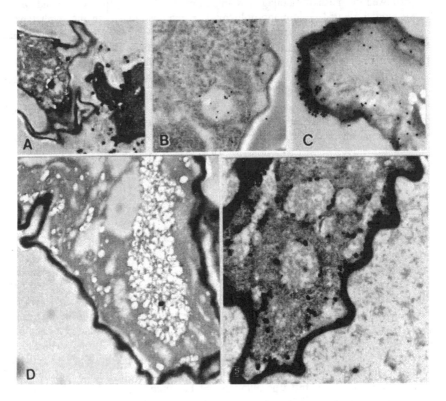

Figure 5. Localization of laccases of *P. radiata* and their transcripts by immunocytochemistry and in-situ hybridization in TEM. A-C, Immunolabeling of Lac: A, before transfer in culture medium added with veratryl alcohol; B; one day after transfer; C, three days after transfer;D-E:localization of Lac transcripts;.D, sense probe: no labeling; E, antisense probe: gold particles indicate the presence of transcripts of Lac.

intensity of the bands increased from day 1 to day 3. A second band suggests that another *lac* transcript could be expressed in our conditions of culture. From these results [33], day 2 after transfer was selected for visualizing *lac* transcripts at the ultrastructural level.

In situ hybridization with antisense Dig-Lac mRNA probe was adjusted to TEM for visualizing Lac transcripts as a function of growth. A positive signal was obtained on day 2 of growth in LN+VA (Fig. 5 D,E).

3. Conclusions

Immunocytochemical investigation of the mode of colonization and degradation of lignocellulosic substrates by white-rot fungi allowed us both to visualize the specific removal of lignin and to identify the enzymes involved in the degradation. Thanks to our antibodies directed against specific structural motives of lignin, it was possible to establish a relationship between the nature of lignin and the morphology of degradation performed by white-rot fungi. A correlation between the degraded zone during the first stage of the attack and the noncondensed nature of lignin indicates that the ligninolytic enzymes oxidize preferentially this type of lignin. Interestingly, the participation of the extracellular sheath as a supporting network for the enzymes and as a material of junction which serves for the attachment of the hyphae to the lignocellulosic substrates, constitutes an important factor in the colonization by fungi.

Among the oxidative enzymes, Lac transcripts could be detected about one day before Lac protein could be identified. This delay could indicate that the protein was secreted as a preprotein not recognized by the antibody. In this report, Lac transcripts have been identified at an early stage of idiophasic growth. This means that the response to LN+VA conditions is rapid. The detection of mRNA in a shorter delay after transfer into LN+VA must be done since mRNA synthesis could be detected after 15 min by PCR techniques [34].

The method we proposed will be currently used to detect early transcripts which are triggered when the hyphal tip contacts its natural lignocellulosic substrate. The concomitant labeling of the proteins and its mRNA could be a powerful experimental tool to investigate the first enzymic events which happen during fungal attack on lignocellulosic cell walls by white rot fungi.

Finally, when growing fungi supplemented with a lignocellulosic substrates are introduced in soil the capacity of the fungi to secrete their oxidizing enzymes seems to be related to the production of extracellular sheath.

References

1.   Kirk, T.K. and Farrell, R.L. (1987) Enzymatic combustion: the microbial degradation of lignin, *Annu. Rev. Microbiol.* **41**, 465-505

2.   Waldner, R., Leisola, M.S.A., and Fletcher, W. (1988) Comparison of lignolytic activities of selected white-rot fungi. *Appl. Microbiol. Biotechnol.* **29**, 400-407.

3.   Wariishi, H., Valli, K., and Gold, M.H. (1991) *In vitro* depolymerization of lignin by manganese peroxidase of *Phanerochaete chrysosporium, Biochem. Biophys. Res. Comm.* **176**, 269-275.

196

4. Eriksson, K.-E.L., Blanchette, R.A. and Ander, P. (1990) *Microbial and enzymatic degradation of wood components*, Springer-Verlag Berlin, Heidelberg.

5. Hatakka, A. (1994) Lignin-modifying enzymes from selected white-rot fungi: production and role in lignin degradation. *FEMS Microbiol. Rev.* **13**, 125-135.

6. Schoemaker, H.E., Lundell, T., Hatakka, A. and Piontek, K. (1994) The oxidation of veratryl alcohol, dimeric lignin models and lignin by lignin peroxidase: the redox cycle revised. *FEMS Microbiol. Rev.* **13**, 314-321.

7. Daniel, G. (1994) Use of electron microscopy for aiding our understanding of wood biodegradation *FEMS Microbiol. Rev.* **13**, 199-234.

8. Bourbonnais, R. and Paice, M.G. (1990) Oxidation of non-phenolic substrates, an expanded role for laccase in lignin biodegradation. *FEBS Lett.* **267**, 99-102.

9. Munoz, C., Guiller, F., Martinez, A.T. and Martinez, M.J. (1997) Laccase isoenzymes of *Pleurotus eryngii*: characterization, catalytic properties, and participation in activation of molecular oxygen and Mn^{+2} oxidation. *Appl. Environ. Microbiol.* **63**, 2166-2174.

10. Eggert, C., Temp, U., Dean, J.F.D., and Eriksson, K.E.L. (1996) A fungal metabolite mediates degradation of non phenolic lignin structures and synthetic lignin by laccase. *FEBS Lett.* **391**, 144-148

11. Eriksson, K.E., Habu, N., and Samajima, M. (1993) Recent advance in fungal cellobiose oxidoreductases. *Enzyme Microb. Technol.* **15**, 1002-1008.

12. Hammel, K.E. (1989) Organopollutant degradation by ligninolytic fungi. *Enzyme Microb. Technol.* **11**, 776-777.

13. Joshi, D.K., and Gold, M.H. (1994) Oxidation of di-benzo-p-dioxin by lignin peroxidase from the basidiamycete *Phanerochaete chrysosporium*. *Biochemistry* **33**, 10969-10975.

14. Sack, U., Hofrichter, M., and Fritsche, W. (1997) Degradation of polycyclic aromatic hydrocarbons by manganese peroxidase of Nematoloma frowardii. *FEMS Microbiol. Lett.* **152**, 227-234

15. Novotny, C., Erbanova, P., Sasek, V., Kubatova, A., Cajthaml, T., Lang, E., Krahl, J. and Zadrazil, F. (1999) Extracellular oxidative enzyme production and PAH removal in soil by exploratory mycelium of white-rot fungi. *Biodegradation* **10**, 159-168.

16. Boyle, C.D. (1995) Development of a practical method for inducing white-rot fungi to grow into and degrade organo-pollutants in soil. *Can. J. Microbiol.* **41**, 345-353.

17. Laine, M.M., and Jorgensen, K.S. (1996) Straw compost and bioremediatiated soil as inocula for bioremediation of chlorophenol-contaminated soil, *Appl. Environ. Microbiol.* **62**, 1507-1513.

18. Lang, E., Nerud, F., and Zadrazil, F. (1998) Production of ligninolytic enzymes by *Pleurotus sp.* and *Dichomitus squalens* in soil and lignocellulose substrate as influenced by soil microorganisms, *FEMS Microbiol. Lett.* **167**, 239-244.

19. Lang, E., Nerud, F., Novotna, E., Zadrazil, F., and Martens, R. (1996) Production of ligninolytic exoenzymes and pyrene mineralization by *Pleurotus sp.* in lignocellulose substrate, *Folia Microbiol.* **41**, 489-493.

20. Vares, T., Kalsi, M., and Hatakka, A. (1995) Lignin peroxidases, manganese-peroxidases, and other ligninolytic enzymes produced by *Phlebia radiata* during solid-stage fermentation of wheat straw. *Appl. Environ. Microbiol.* **61**, 3515-3520.

21. Ruel, K., Faix, O., and Joseleau, J-P. (1994) New immunogold probes for studying the distribution of the different lignin types during plant cell wall biogenesis. *J. Trace Microprobe Techniques* **12**, 247-265.

22. Joseleau, J.-P., and Ruel, K. (1997) Study of lignification by noninvasive techniques in growing maize internodes. An investigation by FTIR, CP/MAS ^{13}C NMR and immunocytochemical transmission electron microscopy, *Plant Physiol.* **114**, 1123-1133.

23. Sarkanen, K.V. (1971) Lignin precursors and their polymerization. *Lignins: occurrence, formation, structure and reactions*, Sarkanen KV. and Ludwig G.H., eds., Wiley-Interscience, New York, pp. 95-163.

24. Ruel, K., Ambert, K. and Joseleau, J.-P. (1994) Influence of the enzyme equipment of white-rot fungi on the patterns of wood degradation. *FEMS Microbiol. Rev.* **13**, 241-254.

25. Ruel, K., Burlat, V. and Joseleau, J-P. (1999) Relationship between ultrastructural topochemistry of lignin and wood properties. *IAWA J.* **20**, 2 : 203-211.

26. Ruel, K., Barnoud, F. and Eriksson, K.E. (1981) Micromorphological and ultrastructural aspects of spruce wood degradation by wild-type *Sporotrichum pulverulentum* and its cellulase-less mutant cel 44. *Holzforschung*, **35**, 157-171.

27. Bes, B., Pettersson, B., Lennholm, H., Iversen, T. and Eriksson, K.E. (1987) Synthesis, structure and enzyme degradation of an extracellular glucan produced in nitrogen-starved cultures of the white rot fungus *Phanerochaete chrysosporium*, *Appl. Biochem. Biotechnol.* **9**, 310-318.

28.   Ruel, K. and Joseleau, J.-P. (1991) Involvement of an extracellular glucan sheath during degradation of *Populus* wood by *Phanerochaete chrysosporium. Appl. Environ. Microbiol.* 57, 374-384.

29.   Srebotnik, E., Messner, K. and Foisner, R. (1988) Penetrability of white rot degraded pine wood by the lignin peroxidase of *Phanerochaete chrysosporium. Appl. Environ. Microbiol.* 54, 2608-2614.

30.   Daniel, G., Petterson, B., Nilsson, T. and Volc, J. (1990) Use of immunogold cytochemistry to detect Mn(II)-dependent and lignin-peroxidases in wood degraded by the white rot fungi *Phanerochaete chrysosporium* and *Lentinula edodes. Can. J. Bot.*, 68, 920-933.

31.   Joseleau, J-P. and Ruel, K. (1992)   Ultrastructural examination of lignin and polysaccharide degradation in wood by white rot fungi. In : *Biotechnology in Pulp and Paper Industry.* M. Kuwahara, M. Shimada eds. Uni Publishers CO, LTD, Ohyn Bidg, 2-6-8 Kayaba-cho, Nihonbashi, Tokyo, Japan, part II, 195-202 .

32.   Saloheimo, M., Niku-Paavola, M-L. and Knowles, J.K.C. (1991) Isolation and structural analysis of the laccase gene from the lignin-degrading fungus *Phlebia radiata. J. Gen. Microbiol.* 137, 1537-1544.

33.   Ruel, K., Zhang, H., Niku-Paavola, M-L ; Saloheimo, M., Moukha, S. and Joseleau, J-P (1999) *In situ* Hybridization and Immunocytochemistry in Electron Microscopy to Study the Expression of Ligninolytic Enzymes in Fungi Growing in Wood. *Proceedings from the 10th International Symposium on Wood and Pulping Chemistry* , Yokohama, Japan, Vol. I, 534-539.

34.   Janse, B.J.H., Gaskell, J., Akhtar, M. and Cullen, D. (1998) Expression of Phanerochaete chrysosporium genes encoding lignin peroxidases, manganese peroxidases, and glyoxal oxidase in wood. *Appl. Environ. Microbiol.* 64, 3536-3538.

ECOTOXICOLOGICAL EVALUATION OF PAH-CONTAMINATED SOIL USING EARTHWORMS

M. BHATT, O. SZOLAR, R. BRAUN AND A.P. LOIBNER
Institute For Agrobiotechnology
Department of Environmental Biotechnology
Konrad Lorenz Str. 20, A-3430 Tulln, AUSTRIA

1. Introduction

Oligochaete earthworms are functionally important in terrestrial ecosystems as primary decomposers, as food organisms for many avian and mammalian organisms and as intermediates in nutrient cycling process [1]. In terms of biomass earthworms are usually the predominant component of the soil fauna with 10-200 g fresh weight per square metre [2]. The potential toxicity of chemicals for earthworms provides a useful parameter when attempting to determine soil quality criteria for ecological risk assessment of contaminated sites. An acute toxicity test using a litter-dwelling species (*Eisenia fetida*) has been accepted as OECD Guideline No. 207 [3]. Individual researchers have also reported chronic toxicity tests on earthworms [4,5].

Polycyclic aromatic hydrocarbons (PAH) have been recognized as one of the most important classes of contaminants in many urban and agricultural soils. They are of great environmental concern because of their biotransformation in organisms, which may produce metabolites with mutagenic and carcinogenic properties [6]. Some evidence indicates that PAH may accumulate in the body tissues of earthworms [7].

The purpose of this study was to evaluate acute and chronic toxicity of PAH-contaminated soil samples for the earthworm (*Eisenia fetida*) for a period of 7, 14 and 42 days. In addition, estimation of bioaccumulation of PAH in earthworms which survived after 42 days was also carried out.

2. Materials and Methods

2.1 SOIL SAMPLES

Soil samples used for the ecotoxicological studies include a noncontaminated control soil and a number of PAH-contaminated soils subjected to remediation techniques or not. All PAH-contaminated soil samples were collected from a biopile remediation and were further treated with either active or passive aeration. Samples were collected in the

V. Šašek et al. (eds.),
The Utilization of Bioremediation to Reduce Soil Contamination: Problems and Solutions, 199–203.

course of remediation after 0, 17, 28, 42, 57, 86, 127 days. Due to biodegradation the PAH-concentration (16 USEPA-PAH) declined from 160 to 100 mg/kg dry soil. No difference in PAH removal was observed with active or passive treatment.

In addition, two output materials of a soil washing plant, namely Filtercake (fine silt type fraction; total organic carbon 1.1 % , PAH concentration 395 mg/kg dry soil) and Finegrain (coarser sand type fraction; total organic carbon 3.0 %, PAH concentration 203 mg/kg dry soil) were investigated for earthworm toxicity. Filtercake and Finegrain were mixed with noncontaminated soil and compost for biopile treatment. All soil samples were sieved (< 2mm) and the amount of water was adjusted to 65 % of their maximum water holding capacity which was maintained throughout the experiment.

2.2 EARTHWORMS

Cultures of earthworms were purchased from a local supplier and identified as *Eisenia fetida* at the Institute of Zoology, University for Agricultural Science, Vienna. The earthworms were immediately maintained in moist field soil filled in plastic containers (70 x 40cm size), that had been mixed thoroughly with compost residues, cow manure, and green grass. The containers were loosely covered with perforated dark plastic sheet and kept for at least a week prior to the start of experiment.

2.3 EXPERIMENTAL SET-UP

All earthworm tests were performed in 700-mL (13-cm height x 10-cm diameter) glass jars with lids. Each glass jar contained 500 g (total solids) soil. Oxygen transfer to the replicates was maintained by perforated lids. The glass jars were left in continuous light at 20 °C. No food substance was provided during the test-incubation period. Each soil sample was tested in triplicate.

Ten adult *Eisenia fetida*, with well developed clitellum were added in each of the jars on the surface of the soil. The glass jars were kept under continuous light to keep the worms in the soil.

For the acute survival tests, motile earthworms were counted after 7 and 14 days of exposure. The nonmotile (dead) worms or their debris were discarded and the remaining live worms were placed back into the same soil for further 28 days chronic exposure. At the end of the experiment (42 days) the live worms were counted again.

2.4 BODY BURDEN ANALYSIS OF PAH IN *Eisenia fetida*

The protocol used to detect uptake of PAH by earthworms was modified from Chung and Alexander [8]. Earthworms that survived after chronic exposure to biopile-treated-soil samples (active aeration), were collected in a glass Petri dish. The organisms were cleaned with distilled water. They were placed on filter paper and kept there for 20 min. The animals were then kept in dry and sterile marine-sand for about 4 h for depuration. Subsequently, the surviving animals were pooled in every jarand frozen using liquid nitrogen (to minimize the decaying activity) and stored at –20 °C until extraction.

The animals were weighed (1.10 to 2.00 g/jar) and then homogenized by grinding with 8 g of Na_2SO_4. The tissues were placed in preweighed extraction thimbles. The homogenized mixture of worm-sodium sulfate was subjected to soxtherm extraction (100 ml acetone-hexane mixture 1:1 V/V; 60 min with soil submersed in boiling solvent, followed by 90 min soxhlet-like extraction). HPLC was used to analyze PAH in extracts of earthworm tissues.

3. Results and Discussion

All earthworms added to noncontaminated control soil samples survived after 42 days. After exposure of 14 days, only three of the 13 biopile-remediated soils showed mortality up to 20 %. All the soil samples with remaining live earthworms were further incubated for four weeks and additional mortality was observed in already acutely toxic biopile-remediated soil samples (10 – 17 % mortality). However, mortality of animals was only found in soil samples collected at the early phase of biopile treatment and not with remaining samples indicating a gradual decrease in lethal toxic effects of the contaminated soil samples due to bioremediation (Fig. 1).

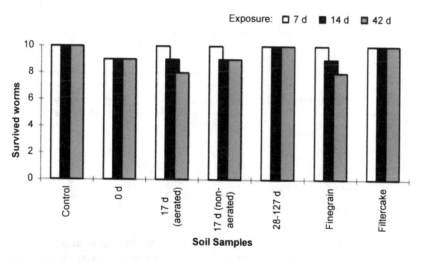

Figure 1: Earthworm survival in PAH-contaminated soil samples: 0 d, 17 d, 28-127 d indicate samples taken from the biopile treatment (aerated and non-aerated) at different times (remediation time is given in days). Finegrain and Filtercake are components of bioremediated soil. Columns of different color indicate exposure times (7, 14, 42 days).

After 42 days of exposure, the organisms in the Filtercake component, which was higher in PAH concentration (395 mg PAH/kg dry soil) did not show any significant effect on earthworm survival while the Finegrain component with lower PAH-concentration (203 mg PAH/kg dry soil) revealed 80 % of earthworms to be motile. These results suggest possible involvement of other toxic substance(s) present in the

Finegrain component, or a lower bioavailability of PAH in Filtercake possibly due to the higher organic matter content of this component [9, 10].

In addition to acute and chronic survival of the earthworms, the body burden of 16 USEPA-PAH was analyzed for worms which survived chronic exposure (42 days) to soil samples from the active aeration biopile treatment. This experiment revealed significant amounts of PAH bioaccumulation; 13 – 48 µg of total PAH per gram earthworm tissue were detected (Fig. 2). Highest bioaccumulation was observed for the 28 d-sample, wherein all worms survived but showed least average weight.

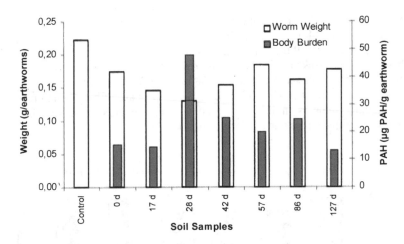

Figure 2. Body burden of total PAH in earthworms (grey columns) after exposure to contaminated soil. For extraction and analysis the surviving worms were pooled per sample. Average weight of pooled worms is presented using white columns. Sample names indicate time (days) of biopile treatment.

4. Conclusion and Summary

Oligochaetes are one of the predominant members of soil fauna and they should be regarded as useful assessment endpoint in terrestrial ecotoxicology. Earthworms may not exhibit lethal toxic effects if exposed to soil samples with low PAH concentrations demanding the inclusion of sublethal toxicity endpoints such as effects on reproduction [4, 11] and the influence on enzymic systems [12].

Concurrent analysis of bioaccumulation of PAH may deliver useful information for relating toxic effects shown by contaminated soil samples, to the bioavailable fraction of the contaminants. However, further research is necessary for a significant correlation of body burden to adverse effects.

Acknowledgements

This work was supported by the European Union, 4th Framework Programme of DGXII (ENV4-CT97-602) and the Kommunalkredit Austria (GZ 98.20015). We are also grateful to Prof. Erhard Christian (Institute of Zoology, Univ. For Agricultural Science, Vienna) for the identification of earthworms.

References

1. Edwards, C.A. and Bohen, P.J. (1996) Biology of earthworms, 3rd edn. Chapman & Hall, London.
2. Dunger, W. and Fielder, H.J. (1989) Methoden der Bodenbiologie, Gustav Fischer Verlag, Jena.
3. OECD (1984) *Guideline for Testing of Chemicals. No. 207. Earthworm, Acute Toxicity Tests,* Organization for Economic Cooperation and Development, Paris.
4. Saterbak, A., Toy, R.J., Wong, D.C.L., McMain, B.J., Williams, M.P., Dorn, P.B., Brzuzy, L.P., Chai, E.Y., and Salanitro, J.P. (1999) Ecotoxicological and analytical assessment of hydrocarbon-contaminated soils and application to ecological risk assessment, *Environ Toxicol Chem* **18,**1591-1607.
5. Potter, C.L., Glaser, J.A., Chang, L.W., Meier, J.R., Dosani, M.A., and Herrmann, R.F. (1999) Degradation of polynuclear aromatic hydrocarbons under bench-scale compost conditions, Environ Sci Technol 33, 1717-1725.
6. Harvey, R.G. (1991) *Polycyclic aromatic hydrocarbons: Chemistry and Carcinogenicity,* Cambridge University Press, New York, NY, USA.
7. Ma, W.C., Immerzeel, J. and Bodt, J. (1995) Earthworms and food interactions on bioaccumulation and disappearence in soil of polycyclic aromatic hydrocarbons: Studies on phenanthrene and fluoranthene, *Ecotoxicol Environ Saf* **32,** 226-232.
8. Chung, N. and Alexander, M. (1999) Effect of concentration on sequestration and bioavailability of two polycyclic aromatic hydrocarbons, *Environ Sci Technol* **33,** 1717-1725.
9. Nam, K., Chung, N., and Alexander, M. (1998) Relationship between organic matter content of soil and the sequestration of phenanthrene, *Environ Sci Technol* **32,** 3785-3788.
10. Hatzinger, P.B. and Alexander, M. (1997) Biodegradation of organic compounds sequestered in organic solids or in nanopores within silica particles, *Environ Toxicol Chem* **16,** 2215-2221.
11. Meier, J.R., Chang, L.W., Jacobs, S., Torsella, J., Meckes, M.C., and Smith, M.K. (1997) Use of plant and earthworm bioassays to evaluate remediation of soil from a site contaminated with polychlorinated biphenyles, *Environ Toxicol Chem* **16,** 928-938.
12.. Achazi, R.K., Flenner, C., Livingstone, D.R., Peters, L.D., Schaub, K., and Scheiwe, E. (1998) cytochrome P450 and dependent activities in unexposed and PAH-exposed terrestrial annelids, *Comp Biochem Physiol C,* **21,** 339-350.

USE OF BIOASSAYS IN DETERMINING THE ECOTOXICITY OF INDUSTRIAL SOILS

M.A.T. DELA CRUZ, O. SZOLAR, R. BRAUN, AND A.P. LOIBNER
Institute for Agrobiotechnology (IFA-Tulln)
Department of Environmental Biotechnology
Konrad Lorenz Strasse 20, A-3430 Tulln, AUSTRIA

1. Introduction

Polycyclic aromatic hydrocarbons (PAH) are ubiquitous contaminants [1] which are of concern due to their mutagenic [2] and epigenetic [3] characteristics. Because of their environmental and biological importance, 16 PAHs have been included in the U.S. Environmental Agency's priority list of pollutants.

The use of microbial bioassays to assess the toxicity of organic pollutants is increasing. Bioassays measure the total toxic potential of a sample, comprising possible synergistic or antagonistic effects of contaminants. They also provide information on the sensitivity of different test organisms exposed to the pollutants. Various endpoints such as growth, reproduction, and survival have been employed [4,5]. In addition, physiological parameters such as bioluminescence [6] and enzymic activities [7,8] are also often used.

The objective of the present study was to assess the toxicity of elutriates from three industrial soils. To evaluate the acute toxicity of soil elutriates to *Vibrio fischeri*, a bioluminescence inhibition test was conducted. The chronic toxicity of soil elutriates to *Selenastrum capricornutum* was also determined using an algal growth inhibition test in microtiter plates.

2. Materials and Method

2.1. SOIL SAMPLES

Soils from a former coke production and manufacturing plant site were used. The degree of PAH contamination was as follows:

Industrial soil E1: low PAH contamination (200 mg/kg dry soil)
Industrial soil E2: moderate PAH contamination (2000 mg/kg dry soil)
Industrial soil E3: high PAH contamination (12 000 mg/kg dry soil)

V. Šašek et al. (eds.),
The Utilization of Bioremediation to Reduce Soil Contamination: Problems and Solutions, 205–209.

2.2. ELUTRIATE PREPARATION

Ecotoxicological investigations were carried out using the aqueous soil extract (elutriate) to determine the potential risks of toxicants leached out of the soil and the risks associated with soil inhabitants.

Preparation of the aqueous soil elutriate was modified after the methodology outlined by Hund and Traunspurger [6]. Soil elutriates from 3 industrial soils were prepared (dry soil to water ratio of 1:2.5) for 24 h at room temperature. The aqueous solution was centrifuged (5000 g) in 15-mL glass test tubes for 20 min at 20 °C and then filtered with a glass-fiber filter to remove particulate matter.

2.3. BIOLUMINESCENCE INHIBITION TEST

2.3.1. Dr. Lange LUMIStox Standard Method
The Dr. Lange LUMIStox test is a screening tool which uses the bioluminescence of the bacterium, Vibrio fischeri. The test was performed following the manufacturer's procedure [9]. Solid sodium chloride crystals were added to the prepared soil elutriates to a final concentration of 2 % W/V. The pH was adjusted to 7.0 ± 0.2.

The bioluminescence inhibition of Vibrio fischeri using the Dr. Lange LUMIStox test kit is determined by combining 0.5 mL of elutriate and 0.5 mL luminescent bacterial suspension. After an exposure of 30 min at 15 °C, the decrease in light emission is measured. The toxicity of elutriates is expressed as percent bioluminescence inhibition ($\%I_B$) relative to a noncontaminated reference. The results were calculated using a correction factor (C_f), which is a measure of changes in intensity within control samples during the 30 min exposure time.

2.4. ALGAL GROWTH INHIBITION TEST

The algal growth inhibition test in microtiter plate is a screening tool following the OECD guideline 201 [10] and the Environment Canada biological test method [11].

Algal cells (Selenastrum capricornutum, PRINTZ) were cultivated for 4 days in an incubation chamber with cool luminescent lights with a light/dark cycle of 13h/11h. Temperature was maintained at 23 ± 2 °C and humidity at 50 %.

Algal cells were transferred to a centrifuge tube and the medium was removed by centrifuging the algal solution for 10 min. The supernatant was discarded and the algal cells were subsequently resuspended by adding 10 mL of buffer solution (15 mg/L $NaHCO_3$). The concentration of algal inoculum cells was adjsuted to obtain the desired initial algal concentration of 1 x 10^4 cells/mL in each well of the microtiter plates.

Sterile white 96-well microtiter plates were used wherein 36 peripheral wells were filled with 220 µL double distilled water to minimize evaporation in the microtiter plate. The test system consisted of 60 inner wells (maximum volume of 300 µL). Each well was filled with 205 µL double distilled water (for the control) or soil elutriate, 5 µL ATCC media, and 10 µL algal suspension. Three wells were filled for each soil elutriate.

Chlorophyll *a* fluorescence (at the beginning of the experiment and after 72 h of incubation) for each well was measured using a microplate reader (Tecan SPECTRAFluor Plus, Tecan GmbH, Austria). Chlorophyll *a* fluorescence of the algae, exposed to the soil elutriate over a period of 72 h, was compared with the fluorescence of the algae in a noncontaminated control.

3. Results and Discussion

3.1. BIOLUMINESCENCE INHIBITION TEST

The bioluminescence inhibition of elutriates from three industrial soils was determined. The mean bioluminescence inhibition for each industrial soil was calculated (Table 1). The lowest bioluminescence inhibition was observed in the elutriate from industrial soil E1, followed by the elutriate from industrial soil E2. The highest bioluminescence inhibition was exhibited by the elutriate from industrial soil E3.

TABLE 1. Acute toxicity to *Vibrio fischeri* of soil elutriates from three industrial soils ($n=3$; $*n=2$).

Industrial soil	Bioluminescence inhibition test	
	Mean inhibition (%)	SD (%)
E1	13.6*	4.6
E2	68.9	0.5
E3	88.5	0.5

When relating the results of bioluminescence inhibition obtained from the elutriates to the degree of PAH contamination of these industrial soils, the data seem to indicate a positive relationship. Industrial soil E1 has the lowest degree of PAH contamination followed by industrial soil E2. Industrial soil E3 has the highest degree of PAH contamination.

However, these industrial soils are from a former coke production and manufacturing plant site and there are organic pollutants other than PAHs and inorganic contaminants that may be present. These contaminants may have contributed to the overall reduction of bioluminescence in the elutriates of the respective industrial soils. Moreover, not only the parent compounds may be present but also metabolites from microbial and chemical degradation can occur. Metabolites can be more toxic than the parent compound. Thus, the bioluminescence inhibition for each elutriate cannot be attributed to a particular type of contaminant but to the toxicity of all pollutants present in each industrial soil, both organic and inorganic.

3.2. ALGAL GROWTH INHIBITION TEST

The fluorescence inhibition of elutriates from three industrial soils were determined. The mean fluorescence inhibition observed for the elutriate from each industrial soil is presented in Table 2.

TABLE 2. Chronic toxicity to *Selenastrum capricornutum* of
soil elutriates from three industrial soils (*n*=3).

Industrial soil	Algal growth (fluorescence) inhibition Test	
	Mean inhibition (%)	SD (%)
E1	17.6	3.3
E2	78.3	1.4
E3	98.7	0.7

As the data suggests, elutriates from the lowest contaminated industrial soil exhibited the lowest fluorescence inhibition whereas elutriates from the highest contaminated industrial soil showed the highest fluorescence inhibition. Moreover, as already observed with the bioluminescence inhibition test, there seems to be a positive relationship between the fluorescence inhibition of the elutriates and the degree of PAH contamination of these industrial soils.

However, as mentioned previously, the fluorescence inhibition cannot be attributed entirely to PAH contamination since there may be other organic and inorganic pollutants contained in the industrial soils. These pollutants may have been leached out into the elutriates. Moreover, the mixture of contaminants in these soils and the influence of soil matrix may have caused other less water soluble pollutants to be further dissolved in the aqueous solution.

3.3. COMPARISON OF TEST ORGANISMS

Responses of the two test organisms to the elutriates were compared (Fig. 1). In general, *Selenastrum capricornutum* was slightly more inhibited than *Vibrio fischeri*.

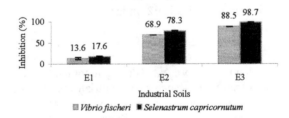

Figure 1. Acute toxicity to *Vibrio fischeri* and chronic toxicity to *Selenastrum capricornutum* of soil elutriates from three industrial soils (mean ± SE).

The difference in responses may be due to the type of contaminants present in the industrial soils. Thus, to assess the toxicity of PAH contamination in industrial soils, further chemical analysis as well as toxicity testing with various test organisms and endpoints are needed. These bioassays may include exposure of elutriates as well as solid soil to test organisms from various ecological levels with different endpoints such as growth, reproduction, survival, and genotoxicity, to name a few.

4. Conclusion

In conclusion, the elutriates from each industrial soil show acute and chronic toxicities to *Vibrio fischeri* and *Selenastrum capricornutum*, respectively. The bioluminescence inhibitions were as follows: 13.6 (SD: 4.6) % for industrial soil E1, 68.9 (SD: 0.5) % for industrial soil E2, and 88.5 (SD: 0.5) % for industrial soil E3. The fluorescence inhibitions were found to be 17.6 (SD: 3.3) % for industrial soil E1, 78.3 (SD: 1.4) % for industrial soil E2, and 98.7 (SD: 0.7) % for industrial soil E3.

Ecotoxicities of the elutriates seem to be positively related with the degree of PAH contamination for each industrial soil. However, bioluminescence and fluorescence inhibitions cannot be attributed solely to PAHs since there are other contaminants present in these industrial soils. Thus, the bioluminescence and fluorescence inhibitions are due to the toxicity of the total pollutants present in each industrial soil that were leached out into the aqueous elutriate.

References

1. Harvey, R.G. (1997) *Polycyclic Aromatic Hydrocarbons*, Wiley-VCH Inc., New York.
2. Hannigan, M.P., Cass, G.R., Penman, B.W., Crespi, C.L., Lafleur, A.L., Busby Jr, W.F., Thilly, W.G., and Simoneit, B.R.T. (1998) Bioassay-directed chemical analysis of Los Angeles airborne particulate matter using a human cell mutagenicity assay. *Environmental Science and Technology* **32**, 3502-3514.
3. Ghoshal, S., Weber Jr, W.J., Rummel, A.M., Trosko, J.E., and Upham, B.L. (1999) Epigenetic toxicity of a mixture of polycyclic aromatic hydrocarbons on gap junctional intercellular communication before and after biodegradation. *Environmental Science and Technology* **33**, 1044-1050.
4. Salanitro, J.P., Dorn, P.B., Huesemann, M.H., Moore, K.O., Rhodes, I.A., Rice Jackson, L.M., Vipond, T.E., Western, M.M., and Wisniewski, H.L. (1997) Crude oil hydrocarbon bioremediation and soil ecotoxicity assessment. *Environmental Science and Technology* **31**, 1769-1776.
5. Sayles, G.D., Acheson, C.M., Kupferle, M.J., Shan, Y., Zhou, Q., Meier, J.R., Chang, L., and Brenner, R.C. (1999) Land treatment of PAH-contaminated soil: Performance measured by chemical and toxicity assays. *Environmental Science and Technology* **33**, 4310-4317.
6. Hund, K. and Traunspurger, W. (1994) Ecotox: Evaluation strategy for soil bioremediation exemplified for a PAH-contaminated site. *Chemosphere* **29**, 371-390.
7. Gaudet, C. (1994) *A framework for ecological risk assessment at contaminated sites in Canada: Review and recommendations*. Environment Canada Scientific Series No. 199, Ottawa.
8. Bogaerts, P., Senaud, J., and Bohatierm J. (1998) Bioassay technique using nonspecific esterase activities of *Tetrahymena pyroformis* for screening and assessing cytotoxicity of xenobiotics. *Environmental Toxicology and Chemistry* **29**, 371-390.
9. Dr. Lange LUMIStox Operating Manual. Düsseldorf, Germany.
10. Organization for Economic Cooperation and Development (OECD) (1984) *Guidelines for testing of chemicals: Alga Growth Inhibition Test*. Document No. 201. Paris, France.
11. Environment Canada. (1992) *Biological test method: Growth inhibition test using the freshwater alga, Selenastrum capricornutum*. Environmental Protection Series Report EPS 1/RM/25, Environment Canada, Ottawa.

GENOTOXICITY ESTIMATION IN SOILS CONTAMINATED WITH POLYCYCLIC AROMATIC HYDROCARBONS AFTER BIODEGRADATION

K. MALACHOVÁ [1], D. LEDNICKÁ [1] and Č. NOVOTNÝ [2]

[1] Faculty of Science, University of Ostrava, 30th April Street 22, 701 03 Ostrava 1, Czech Republic; [2] Institute of Microbiology, Academy of Sciences of the Czech Republic, Vídeňská 1083, 142 20 Prague 4, Czech Republic.

1. Introduction

Polycyclic aromatic hydrocarbons (PAHs) belong to recalcitrant pollutants that resist to decomposition by natural processes. As a result of long term industrial activities and accidents, PAHs and their residues accumulate in soil. Biodegradation processes that are able to decompose PAHs and their mixtures are rather complex and can be affected by many physical, chemical and biological factors [9,5]. Some PAHs are degraded to produce intermediates with a mutagenic activity, such as dihydrodiols, phenols, arene oxides, etc. [3,1]. The purpose of the study was to detect mutagens in soil contaminated with coke plant wastes in the course of many years of production and, on the basis of genotoxicity changes, evaluate the efficiency of biodegradation technologies used for decontamination.

2. Materials and Methods

2.1. SOIL SAMPLES

All soil samples were collected in the former Karolína coke plant in Ostrava, Czech Republic. The following contaminants were detected in the soil: PAHs by gas chromatography (Hewlett-Packard 5971 Series II), nonpolar extractable compounds by infrared spectrophotometry (M80), cyanides by spectrophotometry (LKB Ultraspec II), and heavy metals, such as As, Hg, Zn and Pb, by atomic absorption spectrophotometry (AAS). Two groups of samples were evaluated (Group I and II), originating from two independent, pilot biodegradation tests. Group I samples were exposed to a 5-month biodegradation by indigenous soil microflora capable of enzymic

211

V. Šašek et al. (eds.),
The Utilization of Bioremediation to Reduce Soil Contamination: Problems and Solutions, 211–215.
© 2003 *Kluwer Academic Publishers. Printed in the Netherlands.*

degradation of polycyclic organic substances. Group II samples represented contaminated soil exposed to a 6-month biodegradation by indigenous soil microflora and a special fungal preparation supported by natural compost (EKOHUM).

2.1.1. Soil characteristics of Group I samples

The samples originated from two corridors designated A and B and were removed in regular monthly intervals from June to October. An initial PAH concentration of 784 mg/kg dry matter found in June was reduced to 34 mg/kg after a 5-month bioremediation which represented a biodegradation efficiency of 91 %. The removal efficiency for phenols and BTX (benzene, toluene, xylene) was about 90 % (DEKONTA Co.).

2.1.2. Soil characteristics of Group II samples

The content of PAHs and mutagenicity were evaluated after a) an uncontrolled 6-month remediation [samples Z0 (time 0) and Z6]; b) a 6-month remediation in the presence of components accelerating the degradation, i.e. EKOHUM (liquid manure) (sample ZE), fungi+EKOHUM (sample ZHE) and EKOHUM+fungi (sample ZEH). In the fungi+EKOHUM system, the contaminated soil was first mixed with the fungal inoculum and then, after 14 days of biodegradation, EKOHUM was added and the biodegradation continued for six months. In the EKOHUM+fungi system the order of addition of the two accelerating components was reverse. The contaminated soil used in the experiment contained 221 mg PAHs/kg dry matter before the treatment. The respec-tive PAH removals in ZE, ZHE and ZEH samples were 79, 70 and 61 %. The chemical analysis was carried out by ATE a.s. Co., responsible for   the decontamination.

2.2. PREPARATION OF SOIL EXTRACTS FOR DETECTION OF MUTAGENI-CITY

The soil samples (20 g) were extracted with (3 x 50 ml) dichloromethane (Merck) at 20 °C using sonification (3 x 20 min) as usual. The soil samples containing EKOHUM were extracted using a SOXWAVE 3.6 extractor in a focus, microwave field (MAE). In this case, a mixture of toluene and methanol (50/6.5 V/V) was used.

After extraction, the solvents were evaporated to dryness and the amount of extractable organic matter (EOM) was determined [6]. The residues were redissolved in dimethyl sulfoxide in nitrogen atmosphere.

2.3. MUTAGENICITY BIOASSAY

SOS Chromotest is a quantitative bacterial colorimetric assay for genotoxins based on an induction of the SOS sfiA function. The SOS-inducing potency (SOSIP) is a single parameter representing the induction factor (Ic) determined for the concentration of the compound tested that is read from the linear region of the dose-response curve [4, 8]. In the Ames test (histidine reversion assay), auxotrophic indicator tester strains of Salmonella typhimurium His⁻ were used, i.e. TA100 for the detection of base-pair substitution mutations and TA98 for the detection of frameshift mutations. Mutation potential (MP) represents the index of mutagenicity (IM) for the concentration of the compound tested that is read from the linear region of the dose-response curve [2, 7].

Mutagenicity of all samples was determined in both the presence and absence of metabolic activation *in vitro* by the S9 liver microsomal fraction and a cofactor mixture [7].

The results were expressed as a means of at least three replicates. Similarly, at least two plates or tubes per dose were tested in each assay.

3. Results and Discussion

3.1. BIOREMEDIATION BY SOIL BACTERIA

The results of SOS Chromotest demonstrate that, before the reclamation process, the soil samples A and B contained indirectly acting mutagens inducing SOS repair. A significant decrease in the SOS induction activity of both samples was observed after 1 month of degradation.

Using the Ames test with S9 activation in strain TA98, the presence of mutagens with indirect mutagenic activity of frameshift-mutation type was detected during the whole remediation process (Fig. 1). In both sample groups, a gradual increase in the mutagenic activity could be observed from the beginning of remediation (June to October). Consequently, the biodegradation did not eliminate the pollutants contributing to the indirect acting mutagenicity of frameshift-mutation type. Contrary to SOS Chromotest, the presence of direct acting frameshift mutagens was detected in the Ames test (Fig. 1). This effect is exerted by some PAH derivatives after their adsorption to soil particles and a reaction with compounds contained in the soil or resulting from the activity of soil microorganisms [1]. Base-substitution mutations detectable in strain TA100 were not found in any of the Ames test variants.

Comparing the mutagenicity results obtained we can conclude that promutagens inducing SOS repair were rapidly and effectively decomposed in the biodegradation process whereas the degradation of frameshift promutagens was ineffective.

3.2. BIODEGRADATION BY FUNGAL CULTURE

The sample Z0 removed at time zero exhibited the indirect and direct mutagenic effect. In SOS Chromotest, the mutagenicity was detected only using the test with metabolic activation *in vitro*. However, in the Ames test, both the induction of frameshift and substitution mutations in the tests with metabolic activation *in vitro* and of frameshift mutations in the tests without metabolic activation was detected (Fig. 2).

After a 6-month biodegradation the mutagenicity of 4 soil samples was determined (Fig. 2). Interestingly, sample Z6 collected after a spontaneous degradation by autochthonous microflora showed an increase of the induction activity as measured by SOS Chromotest with metabolic activation. In contrast, using the Ames test, a decrease in mutagenicity in strain TA98 with and without metabolic activation was observed, compared to the initial values of mutation potential. In case of strain TA100, negative results were obtained. Similar results, i.e. a higher induction potential and a

214

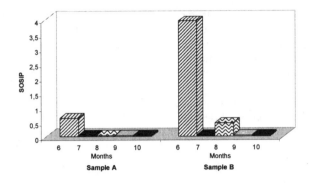

Figure 1. Results of Ames test using strain TA98. The tested mutagen concentrations ranged from 3.1 to 593.3 µg extractable organic matter per plate.

decrease of mutagenicity in TA98, were also found with sample ZE to which only EKOHUM was added.

A decrease of the induction of SOS repairs was observed only in samples ZHE and ZEH that differed in the time for which the fungal culture was applied. The SOS Chromotest results, however, did not correlate with the evaluation of mutagenicity by the Ames test. Using strain TA98, some differences between samples ZHE and ZEH were detected in the variants +S9 and −S9. Whereas ZEH was as mutagenic as ZE using the Ames test and thus no expected decrease in frameshift mutations occurred, a decrease of mutagenicity was found in ZHE. Consequently, at the end of biodegradation, the least mutagenic was ZHE. The results demonstrate that the decontamination of compounds with genotoxic activity was most efficient in the contaminated soil first inoculated with the fungi to which subsequently EKOHUM was added.

The study showed that the bacterial systems for mutagenicity detection can be applied not only to immediate screening of the genotoxic risk of contaminated soils but that they also allow us to assess the course and character of long-term changes occurring in this environment. This might broaden the possibilities of their future use in ecological monitoring.

Acknowledgements
This work was supported by Czech Ministry for Education: CEZ: J09/98:173100002, the Grant of University of Ostrava IGS: 8812 and by the Grant Agency of the Czech Republic: project No. 526/00/1303.

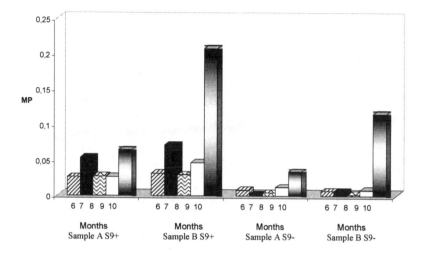

Figure 2. Results of SOS Chromotest and Ames test with metabolic activation *in vitro* (S9+) or without metabolic activation *in vitro* (S9-). Samples Z0, Z6, ZE, ZHE, ZEH, cf. Materials and Methods. The tested mutagen concentrations ranged from 5 to 150 μg extractable organic matter per mL.

References

1. Alexander, M. (1994) *Biodegradation and Bioremediation.*, San Diego, New York, Boston, London, Sydney, Tokyo, Toronto, Academic Press, pp.51-70.
2. Ames, B. J., M. C. Cann, J. and Yamasaki, R. (1975) Methods for detecting carcinogens and mutagens with the *Salmonella* / mammalian-microsome mutagenicity Test. *Mutation Res.*, **31**, 347-364.
3. Carmichael, L. M. and Pfaender, K. (1997) Polynuclear aromatic hydrocarbon metabolism of soils: relationship to soil characteristic and preexposure. *Environ. Tox. Chemistry*, **16**, **4**, 666-675.
4. Hofnung, M. and Quillardet, P. (1984) Use of the terms mutagenicity and genotoxicity. *Mutation Res.*, **132**, 141-142.
5. Holub, Z., Šimonovičová, A. and Banásová, V. (1993) The influence of acidification on some chemical and microbiological properties of soil, determining plant viability. *Biologia*, Bratislava, **48**, 6, 671-675.
6. Malachová, K. and Lednická, D. (1997) Mutagenic effect of the complex mixtures of substances contaminating the air in the city of Ostrava. *Ochrana ovzduší*, **6**, 5–7 (in Czech).
7. Maron, D. M. and Ames, B. N. 1983. Revised methods for the *Salmonella* mutagenicity test. *Mutation Res.*, **113**, 173-215.
8. Quillardet, P., DeBellecombe, C. and Hofnung, M. (1985) The SOS Chromotest, a colorimetric bacterial assays for genotoxins: validation study with 83 compounds. *Mutation Res.*, **147**, 79-95.
9. Wilson, S. C. and Jones, K. C. (1993) Bioremediation of soil contaminated with polynuclear aromatic hydrocarbons (PAHs): A review. *Environ. Pollution*. **81**, 229-249.

ECOTOXICOLOGICAL HAZARD ASSESSMENT OF SOLID-PHASE SAMPLES

L. PÕLLUMAA and A. KAHRU

National Institute of Chemical Physics and Biophysics,

Akadeemia 23, 12618 Tallinn, Estonia

1. Introduction

The legislative control and prediction of hazardous effects in soils and sediments are mainly based on chemical data on the level of selected hazardous key pollutants. This approach has some limitations: inability to account for the bioavailability of the contaminants and foresee the interactive effects of pollutants in complex matrixes. Also there is always the risk of poor selection of the contaminants to be measured. However, additional biological end ecological tests enable to compose a more objective picture of the environmental hazard. Due to the complexity of ecosystems and multifunctional character of toxicity *per se* which can be chemical, species- and end-point dependent variable, the assessment of polluted wastewaters and soils is recommended to be performed by using several organisms of different trophic levels [1,2]. However, the selection of a suitable test battery is not a trivial task. With the appearance of new generation of biotesting means, so-called microbiotests [3], the biological approach to assessing environmental pollution could become a powerful counterpart to the chemical one. Currently ecotoxicological testing (fish, daphnia and algal tests) is introduced in the EC legislation only for the analysis of new chemicals (EC Directive 67/548, 7th amendment 92/32).

2. Materials and Methods

Fourteen contaminated solid samples (soils, sediments and solid wastes; Table I) from Estonia were analyzed. Nine samples belonged to the industrial zone (I) and four to the living zone (L). 300 mL of bidistilled water was added to 100 g of not predried solid sample and shaken on a rotary shaker at 170 rpm in the dark at about 25 °C for 24 h. The resulting slurries were tested for toxicity in the Solid-phase Flash-assay. Aqueous extracts for the toxicity testing were obtained by centrifuging the slurries at 10 500 g for 10 min. The resulting extracts were defined as 100 %.

The concentration of heavy metals (Cd, Cu, Zn, Ni, Cr, Pb, As) was measured with atomic adsorption spectrometry and total oil products gravimetrically, both in the Laboratory of Geological Survey of Estonia; polyaromatic hydrocarbons (PAHs) were determined by HPLC in the Institute of Chemistry, Tallinn.

V. Šašek et al. (eds.),

The Utilization of Bioremediation to Reduce Soil Contamination: Problems and Solutions, 217–220.

Table I. Chemical and ecotoxicological characterization of samples

PLV for the industrial zone, mg/kg dry mass 5000 200

PLV for the living zone, mg/kg dry mass 500 20

Sample	pH	Tot oil, mg/kg dry mass	Tot PAH, mg/kg dry mass	Average toxicity, TU/dry mass	Nr of tests showing toxicity	value of the most sensitive test(s) in TU/dry mass	Chemical ranking	Ecotoxicological ranking
1 Waste-rock (I)	12.1	321	0	11.0*	7	T (18)	ok	VT
2 Aged semi-coke (I)	7.2	1044	39	0.4	2	R (1.2), P (1.3)	**	T
3 Fresh semi-coke (I)	12.4	527	17	10.4*	7	T&D (14)	**	VT
4 Oil-shale region soil (I)	10.3	7231	240	2.7*	6	D (7.5), SF (5.6)	***	T
5 Oil-shale region soil (L)	7.6	640	434	1.2*	4	FS (5.3), R (2.2)	***	T
6 Oil-shale region soil (I)	6.8	2334	0	0.2*	1	FS (1.2)	**	T
7 Semi-coke (L)	7.2	240	0	0.9*	3	FS (3.5)	ok	T
8 Oil-shale region control soil (L)	6.6	486	1	0.16	0	0	ok	NT
9 Control soil (L)	6.5	248	3	0.10	0	0	ok	NT
10 Oil-polluted soil (I)	7.3	2214	4	0.25	0	0	**	NT
11 Oil-polluted soil (I)	7	19294	509	3.7*	8	M (8.7), D (6.5)	***	T
12 Oil-polluted soil (I)	7.2	457	4	0.08	0	0	ok	NT
13 Oil-polluted soil (I)	7	745	1	0.10	0	0	**	NT
14 Sediment	6.8	1219	24	0.40	1	R (2.5)	**	T

I - industrial zone; L - living zone; * - adsorbed toxicity detected; T – Thamnotox; R – Rotox; P – Protox; D – Daphnia; SF – Solid-phase Flash-assay; M – Microtox; underlined values exceed respective living zone PLVs; values in bold exceed respective industrial zone PLVs; ok - <PLVs for the living zone; ** - >PLVs for the living zone; *** - >PLVs for the industrial zone; T - *toxic*; VT - *very toxic*; NT - *not toxic*

Microtox™ bacteria (*Vibrio fischeri* NRRL-B 11177; AZUR Environmental, Carlsbad, CA, USA) were used as test organisms in conventional photobacterial luminescence inhibition test (contact time 15 min) as described in [4] as well as in the Solid-phase Flash-assay [5] that makes it possible to detect particle-bound toxicants. The 72-h growth inhibition of the alga *Selenastrum capricornutum*, the 24-h and 48-h mortality of crustaceans *Thamnocephalus platyurus* and *Daphnia magna*, the 48-h reproduction inhibition of the rotifer *Brachionus calyciflorus*, and the 24-h growth inhibition of the protozoon *Tetrahymena thermophila* were performed according to the respective Manufacturer's Standard Operational Procedures (MicroBio Tests Inc., Belgium). Toxicity data obtained as 50 % effect end-point values (in % of dilution) were converted into toxic units (TU): TU = [1/L (E) C50] x 100. Toxicity of the samples presented in Table I was calculated on dry mass basis, i.e. TU/dry mass.

3. Results and Discussion

3.1 CHEMICAL CHARACTERIZATION OF THE SAMPLES

The chemical ranking of the samples was performed according to Estonian legislation [6] whereas the rank was defined by the results of three pollutant groups measured: heavy metals, oil products and PAHs. Samples were classified as *environmentally safe* (all concentrations below permitted limit values, PLVs, for the living zone), *not suitable for the living zone* (at least one pollutant >PLVs for living zone) and *not suitable for the industrial zone* (at least one pollutant >PLVs for the industrial zone) (Table I). **Five** samples out of fourteen were characterized as *environmentally safe*, **six** samples (2, 3, 6, 10, 13, 14) were *not suitable for the living zone* and **three** samples (4, 5, 11) were too polluted even for the industrial zone (Table I).

The main contaminants in all samples were **oil products** (Table I) exceeding PLV for the living zone (500 mg/kg) in nine samples. In two samples out of these nine (4 and 11) the oil level exceeded even the PLV for the industrial zone (5000 mg/kg). The concentration of **PAHs** in nine samples was low, not exceeding the PLV for the living zone (20 mg/kg) (Table I). However, in three samples (4, 5 and 11) even the PLV for the industrial zone (200 mg/kg) was exceeded.

Generally the concentration of **heavy metals** in the samples was low, not exceeding the PLV for the living zone. Only in the sediment from the Tallinn Old Harbor the concentration of copper (183 mg/kg dwt) slightly exceeded the PLV for the living zone (data not shown). Both **control soils** (8 and 9) were *environmentally safe*, as the concentrations of the measured key pollutants did not exceed the PLV for the living zone.

3.2 TOXICOLOGICAL CHARACTERIZATION OF THE SAMPLES

The samples were analyzed using the battery of biotests (altogether 8 tests): algae (A), *Thamnocephalus* (T), *Daphnia* (D), protozoa (P), rotifers (R), Microtox (M), Solid-phase Flash-assay (FS) & Flash Assay (FE). The sum of the results is shown in Table I. The toxicological ranking of the samples was performed according to [7] whereas the rank was defined by the results of the most sensitive test in the battery (predicting the weakest point in the food web). Samples were classified as follows: *nontoxic* (< 1 TU), *toxic* (1 - 10 TU) and *very toxic* (10-100 TU) (Table I).

Thus, **two** samples were found to be *very toxic* toward test organisms: samples 1 and 3 (both strongly alkaline solid wastes from the oil-shale region) remaining *very toxic* also after neutralization (data not shown). **Seven** samples were *toxic* and **five** (including control soils 8 and 9) were *nontoxic* (Table I). The Flash-Assay detected the particle-bound toxicity in seven samples that mostly originated from the oil-shale region (Table I).

The most toxic samples were oil-shale industry solid wastes (1 and 3), leachate-polluted soil (4) and soil from the former gas station (11) as almost the whole test battery detected the toxicity (TU>1) in these samples.

In most cases the toxicological results pointed in the same direction as the chemical analysis. There were two exceptions: both oil-shale industry solid wastes (1 and 3) that did not contain any of the measured pollutants over the PLVs but proved to be *very toxic*.

In order to choose tests for the optimal test battery, the sensitivity of the tests was evaluated according to the number of the *toxic* (TU>1) samples detected. The most sensitive tests were Solid-phase Flash-assay (photobacterium *Vibrio fischeri*), Rotoxkit chronic (rotifer *Brachionus calyciflorus*), Protoxkit (protozoon *Tetrahymena thermophila*) and Daphtoxkit (crustacean *Daphnia magna*) where, respectively, 50 %, 50 %, 43 % and 43 % of tested samples proved *toxic*. Microtox, Thamnocephalus and algal tests proved less sensitive (29 %, 21 % and 14 % of tested samples proved *toxic*, respectively). According to the representation of the trophic level, cost of the test and simplicity of the test protocol the following battery could be proposed: **Solid-phase Flash-assay** (bacterial acute test for the evaluation of particle-bound toxicity), **Protoxkit** test (protozoan chronic assay), **Daphnia test** (crustacean acute assay) and **Algaltoxkit** (microalgae, chronic test with primary producers). Microtox test could be recommended for the screening as it showed good correlation with crustacean and algal tests.

Acknowledgements

This work was supported by the Estonian Science Foundation grant number 3845 and RSS grant nr. 430/1999. We thank Dr. Marina Trapido for the chemical analysis of PAHs and Alla Maloveryan for skillful technical assistance.

References

1. Dutka, B.J., Jones, K., Kwan, K.K., Bailey, H. & McInnis, R. (1988) Use of microbial and toxicant screening tests for priority site selection of degraded areas in water bodies. *Water Res.* **22**, 503-510.
2. Blaise, C. (1998) Microbiotesting: An expanding field in aquatic toxicology. *Ecotoxicol. Environ. Saf.* **40**, 115-119.
3. Blaise, C. (1991) Microbiotests in aquatic ecotoxicology: Characteristics, utility and prospects. *Tox. Assess.* **6**, 145-155.
4. Kahru, A. (1993) In vitro toxicity testing using marine luminescent bacteria *Photobacterium phosphoreum*: The BiotoxTM test. *ATLA*, **21**, Nr 2, 210-215.
5. Lappalainen, J., Juvonen, R., Vaajasaari, K. & Karp, M. (1999) A new flash method for measuring the toxicity of solid and colored samples. *Chemosphere* **38**, 1069-1083.
6. Environmental Ministry of Estonia (1999) Maximum limits of hazardous substances in soil and ground water. Regulation No. 58 (of the Minister of Environment of 16 June 1999), RTL, **105**, 1319
7. Persoone, G., Goyvaerts, M., Janssen, C., De Coen, W. & Vangheluwe, M. (1993) Cost-effective acute hazard monitoring of polluted waters and waste dumps with the aid of Toxkits, Final Report. Commission of European Communities. Contract ACE 89/BE 2/D3, 600 p.

MONITORING OF POLYCHLORINATED BIPHENYLS IN SLOVAK FRESHWATER SEDIMENTS

USE OF SEMIPERMEABLE MEMBRANE DEVICES

R. TANDLICH[1], B. VRANA[2] and S. BALÁŽ[1]

1 Department of Pharmaceutical Sciences, College of Pharmacy, North Dakota State University, Fargo, ND 58105, USA; 2 Department of Chemical Toxicology, Centre for Environmental Research, Permoserstr. 15, 04301 Leipzig, Germany

Abstract

Semipermeable membrane devices (SPMDs) and low-density polyethylene membranes (LDPEs) were used to estimate the "bioavailable" fraction of polychlorinated biphenyls (PCBs) in freshwater sediments from the area of Strážske, Slovakia. In both techniques, the molecules of analytes must diffuse through the polyethylene membrane, thus particle-associated PCBs are excluded from sampling. The partitioning equilibrium was assessed in sediment slurries upon incubation with gentle shaking for 28 days. The influence on the sampling process of the organic carbon content in the sediments, turbation, and total PCB concentration in the sediment samples was studied.

1. Introduction

Semipermeable membrane devices (SPMDs) have been used for sampling of air and water environments [1,2]. Low-density polyethylene membranes (LDPEs) are hydrophobic in character and constitute an important part of SPMDs. The advantages of SPMD application in environmental sampling include low cost, time-integrated sampling, and easy operation. Polychlorinated biphenyls (PCBs) are hydrophobic organic pollutants that have been used for wide range of industrial applications in the past. Because of their high hydrophobicity and low water solubility, they have become a major threat to the higher stages of the food chains. Former or present sources of pollution are often located close to the rivers making their sediments become sink phases or secondary sources of PCB pollution. Recently it has been suggested that the total PCB concentrations do not constitute a proper toxicological risk assessment basis and the "bioavailable" concentration fractions should be use instead [3]. SPMDs and LDPEs seem ideal for this purpose and their applicability was tested on contaminated sediment samples from the area around a former PCB-producing plant.

V. Šašek et al. (eds.),
The Utilization of Bioremediation to Reduce Soil Contamination: Problems and Solutions, 221–226.
© 2003 Kluwer Academic Publishers. Printed in the Netherlands.

2. Materials and Methods

2.1. SEDIMENT SAMPLES

Sediment samples were collected near the town of Strážske (Eastern Slovakia) in 1997 [4]. Before experiments, the samples were air-dried, passed through a 2 mm sieve and stored in the dark at room temperature until used. The total organic carbon content was determined with high TOC analyzer (Elementar, Merck, Darmstadt, Germany) using sample combustion at 1050 °C.

2.2. CHEMICALS AND MATERIALS

The sources of chemicals and materials are given in parentheses: triolein (Sigma Chemical, St. Louis, MO, USA), hydrochloric acid, n-hexane, acetone, d10-anthracene, 2-propanol (Merck, Leipzig, Germany), PCB (USP 100, Promochem, 100 µg/mL), LDPE layflat tubing (diameter 2.5 cm, wall thickness 51 µm, Polymer-Synthese-Werk GmbH, Rheinberg, Germany).

2.3. SPMD AND LDPE PREPARATION

The polyethylene tubing was cut into 1-meter long segments, soaked in n-hexane overnight, rinsed with acetone and dried under a gentle stream of nitrogen. For SPMD preparation, the segments were partitioned into 10 cm cuts, which were heat sealed at one end. After loading with 0.1 g of triolein, the cuts were tightly rolled to expel any air from inside and spiked with 10 µL of a stock solution (100 µg/mL) of d_{10}-anthracene in n-hexane, as internal standard. Finally, SPMDs were heat-sealed at the other end and stored at 4 °C until use. For LDPE preparation, the triolein loading step was omitted. For extraction efficiency assessment, the SPMDs/LDPEs were spiked the second time with PCBs (USP 100, 100 µg/mL), placed in 50 ml Erlenmeyer flasks and dialyzed against 50 mL of *n*-hexane. Dialyzates were concentrated on a rotary evaporator under a gentle stream of nitrogen to 0.5 mL PCBs were quantified using GC/MS according to Kocan *et al.* [4]. Recovery rates of 95 to 100 % were reached after 24 h.

2.4. SPMD AND LDPE EXPOSURE TO SEDIMENT SUSPENSIONS

5 g (dry mass, 0.01 g accuracy) of the sediment was weighed into a 250-mL amberlite wide-mouth jars and 50 mL of distilled water (with 0.1 % NaN_3 prevent biodegradation) was added. One SPMD/LDPE spiked with d_{10}-anthracene (see above) was placed into each jar, the jars were placed on a rotary shaker (125 rpm) and incubated in the dark at (25 ± 2) °C for 28 days. After the end of the incubation period, individual SPMDs/LDPEs were removed from the jars and mechanically cleansed by brushing in distilled water. Successively, biofouled matter was removed by rinsing in 0.1 M hydrochloric acid for 30 s, the SPMDs/LDPEs were rinsed with acetone, 2-propanol, and dried under stream of nitrogen. PCBs were extracted upon dialysis against *n*-hexane (see SPMD and LDPE preparation) and quantified using GC/MS according to [4].

3. Results and Discussion

Results are summarized in Table I and selected data are presented in Fig. 1.

TABLE 1. SPMD/LDPE accumulated amounts of PCB congeners in ng (identified by the IUPAC number) versus their total sediment concentrations (ng/g, adapted from [4]).

IUPAC #	Substitution pattern	In	Sediment						
			S1	S2	S4	S6	S7	S18	S20
52	2, 2', 5, 5'	Sample	100	130	440	640000	1700	2100	2600
		SPMD	110	35	200	33500	56	ND*	12600
		LDPE	40	42	199	21900	116	1001	14400
101	2, 2', 4, 5, 5	Sample	70	90	406	290000	1000	1462	160000
		SPMD	60	23	91	10600	40	310	5690
		LDPE	23	25	370	8530	ND[1]	1200	7720
105	2, 3, 3', 4, 4'	Sample	19	22	81	21000	190	120	23000
		SPMD	ND[1]	ND[1]	62	950	ND[1]	170	1380
		LDPE	ND[1]	ND[1]	ND[1]	780	ND[1]	39	1136
118	2, 3, 4, 4', 5	Sample	27	53	210	75000	430	310	61000
		SPMD	ND[1]	ND[1]	162	3920	48	498	422
		LDPE	44	ND[1]	65	780	48	119	696
138	2, 2,' 3, 4, 4', 5	Sample	100	130	590	320000	1200	1300	170000
		SPMD	96	38	434	8060	89	1383	7859
		LDPE	31	31	ND[1]	5390	76	355	5858
180	2, 2,' 3, 4, 4', 5, 5'	Sample	75	84	440	250000	570	1212	120000
		SPMD	61	ND[1]	350	7610	168	234	2970
		LDPE	ND[1]	ND[1]	125	5810	72	830	3270

[1]ND – not determined

For all individual PCBs, the SPMD/LDPE accumulated amounts increase with increasing total concentrations in the sediment samples. In the case of 2,2',5,5'-tetrachlorobiphenyl (IUPAC 52) and 2,2',4,5,5'-pentachlorobiphenyl (IUPAC 101), SPMDs and LDPEs have comparable extraction efficiencies. There is little difference between the 2,2',5,5'-tetrachlorobiphenyl amounts accumulated from sediment samples S6 and S20. These results suggest comparable bioavailable fractions in these two sediments, or indicate that the saturation limits of the devices have been reached. There was a linear relationship between the SPMD-accumulated amounts and the total sediment concentration of 2,2',5,5'-tetrachlorobiphenyl and 2,2',4,5,5'-pentachloro-biphenyl (r^2=0.999, data not shown). This means that triolein increases the accumulation capacity of the sampling device for these two congeners. Data for 2,2',3,4,4',5-hexachlorobiphenyl (IUPAC 138) suggest that saturation limits of both sampling devices might have been reached for this congener above the total sediment concentration of 170 µg/g. Recently, the SPMD sampling rates in aquatic environments have been shown to decrease for higher-molar-mass PAHs, due to steric hindrances to their transport through the pores of LDPE [5]. If an analogous situation is encountered during sediment sampling, PCB congeners with higher molar mass diffuse so slowly through the pores of LDPE that, with its saturation limit reached, the presence of triolein does not influence the sampling process, i.e. SPMDs provide comparable

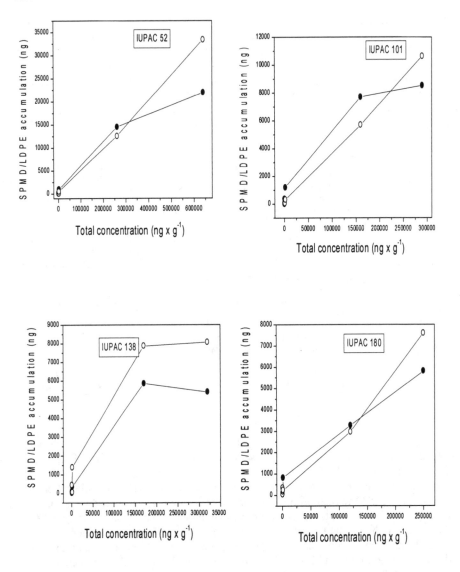

Figure 1. The amounts of specified PCB congeners accumulated in SPMDs (o) / LDPEs (•) after a 28 day incubation at (25 ± 2) °C in the dark are plotted as a function of the total concentrations in the sediment samples as detailed in Kocan *et al.* [4].

results with LDPEs. The accumulation patterns of 2,2',3,4,4',5,5'-heptachlorobiphenyl (IUPAC 180) should be analogous to those observed for 2,2',3,4,4',5-hexachloro-biphenyl. However, as can be seen from Table 1. and Fig. 1, rather the opposite is true.

Further experiments, with a broader spectrum of PCB congeners of similar molar mass and hydrophobicity, are currently under way to account for this discrepancy.

For individual sediments, good linear correlations have been obtained when fitting the total concentration as a function of the amount accumulated in the SPMDs/LDPEs for quantified congeners. The results are summarized in Table 2. Optimized values of adjustable parameters and the amounts accumulated in the sampling devices can be used to estimate the concentration of a particular PCB congener in the sediment sample in question. The slopes of the dependences seem to be directly proportional to the total organic carbon content of the sediment samples for both SPMDs and LDPEs. At the same time, the increasing level of organic carbon leads to an increased value of the intercept of the linear dependence, suggesting higher background interference.

TABLE 2. Linear dependences of the total sediment concentration on the amount of PCB congeners accumulated in SPMDs/LDPEs optimized values of adjustable parameters and statistical indices for linear regressions.

Sediment	TOC %	SPMD				LDPE			
		A	B	N	R	A	B	N	R
S1	1.92	0.71	28	5	0.99	1.36	47	4	0.92
S2	1.88	8.08	315	4	0.81	1.62	63	4	0.81
S4	1.73	0.82	127	6	0.89	1.90	145	5	0.93
S18	1.15	0.62	120	6	0.92	1.86	307	6	0.93
S20	6.60	18.29	41879	6	0.95	13.58	52127	6	0.87

A-regression line intercept, B-regression line slope, N-number of data points, R-value of the regression coefficient.

4. Conclusions

The results suggest that SPMDs/LDPEs can be used to assess PCB contamination over several orders of magnitude of total sediment concentration. Because the sampling devices accumulate contaminant molecules selectively, instrumental analysis of samples is possible without expensive cleanup procedures, which are unavoidable when analyzing sediment extracts using other procedures e.g., liquid-liquid extraction. The accumulated amounts of PCBs are in good mutual agreement between SPMDs and LDPEs for sediment samples but one. This fact proves that triolein is not probably essential for the sampling process, as suggested previously by others [6,7]. LDPEs and SPMDs might deliver different results in assessment of the bioavailable fraction of PCB at very high levels of contamination. On the other hand, leveling-off in accumulation was observed for both SPMDs and LDPEs, when assessing 2,2',3,4,4',5-hexachlorobiphenyl (IUPAC 138). This observation could be explained through the steric hindrances of the transport of 2,2,'3,4,4',5-hexachlorobiphenyl through the matrix of the polyethylene tubing, due to the pore size of the material. In general, the results suggest that LDPEs could well substitute for SPMDs in low-cost first-hand sampling or bioavailable fraction assessment of PCB contaminated habitats with low or moderate level of pollution.

226

Acknowledgements

Although the research described in this article has been funded in part by the United States Environmental Protection (grant number R82-6652-011), it has not been subjected to the Agency's required peer and policy review and therefore does not necessarily reflect the views of the Agency and no official endorsement should be inferred. The authors would like to thank Uwe Schröter and Petra Fiedler for instrumental measurements.

References

1. Litten S., Mead B., Hassett J. (1993) Application of passive samplers (PISCES) to loating a source of PCBs on the Black River, New York. Environmental Toxicology and Chemistry 12, 639-647.
2. Ockenden W., Prest H., Thomas G. O. et al. (1998) Passive air sampling of PCBs. Field calculations of atmospheric sampling rates by triolein-containing semipermeable mebrane devices. Environmental Science and Technology 32, 1538-1543.
3. Sijm D., Kraaij R., Belfroid A. (2000) Bioavailability in soil or sediment: exposure of different organisms and approaches to study it. Environmental Pollution 108, 113-119.
4. Kocan A., Petrik J., Drobna B. et al. (1998) AnonymousZatazenie zivotneho prostredia ludskej populacie v oblasti kontaminovanej polychlorovanymi bifenylmi [sprava za 2. rok riesenia projektu]. Bratislava, Ustav preventivnej a klinickej mediciny 2, 206.
5. Huckins J. N., Petty J. D., Orazio C. E., et al (1999) Determination of uptake kinetics (Sampling rates) by lipid-containing semipermeable membrane devices (SPMDs) for polycyclic aromatic hydrocarbons (PAHs) in water. Environmental Science and Technology 33, 3918-3923.
6. Booij K., Sleiderink H. M., Smedes F. (1998) Calibrating the uptake kinetics of semipermeable membrane devices using exposure standards. Environmental Toxicology and Chemistry 17, 1236-1245.
7. Shea D., Hofelt C. S., Luellen D. R. (1999) Use of low-density as a substitute for the standard semipermeable membrane device.; in Anonymous 9th Annual Meeting of SETAC-Europe. Leipzig, Germany.

PART IV

APPLICATION OF
BIOREMEDIATION TO
ENVIRONMENTAL PROBLEMS

BIOPILES FOR REMEDIATION OF PETROLEUM-CONTAMINATED SOILS: A POLISH CASE STUDY

Polish Refinery Biopile

T. C. Hazen[1], A. J. Tien[2], A. Worsztynowicz[3], D. J. Altman[2], K. Ulfig[3], and T. Manko[3]

[1]*Lawrence Berkeley National Laboratory, MS 70A-3317, One Cyclotron Rd., Berkeley, CA 94720 USA,* [2]*Westinghouse Savannah River Company, Bldg 704-8T, Aiken, SC 29808 USA,* [3]*Institute for Ecology of Industrial Areas, 6 Kossutha St., Katowice, POLAND*

Abstract

The US Department of Energy and the Institute for Ecology of Industrial Areas of Poland demonstrated bioremediation techniques for the cleanup of acidic petroleum sludge impacted soils at an oil refinery in southern Poland. The waste was composed of high-molar mass paraffinic and polynuclear aromatic hydrocarbons. Benzo(a)pyrene and BTEX compounds were identified as the contaminants of concern. Approximately 3 300 m³ of contaminated soil (TPH ~ 30 000 ppm) was targeted for treatment. A biopile design which employed a combination of passive and active aeration in conjunction with nutrient and surfactant application was used to increase the biodegradation of the contaminants of concern. Over the 20 month project, more than 81 % (120 metric tons) of petroleum hydrocarbons were biodegraded. Despite the fact that the material treated was highly weathered and very acidic, biodegradation rates of 121 mg per kg soil per day in the actively aerated side (82 mg per kg soil per day in the passive side) were achieved in this biopile. Microbial counts and dehydrogenase measurements gave the best correlation with the biodegradation rates. Costs were competitive or significantly lower when compared with other *ex situ* treatment processes.

1. Introduction

Biodegradation of petroleum hydrocarbons in soil (petroleum land farming) has been used by the oil industry for more than 30 years as an efficient way to destroy oil sludges [6]. By applying oil to the soil surface, adding fertilizer (P + N), water, and then tilling to aerate (oxygenate), the soil microbes have been shown to completely degrade large quantities of oil. Until recently, the state-of-the-art approach to soil remediation was excavation and disposal at a secure landfill. Changes in liability concerns, increasing costs, and regulatory constraints have decreased the popularity of excavation and landfill disposal as a soil cleanup alternative. Landfill disposal of contaminated soil does not remove the future liability of its generator, who will be held jointly liable with the landfill operator for any future associated contamination. Thus, on site permanent solutions are the preferred method of treatment, especially those that involve the

229

V. Šašek et al. (eds.),
The Utilization of Bioremediation to Reduce Soil Contamination: Problems and Solutions, 229–246.

complete destruction of the contaminant using biological (natural) techniques, i.e. bioremediation.

This bioremediation demonstration project focused on the cleanup technique known as "biopiles". The biopile process is very similar to active bioventing where air, as an oxygen source, and other amendments are forced through the vadose zone sediments either by vacuum extraction or by injection to stimulate the microbial oxidation of the hydrocarbons (for a complete set of definitions see Hazen [10]). As the name implies, biopiling is an *ex situ* process. The contaminated material is excavated and recombined or amended with other materials e.g., nutrients, sand, sawdust, wood chips, compost or other similar bulking agents, as needed, to improve permeability and moisture retention, and then placed in an engineered structure (configuration), to support and stimulate the biological reactions necessary to oxidize the hydrocarbons. Typically, this is a composting process which utilizes forced air via injection or vacuum extraction, moisture control, nutrient addition and environmental monitoring. Using commercially available vacuum pumps, or blowers, leachate pumps, moisture probes, thermocouple temperature probes, and real time soil gas monitoring equipment provides a mature and effective technology base for the operation and monitoring of the biopile.

The US Department of Energy and the Institute for Ecology of Industrial Areas (IETU), Katowice, Poland have been cooperating in the development and implementation of innovative environmental remediation technologies since 1995 (URL: www.iicer.fsu.edu/publications.cfm). A major focus of this program has been the demonstration of bioremediation techniques to clean up the soil and sediment associated with a waste lagoon at the Czechowice Oil Refinery (CZOR) in southern Poland. After an expedited site characterization (ESC), treatability study, and risk assessment study, a remediation system was designed that took advantage of local materials to minimize cost and maximize treatment efficiency. U.S. scientists and engineers worked in tandem with counterparts from the IETU and CZOR throughout this project to characterize, assess and subsequently, design, implement and monitor a bioremediation system. The CZOR was named by PIOS (State Environmental Protection Inspectorate of Poland) as one of the top 80 biggest polluters in Poland. The history of the CZOR dates back more than 100 years to its establishment by the Vacuum Oil Company (a U.S. company and forerunner of Standard Oil). More than a century of continuous use of a sulfuric acid-based oil refining method by the CZOR has produced an estimated 120,000 tons of acidic, highly weathered, petroleum sludge. This waste has been deposited into three open, unlined process waste lagoons, 3 meters deep, now covering 3.8 hectares (Fig. 1). Initial analysis indicated that the sludge was composed mainly of high-molar mass paraffinic and polynuclear aromatic hydrocarbons (PAHs). The overall objective of this full-scale demonstration project was to characterize, assess and remediate one of these lagoons. The remediation tested and evaluated a combination of U.S. and Polish-developed biological remediation technologies. Specifically, the goal of the demonstration was to reduce the environmental risk from PAH compounds in soil and to provide a green zone (grassy area) adjacent to the site boundary. The site was characterized using the DOE-developed Expedited Site Characterization (ESC) methodology. Based on the results of the ESC, a risk assessment was conducted using established U.S. procedures; a 0.3-hectare site, the smallest of the waste lagoons, was selected for a modified aerobic biopile demonstration.

Bioremediation is generally attempted by employing biostimulation, a process in which the conditions for microbial growth are optimized by supplying adequate amounts of electron acceptor(s), water, nutrients, in the form of nitrogen, phosphorus and trace elements, to the contaminated material [10]. Because biodegradation rates for petroleum hydrocarbons are fastest under aerobic conditions, maintaining adequate oxygen levels and moisture control are two of the main objectives associated with this project.

The material selected for the technology demonstration contains petroleum sludges, soils contaminated with crude and processed oil and other petroleum by-products and process waste from the refining of crude oil. The predominant contaminants of concern (COCs) are polycyclic aromatic hydrocarbons (PAHs) including benzo(a)pyrene, a known carcinogen. Also benzene, toluene, ethylbenzene and xylene known as BTEX and very recalcitrant high-molår mass molecules, the remnants and residue from tank bottoms of acid refining of crude oil. Although the high-molar mass molecules represent a portion of the total petroleum hydrocarbons (TPH) present in the waste material, they are of less concern to human health and the environment from a risk assessment standpoint, due to their highly insoluble state and lack of mobility within a soil matrix. The bioremediation processes will, however, reduce even the highly recalcitrant substances found in the waste material over time. The contaminated material presented several unique challenges to the remediation effort. The refinery and its associated lagoons are over one hundred years old, creating highly weathered conditions and material that would require special handling and preparation for the remediation process to be effective. The integrated bioremediation system (biopile), as designed, provided the stimulation needed to support the biological processes required to break down the recalcitrant hydrocarbon complexes to a more innocuous and stable material.

Wood chips (weathered) were selected as a bulking agent for the biopile because they provide the necessary porosity increase while utilizing an inexpensive waste product from a local lumber mill which otherwise would have to be disposed of. [Wood chips are normally sold as feed stock for pressed wood manufacturing. However, weathered (i.e. old and dirty) chips are not usable and must be disposed of separately.] During the construction of the biopile, the refinery took the initiative in utilizing grass clippings, leaf litter, and chipped waste lumber or wood originating from the refinery property, thus eliminating the need and associated costs for transporting wood chips from Kobior, located approximately 25 km from the refinery in Czechowice.

Dolomite was selected over other materials, e.g., gravel, as the leachate collection layer based on several factors including its ease of handling, relative by low cost and availability, pH amelioration, and a direct and inexpensive transportation route via train from the quarry to the refinery. Dolomite was also available in a variety of screen sizes which was incorporated into the process design to ensure effective air distribution throughout the system.

The final site use, proposed by the refinery, for the lagoons is a "green zone" to serve as a buffer and visual barrier between the refinery installations and the city of Czechowice-Dziedzice. The green zone will have limited access by trained refinery and IETU personnel for scientific and research purposes and for continued monitoring of the biopile processes. The area is not intended for recreational use by the general population or the refinery staff. No other regularly scheduled activities associated with the operations of the facility are planned for the site. The removal of the lagoon and the creation of the green zone has great public relations significance and greatly reduces the overall risk of the refinery for the city.

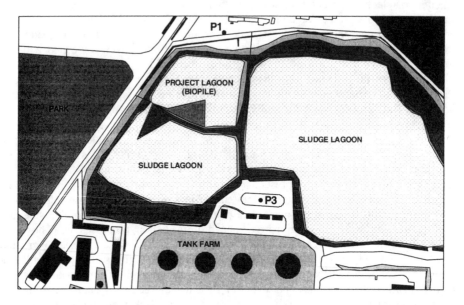

Figure 1. Refinery lagoon

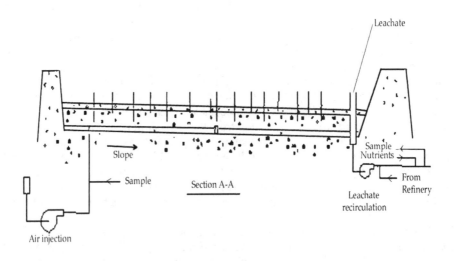

Figure 2. Cross section design drawing

1.1 BIOPILE DESIGN AND CONSTRUCTION

The biopile was constructed utilizing contaminated soil amended with wood chips and other vegetative materials. The pile was constructed in the existing excavated lagoon as seen in Figure 1. The empty lagoon bottom was sloped toward a sump pump, which was

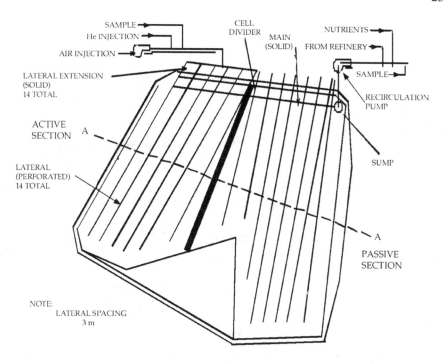

Figure 3. Plan view

connected to the leachate system. A leachate collection system consisting of perforated leachate collection piping was placed at the bottom of a dolomite base, approximately 30 cm deep. A cell divider (constructed of clay) was placed within the dolomite to create a separate active and passive section of the biopile. Fig. 2 provides a cross-sectional perspective.

The sump and its associated pump (Figs. 2 and 3) is used to recirculate any collected leachate to the top of the biopile. In addition, use of make-up water from the existing wastewater treatment facility at the refinery ensured that an adequate supply of moisture was available.

As described by the USEPA (EPA/540/R-95/534a), one driving force behind the development of bioventing was the difficulty of delivering oxygen *in situ* [22]. Many contaminants, especially petroleum hydrocarbons, are biodegradable in the presence of oxygen. Enhanced bioreclamation processes use water to carry oxygen or an alternative electron acceptor to the contaminated zone. This process was common, whether the contamination was present in the groundwater or in the unsaturated zone. Media for adding oxygen to contaminated areas have included pure-oxygen-sparged water, air-sparged water, hydrogen peroxide, and air. In all cases where water is used, the solubility of oxygen is the limiting factor effecting mass transfer. At standard conditions, a maximum of 8 to 10 mg/L of oxygen can be obtained in water when aerated. The stoichiometric equation 1 shown below is an example that can be used to calculate the quantity of water that must be delivered to provide sufficient oxygen for biodegradation.

$$C_6H_{14} + 9.5\ O_2 \Leftrightarrow 6CO_2 + 7H_2O \tag{1}$$

An example of the mass of water that must be delivered for hydrocarbon degradation to occur is shown below. Based on Eq. (1) the stoichiometric molar ratio of hydrocarbon to oxygen is 1:9.5 or, to degrade 1 mole of hydrocarbon, 9.5 moles of oxygen must be consumed. On a mass basis:

$$\frac{1\ \text{mole } C_6H_{14}}{9.5\ \text{moles } O_2} \times \frac{1\ \text{mole } O_2}{32\ \text{g } O_2} \times \frac{86\ \text{g } C_6H_{14}}{1\ \text{mole } C_6H_{14}} = \frac{86\ \text{g } C_6H_{14}}{304\ \text{g } O_2} = \frac{1\ \text{g } C_6H_{14}}{3.5\ \text{g } O_2}$$

Given an average concentration of 9 mg/L of oxygen dissolved in water, the amount of air-saturated water that must be delivered to degrade 1 g of hydrocarbon is calculated as follows:

$$\frac{3.5\ \text{g } O_2\ \text{required}}{\dfrac{9\ \text{mg } O_2}{1\ \text{L } H_2O} \times \dfrac{1\ \text{g}}{1,000\ \text{mg}}} = \frac{390\ \text{L } H_2O}{1\ \text{g } C_6H_{14}}$$

or, to degrade 1 lb:

$$\frac{390\ \text{L } H_2O}{1\ \text{g } C_6H_{14}} \times \frac{1,000\ \text{g}}{2.2\ \text{lb}} = 178,600\ \text{L } H_2O \Big/ 1\ \text{lb } C_6H_{14}$$

Based on the findings from the IETU treatability study, and an understanding of the mass transfer limitations of air-saturated water as an oxygen delivery system and the costs and safety concerns associated with pure oxygen generation, air injection was selected for the biopile electron-acceptor delivery system.

Previous field demonstrations at SRS have shown that direct air injection is an acceptable method of delivering oxygen to the subsurface microbiota. Additionally, a recent demonstration at a local municipal landfill (Columbia County, GA), has shown that air can be delivered via the leachate collection piping without adversely impacting the collection of leachate. This dual use of the leachate collection system was also applied to the construction of the biopile. A regenerative blower was obtained to provide the necessary airflow for the biopile. The Columbia County Landfill demonstration has shown that air injected into the perforated leachate collection piping is distributed to the entire cell via the leachate drainage layer (i.e. dolomite) (Fig. 2).

Approximately 1 meter of amended biopile material was placed above the leachate collection system. The composition of the mixture was approximately 80 % to 90 % contaminated material, and 10 % to 20 % wood chips as a bulking agent. Results from the column study for permeability conducted by IETU indicated that 10 % wood chips

Table 1. Oxygen requirements based on source

Oxygen form	Oxygen concentration in H_2O	Volume to degrade 1 lb hydrocarbon
Air-saturated H_2O	8 – 10 mg/L	180 000 L
Oxygen-saturated H_2O	40 – 50 mg/L	40 000 L
Hydrogen peroxide	up to 500 mg/L	6,100 L
Air	(21 % in air)	4,800 L

(V/V) to be adequate.

Immediately above the amended biopile material a 20 to 30 cm of cover of silty topsoil was used. The cover was planted with grass seed, which provided protection from erosion, support of a green zone and a biofilter for any VOCs that may reach the surface of the biopile. In addition, the refinery planted a mixture of deciduous conifer and beech trees and also some evergreen pines for landscaping purposes.

Atop the biopile, a trickle system for water application was installed to maintain soil moisture at 20–80 % of the biopile's field capacity. The target range is 30–40 %. The water supply came from the leachate collection sump with make-up water coming from the refinery's process water sources including treated wastewater.

2.   Experimental plan

2.1.   CRITERIA FOR SUCCESS

There are three primary criteria by which the overall success of this demonstration was evaluated:

1. Demonstrate the application of bioventing/biosparging as a viable cost-effective process to remediate contaminated sites to reduce risk to humans and environment, resulting in a green zone. The ability of the remediation process to degrade highmolar mass compounds (PAHs) will be evidenced by utilizing state-of-the-art monitoring equipment, analytical techniques and treatability studies to determine the rate and volume reduction in the starting concentrations of the contaminants.

2. Evidence of biological destruction (biodegradation) of petroleum (PAH, TPH and BTEX) from the contaminated material. Since a major advantage of bioremediation is destruction, it is important and significant to demonstrate that biodegradation is occurring. The evidence is expected to come primarily from comparison of the biopile material and soils analysis taken before, during and after the material is subjected to the treatment process (nutrient addition, aeration, pH adjustment and moisture control) to stimulate microbial activity and thus biodegradation of the contaminants.

3. Relatively simple and trouble-free operation. A critical assumption for the successful demonstration of the technology is that the system, as designed, will function with little or no down time and provide operating conditions that minimize fugitive air emissions and maximize biodegradation rates. The proposed project has no precedence in Poland and as such represents new technology for the country. However, since several other nations have demonstrated similar technologies, it represents a relative by low risk and should have high public acceptance. The simplistic design contributes direct benefits associated with the ease of management and operation. A minimal staff will be required to operate the equipment, again adding to the low risk factor by limiting exposure to operations personnel.

2.2.   PROCESS MONITORING

Monitoring of the system was accomplished in a variety of ways. Soil gas piezometers (Fig. 4) were installed in the biopile to monitor carbon dioxide (CO_2), oxygen (O_2), methane (CH_4), volatile organic hydrocarbons (VOCs) and semivolatile organic

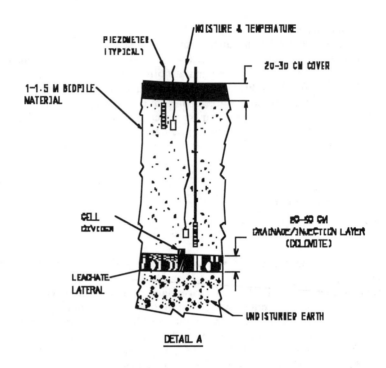

Figure 4. Monitoring point design

hydrocarbons (sVOCs). Water and soil samples were analyzed for polycyclic aromatic hydrocarbons (PAHs), nutrients, metals, and microbiological activity. Water quality of the leachate was tested for pH, dissolved oxygen (DO), specific conductance, temperature and other chemical and physical parameters like BOD and COD. The data gained from the monitoring program were used to calculate the biodegradation rates for PAH and total petroleum hydrocarbons (TPH). Respiration testing was conducted to monitor O_2 utilization during the remediation and helium tracer tests were performed to monitor the air flow, distribution and hydraulic conductivity with respect to the permeability of the contaminant mixture within the biopile during air injection. Moisture and temperature were measured by *in situ* moisture probes and thermocouples set next to each of the vadose zone piezometers.

2.2.1. Monitoring equipment

The Brüel & Kjær (B&K) Multi-gas Monitor, Type 1302, photoacoustic infrared spectrophotometer was used to monitor VOCs in the soil gas by direct analysis from the vadose zone piezometers in the field. A Landtec GEM-500 Gas Extraction Monitor was used to measure CH_4, CO_2 and O_2 concentrations in soil gas. Soil moisture was monitored with multiple gypsum moisture blocks installed throughout the biopile at approximately the same locations as the soil piezometers (Fig. 4). The soil moisture blocks consist of two concentric electrodes cast into gypsum blocks. When the blocks are placed within the biopile, the moisture content of the gypsum approaches equilibrium with the moisture content of the biopile material. The moisture blocks and moisture meter (Model KS-D1) utilized in this project were manufactured by Delmhorst

Instrument Company, Towaco, NJ. Soil temperature was monitored using J Type or K Type thermocouples placed within the biopile at approximately the same locations as the piezometers and moisture blocks. The readout device being used on this project was a Fluke Model 52 K/J Thermometer manufactured by the John Fluke Manufacturing Co. Inc., Everett, WA.

2.3. SAMPLING AND ANALYSIS

Soils were collected during the Expedited Site Characterization (ESC), the initial phase of the remediation demonstration and again for post demonstration characterization for analysis of VOCs, microbial counts, physical parameters, and miscellaneous parameters. Since the biopile contaminated soil was mixed with wood chips, the soil matrix should be much more homogeneous than the native soil, thus random soil sampling of the passive and active zones should provide a reasonable estimate of soil parameter changes during interim sampling intervals. With biopile systems, *in situ* respiration testing can also indicate when the site is clean and, therefore, when to collect final soil samples.

Soils for organic and inorganic analysis were collected using a hand auger, and placed in a Whirl-Pak bag or other clean container. Samples were placed in a cooler on ice and managed according to the hold times as seen in the Test Plan. Prior to sample analysis, samples were weighed to determine the mass of the sample. Core specimens for microbial analysis were obtained directly from the soil sampler. Cores were sectioned with sterile spatulas and the outermost layer scraped off using a sterile scoopula. The sample was then placed in a sterile Whirl-Pak bag and transported to the laboratory on ice for immediate analysis according to the test plan. Laboratory analyses were performed by personnel at the IETU laboratory.

Helium tracer tests and vadose zone respiration measurements were done using a Mark 4 Helium Detector Model 9821 helium detector using the Bioventing Respiration Test protocols of EPA (EPA/540/R-95/534a) [22].

A sampling port, pump or bailer was used to collect leachate samples from the leachate recirculation system. Water was filtered in the field, if required, with 0.45 μm pore filters. Field water parameters including dissolved oxygen, oxidation-reduction potential (ORP), pH, specific conductivity, and temperature, were monitored using a Hydrolab Surveyor.

Hydrocarbons were analyzed as follows: *EPA Method 8010*: 1,2-dichloroethane; *EPA Method 8020*: benzene, toluene, ethylbenzene, total xylenes total VOCs, *n*-propylbenzene; *EPA Method 5030* and *GC-FID California*: aromatics; *EPA Method 9071*: TCLP. Soluble reactive phosphate concentrations were measured by the ascorbic acid colorimetric determination method (EPA 365.2). Total phosphorus was determined by the persulfate digestion and ascorbic acid colorimetric determination (EPA 365.2). Total nitrogen was determined using a thermal-conductivity detector (TCD) which includes free-ammonia plus organic nitrogen, colorimetrically following digestion, distillation and nesslerization (EPA 351.3). Ammonia as distilled ammonia nitrogen was determined colorimetrically following distillation and nesslerization (EPA 350.2). Nitrate, nitrite, and sulfate were determined by ion chromatography (EPA 300.0). BOD and COD was determined by 5-day BOD test 5210 B APHA [3] and the 5220B Open Reflux COD method APHA [4]. For EPA methods see [22-26].

DAPI (4,6-diamino-2-phenylindole) was used to provide a direct estimate of the total number of bacteria in the environment, regardless of ability to grow on any media that might be used [12]. A comparison of acridine orange (AODC)-stained samples and DAPI stained samples of soil obtained earlier from the refinery site showed the DAPI gave superior results [21]. Calcofluor White (CFW) was used to provide a direct estimate of the total number of fungi in the environment. Naphthalene and crude oil enrichment was done with minimal salts media (MSM) [9]. The plates for naphthalene were incubated in an enclosed environment with naphthalene vapors available to the bacteria as a source of carbon for metabolism. For crude oil degraders MSM was placed in 96-well microtiter plates. Samples were diluted 10- fold across 8 wells in triplicate and a drop of crude oil placed in each well. After incubation, each well was scored for turbidity and oil emulsification to obtain a most probable number (MPN) density per gram dry mass or per mL Since the actively aerated parts of the biopile could reach high temperatures due to high rates of biodegradation (composting type conditions). Enrichments from the active aeration side were incubated both at 25 °C and at 45 °C. Oxidation of petroleum by microbes like other types of organic oxidation under aerobic conditions is linked to the electron transport system (ETS) of the cell. The enzymes of the ETS include a number of dehydrogenases, thus dehydrogenase activity can be used as an overall measure of activity in the soil. Triphenyltetrazolium chloride (TTC) is used as an artificial electron acceptor to estimate dehydrogenase activity since the reduction of TTC to triphenylformazan (TTF) causes a color change that can be quantified using a spectrophotometer. Soil samples were incubated with TTC (1.5 g/100 ml) for 24 h. The samples were then extracted with acetone and the extract measured at 546 nm using a spectrophotometer [1]. Values are presented as TPF μg/g dry mass.

For a complete description of all techniques and methods see Altman et al. [2]; and URL: www.iicer.fsu.edu/publications.cfm

3. Results

The project was divided into 5 operating campaigns: OC1: mobilization and air injection startup (9/25/97-1/27/98); OC2: air injection (2/1/98-4/15/98); OC3: air injection + fertilizer (4/16/98-6/30/98); OC4: air injection + fertilizer + leachate recirculation (7/1/98-9/29/98), and OC5. Air injection + fertilizer + leachate recirculation + surfactants (7/4/99-9/29/99). The surfactant used in OC5 was a Triton N-101 analogue (Rokafenol N-8).

The estimated removal of BTEX for the 20 month project was between 90 and 99.9 % with a reduction in TPH of 65–90 % and PAH removal of 50–75 %. These results are similar and comparable to a number of other sites reporting bioremediation of petroleum contaminated sediments. Sims [18] reported a 50–100 % reduction of fossil fuels in soil after only 22 days. St. John and Sikes [19] reported that a prepared bed system, complete with fugitive air emissions control, at a Texas oil field was able to reduce volatile organic carbon by >99 % after 94 days, with semivolatiles being reduced by more than 89 %. In California, Ross et al. [17] reported that four acres of soil 38 cm deep, contaminated with diesel and waste motor oils were decreased from 2800 ppm TPH to less than 380 ppm in only four weeks. He also reported that at another site owned by a heavy equipment manufacturer, 7,500 m³ were reduced to <100 ppm TPH after nine weeks and an additional 9000 m³ with 180 ppm TPH were reduced to <10 ppm after only five weeks. Another site in California had 600 m³ reduced from 1000 ppm TPH to <200 ppm in 35 days. Molnaa and Grubbs [15] report other sites in

California where similar results were obtained, e.g., 2000 m^3 with 2800 ppm TPH were reduced to less than 38 ppm in 74 days, a truck stop where 15 000 m^3 were reduced from 3000 ppm TPH to less than 30 ppm TPH in 62 days, and a site contaminated with lubricating oils where 25 000 m^3 were reduced from 4800 ppm down to 125 ppm in 58 days. Based on the initial concentrations in the biopile (>200 g/kg TPH in litter) reported by Ulfig et al. [20], the rates of removal were predicted to be between 10 to 80 mg/kg of soil per day for TPH and could exceed 120 mg/kg of soil per day, based on similar work by Reisinger et al. [16]. Reisinger experienced a 41 % removal of TPH over the first two quarters (180 days) of biopile operations, based on respiration test data. The overall removal rates were as high as 121 mg TPH/kg soil per day, with a total removal of 120 metric tons in 20 months or 81% of the TPH inventory in the biopile (Tables 2 and 3).

The spatial distribution of contaminants was different between the passive and active sides of the biopile. The active side of the biopile had higher contaminant concentrations initially than the passive side, but the differences between the shallow and deep areas of each side of the biopile were not significant. By the end of the demonstration there were only trace concentrations of TPH and PAHs left in either the passive or active parts of the biopile.

The carbon dioxide concentrations were higher in the passive parts of the biopile and the deep parts of the biopile (data not shown). Where air penetration was the poorest carbon dioxide concentration was the highest, e.g. passive and deep areas of the biopile. This coincides with more anaerobic conditions that would generate more carbon dioxide.

By plotting changes in TPH concentrations together with changes in microbial activity (TTC), the relationship between reduction in contaminant and microbial activity becomes more obvious (Fig. 5). The highest microbial activity areas for the end of OC2 both shallow and deep coincide nicely with the areas in the biopile that showed the greatest reduction in TPH concentration.

Table 2. TPH biodegradation rate by operating campaign and treatment

Campaign	Average	Passive	Active
OC-1	80	44	119
OC-2	88	82	94
OC-3	<33	33	0
OC-4	<37	0	37
OC-5	91	60	121

All values in mg/kg soil per day

Table 3. TPH inventory by operating campaign

	Baseline	OC1	OC2	OC3	OC4	OC5
Metric tons	148	100	68.6	68.6	66.8	28.1
% Remaining	100	68	46	46	45	19

240

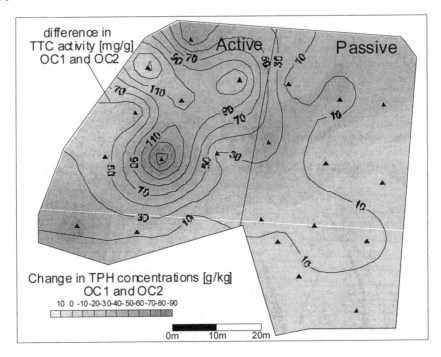

Figure 5. Change in TPH concentration vs. change in TTC activity in the soil of the biopile for OC1 and OC2.

The overall correlations for the soil parameters showed that the direct bacterial counts (DAPI) were significantly inversely correlated with total petroleum hydrocarbons (TPHTOT), polar total petroleum hydrocarbons (TPHPOL), and fluoranthene (FLORAN) (Table 4). This shows that, over the entire demonstration, as the bacterial numbers increased the contaminant concentrations decreased, suggesting a direct relationship. Further inspection of the matrix also shows that as DAPI numbers increases the fungal numbers (CFW) decrease, this suggesting that the microbial community shifted due to zhe inability of fungi to compete with the smaller and more metabolically active bacteria. The fungi also apparently played little if any role in the reduction in contaminant numbers since CFW was not significantly correlated with any of the contaminant parameters (TPHTOT, TPHPOL, FLORAN, BBFLUORA, BKFLUORA, BAPYRE, BOPERY, and I123CDY). DAPI numbers were also significantly correlated with enzymic activity (TTC), so that as the bacterial numbers increased the total dehydrogenase enzyme activity in the soil also increased. The dependence of bacterial density on adequate sources of phosphate was also indicated by the significant positive correlation between DAPI and PO_4. The petroleum degrader enrichments at 20 and 37 °C and the naphthalene degrader enrichments at 20 and 37 °C appeared to be indicating the density of contaminant-tolerant microbial populations in the soil rather than degraders. This is suggested by the significant direct correlations between these parameters and the contaminant parameters (TPHTOT and TPHPOL), thus the higher the concentration of the contaminant the higher the density of these types of microbes. Both enrichments at 37 °C had very few correlations to any of the other soil parameters, in part due to few measurements. The enrichment assays seem to be a poor index of biodegradation activity during the demonstration. Soil pH was also significantly inversely correlated with nearly all the contaminant parameters. This suggests that either the biostimulation process was increasing the pH as the contaminants were being degraded and/or that the biodegradation of the contaminants was causing the soil pH to increase. It is likely that to some extent both processes were occurring since it is well known that oxidation processes tend to increase pH when anaerobic acidic environments are driven aerobic. The validity of the soil data matrix is supported by highly significant direct correlations between all the contaminant parameters and between the contaminants and total Kjeldahl nitrogen (TKN), a normal expectation since the contaminants are normal components of petroleum and that petroleum is high in organic nitrogen.

A simple model that was dependent of non aqueous phase liquid (NAPL) partitioning was found to describe the bioremediation process in the biopile. The mass of contaminants removed during time (t) is given by the following equation:

$$m(t) = M/R^3(R^2-2Da\Delta ct/\gamma)^{3/2} \tag{2}$$

where::$m(t)$ is the mass of NAPL removed in time t, M is the mass of NAPL at t=0, R is the average radius of NAPL particles at t=0, D_a denotes the diffusivity of NAPL in water, c is the saturation concentration of NAPL in water, and γ represents the density of NAPL

Table 4. Correlation matrix of soil analysis parameters

	Temp	DAPI	CFW	NA20	NA37	PE20	PE37	TTC
Temp	1							
DAPI	**-0.652**	1						
CFW	-0.2025	-0.1249	1					
NA20	**-0.3153**	**-0.3254**	**0.3995**	1				
NA37	-0.001	-0.0344	-0.0322	-0.0692	1			
PE20	**-0.3462**	**-0.3222**	**0.398**	**0.9971**	-0.0597	1		
PE37	-0.2939	0.1105	-0.0282	-0.0638	0.0157	-0.0551	1	
TTC	**-0.4051**	**0.227**	-0.1388	**-0.2665**	0.0037	**-0.1306**	-0.057	1
pH	**0.3368**	0.0605	0.0159	-0.0961	0.0961	-0.0143	-0.0255	0.1017
MOIS	-0.2185	0.0941	-0.1241	-0.1513	**0.1592**	-0.1087	0.0078	**0.1801**
NO3	**0.2302**	-0.0361	0.0168	-0.0536	-0.0158	-0.0398	-0.0219	-0.0513
NO2	0.0728	**0.2026**	-0.1031	**-0.2451**	0.0286	**-0.2313**	-0.0496	0.0176
PO4	**-0.5656**	**0.1594**	0.0306	0.0289	-0.0426	0.085	**0.3533**	0.0272
NH4	**-0.3177**	-0.014	-0.1214	0.0918	-0.0583	-0.0937	-0.0403	0.0122
PTOT	**-0.775**	-0.0733	0.1182	-0.018	-0.1356	-0.0169	-0.066	-0.0633
TKN	**0.2623**	0.0943	-0.0921	-0.1252	**0.147**	**-0.1224**	0.055	0.0376
FLORAN	0.1409	**-0.1904**	-0.0431	**0.2171**	0.0254	**0.2172**	0.0011	0.013
BBFLUORA	**0.2162**	-0.1067	-0.0703	0.048	0.0472	0.0409	-0.0092	**0.1619**
BKFLUORA	0.1372	-0.0279	-0.0831	-0.0606	0.0654	-0.0589	0.1105	**0.1596**
BAPYRE	0.0854	-0.0425	-0.0744	0.0019	0.0361	0.0026	**-0.1535**	**0.1382**
BOPERY	**0.2512**	**-0.3095**	-0.0636	**0.1776**	**0.1451**	0.0277	-0.0025	0.0855
I123CDY	0.1632	**-0.141**	-0.0905	0.0814	0.1098	0.0848	0.1046	**0.2188**
TPHTOT	**-0.1943**	**-0.3574**	-0.0304	**0.2532**	-0.1079	**0.2267**	**-0.2095**	**0.1627**
TPHPOL	-0.1124	**-0.3187**	-0.053	**0.1932**	-0.0971	**0.1723**	**-0.2012**	**0.1816**

Temp=temperature, DAPI= direct counts, CFW= fugal element count, NA20 = naphthalene degraders at 20 °C, NA37 = naphthalene degraders at 37°C, PE20 = petroleum degraders at 20 °C, PE37 = petroleum degraders at 37 °C, TTC = dehydrogenase activity, MOIS = moisture content, NO_3= nitrate, NO_2 = nitrite, PO_4= phosphate, NH_4 = ammonia, PTOT = total phosphorus, TKN = total kjeldahl nitrogen, FLORAN = flouranthene, BBFLUORA = benzo(b)flouranthene, BKFLUORA = benzo(k)flouranthene, BAPYRE = benzo(a)pyrene, BOPERY = benzo(g,h,i)perylene, I123CDY = indeno(1,2,3-cd)pyrene, TPHTOT = total petroleum hydrocarbons, TPHPOL = total polar petroleum hydrodarbons. All bold figures are significant at P < 0.01.

It is assumed that the all of the TPH in the bioremediated soil inventory can be divided into three categories: (A) NAPL dispersed in the form of small droplets or aggregates throughout the whole biopile, (B) a fraction of TPH contained in macropores and/or weakly adsorbed to soil structure and thus readily available to microbial attack, and (C) a fraction of TPH which is strongly adsorbed to soil structures and prior to being degraded has to desorb and diffuse to the place where it is available to a microbial community. Values of parameters which characterize each individual fraction were based on our own observations and literature data. The following parameters were assumed as characteristic for the contaminated soil bioremediated in the biopile:

- NAPL (fraction A) content   ~ 40 % of total TPH inventory in soil
- Readily available fraction content   ~ 45 % of total TPH inventory in soil
- Adsorbed fraction content   ~ 15 % of total TPH inventory in soil
- Soil porosity   ~ 0.3
- Characteristics of NAPL fraction (Fraction A):
-     Average radius of aggregates (droplets) R 1.0 cm
-     Solubility in water c 10 mg/l before the surfactant was added; 100 mg/l after the surfactant was added
- Characteristics of readily available fraction (Fraction B):
-     Average radius of soil aggregate π 1.0 cm
-     Desorption coefficient K_d 100
-     Pore diffusivity of contaminant D_{eff} 5×10^{-11} cm^2/s
-     Liquid mass transfer coefficient k_1 1×10^{-5} cm/s
- Characteristics of sorbed fraction (Fraction C):
-     Average radius of soil aggregates r_0 30 m
-     Desorption coefficient K_d 1×10^5
-     Pore diffusivity of contaminant D_{eff} 5×10^{-13} cm^2/s
-     Liquid mass transfer coefficient k_1 1×10^{-5} cm/s
-

As can be seen from Fig. 6., the model results are in quite good agreement with experimental data form the biopile.

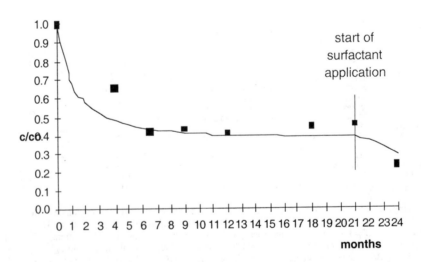

Figure 6. Comparison of model with actual TPH concentrations over time in the biopile.

4.   Cost analysis

The cost of using this technology to remediate the acidic and highly recalcitrant petroleum contaminated soil at the CZOR is estimated and compared with the actual costs for remediation of petroleum contaminated soil reported in the literature. A conservative approach was taken in this comparison by using the actual costs of this demonstration. Since this was a demonstration, the analytical costs, and the specialized equipment costs were much higher than they would be in an actual deployment. The calculations use the average hourly rate for a Polish worker and a US worker (DOE site contractor, fully loaded). Equipment and materials costs were converted from $US to Polish Z (4.00 Z/$1.00). All equipment costs are actual purchase prices. The demonstration remediated 120 metric tons of TPH in 5000 metric tons of soil. Thus 4167 cubic yards of contaminated soil were remediated to cleanup standards. The cost per cubic yard was $96.33/Z142.76 (Table 5). If the specialized equipment was removed, the cost was $86.98/Z105.37. If the fully optimized system had been used from the beginning the remediation time would have been cut in half and the cost per cubic yard would have been reduced to $62.21/Z91.69. If we compare these results with published costs for conventional treatment technologies, we see that even in the worst case scenario, this bioremediation application is better than all costs except the soil washing costs reported by Davis et al. [7]. Thus the biopile remediation costs are lower than incineration, landfilling, stabilization, and asphalting. Considering the reduced time scenario this application even beats the cost of soil washing (Table 5). If we compare this demonstration with other biopile and bioremediation techniques we can see that it falls within the same range but is higher than the prepared bed bioreactor and biopile deployments reported by Kastner et al. [10]. Given that the material being remediated was extremely acidic, more than 100 years old and had heavy metal contamination it also compares well. The costs reported by Kastner et al. [11] were for

Table 5. Cost analysis of the biopile project.

	Polish Z	US $
Total costs	594,882	401,405
Cost/cy remediated 5000 metric tons or 4167 cy	142.76	96.33
Cost/cy less special equipment	105.37	86.98
Bioremediation treatment costs	Cost/cy	Reference
Biopile costs at ISL <100 ppm	30.75	Kastner et al. [11]
Prepared Bed bioreactor <100 ppm	46.07	Kastner et al. [11]
Land treatment	10 -100	EPRI [8]
Biotreatment	40 -100	Levin and Gealt[13]
Bioremediation	39 -111	Molnaa and Grubbs [15]
Conventional treatment costs		
Incinerate (Tennessee) 3 units	309.00	Davis et al [7]
Incineration	>100	EPRI [8]
Incineration	250 - 800	Levin and Gealt [13]
Incineration (mobile) [15]	195 - 520	Molnaa and Grubbs
Soil Wash (Tennessee) 2 units	70.00	Davis et al [7]
Landfill	150 - 250	Levin and Gealt [13]
Landfill	>100	EPRI [8]
Landfill	176 - 202	Molnaa and Grubbs [15]
Asphalt	10 - 100	EPRI [8]
Stabilization/fixation	130 - 260	Molnaa and Grubbs [15]

fresh diesel-contaminated soil, and inherently biodegradable PCS. This demonstration showed that biopile type bioremediation for TPH and PAH contaminated soil can be a highly effective and cost effective solution for waste lagoon cleanup.

5. Discussion

The Polish petroleum refinery biopile field demonstration produced a number of significant findings. The first criterion of success was to demonstrate the application of bioventing/biosparging as a viable cost-effective process to remediate contaminated sites to reduce risk to man and environment and resulting in a green zone. Over the entire field demonstration more than 120 metric tons or 81 % of the total petroleum hydrocarbons that were present were destroyed (Table 3). By the end of the 20 month biopile demonstration, concentrations of TPH and all PAHs were below the Polish and US risk guidelines for even shallow soils (0.3 –15 m) for sites with multiple uses (including residential).This full-scale demonstration simultaneously remediated the contaminants present in the soil to acceptable risk levels and created a permanent green zone with a park-like atmosphere in less than 20 months. The comparison of passive with active aeration demonstrated that the Baroballs, for passive aeration, could be used effectively to provide aeration of the biopile via barometric pumping. This comparison showed that passive air injection required 3 –5 months longer to reach the same end point as blower injection of air. Passive injection could thus provide significant cost savings whenever there is no urgency for remediation due to immediate risk to human health or the environment. New field instruments that were used to monitor physical and chemical parameters in the field had variable success. The landfill gas analyzer proved extremely robust for measuring changes in carbon dioxide, oxygen, and methane in the soil gas. These measurements helped verify the respiration rates, the degree of injected air penetration into the biopile and a measure of aerobic conditions in the biopile. The installed temperature and moisture blocks were not sensitive enough to indicate significant changes in either parameter, thus were of minimal value. However, the moisture blocks did provide evidence that the biopile was drying out in the later part of operating campaign The photoacoustic infrared spectrophotometer proved difficult to operate and did not provide reliable data on soil gas concentrations in the later part of the demonstration. This was primarily due to humidity interference from water vapor becoming entrained in the instrument. The five operating campaigns showed that initially air injection alone (OC1 and OC2) stimulated dramatic reductions of contaminants (>50 %) in the biopile in less than 7 months (Table 3). Subsequent operating campaigns using the addition of fertilizer with the air (OC3) and leachate recirculation (OC4) removed only an additional 1% of the contaminant inventory in a similar period of time. However, the final operating campaign (OC5), which added surfactants, decreased the contaminant inventory by an additional 30 % and caused a significant reduction of all of the metals in the soil. These findings are verified by the biodegradation rates observed during the different operating campaigns (Table 2). The first two operating campaigns had high rates of biodegradation in the active injection areas and these fell off during OC3 and OC4, but were increased to their highest levels in any area during OC5. This suggests that surfactants make more of the strongly sorbed contaminants bioavailable. The passive section of the biopile responded in a similar manner but with about a 3–5-month lag and did not achieve the rates seen in the active aeration sections. The rates observed are similar for bioremediation of other petroleum-contaminated soils, and quite good for similar biopile studies, e.g., prepared beds (52–641 mg/kg soil per day), biopiles (20–60 mg TPH/kg soil per day), bioventing (2.5–10 mg TPH/kg soil per day)} [4, 11,14]. This demonstration suggests the combination of active aeration, fertilizer, and surfactants with leachate recirculation will provide the fastest site remediation and substituting passive aeration will reduce the cost but increase the time to reach the endpoint.

The second criterion of success was to demonstrate evidence of biological destruction (biodegradation) of petroleum (PAH, TPH and BTEX) from the contaminated material. Since a major advantage of bioremediation is destruction, it is important and significant to demonstrate that biodegradation is occurring. Multiple lines of evidence show that biodegradation of PAH and TPH occurred during the demonstration. BTEX compounds were undetectable in the soil after the first samples. First, the contaminant inventory changes showed that large concentrations of nonvolatile petroleum contaminants were being removed at rates that could only have been attributed to rapid biodegradation (Table 2). Second, respiration studies showed that the oxygen demand in the active area corresponded to the rates of contaminant reduction observed. Third, over the entire demonstration there was a highly significant inverse correlation between bacterial density and contaminant concentration. Fourth, limiting nutrients such as phosphate concentrations, were directly correlated to bacterial density, as were enzymic activity measurements of soil (TTC). This suggests that bacteria were being biostimulated by the nutrient amendments being applied during the operating campaigns. The treatability and column simulation studies also verified in a like manner that biodegradation was responsible for reduction of contaminants and that the nutrient and surfactant amendments being applied in the field caused similar responses in the laboratory. Fifth, the aeration of the biopile soil created an aerobic environment conducive to aerobic biodegradation of petroleum contaminants. Sixth, all PAHs measured were also being degraded. Seventh, very active, low-pH-tolerant petroleum and PAH degraders could be isolated from the biopile that used petroleum and PAHs as their sole carbon and energy source. And eighth, the modeling studies of the field data showed that the biodegradation of the contaminants observed during the demonstration could be simulated by a kinetic model of contaminant biodegradation.

The third criterion of success was to demonstrate a relatively simple and trouble-free operation. A critical assumption for the successful demonstration of the technology is that the system, as designed, will function with little or no downtime and provide operating conditions that minimize fugitive air emissions and maximize biodegradation rates. The equipment and operation went through a number of delays during the initial three months of operation, due to differences in voltage, planning, weather and manpower delays from a variety of sectors in this multi-institutional and multi-national demonstration. However, once the initial problems were solved, operation was relatively simple and maintenance free. The simplistic design served the project well and helped make the project the success it was.

References

1. Alef, K., and P. Nannipieri (1995) Methods in Applied Soil Microbiology and Biochemistry. Academic Press, New York. 576 pp.
2. Altman, D. J., Hazen, T. C., A. Tien, K. H. Lombard, and A. Worsztynowicz (1997) Czechowice Oil Refinery Bioremediation Demonstration Test Plan. WSRC-MS-97-214. Westinghouse Savannah River Company, Aiken, SC DOE - NITS.
3. APHA (1992) Standard Methods for the Examination of Water and Wastewater. 18th Ed. American Public Health Association. Washington, DC.

4. APHA (1995) Standard Methods for the Examination of Water and Wastewater. 19th Ed. American Public Health Association, Washington, D.C

5. Bartha, R. (1986) Biotechnology of petroleum pollutant biodegradation. *Microbial Ecology* 12, 155-172.

6. Bartha, R., and I. Bossert (1984) Treatment and Disposal of Petroleum Refinery Wastes. In: R. M. Atlas (ed.), Petroleum Microbiology. Macmillan Publishers, New York, NY.

7. Davis, K. L., G. D. Reed, and L. Walter (1995) A comparative cost analysis of petroleum remediation technologies. p. 73- 80. IN: Applied Bioremediation of Petroleum (ed.) R. E. Hinchee, J. A. Kittel, and H. J. Reisinger. Battelle Press, Columbus, Ohio.

8. EPRI (1998) Evaluation of In Situ Remedial Technologies for Sites Contaminated With Hydrocarbons Report TR-110230, EPRI, Palo Alto, CA

9. Fogel, M. M., A. R. Taddeo, and S. Fogel (1986) Biodegradation of chlorinated ethenes by a methane-utilizing mixed culture. *Appl. Environ. Microbiol.* 51, 720-724.

10. Hazen, T. C. (1997) Bioremediation. In: Microbiology of the Terrestrial Subsurface (eds) P. Amy and D. Haldeman, p 247-266. CRC Press, Boca Raton, Florida.

11. Kastner, J. R., K. H. Lombard, J. A. Radway, J. Santo Domingo, G. L. Burbage, and T. C. Hazen (1997) Characterization using laser induced fluorescence and nutrient injection via barometric pumping, p. 385-392. In Situ and On-Situ Bioremediation: Volume 1. Battelle Press.

12. Kepner, R. L., and J. R. Pratt (1994) Use of fluorochromes for direct enumeration of total bacteria in environmental samples: past and present. *Microbiol. Rev.* 58, 603-615.

13. Levin, M. A. and M. A. Gealt (1993) Biotreatment of Industrial and Hazardous Waste. McGraw Hill New York.

14. Lombard, K. H., and T. C. Hazen (1994) A Petroleum Contaminated Soil Bioremediation Facility. Proceedings American Nuclear Society Spectrum'94: 952-959.

15. Molnaa, B. A., and R. B. Grubbs (1989) Bioremediation of Petroleum Contaminated Soils Using a Microbial Consortia as Inoculum. In: E. J. Calabrese & P. T. Kostecki (eds.), Petroleum Contaminated Soils, vol. 2. Lewis Publishers, Chelsea, MI.

16. Reisinger, H. J., S. A. Mountain, G. Andreotti, G. DiLuise, A. Porto, A. Hullman, V. Owens, D. Arlotti, and J. Godfrey (1996) Bioremediation of a Major Inland Oil Spill Using a Comprehensive Integrated Approach. Proceedings: Third International Symposium and Exhibition on Environmental Contamination in Central and Eastern Europe.

17. Ross, D., T. P. Marziarz, and A. L. Bourquin (1988) Bioremediation of Hazardous Waste Sites in the USA: Case histories. In: Superfund '88, Proc. 9th National. Conference. Hazardous Materials Control Research Institute, Silver Spring, MD.

18. Sims, R. C. (1986) Loading Rates and Frequencies for Land Treatment Systems. In: R. C. Loehr and J. F. Malina, Jr., (eds.) Land Treatment: A Hazardous Waste Management Alternative. Water Resources Symposium No. 13, Center for Research in Water Resources, The University of Texas at Austin, Austin, TX.

19. St. John, W. D., and D. J. Sikes (1988) Complex Industrial Waste Sites. In: G. S. Omenn (ed.), Environmental biotechnology. Reducing Risks from Environmental Chemical through Biotechnology. Plenum Press, New York, NY.

20. Ulfig, K., G. Plaza, T. C. Hazen, C. B. Fliermans, M. M. Franck, and K. H. Lombard (1997) Bioremediation Treatability and Feasibility Studies at a Polish Petroleum Refinery. Proceedings: Third International Symposium and Exhibition on Environmental Contamination in Central and Eastern Europe. September, 1996.

21. Ulfig, K. (1997) Proposed Method for Evaluating Toxicity with Keratinolytic Fungi, Institute of Ecology For Environment for Industrial Areas Year-End Report. IETU, Katowice, Poland.

22. U.S. Environmental Protection Agency (USEPA) (1995) Manual: Bioventing Principles and Practice. Vol 1 and 2. EPA/540/R-95/534a. Office of Research and Development Washington, DC 20460.

23. U.S. Environmental Protection Agency (USEPA) (1985) Compilation of Air Pollutant Emission Factors. AP-42, Fourth Edition. Office of Air Quality, Research Triangle Park, NC 27711.

24. U.S. Environmental Protection Agency (USEPA) (1986) Test Methods for Evaluating Solid Wastes, Physical/Chemical Methods. SW-846, 3rd edition. Superintendent of Documents, U.S. Government Printing Office, Washington D.C. 20402.

25. U.S. Environmental Protection Agency (USEPA) (1987) Industrial Source Complex (ISC) User's Guide; Second Edition (Revised). EPA-450/2-78-027R. Office of Air Quality, Research Triangle Park, NC 27711.

26. U.S. Environmental Protection Agency (USEPA) (1988) A Workbook of Screening Techniques for Assessing Impacts of Toxic Air Pollutants. EPA-450/4-88-009. Office of Air Quality, Research Triangle Park, NC 27711.

WHY MYCOREMEDIATIONS HAVE NOT YET COME INTO PRACTICE

V. ŠAŠEK

Institute of Microbiology, Academy of Sciences of the Czech Republic, Videňská 1083, CZ-142 20 Prague 4, Czech Republic

Abstract

Application of fungal technology (mycoremediation) for the cleanup of polluted soils holds promise since 1985 when the white rot fungus *Phanerochaete chrysosporium* was found to be able to metabolize a number of important environmental pollutants. This ability is generally attributed to the lignin-degrading enzymic system of the fungus. A similar degrading capability was later described with other white-rot fungal species. Most of the experiments were performed using liquid culture media. In soil conditions, where besides the fungus-degrading capability also other factors affect the process, our knowledge is rather limited. Many of the factors are similar to those generally influencing any soil bioremediation process (properties of the environmental matrix, bioavailability, temperature and other physical parameters, pollutant toxicity). Optimum performance of white rot fungal mycelium introduced into soil depends especially on its survival, colonization of the soil matrix and relation to autochthonous soil microflora. The development of fungal technology for decontamination of polluted soil has also been retarded by limited basic research knowledge; most of the results were obtained with the single fungal species *Phanerochaete chrysosporium*. However, the data were often generalized for all other white-rot fungal species without considering their physiological and ecological diversity. The goal of the presentation is the evaluation of the above aspects influencing the development of mycoremediation.

1. Introduction

Bioremediation is often characterized as a cost-effective and environmentally friendly approach to removal of contaminants from the environment. As cleanup tools, various types of organisms are employed. The bioremediation method has been established to exploit mostly bacterial microorganisms [1] and its limits have been repeatedly reviewed [2,3]. The most simple (and sometimes efficient) way of reduction of xenobiotics in polluted soils is the enhancement of growth and metabolic activity of

247

V. Šašek et al. (eds.),
The Utilization of Bioremediation to Reduce Soil Contamination: Problems and Solutions, 247–266.
© 2003 *Kluwer Academic Publishers. Printed in the Netherlands.*

autochthonous microflora by addition of mineral and/or organic nutrients, and supplying oxygen to ensure aerobic processes. A promising cleanup tool is represented by plant-based technology – phytoremediation – which exploits both plant sorption and degradative capacity and biodegradation potential of rhizosphere microorganisms. A technology that employs fungi as remediative agents is mycoremediation.

A group of lignin-degrading fungi belonging to the class Basidiomycetes have received considerable attention for their bioremediation potential. In nature these fungi colonize lignocellulosic materials, playing a role of litter decomposers, or they inhabit living or dead wood causing white rot of wood. Lignin is a complex three-dimensional polymer and ligninolytic fungi possess a very nonspecific mechanism for its degradation. Initial attack of lignin macromolecule is caused by an extracellular degradative system that consists of heme peroxidases, lignin peroxidase (LiP) and manganese-dependent peroxidase (MnP), as well as an H_2O_2 generative system. Phenol oxidases such as laccase, which is present in many ligninolytic fungi, are also believed to particiate in the degradation of lignin. The same nonspecific mechanism that white-rot fungi use to degrade lignin, also allows them to degrade a wide range of pollutants.

Originally, the ability of ligninolytic fungi to degrade numerous recalcitrant aromatic organopollutans was attributed to ligninolytic system and many of those aromatics were degraded by using isolated ligninolytic enzymes [4,5] (see also the list summarized by Field et al. [6], and references therein). However, several papers have been recently published [7–12], that put in doubt a direct correlation between the activity of ligninolytic enzymes and metabolism of xenobiotic aromatics and thus support conclusions of Field et al. [6] that "other, unidentified enzymes are implicated". This view has recently been supported also by Bogan et al. [13] who found that fluorene oxidation in vitro was a consequence of lipid peroxidation mediated by the P. chrysosporium Mn-peroxidase. Contrary to polycyclic aromatic hydrocarbons (PAHs) and chlorophenols, where the involvement of ligninolytic enzymes in their biodegradation was elucidated and demonstrated in vitro using isolated LiP, MnP and laccase, the role of ligninolytic enzymes in degradation of polychlorinated biphenyls (PCBs) has been unclear. On the basis of observations that different fungal species transformed and mineralized PCBs to different extent, Beaudette et al. [14] concluded that „different white rot fungi had enzymes of different specificity or different mechanism for degradation of PCB congeners". These authors support the conclusion of Thomas et. al. [8] that other enzymes besides LiP and MnP, enzymes related to lignin degradation, such as laccase or aryl alcohol oxidase, take part in the oxidation of PCBs.

Despite the fact that the role of ligninolytic enzymes in the destruction of aromatic pollutants has not been elucidated completely, the ligninolytic fungi undoubtedly represent a powerful prospective tools in soil bioremediation. This was well demonstrated by Lamar and coworkers [15,16] using inoculum of P. chrysosporium and P. sordida in field trials with soil polluted with pentachlorophenol and PAHs. Pollutant removal from the soil reaching up to 90 % within a few months seems very promising. The process using these fungi has been patented [17,18] and a few companies included the use of ligninolytic fungi for soil remediation into their program (EarthFax Development Corp., Utah, USA [19] or Gebruder Huber Bodenrecycling in Germany [20] but a broader application has not yet been put to effect.

2. Biodegradation capacity of lignin-degrading fungi

2.1. *PHANEROCHAETE CHRYSOSPORIUM*

During mid-1980s the first demonstration of organopollutant degradation by white-rot fungi was published [21,22]. The fungus was *Phanerochaete chrysosporium* Burds. The reason was that this fungus was intensively studied as a model microorganism in the research on the mechanism of lignin degradation. Since optimum conditions for the activity of ligninolytic enzymes were known, and a well-defined nitrogen-limited medium [23] was described, the fungus has been the subject of extensive investigation with a view to use the ligninolytic system in the degradation of xenobiotics. Many structurally diverse environmental pollutants were found to be degraded by this fungus. The fungus in its telemorphic form was described only in 1974 [24]. At the beginning, the biodegradation experiments were performed using N-limited liquid medium but later on also the degradation under non-ligninolytic conditions was described [25].

2.1.1. Degradation of chlorophenols

Of the chlorophenols, pentachlorophenol (PCP) degradation was investigated most often because of its former extensive use as a wood preservative, fungicide and insecticide, which led to the contamination of soil and aquatic ecosystems worldwide. Under laboratory conditions, $[^{14}C]PCP$ reduction of 97 % and 20–50 % mineralization was observed in a 30-d experiment [26]. Using different types of bioreactors, extensive removal of PCP from liquid media was demonstrated [27,28]. Efficient removal of PCP using ammonium lignosulfonates, a waste product of paper-mill industry, was also demonstrated [29]. Armenante *et al.* [30], following the degradation of 2,4,6-TCP, found that both extra- and intracellular enzymes were involved in the degradation. Gold and coworkers [31,32] demonstrated that fungal peroxidases are directly involved in chlorophenol degradation. These biodegradations evidently result from a simultaneous operation of several extra- and intracellular processes, in which both LIP and MnP are involved [33].

2.1.2. Degradation of polychlorinated biphenyls and dioxins

Polychlorinated biphenyls (PCBs) used extensively in industrial application until the mid-1970s are extremely nonreactive and heat stable chemicals. Mineralization of PCBs by *P. chrysosporium* was first reported by Eaton [22]. Thomas and coworkers [8] demonstrated mineralization of 2-chlorobiphenyl and 2,2′,4,4′-tetrachlorobiphenyl to $^{14}CO_2$. After 32-day incubation only 1% of tetrachlorobiphenyl was mineralized but 40 % was found to be incorporated into fungal biomass. Also Dietrich and coworkers [34] found that low-chlorinated PCB congeners are degraded significanlty, while higher chlorination is a limiting factor. Evidence for substantial degradation of Arochlors 1242, 1254, and 1260 was documented by Yadav and coworkers [11] who, contrary to a former observation with N-limited media, found degradation also in a high-N medium, and most extensive in malt-extract medium. Using congener-specific analysis, they documented that the degree of degradation is affected by the number of chlorine atoms but not by their position on the biphenyl ring. Based on the congener nonspecificity and the fact that the degradation takes place also under non-ligninolytic conditions, these authors put in doubt the role of ligninolytic enzymes, and supposed that the degradation might be due to a free radical mechanism of attack. These results were confirmed by Krčmář and Ulrich [35] who described degradation of a PCB mixture under non-ligninolytic conditions but not under ligninolytic ones. In our laboratory [36] we observed *in vitro* degradation of PCB mixtures containing low- and high-chlorinated biphenyls using intact mycelium, crude extracellular liquid and enriched MnP and LIP. A decrease in PCB concentration during a 44-h treatment with mycelium (74 %) or crude extracellular liquid (60 %) was observed. In contrast, MnP and LIP isolated from the extracellular liquid did not catalyze any degradation. In conclusion, the ability of *P. chrysosporium* to degrade PCBs was proven but the enzymic background and mechanism of the attack are still to be explained.

Polychlorinated dibenzo-*p*-dioxins (PCDDs) are recognized as environmental hazards due to their acute toxicity to animals and humans. PCCDs have been identified in paper-pulp mill effluents and in ash generated by a variety of combustion processes. These compounds are degraded very slowly by soil microflora which increases their environmental impact. The capability of *P. chrysosporium* to degrade and also to mineralize PCDDs was demonstrated in 1985 [21]. Later on, the metabolism of a model compound, 2,7-dichlorodibenzo-*p*-dioxin, was elucidated, the oxidation products were characterized, and the involvement of fungal ligninolytic system was documented [37].

2.1.3. Degradation of aromatic hydrocarbons

Because of their toxicity and wide-spread occurrence, polycyclic aromatic hydrocarbons (PAHs) represent one of the most important groups of environmental pollutants. Soon after the initial observation that white-rot fungi possess the ability to oxidize PAHs, the involvement of the extracellular ligninolytic enzyme system was indicated. Direct oxidation of PAHs was first observed for LIP [38, 39] and later on for MnP [40,41,42] and laccase [43,44]. Ligninolytic enzymes catalyze one-electron oxidation to produce PAH quinones that can be further metabolized via ring-fission [45]. Some PAHs with up to six rings were shown to be degradable via MnP-dependent lipid peroxidation reactions both *in vitro* and *in vivo* [39,46]. Intracellular cytochrome P450 monooxidase activity followed by epoxide hydrolase-catalyzed hydration resulting in hydroxylation of 3-, 4-, and 5-ringed PAHs are believed to initially metabolize PAH molecules including phenanthrene having an IP of 8.03 eV [47-52]. Contrary to PCBs, the metabolism of PAHs by both white-rot fungi and other filamentous fungi and yeasts is relatively well understood (see review by Cerniglia and coworkers [53] and references therein). *P. chrysosporium* was demonstrated to degrade a wide array of individual PAHs as well as the PAHs present in anthracene oil [54-57]. The fungus also degrade simple aromatic hydrocarbons – benzene, toluene, ethylbenzene, and xylenes (BTEX), a group of pollutants derived from gasoline and aviation fuel [58]. Degradation of these compounds occurred under non-ligninolytic conditions. Noninvolvement of ligninolytic enzymes wasobserved in degradation of chlorobenzenes [59].

2.1.4. Decolorization of dyes

Synthetic dyes are extensively used for industrial dyeing and printing. A significant proportion appears in the form of wastewaters and is spilled into the environment.

Synthetic azo dyes are either toxic or can be modified biologically to toxic or carcinogenic compounds. Polymeric dyes like Poly R-478, Poly B-411 or Poly Y-606, which are not taken up by cells, serve as suitable substrates for the detection of some enzymic components of the fungal ligninolytic system. Degradation of those dyes correlates with the start of lignin metabolism in white rot fungi and probably reflects a combined effect of peroxidases and H_2O_2-producing oxidases [60-63]. Therefore, the dye-decolorizing test has been used as a possible, practicable and inexpensive alternative to radiolabeled lignin as a substrate in lignin biodegradation studies. According to Field and coworkers [64] decolorization of polymeric dyes has proved to be a good indicator of the initial transformation of xenobiotics mediated by peroxidase activity of fungi.

Since synthetic dyes themselves are environmental pollutants and *P. chrysosporium* was the organism used in the above studies, naturally, the screening on decolorization of dyes of different chemical structures started with this fungus. It was demonstrated that the fungus degraded triphenylmethane, anthraquinone, heterocyclic, metal-complex, polymeric, azo, and diazo dyes [60,65-73] and the role of MnP, LiP and the presence of veratryl alcohol in the degradation was described (see [74-76] and reference therein). *P. chrysosporium* is also able to decolorize humic acids [77], bleach plant effluent from paper mills [78-81], decompose lignosulfonates [82], degrade colorants in refinery effluents [83] and decolorize many other complex colored organic effluents, such as olive mill wastewaters [84-85].

2.1.5. Degradation of explosives and other nitroaromatics

The contamination of soil and water with residues of explosives and related compounds is a widespread environmental problem. 2,4,6-Trinitrotoluene (TNT) is acutely toxic to humans and animals and is mutagenic in the Ames test.

A number of studies show degradation or mineralization of TNT by *P. chrysosporium* using liquid cultures [86-90]. The initial intermediate products of TNT biodegradation (2-amino-4,6-dinitrotoluene and 4-amino-2,6-dinitroluene) were also metabolized by the fungus, though the rates of degradation were slow. This is important because many microorganisms (other than ligninolytic fungi) are able to mediate only the initial reduction of TNT but the mentioned intermediates are not further degraded or undergo dimerization to azo and azoxy dimers, which are also environmentally persistent. *P. chrysosporium* was reported to be able to remediate water contaminated with TNT (120-175 mg/L) and cyclotrimethylene trinitroamine (RDX) (25 mg/L) with an efficiency of over 90 % [91].

Also the metabolism of nitroglycerin by *P. chrysosporium* was demonstrated and the enzymic mechanism involved in this degradation was elucidated [92]. Another nitroaromatic used in the production of polyurethane and explosives, 2,4 dinitrotoluene, was found to be metabolized by *P. chrysosporium*: its multistep degradation pathway was elucidated and involvement of MnP and LIP was documented [93].

2.1.6. Degradation of pesticides

Many types of pesticides of different chemical nature are produced in large quantities and have been widely used in the past as well as at present. Some of them persist in the environment; their residues are often detected in food, living organisms, soils and surface waters. *P. chrysosporium* was found to degrade herbicides based on 2,4-dichlorophenoxyacetic acid and 2,4,5-trichlorophenoxyacetic acid [94,95], alkyl halide insecticides (chlordane, heptachlor, lindane, dieldrin, mirex) [96], and the herbicide diuron [97]. DDT [1,1,1-trichloro-2,2-bis-(4chlorophenyl)ethane] was subjected to a more detailed study, since this compound was among the first persistent organopollutants demonstrated to be degraded by *P. chrysosporium* [21]. It is one of the most persistent insecticides in the environment. In early studies, the rate of disappearance of DDT in liquid cultures was 50 % within 30 days [21] and mineralization was around 13.5 % for the same time period [98]. However, the degradation of DDT was incomplete even under optimum laboratory conditions and thus it was not possible to move this technology into field conditions.

2.2. OTHER LIGNINOLYTIC FUNGI

White-rot fungi, which are predominantly basidiomycetes, are very abundant in nature and include more than 2000 species [99] which represents approximately 90 % of the total of wood-rot fungi [100]. Thus it is not surprising that, encouraged by the impressive biodegrading capability of *P. chrysosporium,* a screening for other basidiomycete fungi that could compete with this fungus started in several laboratories.

2.2.1. Screening
A complex approach was applied by Field and coworkers at the Wageningen Agriculture University, the Netherlands, using basidiomycete cultures from fungal collections, their own isolates from rotten wood, litter or soil from forest habitats. After screening for ligninolytic activity using decolorization of Poly R-478, they tested preselected strains for anthracene degradation. Out of the tests the fungus *Bjerkandera* sp. strain BOS55 resulted that became their model organism for further research [101]. Testing five fungal species for mineralization of 3,4-dichloroaniline, dieldrin and phenanthrene. Morgan and coworkers [102] found *Trametes versicolor* to be a powerful degrader. At the Jena University, Germany, Fritsche and coworkers, by screening as to fungal strains both for ligninolytic activity and PAH-degradation ability selected several new basidiomycetous species – *Agrocybe aegerita, Stropharia rugosoannulata, Kuehneromyces mutabilis, Trametes versicolor, Pleurotus ostretus, Nematoloma frowardii,* as possible degraders, and the last mentioned species was studied most extensively [103-106].
Zadrazil in Braunschweig, Germany screened 49 strains of wood-rotting fungi using labeled PAHs in soil conditions which enabled simultaneous estimation of the ability to colonize soil [107]. Contrary to expectation, more intense mineralization of PAHs was detected in the controls and with fungi that were not able to colonize the soil. A collection of 100 fungal strains isolated from soil (84 belonging to Basidiomycetes) was tested for their ability to degrade three phenylurea-based herbicides [108]. The best results were obtained with *B. adusta* and *Oxysporus* sp. Colombo and coworkers [109] compared biodegradation of hydrocarbons in soil using natural soil microflora, imperfect fungi isolated from oil-polluted areas, and three basidiomycetous ligninolytic species. Basidiomycetes degraded both aliphatic and aromatic hydrocarbons but some imperfect fungi isolated from polluted soil showed a higher degradation potential.

Capability of four ligninolytic species to degrade TNT in liquid cultures was tested [110]. *Phanerochaete sordida* and *Cyathus stercoreus* showed a faster rate of TNT degradation in comparison with *P. chrysosporium*. Combined test (effect of PCP dose on the growth and biodegrading ability) with six white-rot fungi demonstrated that not the concentration in the medium but the dose of PCP per mycelium biomass regulated the growth and biodegradation [111].

In our laboratory [112] the preselection of ligninolytic basidiomycetes isolated in the Czech Republic was based on the ability to decolorize model synthetic dyes (RBBR and Poly R-478). Besides the strains that have already been used in biodegradation experiments (*P. ostreatus, T. versicolor, B. adusta*), other robust decolorizers (*Irpex lacteus, Mycoacia* sp., *Tyromyces chioneus, Cerioporia metamorphosa, Daedalopsis confragosa*) were detected as prospective for biodegradation studies. Similarly Song [113] tested the fungal degraders isolated from soil in Korea. Six strains were evaluated for the ability to mineralize labeled pyrene in liquid complex medium. The highest mineralization was found with *Irpex lacteus* and *P. chrysosporium*. A set of 30 fungal strains of different taxonomic groups was tested for sensitivity and ability to degrade fluorene [114]. Surprisingly, three representatives of Zygomycetes showed also the highest resistance and biodegradation capability. White-rot fungi were also screened for degradation of other compounds, like styrene in the gaseous phase [115] or the synthetic polymer Nylon [116]. *P. chrysosporium, P. ostreatus, T. versicolor* and *B. adusta* were found to be potent styrene degraders. The addition of lignocellulosic materials to the growth medium increased the fungus-degrading potential; styrene was almost completely degraded during 48 h. In fact, the induction of degrading capacity of white-rot fungi using lignocellulosic material had already been applied to a biofilter based on a lignocellulosic substrate colonized by a white-rot fungus and used for filtration of polluted air [117].

By screening, several species of ligninolytic basidiomycetes were selected and studied in detail both in liquid culture media and in soil.

2.2.2. Other Phanerochaete *species*

Several *Phanerochaete* species (*P. chrysorhiza, P. laevis, P. sanguinea, P. filamentosa, P. sordida*) were included in a screening test for sensitivity to PCP and ability to degrade this pollutant in soil [118] and the last-mentioned species, due to its resistance to PCP, was successfully used in soil remediation trials [119] (for details see below). *P. sordida* was also demonstrated as an efficient degrader of TNT of polychlorinated dibenzo-*p*-dioxins and ofpolychlorinated dibenzofurans [110,120]. *P. laevis* was demonstated to degrade PAHs [121] and *P. flavido-alba* was successfully applied to decolorization paper-mill and olive oil-wastewaters [122,123].

2.2.3. Trametes (Coriolus) versicolor (L.:Fr.) Pilát

This fungus has been used in bioremediation research because of its strong extracellular laccase production, high resistance to PCP and the fact that, unlike *P. chrysosporium* that metabolizes PCP to chloroanisoles, *T. versicolor* produces only insignificant amounts of this metabolite [124]. Besides, it was shown that the organochlorine fraction of the paper-bleaching process was selectively attacked by *T. versicolor*, its laccase rapidly dechlorinated a large number of polychlorinated phenols and guaiacols [125,126]. Recently, oxidation of PAHs and aromatic alcohols by laccase of *T. versicolor* in laccase-mediator systems was demonstrated [43,127-129]. The degradation capacity of the fungus and its laccase was extensively studied with chlorophenols [129-131], synthetic dyes [132,133], and only scarcely was the fungus studied under soil conditions [134] (for details see below) because of it indisposition to colonize soil [135]. Degradation of anthracene [57] and PCBs [136] by the fungus was also investigated but the degrading potential was found to be limited.

2.2.4. Pleurotus ostreatus *(Jacq.:Fr.) Kumm. and* other Pleurotus *species*

P. ostreatus (oyster mushroom) is a cultivated edible mushroom. Its great advantage is that a large-scale production of fungal biomass grown on lignocellulosic substrates has been already developed and is economically feasible because the substrates need not be sterilized (pasteurization is sufficient). The fungal mycelium well colonizes soil [135], the degradation studies were performed on solid matrices – soil or lignocellulosic materials (see below).

Bezalel and coworkers [48-50,52] studied the degradation of PAHs by *P. ostreatus* and involvement of ligninolytic and other enzymes. They provided evidence that, in the initial hydroxylation of phenanthrene, cytochrome P-450 and epoxide hydrolase were involved (like in non-ligninolytic fungi).

Pleurotus pulmonarius (Fr.) Quél. is another cultivated edible oyster mushroom used in biodegradation research. Masaphy and coworkers [51] described transformation of atrazine (triazine-based herbicide) by the fungus both in liquid culture and on straw and demonstrated involvement of cytochrome P-450 in atrazine degradation.

Several *Pleurotus* strains were successfully tested using labeled PAHs applied on sand, straw or soil [137,138]. Similar potential to mineralize fluorene and anthracene was documented for *P. ostreatus* and *Bjerkandera adusta* [139].

2.2.5. Bjerkandera adusta *(Willd.:Fr.) P. Karst. and* Bjerkandera sp. *strain BOS55*

From an extensive screening program in Dr. Field's laboratory [6,64] two Bjerkandera strains were selected for detailed biodegradation studies. Oxidation of PAHs by the extracellular ligninolytic system of the strain BOS55 was stimulated by adding nonionic surfactants indicating that bioavailability was the limiting factor [140]. Oxidation of anthracene increased after addition of solvents (acetone or ethanol) [141]. Successive mineralization and detoxication of benzo[a]pyrene was demonstrated using a combination of Bjerkandera sp. BOS55 and natural mixed populations of microorganisms originating either from activated sludge or from forest soil [142]. Since this fungal strain also removed efficiently PAHs in the organic solvent extract from a polluted soil [143] the fungus became prospective for remediation practice.

2.2.6. Nematoloma frowardii E. Horak

Within an extended screening for TNT mineralizing fungi [144], *N. frowardii* was selected as one of the best degraders. Involvement of MnP in the TNT conversion and mineralization was documented [145,146] as well as mineralization of amino-dinitrotoluenes [150]. Using MnP isolated from the fungus, PAHs and synthetic humic substances were degraded [42,147].

TABLE 1. Degradation of typical environmental pollutants by ligninolytic basidiomycetes

Fungus	Type of pollutant	Reference
Phanerochaete chrysosporium Burds. and other *Phanerochaete* spp.		
	chlorinated aromatics	8,9,11,14-16,21,22,26, 30-32,34-37,59,111, 117-120,136,139,171, 178,174-176
	PAHs	8,13,21,38-52,54-59, 101,102,113,114,121, 135,167-169,170,172, 173
	dyes	60,65-76
	TNT and other nitroaromatics	21,86-93,110
	pesticides	7,21,94-98,102,108, 181,182
	other compounds	29,77-85,115,122,123
Trametes versicolor (L.:Fr.) Pil.		
	chlorinated aromatics	14,111,124,131-133, 134,136,160,175,176
	PAHs	44,48-50,52,57,101, 107,112,113,127- 129,135, 136
	dyes	112,132,133
	pesticides	102,108
	other compounds	115,125,126,129
Pleurotus ostreatus (Jacq.:Fr.) P. Kumm. and other *Pleurotus* spp.		
	chlorinated aromatics	14,171,184,178,179
	dyes	112
	pesticides	108
	PAHs	10,103,107,109,113, 135,137-139,160,170, 171,184,185
	Other compounds	115,161
Bjerkandera adusta (Willd.:Fr.) P. Karst. and Bjerkandera sp. BOS55		
	dyes	101,112,139
	pesticides	108
	PAHs	101,106,107,114,139- 143
	other compounds	115
Nematoloma frowardii E. Horak		
	PAHs	42,106
	TNT	144-146
	other compounds	150

2.2.7. Other ligninolytic fungi in liquid culture

Several other fungal species have been studied as possible organopollutant degraders. In a search for new filamentous fungi with a potential for degrading benzo[a]pyrene, a collection of basidiomycetes were screened [148]. *Marasmiellus troyanus,* the most powerful degrader in the screening, compared to *P. chrysosporium* showed a much higher mineralization and biotransformation of the compound [149]. The white-rot basidiomycete *Phlebia radiata* was shown to transform and mineralize the first identified reduction product of TNT – 4-hydroxylamino-2,6,-dinitrotoluene [150,151]. *Grifola frondosa,* a popular edible mushroom that is cultivated extensively in Japan using a straw matrix, was shown to degrade 40 components in the 41 major peaks of PCBs [152]. The organism accumulated dichloromethoxyphenols during transformation of the PCB mixture. The idea of the work was to evaluate the possibility of using spent sawdust matrix containing fungal mycelium as a cheap agent for the treatment of PCB contamination. Biotransformation of benzo[a]pyrene by purified extracellular laccase of *Pycnoporus cinnabarinus* was investigated in a bench scale reactor [153]. Most of the substrate was converted within 24 hrs if the exogenous mediator ABTS was present. The enzyme preparation oxidized the substrate mainly to benzo[a]pyrene quinones. Laccase isolated from another white rot fungus – *Coriolopsis gallica* was screened for oxidation of seven PAHs, including benzo[a]pyrene [154]. Addition of hydroxybenzotriazole to the reaction mixture increased the oxidation rate nine-fold compared with the oxidation rate in the presence of ABTS alone. Recently the white-rot fungus *Irpex lacteus* (Fr.:Fr.) Fr. has become a research subject because of its ability to intensively decolorize synthetic dyes [112,155] and degrade PAHs [113,156] both in laboratory nutrient media and in soil (see below).

2.3. APPLICATION OF LIGNINOLYTIC FUNGI IN SOIL REMEDIATION

2.3.1. Background

Soil is an extremely variable matrix, diverse in many aspects (chemical, physical, biological) and, up to now, no universally usable organism or a universal way of soil remediation have been described because of the influence of many environmental factors. Large-scale bioremediation can be performed only in nonsterile soil. The bacterial flora present may inhibit the growth of the introduced fungus [157] or the introduced fungus may inhibit the activity of soil microflora [155].

Bioavailability of the respective xenobiotic in soil is an important factor because the sorption of xenobiotics to soil particles competes with biodegradation [3]. On the other hand, sorption and humification of the xenobiotic compound during soil bioremediation can play a positive role in the remediation process. Several authors [159-164] demonstrated formation of nonextractable PAH residues in soil treated with ligninolytic fungus grown on a lignocellulosic substrate. Extracellular fungal oxidoreductases catalyze covalent incorporation of pollutant-derived material into soil organic matter which presumably occurs via free-radical-mediated copolymerization of the pollutant or its degradation product and precursors of humic and fulvic acids [164]. The formation of non-bioavailable organic residues in soil is accepted in EU legislation for pesticides and non-bioavailable pollutants as representing no direct harm to the environment [174].

2.3.2. Bench-scale solid-phase degradation experiments

The promising capability of *P. chrysosporium* to degrade many important soil pollutants in liquid culture provoked interest to apply this fungus in soil conditions. Simultaneously, other white rot fungal species started to be screened for growth and degrading potential in soil. Only rarely were the experiments documenting degradation of a pollutant in both liquid medium and soil directly compared. Ryan and Bumpus [94] found that *P. chrysosporium* mineralized 62 % of 2,4,5-trichlorophenoxyacetic acid after 30-d incubation, and about 30 % of the same compound was mineralized in soil after the same period. Lamar *et al.* [118] compared *P. chrysosporium* and *P. sordida* with respect to removing PCP from liquid medium and from soil. *P. chrysosporium* removed PCP from soil more efficiently and faster whilst *P. sordida* possessed a higher capacity to mineralize PCP in liquid medium.

Zaddel *et al.* [175] compared growth and PCB-mixture degradation by *P. chrysosporium, P. ostreatus* and *T. versicolor* on wood chip substrate and tested the effect of soil addition on the degradation of PCBs. Both the growth and degradation ability were lowest with *P. chrysosporium,* the other two species grew and degraded comparably. An interesting fact was that *P. ostreatus* showed a congener specificity in the degradation of several PCB congeners, which was inconsistent with both former and later observations [11,36] showing that the degradation of PCBs by *P. chrysosporium* was not congener-specific. However, in our laboratory we confirmed the difference between PCB degradation by *P. chrysosporium* and *P. ostreatus*, as well as the low potential of *P. chrysosporium* to degrade PCBs in soil [176].

Degradation of PAHs in soil by straw-grown mycelium of *P. ostreatus* was studied by Eggen and coworkers [163,177,178] who demonstrated the ability of *P. ostreatus* to degrade both aged and artificially added (^{14}C-labeled) benzo[*a*]pyrene in aged creosote-contaminated soil. Using fungal substrate from a commercial oyster mushroom production they demonstrated that such a treatment after 7 weeks at an ambient temperature resulted in the following removal: 86 % of total 16 PAHs, 89 % of 3-ring PAHs, 87 % of 4-ring PAHs, and 48 % of 5-ring PAHs.

T. versicolor was a model fungus in the study of the fate of PCP in autoclaved soil [134]. During a 42-d incubation the fungus mineralized 29 % of artificially added ^{14}C-labeled PCP. The concentration of nonlabeled PCP decreased to 4 % of its original value during the incubation period. Only trace amounts of anisoles were formed during the incubation and a part of ^{14}C-labeled PCP was found to be bound to humic substances. The authors concluded that a treatment of PCP-polluted soils with *T. versicolor* was a possibility.

2.3.3. Field trials

In 1988, the group of Prof. Hüttermann at the University of Göttingen, Germany, started an experiment testing the straw-grown white-rot fungi under field conditions for remediation of PAH-contaminated soil [158,186]. About 23 m^2 of soil from a former gas work (average contamination – sum of 16 measured PAHs – was about 20 mg/kg soil) was treated. The soil was mixed with *Pleurotus ostreatus* grown on straw from a professional mushroom grower. After 8-week treatment the average degradation was 40 % and after 16 weeks in one of the clamps, a degradation of 80 % was measured.

The ability of *P. chrysosporium* and *P. sordida* to deplete PCP from soil contaminated with a commercial wood preservative (initial concentration of PCP was 250-400 ppm), was examined in a field study [119]. Inoculation of soil with either *P. chrysosporium* or *P. sordida* resulted in an overall decrease of about 90 % in the soil in 6.5 weeks. Only a small percentage of the decrease of PCP was the result of fungal methylation to pentachloroanisole. The authors assumed that if the losses of PCP via mineralization and volatilization were negligible (contrary to laboratory-scale studies [15]), most of the PCP was to be converted to non-extractable soil-bound product.

3. Conclusions and perspectives

There is no doubt that ligninolytic fungi possess a great potential for application to soil remediation. Their ability to degrade many xenobiotic organopollutants of different chemical nature was demonstrated in a number of experiments, both in liquid nutrient media and under soil conditions. However, it is not possible to say that our knowledge is satisfactory. When 15 years ago this ability was described in *P. chrysosporium*, the fungus became a model organism of extensive research and we were optimistic that such a powerful fungus represented a universal bioremediation tool. Why our visions have not been fulfilled? There are two reasons – one, that is valid for bioremediations generally, the other is more or less specific for ligninolytic fungi.

Despite valuable basic knowledge on the mechanisms of pollutant biodegradation, bioremediation has not yet been accepted as a routine treatment technology and the environmental industry is wary in applying bioremediation for the treatment of contaminated sites. It is simple to demonstrate biodegradation of specific compounds in a contaminated environmental sample by spiking the compound into soil and estimating its loss in comparison with sterilized control samples. In practice, many factors influence the biodegradation process. Reduced bioavailability results from the interaction of pollutants with both organic and inorganic components in the soil, and with increased contact time (aging), the proportion of the pollutant that becomes biologically available decreases. The mechanism of this effect results from a complex of several physical and chemical phenomena: sequestration of the pollutant into soil micropores, binding to soil minerals by ionic or electrostatic interactions, oxidative covalent coupling with soil organic matter via enzymic or chemical catalysis, and partition/dissolution into the soil organic matter. These mechanisms often operate in consort but individually predominate in specific soils.

One of the specific problems connected with the application of ligninolytic fungi is the fact that mostly the so-called white rot fungi, that normally grow on tree wood, were the research objects; the introduction of their biomass into soil condition was the introduction into a foreign environment. By far the most studied lignin-degrading fungal species in this respect has been *Phanerochaete chrysosporium* and, of the many hundreds of species possessing ligninolytic activity, only few became research objects. Another group of ligninolytic fungi – the litter-decomposing species – have almost been neglected. Since the litter decomposers are physiologically close to soil conditions, and the establishment of robust mycelium in soil is as important for soil remediation as the ability to degrade the pollutants, these fungi deserve more attention.

Little interest has also been focused on the relationship between the ligninolytic fungus mycelium and indigenous soil microflora. This is an important factor since soil microorganisms can inhibit the growth of the introduced fungal mycelium or cooperate in further metabolism of pollutants oxidized by the fungus. In this connection, more attention should be paid to simultaneous (or successive) application of ligninolytic fungi and known bacterial degraders. The action of ligninolytic fungi is extracellular and a free radical mechanism is involved; thus these fungi can play a role in the first attack on the pollutant molecule being oxidized and, often becoming more water- soluble, it can serve as a substrate for collaborating bacterial species.

The application of mycoremediation needs well trained personnel with a sufficient knowledge of microbiology. If fungal inoculum is handled as a chemical compound and not as a living organism, the proper humidity, temperature and access of oxygen are not controlled carefully during the remediation process, the technology is often perceived as one that does not work. Thus, both more extensive (especially searching for and exploiting new fungal species) and intense (deeper study of the influence on the bioremediation process by specific soil parameters) research is needed, as well as improvement of practical application, to establish mycoremediation as an effective and reliable soil-remediation technology.

References

1. Chaudhry, G.R., and Chapalamadugu, S. (1991) Biodegradation of halogenated organic compounds, *Microbiol. Rev.* **55**, 59-79.
2. Morgan, P., and Watkinson, R.J. (1989) Microbiological methods for the cleanup of soil and ground water contaminated with halogenated organic compounds, *FEMS Microbiol. Rev.* **63**, 277-300.
3. Providenti, M.A., Lee, H. and Trevors, J.T. (1993) Selected factors limiting the microbial degradation of recalcitrant compounds, *J. Ind. Microbiol.* **12**, 379-395.
4. Hammel, K.E. (1995) Organopollutant degradation by ligninolytic fungi, In: Young, L., Cerniglia, C. (Eds.). *Microbial Transformation and Degradation of Toxic Chemicals.* Wiley-Liss, New York, USA, pp. 331-336.
5. Cerniglia, C.E. (1997) Fungal metabolism of polycyclic aromatic hydrocarbons: past, present and future application in bioremediation, *J. Industrial. Microbiol. Biotechnol.* **19**, 324-333.
6. Field, J.A., de Jong, E., Costa, G.F., de Bont, J.A.M. (1993) Screening for ligninolytic fungi applicable to the biodegradation of xenobiotics, *Trends Biotechnol.* **11**, 44-49.
7. Kohler, A., Jager A., Willerhausen, H., Graf, H. (1988): Extracellular ligninase of *Phanerochaete chrysosporium* Burdshall has no role in the degradation of DDT, *Appl. Microbiol. Biotechnol.* **29**, 618-620.
8. Thomas, D.R., Carswell, K.S., and Georgiou, G. (1992) Mineralization of biphenyl and PCBs by the white rot fungus *Phanerochaete chrysosporium, Biotechnol. Bioeng.* **40**, 1395-1402.
9. Yadav, J.S., and Reddy, C.A. (1992) Non-involvement of lignin peroxidases and manganese peroxidases in 2,4,5-trichlorophenoxyacetic acid degradation by *Phanerochaete chrysosporium, Biotechnol. Lett.* **14**, 1089-1092.
10. Sack, U., and Gunther, T. (1993) Metabolism of PAH by fungi and correlation with extracellular enzymatic activities, *J. Basic Microbiol.* **33**, 269-277.
11. Yadav, J.S., Quensen, J.F., Tiedje, J.M., Reddy, C.A. (1995) Degradation of polychlorinated biphenyl mixtures (Aroclor 1242, 1254, and 1260) by the white rot fungus *Phanerochaete chrysosporium* as evidenced by congener-specific analysis, *Appl. Environ. Microbiol.* **61**, 2560-2565.
12. Novotný, Č., Vyas, B.R.M., Erbanová, P., Kubátová, A., Šašek, V. (1997) Removal of PCBs by various white rot fungi in liquid culture, *Folia Microbiol.* **42**, 136-140.

260

13. Bogan, B.W., Lamar, R.T., Hammel, K.E. (1996) Fluorene oxidation in vitro by *Phanerochaete chrysosporium* and in vitro during manganese peroxidase-dependent lipid peroxidation. *App. Environ. Microbiol.* **62**, 1788-1792.

14. Beaudette, L.A., Davies, S., Fedorak, P.M., Ward, O.P., and Pickard, M.A. (1998) Comparison of gas chromatography and mineralization experiments for measuring loss of selected polychlorinated biphenyls congeners in cultures of white rot fungi, *Appl. Environ. Microbiol.* **64**, 2020-2025.

15. Lamar, R.T., Glaser, J.A., and Kirk, T.K.(1990) Fate of pentachlorophenol (PCP) in sterile soil inoculated with white-rot basidiomycete *Phanerochaete chrysosporium*: mineralization, volatilization and depletion of PCP, *Soil. Biol. Biochem.* **22**, 433-440.

16. Lamar, R.T., Davis, M.W., Dietrich, D.M., and Glaser, J.A. (1994) Treatment of a pentachlorophenol- and creosote-contaminated soil using the lignin degrading fungus *Phanerochaete sordida*: a field demonstration, *Soil Biol. Biochem.* **26**, 1603-1611.

17. Aust, S.T., Tien, M., and Bumpus, J.A. (1986) Process for the degradation of environmentally persistent organic compounds, *Eur. Pat. Appl.* No. 86102067.5, 18.02.86.

18. Volfová O., Šašek V., Krumphanzl V., Přikryl J., Erbanová P., Pilátová J. (1995): Biodegradation of chlorinated aromatic compounds and hydrocarbons, Czech. Pat. No. 280091, 29.8.1995.

19. Aust, S.T., and Benson, J.T. (1993) The fungus among us: use of white rot fungi to biodegrade environmental pollutants, *Environ. Health Prospect.* **101**, 232-233.

20. May, R., Schröder, P., and Sandermann, H., Jr. (1997) Ex-situ process for treating PAH-contaminated soil with *Phanerochaete chrysosporium*, *Environ. Sci. Technol.* **31**, 2626-2633.

21. Bumpus, J.A., Tien, M., Wright, D., and Aust, S.D. (1985) Oxidation of persistent environmental pollutants by a white rot fungus, *Science* **228**, 1434-1436.

22. Eaton, D.C. (1985) Mineralization of polychlorinated biphenyls by *Phanerochaete chrysosporium*: a ligninolytic fungus, *Enzyme Microb. Technol.* **7**, 194-196.

23. Tien, M., and Kirk, T.K. (1988) Lignin peroxidase of *Phanerochaete chrysosporium*, *Methods Enzymol.* **161**, 238-249.

24. Burdshall, H.H., and Eslyn, W.E. (1974) A new *Phanerochaete* with a *Chrysosporium* imperfect state, *Mycotaxon* **1**, 123-133.

25. Dhawale, S.W., Dhawale, S.S., and Dean-Ross, D. (1992) Degradation of phenanthrene by *Phanerochaete chrysosporium* occurs under ligninolytic as well as non-ligninolytic conditions, *Appl. Environ. Microbiol.* **58**, 3000-3006.

26. Mileski, G., Bumpus, J.A., Jurek, M., and Aust, S.D. (1988) Biodegradation of pentachlorophenol by the white-rot fungus *Phanerochaete chrysosporium*, *Appl. Environ. Microbiol.* **54**, 2885-2889.

27. Kang, G., and Stevens, D.K. (1994) Degradation of pentachlorophenol in bench scale bioreactors using the white rot fungus *Phanerochaete chrysosporium*, *Hazard Waste Hazard Mat.* **11**, 397-410.

28. Alleman, B.C., Logan, B.E., and Gilbertson R.L. (1995) Degradation of pentachlorophenol by fixed films of white rot fungi in rotating tube bioreactors, *Wat. Res.* **29**, 61-67.

29. Aitken, B.S., and Logan, B.E. (1996) Degradation of pentachlorophenol by the white rot fungus *Phanerochaete chrysosporium* grown in ammonium lignosulphonate media, *Biodegradation* **7**, 175-182.

30. Armenante, P.M., Pal, N., and Lewandowski, G. (1994) Role of mycelium and extracellular protein in the biodegradation of 2,4,6-trichlorophenol by *Phanerochaete chrysosporium*, *Appl. Environ. Microbiol.* **60**, 1711-1718.

31. Valli, K., and Gold, M.H. (1991) Degradation of 2,4-dichlorophenol by the lignin-degrading fungus *Phanerochaete chrysosporium*, *J. Bacteriol.* **173**, 345-352.

32. Joshi, D.K., and Gold, M.H. (1993) Degradation of 2,4, 5-trichlorophenol by the lignin-degrading basidiomycete *Phanerochaete chrysosporium*, *Appl. Environ. Microbiol.* **59**, 1779-1785.

33. Hammel, K.E., and Tardone, P.J. (1988) The oxidative 4-dechlorination of polychlorinated phenols is catalyzed by extracellular fungal lignin peroxidase, *Biochemistry* **27**, 6563-6568.

34. Dietrich D., Hickey, W.J., and Lamar, R. (1995) Degradation of 4,4′-dichlorobiphenyl, 4,3′,4,4′-tetrachlorobiphenyl, and 2,2′,4,4′,5,5′-hexachlorobiphenyl by the white rot fungus *Phanerochaete chrysosporium*, *Appl. Environ. Microbiol.* **61**, 3904-3909.

35. Krčmář, P., and Ulrich, R. (1998) Degradation of polychlorinated biphenyl mixture by the lignin-degrading fungus *Phanerochaete chrysosporium*, *Folia Microbiol.* **43**, 79-84.

36. Krčmář, P., Kubátová, A.,Votruba, J., Erbanová, P., Novotný, Č., and Šašek, V. (1999) Degradation of polychlorinated biphenyls by extracellular enzymes of *Phanerochaete chrysosporium* produced in a perforated plate bioreactor, *World J. Microbiol. Biotechnol.* **15**, 237-242.

37. Valli, K., Warishi, H., and Gold, M.H. (1992) Degradation of 2,7-dichlorodibenzo-*p*-dioxin by the lignin-degrading basidiomycete *Phanerochaete chrysosporium*, *J. Bacteriol.* **174**, 2131-2137.

38. Haemmerli, S.D., Leisola, M.S.A., Sanglard, D., and Fiechter , A. (1986) Oxidation of benzo[a]pyrene by extracellular ligninases of *Phanerochaete chrysosporium*: veratryl alcohol and stability of ligninase, *J. Biol. Chem.* **261**, 6900-6903.

261

39. Hammel, K.E., Kalyanaraman, B., and Kirk, T.K. (1986) Oxidation of polycyclic aromatic hydrocarbons and dibenzo[p]-dioxins by *Phanerochaete chrysosporium, J. Biol. Chem.* **261**, 16948-16952.

40. Moen, M.A., and Hammel, K.E. (1994) Lipid peroxidation by the manganese peroxidase of *Phanerochaete chrysosporium* is the basis for phenanthrene oxidation by the intact fungus, *Appl. Environ. Microbiol.* **60**, 1956-1961.

41. Field, J.A., Vledder, R.H., van Zelst, J.G., and Rulkens, W.H. (1996) The tolerance of lignin peroxidase and manganese dependent peroxidase to miscible solvents and the *in vitro* oxidation of anthracene in solvent:water mixtures, *Enzyme Microbiol. Technol.* **18**, 300-308.

42. Sack, U., Hofrichter, M., Fritsche, W. (1997) Degradation of polycyclic aromatic hydrocarbons by manganese peroxidase of *Nematoloma frowardii, FEMS Microbiol. Lett.* **152**, 227-234.

43. Johannes, C., Majcherczyk, A., and Huttermann, A. (1996) Degradation of anthracene by laccase of *Trametes versicolor* in the presence of different mediator compounds, *Appl. Microbiol. Biotechnol.* **46**, 313-317.

44. Collins, P.J.J., Kotterman, M.J.J., Field, J.A., and Dobson, A.D.V. (1996) Oxidation of anthracene and benzo[a]pyrene by laccase from *Trametes versicolor, Appl. Environ. Microbiol.* **62**, 4563-4567.

45. Hammel, K.E., Green, B., and Gai, W.Z. (1991) Ring fission of anthracene by a eukaryote, *Proc. Natl. Acad. Sci.* **88**, 10605-10608.

46. Bogan, B.W., and Lamar, R.T. (1995) One-electron oxidation in the degradation of creosote polycyclic aromatic hydrocarbons by *Phanerochaete chrysosporium, Appl. Environ. Microbiol.* **61**, 2631-2635.

47. Sutherland, J.B., Selby, A.L., Freeman, J.P., Evans, F.E., and Cerniglia, C.E. (1991) Metabolism of phenanthrene by *Phanerochaete chrysosporium, Appl. Environ. Microbiol.* **57**, 3310-3316.

48. Bezalel, L., Hadar, Y., Freeman, J.P., and Cerniglia, C.E. (1996a) Initial oxidation products in the metabolism of pyrene, anthracene, fluorene, and dibenzothiophene by the white rot fungus *Pleurotus ostreatus, Appl. Environ. Microbiol.* **62**, 2554-2559.

49. Bezalel, L., Hadar, Y., Fu, P.P., Freeman, J.P., and Cerniglia, C.E. (1996b) Metabolism of phenanthrene by the white rot fungus *Pleurotus ostreatus, Appl. Environ. Microbiol.* **62**, 2547-2553.

50. Bezalel, L., Hadar, Y., and Cerniglia, C.E. (1996c) Mineralization of polycyclic aromatic hydrocarbons by the white rot fungus *Pleurotus ostreatus, Appl. Environ. Microbiol.* **62**, 292-295.

51. Masaphy, S., Levanon, D., Henis, Y., Venkateswarlu, K., and Kelly, S.L. (1996) Evidence for cytochrome P450 and P450-mediated benzo[a]pyrene hydroxylation in the white rot fungus *Phanerochaete chrysosporium, FEMS Microbiol. Lett.* **135**, 51-55.

52. Bezalel, L., Hadar, Y., and Cerniglia, C.E. (1997) Enzymatic mechanisms involved in phenanthrene degradation by the white rot fungus *Pleurotus ostreatus, Appl. Environ. Microbiol.* **63**, 2495-2501.

53. Cerniglia, C.E., Sutherland, J.B., and Crow, S.A. (1992) Fungal metabolism of aromatic hydrocarbons. In: G. Winkelmann (ed.), *Microbial Degradation of Natural Products.* WCH Press, Weinheim, pp. 193-217.

54. Bumpus, J.A. (1989) Biodegradation of polycyclic aromatic hydrocarbons by *Phanerochaete chrysosporium, Appl. Environ. Microbiol.* **55**, 154-158.

55. Sanglard, D.S., Leisola, A., and Fiechter, A. (1986) Role of extracellular ligninases in biodegradation of benzo[a]pyrene by *Phanerochaete chrysosporium, Enzyme Microb. Technol.* **8**, 209-212.

56. Hammel, K.E., Gai, W.Z., Green, B., Moen, M.A. (1992) Oxidative degradation of phenanthrene by the ligninolytic fungus *Phanerochaete chrysosporium, Appl. Environ. Microbiol.* **58**, 1832-1838.

57. Vyas, B.R.M., Bakowski, S., Šašek, V., and Matucha, M. (1994) Degradation of anthracene by selected white rot fungi, *FEMS Microbiol. Ecol.* **14**, 65-70.

58. Yadav, J.S., and Reddy, C.A. (1993) Degradation of benzene, toluene, ethylbenzene, and xylenes (BTEX) by the lignin-degrading basidiomycete *Phanerochaete chrysosporium, Appl. Environ. Microbiol.* **59**, 756-762.

59. Yadav, J.S., Walace, R.E. and Reddy, C.A. (1995) Mineralization of mono- and dichlorobenzenes and simultaneous degradation of chloro- and methyl-substituted benzenes by the white rot fungus *Phanerochaete chrysosporium, Appl. Environ. Microbiol.* **61**, 667-680.

60. Glenn, J.K., and Gold, M.H. (1983) Decolorization of several polymeric dyes by the lignin-degrading basidiomycete *Phanerochaete chrysosporium, Appl. Environ. Microbiol.* **45**, 1741-1747.

61. Glenn, J.K., and Gold, M.H. (1985) Purification and characterization of an extracellular Mn (III)-dependent peroxidase from the lignin-degrading basidiomycete *Phanerochaete chrysosporium, Arch. Biochem. Biophys.* **242**, 329-341.

62. Kuwahara, M., Glenn, J.K., Morgan, M., and Gold, M.H. (1984) Separation and characterization of two extracellular H_2O_2-dependent peroxidases from ligninolytic cultures of *Phanerochaete chrysosporium, FEBS Lett.* **169**, 247-250.

63. Paszcynski, A., Pasti, M.B., Goszczynski, S., Crawford, D.L., and Crawford, R.L. (1991) New approach to improve degradation of recalcitrant azo dyes by *Streptomyces* spp. and *Phanerochaete chrysosporium, Enzyme Microb. Technol.* **13**, 378-384.

64. Field, J.A., de Jong, E., Feijoo-Costa, G., and de Bont, J.A.M. (1993) Screening for ligninolytic fungi applicable to biodegradation of xenobiotics, *Trends in Biotechnol.*, TIBTECH February 1993, **11**, 44-49.

65. Bumpus, J.A., and Brock, B.J. (1988) Biodegradation of crystal violet by the white-rot fungus *Phanerochaete chrysosporium*, *Appl. Environ. Microbiol.* **54**, 1140-1150.

66. Crips, C., Bumpus, J.A., and Aust, S.A. (1990) Biodegradation of azo and heterocyclic dyes by *Phanerochaete chrysosporium*, *Appl. Environ. Microbiol.* **56**, 1114-1118.

67. Spadaro, J.T., Gold, M.H., and Renganathan, V. (1992) Decolorization of azo dyes by the lignin-degrading fungus *Phanerochaete chrysosporium*, *Appl. Environ. Microbiol.* **58**, 2397-2401.

68. Ollikka, P., Alhonnaki, K., Leppaen, V., Glumoff, T., Raiola, T., and Suonimane, L. (1993) Decolorization of azo, triphenyl methane, heterocyclic and polymeric dyes by lignin peroxidase isoenzymes from *Phanerochaete chrysosporium*, *Appl. Environ. Microbiol.* **59**, 4010-4016.

69. Paszczynski, A., Pasti-Grigsby, M.B., Goszczynski, S., Crawford, R.L., and Crawford, D.L. (1992) Mineralization of sulfonated azo dyes and sulfanilic acid by *Phanerochaete chrysosporium* and *Streptomyces chromofuscus*, *Appl. Environ. Microbiol.* **58**, 3598-3604.

70. Paszczynski, A., and Crawford, R.L. (1991) Degradation of azo compounds by ligninase from *Phanerochaete chrysosporium*: Involvement of veratryl alcohol, *Biochem. Biophys. Res. Commun.* **178**, 1056-1063.

71. Goszczynski, S., Paszczynski, A., Pasti-Grigsby, M.B., Crawford, R.L., and Crawford, D.L. (1994) New pathway for degradation of sulfonated azo dyes by microbial peroxidases of *Phanerochaete chrysosporium* and *Streptomyces chromofuscus*, *J. Bacteriol.* **176**, 1339-1347.

72. Paszczynski, A., Goszczynski, S., Crawford, R.L., Crawford, D.L. (1997) Biodegradation of diazo dyes by *Phanerochaete chrysosporium*, pp. 505-510 in *In Situ and On-Site Bioremediation 4*, **2**, Battele Press, Columbus – Richland.

73. Young, L., and Yu, J. (1997) Ligninase catalysed decolorization of synthetic dyes, *Wat. Res.* **31**, 1187-1193.

74. Paszczynski, A., and Crawford, R.L. (1995) Potential for bioremediation of xenobiotic compounds by the white-rot fungus *Phanerochaete chrysosporium*, *Biotechnol. Prog.* **11**, 368-379.

75. Swamy, J., and Ramsay, J.A. (1999) The evaluation of white rot fungi in the decoloration of textile dyes, *Enzyme Microb. Technol.* **24**, 130-137.

76. Rodrigues, E., Pickard, M.A., and Vazgues-Duhalt, R. (1999) Industrial dye decolorization by laccases from ligninolytic fungi, *Current Microbiol.* **38**, 27-32.

77. Blondeau, R. (1989) Biodegradation of natural and synthetic humid acids by the white rot fungus *Phanerochaete chrysosporium*, *Appl. Environ. Microbiol.* **55**, 1282-1285.

78. Fukai, H., Presnell, T.L., Joyce, T.W., and Chang, H.M. (1992) Dechlorination and detoxification of bleach plant effluent by *Phanerochaete chrysosporium*, *J. Biotechnol.* **24**, 267-275.

79. Joyce, T.W., Chang, H., Campbell, A.G., Gerrard, E.D., and Kirk, T.K. (1984) A continuous biological process to decolorize bleach plant effluents, *Biotechnol. Adv.* **2**, 301-308.

80. Michel, F.C., Jr., Dass, S.B., Grulke, E.A., and Reddy, C.A. (1991) Role of manganese peroxidases and lignin peroxidases in decolorization of kraft bleach plant effluent, *Appl. Environ. Microbiol.* **57**, 2368-2375.

81. Garg, S.K., and Modi, D.R. (1999) Decolorization of pulp-paper mill effluents by white rot fungi, *Critic. Rev. Biotechnol.* **19**, 85-112.

82. Ritter, D., Jaklin-Farcher, S., Messner, K., and Stachelberger, H. (1990) Polymerization and depolymerization of lignosulfonates by *Phanerochaete chrysosporium* immobilized on foam, *J. Biotechnol.* **13**, 229-241.

83. Guimaraes, C., Bento, L.S., and Mota, M. (1999) Biodegradation of colorants in rafinery effluents – Potential use of the fungus *Phanerochaete chrysosporium*, *Internat. Sugar J.* **101**, 246.

84. Sayadi, S., and Ellouz, R. (1992) Decolourization of olive mill waste-waters by *Phanerochaete chrysosporium*: involvement of the lignin degrading system, *Appl. Microbiol. Biotechnol.* **37**, 813-817.

85. Sayadi, S., and Ellouz, R. (1995) Role of lignin peroxidase and manganese peroxidase from *Phanerochaete chrysosporium* in the decolorization of olive mill wastewaters, *Appl. Environ. Microbiol.* **61**, 1098-1103.

86. Fernando, T., Bumpus, J.A., and Aust, S.D. (1990) Biodegradation of TNT (2,4,6-trinitrotoluene) by *Phanerochaete chrysosporium*, *Appl. Environ. Microbiol.* **56**, 1666-1671.

87. Bumpus, J.A., and Tatarko, M. (1994) Biodegradation of 2,4,6-trinitrotoluene by *Phanerochaete chrysosporium*: identification of initial degradation products and the discovery of a TNT metabolite that inhibits lignin peroxidase, *Curr. Microbiol.* **28**, 185-190.

88. Michels, J., and Gottshal, G. (1994) Inhibition of lignin peroxidase by hydroxyl-amine-dinitrotoluene, an early intermediate in the degradation of 2,4,6-trinitrotoluene, *Appl. Environ. Microbiol.* **60**, 187-194.

89. Spiker, J.K., Crawford, D.L., and Crawford, R.L.(1992) Influence of 2,4,6-trinitrotoluene (TNT) concentration on the degradation of TNT in explosive-contaminated soil by the white rot fungus *Phanerochaete chrysosporium*, *Appl. Environ. Microbiol.* **58**, 199-202.

90. Hawari, J., Halasz, A., Beaudet, S., Paquet, L., Ampleman, G., and Thiboutot, S. (1999) Biotransformation of 2,4,6-trinitrotoluene with *Phanerochaete chrysosporium* in agitated culture, *Appl. Environ. Microbiol.* **65**, 2977-2986.

91. Sublette, K .L., Ganapathy, E. V., and Schwartz, S. (1992) Degradation of munitions wastes by *Phanerochaete chrysosporium, App .Biochem .Biotechnol.* **34/35**, 709-723.

92. Servent, D., Ducrocq, C., Henry, Y., Guissani, A., and Lenfant, M. (1991) Nitroglycerin metabolism by *Phanerochaete chrysosporium*: evidence for nitric oxide and nitrite formation, *Biochem. Biophys. Acta* **1074**: 320-325.

93. Valli, K., Brock, B.J., Joshi, D.K., and Gold, M.H. (1992) Degradation of 2,4-dinitrotoluene by the lignin-degrading fungus *Phanerochaete chrysosporium, Appl. Environ. Microbiol.* **58**, 221-228.

94. Ryan, T.P., and Bumpus, J.A. (1989) Biodegradation of 2,4,5-trichlorophenoxyacetic acid in liquid culture and in soil by the white rot fungus *Phanerochaete chrysosporium, Appl. Microbiol. Biotechnol.* **31**, 302-307.

95. Yadav, J.S., and Reddy C.A. (1993) Mineralization of 2,4-dichlorophenoxyacetic acid (2,4-D) and mixtures of 2,4-D and 2,4,5-trichlorophenoxyacetic acid by *Phanerochaete chrysosporium, Appl. Environ. Microbiol.* **59**, 2904-2908.

96. Kennedy, D.W., Aust, S.D., and Bumpus, J.A. (1990) Comparative biodegradation of alkyl halide insecticides by the white rot fungus, *Phanerochaete chrysosporium* (BKM-F-1767), *Appl. Environ. Microbiol.* **56**, 2347-2353.

97. Fratila-Apachitei, L.E., Hirst, J.A., Siebel, M.A., and Gijzen, H.J. (1999) Diuron degradation by *Phanerochaete chrysosporium* BKM-F-1767 in synthetic and natural media, *Biotechnol. Lett.* **21**, 147-154.

98. Bumpus, J.A., and Aust, S.D. (1987) Biodegradation of DDT (1,1,1-trichloro-2,2-bis-(4-chlorophenyl) ethane by the white rot fungus *Phanerochaete chrysosporium, Appl. Environ. Microbiol.* **53**, 2001-2008.

99. Golovleva, L.A., and Leontievskii, A.A. (1998) Ligninolytic activity of wood-decaying fungi, *Microbiology* **67**, 581-587.

100. Gilbertson, R.L., and Ryvarden, L. (1986) North American Polypores. Vol. 1, Fungiflora, Oslo, Norway.

101. Field, J.A., de Jong, E., Costa, G.F., and de Bont, J.A.M. (1992) Biodegradation of polycyclic aromatic hydrocarbons by new isolates of white rot fungi, *Appl. Environ. Microbiol.* **58**, 2219-2226.

102. Morgan P., Lewis, S.T., and Watkinson, R.J. (1991) Comparison of abilities of white-rot fungi to mineralize selected xenobiotic compounds, *Appl. Microbiol. Biotechnol.* **14**, 691-696.

103. Sack, U., Hofrichter, M., and Fritsche, W. (1997) Degradation of phenanthrene and pyrene by *Nematoloma frowardii, J. Basic Microbiol.* **37**, 287-293.

104. Sack U., and Fritsche, W. (1997) Enhancement of pyrene mineralization in soil by wood-decaying fungi, *FEMS Microbiol. Ecol.* **22**, 77-83.

105. Sack, U., Heinze, T.M., Deck, J., Cerniglia, C.E., Martens, R., Zadrazil, F., Fritsche, W. (1997) Comparison of phenanthrene and pyrene degradation by different wood-decaying fungi, *Appl. Environ. Microbiol.* **63**, 3919-3925.

106. Gramss, G., Kirsche, B., Voigt, K.-D., Gunther, Th., and Fritsche, W. (1999) Conversion rates of five polycyclic aromatic hydrocarbons in the liquid cultures of fifty-eight fungi and the concomitant production of oxidative enzymes, *Mycol. Res.* **103**, 1009-1018.

107. Martens, R., and Zadrazil, F. (1998) Screening of white-rot fungi for their ability to mineralize polycyclic aromatic hydrocarbons in soil, *Folia Microbiol.* **43**: 97-103.

108. Khadrani, A., Siegle-Murandi, F., Steiman, R., and Vroumsia, T. (1999) Degradation of three phenylurea herbicides (chlortorulon, isoproturon and diuron) by micromycetes isolated from soil, *Chemosphere* **38**, 3041-3050.

109. Colombo, J.C., Cabello, M., and Arambarri, A.M. (1996) Biodegradation of aliphatic and aromatic hydrocarbons by natural soil microflora and pure cultures of imperfect and ligninolytic fungi, *Environ. Poll.* **94**, 355-362.

110. Donnelly, K.C., Chen, J.C., Huebner, H.J., Brown, K.W., Autenrieth, R.L., and Bonner, J.S. (1997) Utility of four strains of white-rot fungi for the detoxification of 2,4,6-trinitrotoluene in liquid culture, *Environ. Toxicol. Chem.* **16**, 1105-1110.

111. Alleman, B.C., Logan, B.E., and Gilbertson, R.L. (1992) Toxicity of pentachlorophenol to six species of white rot fungi as a function of chemical dose, *Appl. Environ. Microbiol.* **58**, 4048-4050.

112. Šašek, V., Novotný, Č., and Vampola, P. (1998) Screening for efficient fungal degraders by decolorization, *Czech Mycol.* **50**, 303-311.

113. Song, H.-G. (1999) Comparison of pyrene biodegradation by white rot fungi, *World J. Microbiol. Biotechnol.* **15**, 669-672.

114. Garon, D., Krivobok, F., Siegle-Murandi, F. (2000) Fungal degradation of fluorene, *Chemosphere* **40**, 91-97.

115. Braun-Lullemann, A., Majcherczyk, A., Huttermann A. (1997) Degradation of styrene by white-rot fungi, *Appl. Microbiol., Boitechnol.,* **47**, 150-155.

116. Deguchi, T., Kakezawa, M., and Nishida, T. (1997) Nylon biodegradation by lignin degrading fungi, *Appl. Environ. Microbiol.* **63**, 329-331.

117. Majcherczyk, A., Braun-Lullemann, A., and Huttermann, A. (1990) Biofiltration of a polluted air by a complex filter based on white-rot fungi growing on lignocellulosic substrates. In: M.P. Coughlan, and M.T. Amaral Collaco (eds.) *Advances of Biological Treatment of Lignocellulosic Materials.* Elsevier Applied Science , London, pp. 233-329.

118. Lamar, R.T., Larsen, M.J., and Kirk, T.K. (1990) Sensitivity to and degradation of pentachlorophenol by *Phanerochaete* spp, *Appl. Environ. Microbiol.* **56**, 3519-3526.

119. Lamar, R.T., and Dietrich, D.M. (1990) In situ depletion of pentachlorophenol from contaminated soil by *Phanerochaete* spp, *Appl. Environ. Microbiol.* **56**, 3093-3100.

120. Takada, S., Nakamura, M., Matsueda, T., Kondo, R., and Sakai, K. (1996) Degradation of polychlorinated dibenzo-*p*-dioxins and polychlorinated dibenzofurans by the white rot fungus *Phanerochaete sordida* XK-624, *Appl. Environ. Microbiol.* **62**, 4323-4328.

121. Bogan, B.W., and Lamar, R.T. (1996) Polycyclic aromatic hydrocarbon-degrading capabilities of *Phanerochaete laevis* HHB-1625 and its extracellular ligninolytic enzymes, *Appl. Environ. Microbiol.* **62**, 1597-1603.

122. Pérez, J., Saez, L., de la Rubia, T., and Martínez, J. (1997) *Phanerochaete flavido-alba* ligninolytic activities and decolorization of partially bio-depurated paper mill wastes, *Wat. Res.* **31**, 495-502.

123. Pérez, J., de la Rubia, T., Hamman, B., and Martínez, J. (1998) *Phanerochaete flavido-alba* laccase induction and modification of manganese peroxidase isoenzyme pattern in decolorized olive oil mill wastewaters, *Appl. Environ. Microbiol.* **64**, 2726-2729.

124. Lamar, R.T., Evans, J.W., and Glaser. J.A. (1993) Solid-phase treatment of a pentachlorophenol-contaminated soil using lignin-degrading fungi, *Environ. Sci. Technol.*. **27**, 2566-2571.

125. Roy-Arcand, L., and Archbald, F.S. (19991) Direct dechlorination of chlorophenolic compounds by laccase from *Trametes (Coriolus) versicolor, Enzyme Microb. Technol.* **13**, 194-203.

126. Iimura, Y., Hartikainen, P., and Tatsumi, K. (1996) Dechlorination of tetrachloroguaiacol by laccase of white rot basidiomycete *Coriolus versocolor, Appl. Microbiol. Biotechnol.* **45**, 434-439.

127. Johannes, C., Majcherczyk, A., and Huttermann, A. (1998) Oxidation of acenaphthene and acenaphthylene by laccase of *Trametes versicolor* in a laccase-mediator system, *J. Biotechnol.* **61**, 151-156.

128. Majcherczyk, A., Johannes, C, Huttermann, A. (1998) Oxidation of polycyclic aromatic hydrocarbons (PAH) by laccase of *Trametes versicolor, Enzyme Microbial Technol.* **22**, 335-341.

129. Majcherczyk, A., Johannes, C., and Hutterman, A. (1999) Oxidation of aromatic alcohols by laccase from *Trametes versicolor* mediated by the 2,2´-azino-bis-(3-ethylbenzothiazoline-6-sulphonic acid) cation radical and dication, *Appl. Microbiol. Biotechnol.* **51**, 267-276.

130. Ricotta, A., Unz, R.F., and Bollag, J.-M. (1996) Role of laccase in degradation of pentachlorophenols. *Bull. Environ. Contam. Toxicol.* **57**, 560-567.

131. Grey, R., Hofer, C., and Schlosser, D. (1998) Degradation of 2-chlorophenol and formation of 2-chloro-1,4-benzoquinone by mycelia and cell-free crude culture liquids of *Trametes versicolor* in relation to extracellular laccase activity, *J. Basic Microbiol.* **38**, 371-382.

132. Wang, Y., and Yu, J. (1998) Adsorption and degradation od synthetic dyes on the mycelium of *Trametes versicolor, Wat. Sci. Tech.* **38**, 233-238.

133. Swamy, J., and Ramsay, J.A. (1999) Effect of glucose and NH4$^+$ concentrations on sesquential dye decolorization by *Trametes versicolor, Enzyme Microbial Technol.* **25**, 278-284.

134. Tuomela, M., Lyytikainen, M., Oivanen, P., and Hatakka, A. (1999) Mineralization and conversion of pentachlorophenol (PCP) in soil inoculated with the white-rot fungus *Trametes versicolor, Soil Biol. Biochem.* **31**, 65-74.

135. Novotný, Č., Erbanová, P., Šašek, V., Kubátová, A., Cajthaml, T., Lang, E., Krahl, J., and Zadražil, F. (1999) Extracellular oxidative enzyme production and PAH removal in soil by exploratory mycelium of white rot fungi, *Biodegradation* **10**, 159-168.

136. Vyas, B.R.M., Šašek, V., Matucha, M., and Bubner, M. (1994) Degradation of 3,3´,4,4´-tetrachlorobiphenyl by selected white rot fungi, *Chemosphere* **28**, 1127-1134.

137. in der Wiesche, C., Martens, R., and Zadrazil, F (19956.: Two-step degradation of pyrene by white-rot fungi and soil microorganisms, *Appl. Microbiol. Biotechnol.* **46**, 653-659.

138. Wolter, M., Zadrazil, F., Martens, R., and Bahadir, M. (1997) Degradation of eight highly condensed aromatic hydrocarbons by *Pleurotus* sp. Florida in solid wheat straw substrate, *Appl. Microbiol. Biotechnol.* **48**, 398-404.

139. Schutzendubel, A., Majcherczyk, A., Johannes, C., and Huttermann, A. (1999) Degradation of fluorene, anthracene, phenanthrene, and pyrene lacks connection to the production of extracellular enzymes by *Pleurotus ostreatus* and *Bjerkandera adusta, Internat. Biodeter. Biodegrad.* **43**, 93-100.

140. Kotterman, M.J.J., Rietberg, H.-J., Hage, A., and J.A. Field (1998) Polycyclic aromatic hydrocarbon oxidation by the white-rot fungus *Bjerkandera* sp. strain BOS55 in the presence of nonionic surfactants, *Biotechnol. Bioeng.* **57**, 221-227.

141. Field, J.A., Boelsma, F., Baten, H., and Rulkens, W.H (1995). Oxidation of anthracene in water/solvent mixtures by the white-rot fungus *Bjerkandera* sp. strain BOS55, *Appl. Microbiol. Biotechnol.* **44**, 234-240.

142. Kotterman, M.J.J., Vis, E.H., and Field, J.A. (1998) Successive mineralization and detoxification of benzo[a]pyrene by the white rot fungus *Bjerkandera* sp. strain BOS55 and indigenous microflora, *Appl. Environ. Microbiol.* **64**, 2853-2858.

143. Field, J.A., Baten, H., Boelsma, F., and Rulkens, W.H. (1996) Biological elimination of polycyclic aromatic hydrocarbons in solvent extracts of polluted soil by the white rot fungus, *Bjerkandera* sp. strain BOS55, *Environ. Technol.* **17**, 317-323.

144. Scheibner, K., Hofrichter, M., Herre, A., Michels, J., and Fritsche, W. (1997) Screening for fungi intensively mineralizing 2,4,6-trinitrotoluene, *Appl. Microbiol. Biotechnol.* **47**, 452-457.

145. Scheibner, K., Hofrichter, M., and Fritsche, W. (1997) Mineralization of 2-amino-4,6-dinitrotoluene by manganese peroxidase of the white rot fungus *Nematoloma frowardii, Biotechnol. Lett.* **19**, 835-839.

146. Scheibner, K., and Hofrichter, M. (1998) Conversion of aminonitrotoluenes by fungal manganese peroxidase, *J. Basic Microbiol.* **38**, 51-59.

147. Hofrichter, M., Scheibner, K., Schneegass, I., Ziegenhagen, D., and Fritsche, W. (1998) Mineralization of synthetic humic substances by manganese peroxidase from the white-rot fungus *Nematoloma frowardii, Appl. Microbiol. Biotechnol.* **49**, 584-588.

148. Wunch, K.G., Feibelman, T., and Bennet, J.W. (1997) Screening for fungi capable of biodegrading benzo[a]pyrene, *Appl. Microbiol. Biotechnol.* **47**, 620-624.

149. Wunch, K.G., Alworth, W.L., and Bennet, J.W. (1999) Mineralization of benzo[a]pyrene by *Marasmiellus troyanus*, a mushroom isolated from a toxic waste site, *Microbiol. Res.* **154**, 75-79.

150. Van Aken, B., Skubisz, K., Naveau, H., and Agathos, S.N. (1997) Biodegradation of 2,4,6-trinitrotoluene by the white-rot basidiomycete *Phlebia radiata, Biotechnol. Lett.* **19**, 813-817.

151. Van Aken, B., Godefroid, L.M., Peres, C.M., Naveau, H., and Agathos, S.N. (1999) Mineralization of 14C-U.ring labeled 4-hydroxylamino-2,6-dinitrotoluene by manganese-dependent peroxidase of the white-rot basidiomycete *Phlebia radiata, J. Biotechnol.* **68**, 159-169.

152. Seto, M., Nishibori, K., Masai, E., Fukuda, M., and Ohdaira, Y. (1999) Degradation of polychlorinated biphenyls by a "Maiake" mushroom, *Grifola frondosa, Biotechnol. Lett.* **21**, 27-31.

153. Rama, R., Mougin, C., Boyer, F.-D., Kollmann, A., Malosse, C., and Sigoillot, J.-C. (1998) Biotransformation of benzo[a]pyrene in bench scale reactor using laccase of *Pycnoporus cinnabarinus, Biotechnol. Lett.* **20**, 1101-1104.

154. Pickard, M.A., Roman, R., Tinoco, R., and Vazquez-Duhalt, R. (1999) Polycyclic aromatic hydrocarbon metabolism by white rot fungi and oxidation by *Coriolopsis gallica* UAMH 8260 laccase, *Appl. Environ. Microbiol.* **65**, 3805-3809.

155. Bhatt, M., Patel, M., Rawal, B., Novotný, Č., Molitoris, H.P., and Šašek, V. (1999) Biological decolorization of the synthetic dye RBBR in contaminated soil, *World J. Microbiol. Biotechnol.* **16**, 195-198.

156. Šašek, V., Novotný, C., Erbanová, P., Bhatt, M., Cajthaml, T., Kubátová, A., Dosoretz, C., Rawal, B., and Molitoris, H.P. (1999) Selection of ligninolytic fungi for biodegradation of organopollutants, in A. Leason and B.C. Alleman (eds.), Phytoremediation and Innovative Strategies for Specialized Remedial Applications, The Fifth International In Situ and On-Site Bioremediation Symposium, San Diego, California, April 19-22, 1999, Battelle Press, Columbus, Richland, pp. 69-74.

157. Radtke, C., Cook, W.S., and Anderson, A. (1994) Factors affecting antagonism of the growth of *Phanerochaete chrysosporium* by bacteria isolated from soil, *Appl. Microbiol. Biotechnol.* **41**, 274-280.

158. Loske, D., Hüttermann, A. Majcherzyk, A., Zadrazil, F., and Lorsen, H. (1990) Use of white rot fungi for the clean-up of contaminated sites, in M.P. Coughlan and M.T.A. Collaco (eds.), *Advances of Biological Treatment of Lignocellulosic Materials,* Elsevier Applied Science, London , pp. 311-321.

159. Eschenbach, A., Kästner, M., Wienberg, R., and Mahro, B. (1995) Microbial PAH degradation in soil material from a contaminated site - mass balance experiments with *Pleurotus ostreatus* and different 14C-PAH, in W.J. van der Bring, R. Bosman, and F. Arendt, (eds.), *Contaminated Soil '95.* Kluwer Academic Publishers, the Netherlands, pp. 377-378.

160. Qiu, X.J., and McFarlan, M.J. (1991) Bound residue formation in PAH contaminated soil composting using *Phanerochaete chrysosporium, Hazard. Waste Hazard. Mater.* **8**, 115-126.

161. McFarlan, M.J., Qiu, X.J., Sims, J.L., Randolph, M.E., and Sims, R.C. (1992) Remediation of petroleum impacted soils in fungal compost bioreactors, *Water Sci, Technol.* **25**, 197-206.

162. McFarlan, M.J., and Qiu, X.J. (1995) Removal of benzo[a]pyrene in soil composting systems amended with the white rot fungus *Phanerochaete chrysosporium, J. Hazard. Mater.* **42**, 61-70.

163. Eggen, T., and Majcherczyk A, (1998) Removal of polycyclic aromatic hydrocarbons (PAH) in contaminated soil by white rot fungus *Pleurotus ostreatus*, *Internat. Biodet. Biodegrad.* **41**, 111-117.

164. Bogan, B.W., Lamar, R.T., Burgos, W.D., and Tien, M, (1999) Extent of humifiation of anthracene, fluoranthene, and benzo[a]pyrene by *Pleurotus ostreatus* durin growth in PAH-contaminated soil, *Lett. Appl. Microbiol.* **28**, 250-254.

165. Ruttimann-Johnson, C., and Lamar, R.T. (1997) Binding of pentachlorophenol to humic substances in soil by the action of white rot fungi, *Soil Biol. Biochem.* **29**, 1143-1148.

166. Dawel, G., Kastner, M., Michels, J., Poppitz, W., Gunther, W., and Fritsche, W. (1997) Structure of the laccase-mediated product of coupling of 2,4-diamino-6-nitrotoluene to guiacol, a model for coupling of 2,4,6-trinitrotoluene metabolites to a humic organic soil matrix, *Appl. Environ. Microbiol.* **63**, 2560-2565.

167. Verstraete, W., and Devliegher, W. (1996) Formation of non-bioavailable organic residues: Perspectives for site remediation, *Biodegradation* **7**: 471-485.

168. Mouin, C., Pericaud, C., Dubroca, J., and Asther, M. (1997) Enhanced mineralization of lindane in soil supplemented with the white rot basidiomycete *Phanerochaete chrysosporium*, *Soil. Biol. Biochem.* **29**, 1321-1324.

169. Entry, JA., Donnelly, P.K., and Emmingham, W.H. (1996) Mineralization of atrazine and 2,4-D in soils inoculated with *Phanerochaete chrysosporium* and *Trappea darkeri*, *Appl. Soil. Ecol.* **3**, 85-90.

170. George, E.J., and Neufeld, R.D. (1988) Degradation of fluorene in soil by fungus *Phanerochaete chrysosporium*, *Biotechnol. Bioeng.* **33**, 1306-1310.

171. Šašek, V., Volfová, O., Erbanová, P., Vyas, B.R.M., and Matucha, M. (1993) Degradation of PCBs by white rot fungi, methylotrophic and hydrocarbon utilizing yeasts and bacteria, *Biotechnol. Lett.* **15**, 521-526.

172. Morgan, P., Lee, A.S., Lewis, S.T., Sheppard, A.N., and Watkinson, R.J. (1993) Growrh and biodegradation by white rot fungi inoculated into soil, *Soil Biol. Biochem.* **25**, 279-287.

173. Bogan, B.W., Shoenike, B., Lamar, R.T., and Culen, D. (1996) Manganese peroxidase mRNA and enzyme activity levels during bioremediation of polycyclic aromatic hydrocarbon-contaminated soil with *Phanerochaete chrysosporium*, *Appl. Eniron. Microbiol.* **62**, 2381-2386.

174. Chung, N., and Aust, S.D. (1995) Degradation of pentachlorophenol in soil by *Phanerochaete chrysosporium, J. Hazard. Mat.* **41**, 177-183.

175. Zaddel, A., Majcherczyk, A., and Huttermann, A. (1993) Degradation of polychlorinated biphenyls by white-rot fungi *Pleurotus ostreatus* and *Trametes versicolor* in a solid state systém, *Ecotoxic. Environment. Chem.* **40**, 255-266.

176. Kubátová, A., Erbanová, P., Eichlerová, I., Homolka, L., Nerud, F., and Šašek, V. (2001) PCB congener selective biodegradation by the white rot fungus *Pleurotus ostreatus* in contaminated soil, *Chemosphere* **43**, 207-215.

177. Eggen, T., and Sveum, P. (1999) Decontamination of aged creosote polluted soil: the influence of temperature, white rot fungus *Pleurotus ostreatus*, and pretreatment, *Internat. Biodeter. Biodegrad.* **43**, 125-133.

178. Eggen, T. (1999) Application of fungal substrate from commercial mushroom production –*Pleurotus ostreatus* – for bioremediation of creosote contaminated soil, *Internat. Biodeter. Biodegrad.* **44**, 117-126.

179. Okeke, B.C., Paterson, A., Smith, J.E., and Watson-Craik, I.A. (1997) Comparative biotransformation of pentachlorophenol in soil by solid substrate cultures of *Lentinula edodes*, *Appl. Microbiol. Biotechnol.* **48**, 563-569.

FROM LABORATORY TO INDUSTRIAL SCALE: COMPOSTING OF POLLUTED SOILS FROM FORMER COAL INDUSTRY AND GAS PLANTS: FUTURE RESEARCH NEEDS

H.C. DUBOURGUIER
Institut Supérieur d'Agriculture
41, rue du Port, 59046 Lille Cedex, France

1. Introduction

Activities related to coking and, more generally, to transformation of coal, have been quite varied and underwent an intense evolution from the 18th to the 20th century. The pollutants which are likely to be present in the soils at these sites depend on the type of products being made at the site, the technical knowledge at the time of production, and the environmental policy of the company. More than thirty techniques of treatment of polluted soil from similar sites have been described and summarized in reviews published on behalf of the Environmental Protection Agency (EPA) [1–4]. The various treatment techniques can be classified into three main categories according to their action on the pollutants:

— chemical or physical immobilization
— extraction from the soil matrix
— destruction

Some of these techniques can be applied *in situ* or after excavation (*ex situ*). After excavation, the polluted material can be treated on site in a confined area, or be transported off site to a center specialized in the selected technique. The environmental risk and the financial cost inherent in transport made *in situ* and on-site treatments often advantageous. However, the application of this type of treatment is often severely limited, in particular by the nature of the soil matrix and of the contaminants [5, 6].

Destructive techniques consist mainly of various thermal treatments and bioremediation. Biological techniques are generally less expensive than the thermal treatments and have the further advantage of a positive "green image" for the general public. Bioremediation techniques do not modify the intrinsic nature of the soil, and thus revegetation can be considered after treatment. It is thus not surprising that these techniques are receiving much attention.

The use of microorganisms for wastewater treatment is not a new concept and the bioremediation of polluted soils developed from the principles established in this

V. Šašek et al. (eds.),
The Utilization of Bioremediation to Reduce Soil Contamination: Problems and Solutions, 267–283.

field. Bioremediation has been the subject of many publications [1; 7−14]. Bioremediation exploits the capacity of some microorganisms, like heterotrophic bacteria and fungi, to degrade the complex organic compounds to simple molecules such as water, carbon dioxide, and methane. The microorganisms can either use the organic pollutant like source of carbon and energy for their growth, or to transform indirectly by cometabolism [1]. The techniques of biotreatment have to meet all the environmental conditions necessary for the biological breakdown of the pollutant as fast and complete as possible. Thus, biological treatments require the optimization of the following parameters:

— existence of a microbial population able to degrade the pollutants
— biodegradability of the pollutants
— satisfaction of the metabolic needs for the microorganisms (terminal electron acceptors, C/N/P ratios, micronutrients, substrate for growth in the event of cometabolism)
— physical conditions compatible with the microorganisms, e.g. temperature, pH and moisture
— the absence of inhibitors of the microorganisms [4, 6]

The techniques of bioremediation are applicable to a broad range of organic contaminants. On the other hand, the application of these techniques to inorganic compounds is not possible except for nitrates. Biological methods nevertheless are studied for immobilization of heavy metals or, on the contrary, for their leaching out of the matrix. Bioremediation may be used to treat sludges, sediments, soils and groundwaters. Bioremediation techniques for polluted soils can be used alone, or in combination or in association with other types of techniques, such as washing.

2. Solid-Phase Bioremediation

The techniques of land farming, biopiles and composting are often gathered under the designation of solid-phase treatment as opposed to slurry-phase treatment in bioreactors. Indeed, these techniques treat the soil in its present state, possibly after screening, but without mixing it with a liquid phase. These techniques in general require quite large areas and long-term treatment, i.e. from a few months to two or three years.

2.1. LAND FARMING

This technique has been used for over forty years in oil industry for solid oily waste processing [1]. It consists of spreading out, in thin layers, the material to be cleaned over a soil which undergoes the then traditional agricultural practices of fertilization and irrigation. Ploughing ensures aeration and mixing of the soils. However, due to the risk of leaching of pollutants, this basic technique has been modified. Thus, the polluted soil is now placed in a confined area, built on site. Microorganisms, surfactants or bulking agents, such as straw, are sometimes used to improve the performance of the treatment. This technique is particularly adapted to the biological breakdown of complex compounds, such as fuels, diesel, some pesticides, residues of the treatment of wood (creosote and pentachlorophenol) and waste of coking plants (PAHs) [3]. The principal advantages of this technique lie in the simplicity of its setup and operation, in the improvement of physical qualities of the soil and finally in its low cost [15]. The principal disadvantages are [3]

— the difficulty in controlling parameters like the outside temperature, *i.e.*, making the treatment difficult in winter
— losses by volatilization or generation of dust during mixing.

2.2. COMPOSTING

Historically, composting has been used for organic wastes, wastewater sludges and green wastes from food and feed industry. More recently, this technique was applied to toxic wastes and polluted soils [9]. Composting consists in excavating the polluted soil, then mixing it with organic fertilizers and bulking agents. Various amendments may be used, such as straw, wood shavings or bark, sawdust, plant wastes, such as stalks of corn, or even manures which also constribute microbial inocula. The soil–structuring agent ratio may be high (30–50 % in mass), which often involves, during initial mixing, an important increase in volume of the material to be cleaned. After mixing, the material is laid out in regularly spaced windrows, approximately 1 meter in height, which are mechanically homogenized to allow their aeration. Composting is less often carried out in reactors or in static heaps, aerated by a network of drains. The aerobic composting process proceeds in four different phases: the mesophilic phase (35–55 °C), the thermophilic phase (55–75 °C), the cooling phase, and a final phase of maturation. During these phases, a particular microflora develops: sporulating bacteria and thermophilic fungi during the thermophilic phase, and fungi with bacteria with heat-resistant spores during the cooling phase. Aerobic composting is used for the treatment of soils polluted by petroleum hydrocarbons, pesticides, explosives and PAHs. It can also be considered for cyanides. Compared to land-farming, this technique is not limited by the outside temperature in winter but control of fugitive gas emissions, in particular during the thermophilic phase, may be necessary [3].

2.3. BIOPILES

The excavated soil is mixed with fertilizers and bulking agents, and sometimes inoculated with adapted microflora. The soil–structuring agent ratio is generally lower than in composting. After mixing, the soil is laid out in confined piles a few meters high. These piles may range from approximately 5000–20 000 m^3. The soils are isolated from the ground and are covered by an impermeable cover. Aeration is ensured by a network of drains within the pile. The gas emissions can thus be collected and treated if necessary, *e.g.* for volatile organic compounds. Humidity is often controlled by water sprinkling, which also allows the addition of nutrients during the treatment. Leachates are collected by a drainage system. In the most sophisticated systems, they are treated in a bioreactor before being recycled. The effectiveness of this technique is recognized for light petroleum hydrocarbons and the nonhalogenated volatile organic compounds. Its application is more restricted and discussed with regard to halogenated compounds, pesticides, explosives and semi-volatile compounds [3]. In contrast to land farming and composting, the absence of homogenization during the treatment can lead to a nonhomogeneous removal of pollutants, located preferentially around the ventilation drains. Consequently, the control of the residual concentrations at the end of the treatment must be rigorous.

2.4. APPLICATION OF SOLID-PHASE BIOREMEDIATION

In a general way, the choice of a technique is based on the characteristics of the polluted site and its contamination, the residual concentrations that must be achieved, and the cost. The choice of a technique of remediation and its use is integrated in a global strategy of site restoration that public authorities (in France: *Department of Environment, D.R.I.R.E.*) get busy to order and codify. Thus, this choice is done only after a thorough diagnosis of the site and its pollution. Feasibility studies are often necessary to confirm the relevance of selected techniques of restoration [16, 17].

The limitations of bioremediation lie mainly in the proportion of pollutant which can be degraded by the microorganisms. It is generally considered that bioremediation is not adapted for treatment of soils that are highly polluted by contaminants for which the maximum permissible concentrations are low, and particularly when they are difficult to degrade [1]. The effectiveness of bioremediation is also limited by the proportion of pollutants available to the microorganisms [18, 19]. For organic compounds, this bioavailability depends on the nature of the soil, the capacity of the pollutant considered to bind to the elements of the soil and the age of pollution. Lastly, bioremediation never leads to a complete mineralization of the pollutants. The reduction of soil toxicity after treatment is thus a primary objective [17].

2.5. EFFICIENCY OF SOLID-PHASE BIOREMEDIATION APPLIED TO PAHs

The application of bioremediation to treat soils of abandoned lands from coal-chemical industry was reviewed by Mueller *et al.* [20] for the sites contaminated by creosote, by Thomas and Lester [21] for gasworks sites and by Wilson and Jones [17] for soils polluted by PAHs in general. In this field, and particularly with regard to pilot-scale or industrial-scale tests, many papers report good results without providing numerical criteria supporting their scientific validity. Shannon and Unterman [22] underlined in detail the existence of publications that were more commercial than scientific.

A few reports concern industrial soils highly polluted by PAHs. Ellis *et al.* [23] reported on the treatment of 3500 m^3 soil containing initially 1024 mg PAHs/kg dry mass. The excavated soil was treated on a confined surface 0.75 m thick. The collected leacheates were used to maintain a water content of 20 %. The soils were plowed every week and received fertilizers, microorganisms and surfactants of the same type as those previously used for the *in situ* treatment of the same soils. After 88 days of treatment, the removal yield, based on the sum of the total PAHs, reached 68 %. These authors pointed out that the removal yield of PAHs decreased with the number of rings. They also observed that the solid-phase treatment and the *in situ* treatment were completely inhibited when the average monthly temperatures fell below 5 °C.

Warith *et al.* [24] evaluated a solid-phase treatment of soils from a hydrocarbon gasification plant, containing initially 335 mg PAHs/kg. The treatment, applied to 5000 m^3 of soil, consisted of maintaining a C : N : P ratio of 100 : 10 : 2 by addition of ammonium nitrate and orthophosphate. The soils were plowed once a day, and were inoculated with microorganisms four times a week. The moisture content of the soils was maintained at $\approx$23 %. After 48 days, the removal yields were 86 % for the total PAHs, 92 % for the 2- and 3-ring PAHs, 80 % for the 4-ring PAHs and 65 % for the 5-ring PAHs.

Hund and Traunspurger [25] followed the reduction in the soil toxicity containing initially $\approx$4.5 g PAHs/kg during their treatment in biopiles. After 13 months, a removal yield of 65 % of the total PAHs was obtained. This yield was mainly due to the degradation of 3- and 4-ring PAHs. No significant reduction of the 5- and 6-ring PAHs was observed. The reduction of the ecotoxicity of the treated soil was evaluated by three types of tests: toxicity of the aqueous extracts for microorganisms and fish, survival of plants or animals introduced into the soil, and study of the soil microflora and microfauna. All these tests showed that the toxicity of the soil decreased with the concentration in total PAHs.

Mueller *et al.* [26] studied laboratory-scale solid-phase treatment of soils and sediments contaminated by creosote and pentachlorophenol. The experiments were done with 3 kg soil in closed jars to determine losses by leaching and volatilization. These losses proved to be extremely small ($\approx$100 ppm). The fertilizer addition (in particular the 17 mg of ammonium nitrate and potassium phosphate per kg soil) had a positive effect on the biological breakdown of the heavy PAHs, but little effect on the biological breakdown of the light ones. However, the concentration of light PAHs was significantly reduced whereas the heavy PAHs were not significantly degraded.

From what we have been able to determine, the feasibility of bioremediation of soils highly polluted by coal-processing plants was still a possiblitiy in 1995. Indeed, if the major part of light residues likely to be found on former coal-industry sites are biodegradable under aerobic conditions, their soils present acute pollution from highly polymerized tar and pitch. However, the biological breakdown of PAHs of more than 5 rings was generally regarded as difficult and incomplete. Moreover, the biodegradability of PAHs is likely to be very limited by their notorious insolubility and their capacity to be bound with organic material and clays. Just a few studies were available on the biotreatment of very highly polluted soils (PAH > 5 g/kg dry mass). Thus, this research was undertaken in our laboratory in 1993, working only with highly polluted soils, often found in the Northern region of France.

3. Characteristics of the Soils Used in This Report

Four soils were used during this five-year research project. They were sampled or excavated from former coal industry sites and from a former gas site. The physico-chemical analyses done on the soils before treatment not only aimed to evaluate their extent of pollution, but also to determine the amendments that would be neces-sary to allow the biological breakdown of the pollutants. To estimate the extent of pollution of these soils, they were compared with the agricultural soils located near the sites from which they were sampled, as well as with the Dutch and Canadian standards for the quality of soils. These comparisons emphasized a primarily organic pollution. Although cyanide and heavy-metal concentrations are often high, they do not exceed the intervention guide levels, except for mercury in soil C. The inhibition of the mineralization of nitrogen and nitrification has been demonstrated for con-centrations of ≈ 1 g/kg of Zn, Cu, Ni; 100−500 mg/kg Pb, Cr; 10−100 mg/kg Cd [27]. Only soil A contains lead concentrations that could be limiting for the biore-mediation efficiency.

The extent of pollution by the PAHs was different: very high for soils C and A (total PAHs > 25 g/kg dry mass) and lower for soils D and E (total PAH < 5 g/kg dry mass). Such concentrations were already mentioned for soils of coal wastelands [17]. However, to our knowledge, few published reports have dealt with the biore-mediation of soils at concentration >5 g PAH/kg dry mass.

Moreover, organic matter extractable with an organic solvent (EOMS) repre-sents only a small proportion (8−30 %) of the total organic matter (TOM), esti-mated by mass losses at 550 °C. Thus, the organic pollution of these soils seems to be made up of highly polymerized organic matter which cannot be extracted with a solvent. This may prohibit their analysis by chromatographic techniques. Conse-quently, it was important to follow the evolution of EOMS and TOM during the bioremediation process, in order to evaluate the effect of the treatment on the unre-solved organic compounds.

The PAHs on the EPA list represent 15−38 % of the mass of the EOMS. GC−MS analyses of extracts of one of the most polluted soils identified 169 other organic compounds besides the 16 PAHs. In this extract, the 16 PAHs account for only 9.5 % of the total number of compounds located on the chromatograms. Among those identified, compounds considered carcinogenic were emphasized, *e.g.* carbazol or acridine and their derivatives. The presence of these compounds could also have a direct influence on the bioremediation process.

4. Laboratory Composting

Initial experiments were done in aerated 1-kg mini-windrows. With soil A, the kinetics showed a biodegradation of most of the PAHs, except four that persisted after the eighth month of treatment. After 8 months, the residual levels were dependent on the initial concentrations of the PAHs in the sample. This could be due either to the slow degradation rates (and thus with the premature stopping of the experiment), or to a threshold of bioavailability of the PAHs whose values would be related to the PAH being considered, the age of the pollution and the nature of the soils. These assumptions will be addressed in detail later. These experiments also show an improvement of the degradation of the PAHs by the initial addition of complex organic by-products from the food industry. The analysis of P, N, and organic C of each added amendment suggested that the origin and concentration of N and P had a major influence on the PAH degradation yields. An increase in the concentrations of these two elements resulted in an increase of the PAH degradation rates. However, there was a threshold beyond which no additional increase of degradation could be observed.

Factorial designed experiments were then done in 10 kg mini-windrows with soil D. Average degradations of 58 % were obtained for the total PAHs, 80 % for the light PAHs and 67 % for the 4-ring PAHs, after 130 days in mini-windrows. This relatively short experimental duration may explain why the PAHs with more than 4 rings, which were degraded after eight months with soil A, were not degraded, except for benzo[a]pyrene and indeno[c,d]pyrene. The initial inoculation of the soils by selected microorganisms was not a significant factor in these tests. This can be explained by the incapacity of the bacterial and fungal inocula to compete with the soil natural microflora. However, on this experimental scale ($\approx$10 kg of soil), the transfers of heat (soil to air) are increased and no rise in composting temperature was observed, contrary to the pilot or industrial experiments. The inoculum effect can thus be masked. With regard to the role of the amendments, the effect of N and P was observed as previously. The five studied factors do not have the same influence on the degradation of PAHs. The particularly marked effect of straw on the rate of disappearance of the PAH with 4, 5 and 6 cycles like chrysene and indeno[c,d]pyrene is due to a preferential growth of filamentous fungi on this amendment. The fungi, in particular the lignolytic ones, were stongly suggested in the degradation of the high-molar-mass PAHs. Indeed, the lignolytic enzymes are extracellular and are not specific. The statistical tool used (factorial design) confirmed the effectiveness of straw.

However, this factorial design technique does not allow a determination of the optimum conditions for treatment, that required a further study with another type of experimental plan. Indeed, all the studied parameters had a nonlinear effect on the rate of disappearance of the PAHs, which was also clearly seen in the previous studies with mini-windrows for N and P. Moreover, homogenization of large quantities of the polluted soil (400 kg in this experiment) is difficult in the laboratory. Thus, working with very heterogeneous soils like those studied, the number of repetitions should be increased to reduce the uncertainty.

5. Pilot-Scale Composting

Because of the regulations in France, the pilot experiments were done in a special controlled greenhouse, specially built for these tests, in collaboration with *APINOR Company* (Rieulay, France). Windrows of 3.5 tons of polluted soil were used. These experimental conditions approached industrial conditions, *i.e.* initial preparation and mixing of the windrows with a mechanical engine, regularly spaced reversals of the windrows, monitoring and correction of the water content. Various treatment conditions were studied including the source of nitrogen and the inoculum employed. The first experiments ran for five months. In a second phase, the best windrows were followed over two years in order to evaluate the limits of this bioremediation technique.

In the first five months of treatment, high rates of disappearance (>70 %) could be reached in the treated soils, including PAHs with 4 and even 5 rings like pyrene, fluoranthene or benzo[*a*]pyrene. On the other hand, whatever the conditions of treatment, the MOES did not decrease significantly in five months. All the PAHs were not degraded in similar proportion. The median of the individual degradation yields of each PAH was the best estimate of the total yield of cleanup. On this basis, the various conditions of treatment were compared. The best results, *i.e.* total yield of cleanup exceeding 80 %, were obtained when the soil was supplemented and inoculated. The use of liquid amendments like some soluble fractions from potato industry, although interesting for their high content in N and P and for good results during these tests, induce a compaction of the soils during the treatment. On the other hand, the solid amendments, like dry pulps, supported the maintenance of a soil structure that improved aeration. The other amendments employed and, to a lesser extent, the inoculation of the soils were responsible for a phenomenon important for composting, *i.e.* increase in temperature to >50 °C. This increase in temperature had a negative effect on the mesophilic microflora. The inhibition of the biological activity noticed after 90 days of treatment was attributed not only to the temperature increase, but also to the reduction in the concentration of ammonia nitrogen. At the end of the five month treatment, the residual concentration obtained was ≈5 g PAHs/kg dry mass in the best windrows. Compared with the Dutch guidelines, these residual concentrations were more than one-hundred times higher than the intervention value (40 mg PAHs/kg dry mass), and thus not acceptable.

Windrows 4 and 5 gave the best yields, and thus further treatment was done on them. The composting technique was modified in order to limit the temperature increases. This was done by the splitting of N and P supplementation. Ammonium nitrate was also used together with the by-products of the potato industry to bring nitrogen in a form directly usable by the microorganisms. So, windrows 4 and 5 were treated further. They were re-inoculated and complemented with dry potato by-products, ammonium nitrate and straw. This supplementation was carried out several times during the treatment period. The calculation of supplementation carried out throughout the treatment was done according to EOMS and total PAHs content.

During two and a half years of treatment, the pH of the windrows gradually decreased to 6.2 at the end of the treatment. The disturbances of the pH, observed following the amendments, were probably due to ammonium nitrate, then to ammonium consumption. After each amendment addition, temperature remained <45 °C, due to fractionation. These temperature increases corresponded to the fermentation of organic matter that was in the supplements and to the phenomenon of composting. At the end of the treatment, the important increase of total N, P and S, was higher than the input due to amendments. This accumulation may result from mineralization of unknown organic matter during the treatment. Thus, standardized methods do not take into account the real concentrations of other organics in industrial soils polluted by tar and pitch.

The preferential source of nitrogen used for the biological activity appears to be ammonium. A partial nitrification was also noticed at the end of the treatment. After 2.8 years of treatment, cleanup yields from 80 to 95 % were reached for soils initially containing near 30 g PAHs/kg dry mass. The residual concentrations obtained were respectively 1.49 ± 0.32 g PAHs/kg dry mass for windrow 4 and 3.1 ± 0.12 g PAHs/kg dry mass for windrow 5. The 2-ring PAHs and heterocyclic compounds were not detected at the end of the treatment. The persistent PAHs are primarily compounds with 4 rings, in particular fluoranthene, pyrene and benzo-[b]fluoranthene. Nevertheless, the yield of disappearance of some heavy compounds, such as benzo[a]pyrene, benzo[a]anthracene and pyrene, was remarkably high (>75 %). These results were in apparent contradiction with work of Mueller et al. [26] which showed total recalcitrance of PAHs with more than 4 rings. This is probably due to differences in treatment duration and in aeration. The kinetics of disappearance of the PAH approached first order.

The majority of cleanup was completed during the first five months. After that period, the yields of PAH disappearance evolved more slowly. This was not related to a deficit of N, P or any other essential element since a progressive accumulation of the majority of the proportioned elements was noticed. It was not due to the accumulation of toxic products since the heterotrophic total flora was almost constant during the treatment. Thus, the rate of PAH disappearance was probably limited by the bioavailability of these compounds. However, other unknown limitations may be involved since important degradation yields were observed for some very insoluble non-volatile compounds.

6. Industrial-Scale Composting

The results obtained in pilot-scale experiments justified scaling-up to commercial size. Detailed studies were done on two industrial scale remediations: the first one on 1500 tons of a highly contaminated soil from a former coking plant (soil C), the other on 20 tons of a slightly polluted soil from a former gas site (soil E).

The composting technique was identical for these remediations. It consisted of inoculating the soils at the beginning of treatment with an enrichment of thermophilic indigenous microflora of the corresponding soil. By-products of agro-alimentary industry, agricultural amendments and straw were used. The additions of amendments were done periodically during the treatment period. The parameters that were monitored during the treatment were similar to pilot tests.

6.1 TREATMENT OF 1500 TONS STRONGLY POLLUTED SOILS FROM A COKING PLANT

The treatment was done on site, in the open air, over a period of 625 days. The total quantity of soil C was distributed in 4 confined windrows. The main windrow contained 600 tons of soil, the other three each had 300 tons. The leachates collected from the windrows were recycled to sprinkle the soils during the treatment period. Homogenization and aeration were done by periodic mixing of the soils. The addition of the amendments induced a heating of the soils as during the pilot experiments. This allowed the treatment to proceed at a high rate outside even during winter.

According to the oxidizable OM concentrations at the beginning and at the end of the treatment, the added amendments appear to be almost completely degraded. The effect of dilution of the soil, caused by these additions, was negligible. The stability of heavy-metal concentrations reinforced this assumption.

During the first two months of treatment, an increasing homogeniety of the soils was noticed. This homogeniety caused an increase in the average concentration of PAHs, and thus masked the starting point of the biodegradation. Mixing with a mechanical shovel appeared to be slower than in the pilot-scale windrows.

After 625 days of treatment, a degradation yield of 75 % was obtained with a residual concentration of 5.6 ± 0.7 g total PAHs/kg dry mass. The lightest compounds, such as dibenzofuran or acenaphthene, were not detected at the end of the treatment. The lowest yields of degradation ($\approx$40−50 %) were obtained for compounds with 5- and 6-ring compounds (benzo[b]fluoranthene, benzo[k]fluoranthene and indeno[c,d]pyrene). However, benzo[a]pyrene and benzo[ghi]perylene were degraded at rates equal to or greater than the 3- and 2-ring compounds. The kinetics of disappearance of the various compounds was almost exponential. As in pilot-scale experiments, a great part of the cleanup was carried out in the first six months. The degradation constants of the majority of the PAHs were strongly correlated to their aqueous solubility. This suggests a limiting step in the biological breakdown caused by mass transfer to the aqueous phase. However, such a correlation was not observed for anthracene, benzo[a]pyrene and benzo[ghi]perylene. These results are similar to those of the pilot tests which were carried out with soil A which had a level of contamination close to soil C. As seen in the pilot-scale test, a decrease in the total PAHs concentration <1.5 g/kg dry mass was not observed, even after periods of treatment greater than one year. In order to determine if these high residual concentrations were due to the concentration of the initial pollution or if they correspond to a real limitation of the bioremediation process, the same technique was applied to a slightly polluted soil.

6.2. TREATMENT OF 20 TONS OF SOIL
FROM A FORMER GAS-PRODUCTION SITE

Soil E used in this experiment differed from the previous soils in its industrial origin (a former gas production site in alluvial zone) and by its level of pollution (<5 g total PAHs/kg dry mass). The treatment was done in the control station of *APINOR Company*. The soil (18 tons) was treated in a 1-meter-deep windrow in a metal vat. A small quantity of soil (2 tons) was used as control. This control windrow was humidified and mixed at the same frequency as the treated soil but did not receive any addition of inoculum or amendments. Sampling and analysis were done simultaneously on both windrows.

After 445 days of treatment, the concentration in the control windrow decreased from 2.15 ± 0.5 to 1.29 ± 0.74 g PAHs/kg dry mass. In the treated windrow, the concentration in total PAH followed the same evolution and decreased from 1.73 ± 0.45 to 1.08 ± 0.35 g PAHs/kg dry mass. In the two windrows, the PAH disappearance was mainly due to 2- and 3-ring PAHs.

After 916 days of treatment, the lightest compounds, such as acenaphthylene, dibenzofuran or naphthalene disappeared completely. On the other hand, the 4-ring and heavier PAHs had lower yields of disappearance, except for fluoranthene and benzo[a]pyrene. Thus, at the end of the treatment, thes heaviest PAHs were relatively more abundant, whereas the dominant compounds were initially 4-ring PAHs and light PAHs. The residual concentration of total PAHs was 374 mg/kg dry mass. The yields of PAH disappearance obtained were high: 98 % for 2-ring, 96 % for 3-ring, 82 % for 4-ring and 49 % for 5−6-ring PAHs. The overall yield was 78.6 %. In the control windrow, the overall yield was 66 %.

The reduction of PAH concentrations observed in the control windrow could be due to abiotic phenomena. In fact, N and P concentrations in the initial soil E were not limiting. Thus, stimulation of the indigenous microflora by aeration and humidification might occur, even in the control windrow. Thus, disappearance of 2- and 3-ring PAHs may be due to some abiotic losses, but mainly to biodegradation. Moreover, analysis of PAHs in the gas phase was done after 61 and 437 days. Losses of PAHs by volatilization during the treatment were extremely low and did not have a significant influence on the yield of disappearance after 437 days.

7. Existence of a Concentration Threshold in PAH for Efficacy of Bioremediation?

In all the experiments in this study, it was not possible to obtain residual concentrations of total PAHs <370 mg PAHs/kg mass. This included the best results and in spite of yields ranging from 75 to 95 % after more than two years of treatment. This limitation of yields of clean up may be due to the technique employed itself, or may be dependent on the non-bioavailable fraction of PAHs. The slurry-phase bioremediation is often regarded as a technique more effective than the solid-phase bioremediation because of better control of environmental parameters, particularly temperature and oxygen transfer [21, 28]. Moreover, the soil slurry technique improves the transfer of the pollutants *via* the aqueous phase. Thus, a slurry-phase bioremediation was applied to soils recovered from the previous pilot and industrial windrows. The experiments were done in laboratory fermentors at 37 °C during 4 weeks under strong aeration (2 VVM). The control was done using an agricultural soil artificially polluted by extracts from polluted soils. In these controls, the concentration in total PAHs decreased by 41 %. This disappearance was mainly due to a net decrease of 2-, 3- and 4-ring PAHs. However, the concentrations of higher PAHs remained close to their initial values. The abiotic losses were negligible, except for fluorene and, to a lesser extent, anthracene.

In soils C and E already treated by composting, the slurry-phase bioremediation for 4 weeks did not lower the concentration of PAHs. No abiotic losses were observed. These results confirm that a short-term bioremediation will not improve the removal of PAHs. Soils C and E had initially very different PAH concentrations, *viz.* 24 and 1.8 g/kg dry mass. After more than 2 years of composting, the residual concentrations of PAH were respectively close to 1 g and 370 mg/kg dry mass. Thus, these residual concentrations are probably not related to the initial ones. In addition, comparison of artificially polluted soil and industrial soils indicates that the degradation of PAHs is influenced by the age of the contaminantion or the degree of weathering that has occurred. The longer contact time between pollutants and the soil matrix may lead to a decrease of PAH bioavailability. This is in agreement with a report of Weissenfels *et al.* [29]. Thus, extrapolations of results obtained in the laboratory on simple models (ISO soil, single pollutant) may be erroneous.

8. Conclusions and Discussion

The degradation of PAHs during composting of polluted soils is primarily due to microorganisms; however, it is necessary to stimulate them by making sure their nutritional requirements are met. There are few studies evaluating the influence of various amendments on the cleanup of soils contaminated by tar residues. Many published studies refer to landfarming, mainly for aliphatic petroleum hydrocarbons. In the case of PAHs, an estimate of nutritional requirements of the microflora for C/N and C/P ratios lies close to 120/10 and 120/1, respectively [21]. However, the aliphatic hydrocarbons are generally considered easier to degrade than aromatic and heterocyclic compounds. Consequently, the amendments used for soils polluted by tar and pitch may appear different.

The lab-scale experiments emphasized the importance of N and P for the degradation of PAHs. Increasing amounts of these compounds induce an increase in the degradation yields of PAHs. This observation was confirmed in a pilot scale test. The influence of various types of amendments (organic by-products, inorganic N and P, urea, straw) was studied. The quantity of added N and P is not the only parameter to be taken into account. The nature of the amendments also plays a major role.

Among the by-products of agro-industries, by-products of wheat do not give similar results as by-products of corn. Ammonium nitrate proved to be a better source of nitrogen than urea. This fact had been shown for oil hydrocarbons. Indeed, hydrocarbons, and particularly refined oil, can inhibit the hydrolysis of urea [15]. Our results suggest that pitch and tar could have a similar inhibiting effect. Pilot-scale experiments showed that the physical form (liquid or solid) of amendments also has to be taken into account. Indeed, liquids, interesting by-products by their high content in N and P, induced a compaction of the soils. Thus, liquid by-products are not applicable on a large scale.

Our results suggested the uptake of nitrates at the beginning of the treatment. This period corresponds also to the degradation of amendments and to the fast reduction in the concentrations of PAHs. The degradation of PAHs is generally considered as an aerobic process; however, several authors have shown that naphthalene and acenaphthene can be degraded under denitrifying condition [30−32]. The uptake of nitrates at the beginning of the treatment indicates an insufficient aeration of the soils. Lee *et al.* [33] observed that repeated organic fertilizer additions, with intertidal sediments polluted by aliphatic and ring-fused aromatic hydrocarbons, induced the development of anoxic conditions and limited the disappearance of PAHs. On the contrary, repeated additions of inorganic fertilizers (ammonium nitrate and superphosphate) gave better results. Our lab and pilot-scale results showed that the association of ammonium nitrate, organic by-products and straw proved to be most effective in terms of degradation yields compared to inorganic fertilization. Thus, on a commercial scale, aeration of soils must be improved to limit the extent of anoxic conditions due to the degradation of organic by-products, which are easily fermentable substrates. This can be done by more frequent mixing, and by using other types of mechanical engines.

The uptake of nitrates is observed during the first 6 months. After this initial period, for all experiments, an accumulation of nitrates was observed concurrently with a fast disappearance of ammonium in the treated soil. Mass balances cannot be explained by exogenous nitrates due to amendments. This suggests a partial nitrification of the ammonium added. This is quite remarkable since the nitrifying bacteria are generally regarded as very sensitive to the presence of xenobiotics. The evaluation of nitrification is proposed even as an ecotoxicological test [25].

On the other hand, ammonium was present only in small quantities at all polluted soils. After amendment, ammonium disappeared quickly. The evolution of magnesium and potassium concentrations shows that this disappearance cannot be due to losses in leachates. This suggests that ammonium is the preferential source of nitrogen for the microflora. Surprisingly, the total Kjeldahl nitrogen increased significantly in the pilot-scale experiments. Other elements like S and P followed a similar pattern. The sulfur concentration increased from 1−2 to 10−11 g/kg dry mass after 1135 days. This increase exceeded the amounts that were added by the amendments. Thus, their determination by the conventional standardized methods is probably wrong in soils strongly polluted by tar and pitch. And they are released during the breakdown of polymerized organic matter. This increase in total mineral content, such as S and P, was also observed during the industrial-scale treatment of soil C which was highly polluted by tar.

The use of organic by-products, such as potato fractions, and straw, are responsible for a composting environment in both the pilot and the industrial-scale experiments. This phenomenon was not observed in the lab, although similar amendment ratios were used. This phenomenon of composting induced important modifications to the physicochemical parameters of soil. An alkalinization of the soils was observed, mostly when the increase of temperature was more intense and prolonged. In order to limit this negative effect, the total amount of amendments was split. This fractionation did not have a significant effect on the maximum temperature but reduced the duration of the thermophilic period. As expected, the development of alkaline pH conditions was not observed. Thus, the supply of amendments in weaker but repeated quantities, limits the establishment of adverse conditions to the microbial activity. Estimated dilution based on heavy metals was $\approx 3-11$ % and was negligible compared to the degradation yield of PAHs. Even with strongly polluted soils (soils A and C), whatever the scale, degradation yields of PAHs reached 94 and 74 % on the pilot scale, and 78 % on the industrial scale. These differences are due to different treatment times, *viz.* 1135 days instead of 625 days. For individual PAHs, degradation yields were negatively correlated to their molar mass. Several authors reported similar results [23, 26, 34−37]. However, some PAHs, such as benzo-[*a*]pyrene, or benzo[*ghi*]perylene had high disappearance yields, even higher than yields of 3-ring PAHs, such as phenanthrene. Obtaining such results with compounds, which are generally considered as recalcitrant, can be related to the phenomenon of composting. Indeed, Martens [38] pointed out that a mineralization of PAHs with 4 and 6 rings occurred during composting of household refuse. In addition, composting can stimulate the incorporation of the PAHs into organic material. Qiu and Mcfarland [39] showed that *Phanerochaete chrysosporium* improved the humification of benzo[*a*]pyrene. By using labeled benzo[*a*]pyrene, they found that humification constituted the major pathway of disappearance of this compound in the soil compared to its mineralization. Richnow *et al.* [40] showed that metabolites resulting from the degradation of the PAHs are likely to establish stable chemical bonds (ester bonds) with humic substances. Some authors consider these phenomena as a valid way of degradation since the PAHs are not bioavailable any more [29, 41]. However, there are no research reports on the long-term stability of these humus−xenobiotic complexes.

In pilot and industrial scale, the kinetics of disappearance of PAHs had two phases: a rapid disappearance in the first six months of the treatment, then rates of degradation significantly decreased. An exponential curve can be used to model this kinetics. According to this model, the half-life of PAHs is independent of their initial concentration. The values for some 2- and 3-ring PAHs, such as naphthalene, fluorene or phenanthrene, are higher than those found in the literature which were all obtained under laboratory conditions. As the residual levels of PAHs cannot be reduced by four additional weeks of slurry-phase tests, the composting process is likely to be limited in soils where organic pollution is old, which has also been emphasized by other investigators [29, 42]. Moreover, the use of artificially polluted soil does not allow an accurate prediction of biodegradation of organic pollutants in polluted soils from former industrial sites.

With the slightly polluted soil (soil E), the yields of PAH disappearance obtained after 445 days were relatively low ($\approx$40 % for the total PAHs), compared with highly polluted soils. Several investigators studied the solid-phase treatment of soils having a similar low level of pollution. Results of these studies are rather disparate. In spite of many experimental conditions (fertilizer additions, inoculum), Erickson et al. [43] could not obtain any disappearance of PAHs with soils of from a gas works polluted by coal tars (430 mg PAHs/kg dry mass). They showed that the PAHs were not bioavailable for the microorganisms. On the contrary, Warith et al. [24] obtained yields of PAH degradation of 86 % after 48 days of treatment by landfarming of a soil from a gasification site and containing initially 335 mg PAHs/kg dry mass. With soils polluted by creosote, initially containing 1025 mg PAHs/kg dry mass, Ellis et al. [23] obtained 68 % disappearance after 88 days of solid-phase bioremediation. Lastly, Lamar et al. [44] studied the effect of an inoculum of Phanerochaete sordida with a soil polluted by creosote and pentachlorophenol, initially containing 985−1150 mg PAHs/kg dry mass. After 5 months of treatment, they obtained PAH disappearance of 49.5 % for the inoculated soil and 47.5 % for the non-inoculated soil. Only the 4-ring PAHs (pyrene, benzo[a]anthracene, chrysene) were significantly degraded by the treatment compared with the control. According to this work, it seems that the soils polluted by coal tars are more difficult to treat than soils polluted by creosote.

The study of MOES allows a more complete estimate of the organic matter present in the soil together with the PAHs. During the industrial test on soil C, a transitory accumulation of unknown organic compounds was highlighted concurrently with the first period of rapid PAH degradation. After 370 days of treatment, the concentrations of these compounds decreased significantly. At the end of the experiment, the MOES decreased by 32.6 %, corresponding to the degraded quantity of PAHs. Few authors include MOES together with PAHs in their studies. These MOES would be a relevant criterion to estimate composting efficiency.

Microbial inoculation plays a minor part in the treatment efficiency. The studied soils used in this report were not sterile and contained an indigenous microflora, which is stimulated by amendments. Some authors [28, 37, 44, 45] showed the relevance of an inoculation, but in laboratory tests with high levels of inoculation. Davis et al. [36] obtained less convincing results in the field than in the laboratory with lignolytic fungi. The mixing which may reduce the survival of the inoculated mycelium was an explanation. They also pointed out the difficulty of efficently producing an inoculum for large-scale applications. Indeed, inoculated microorganisms were not recovered from the dominant flora after 1 or 2 weeks of treatment. Temperature increases during composting, and indigenous microflora represent two selective unfavorable factors for exogenous inoculum [46].

282

References

1. Anderson, W.C. (Ed.) (1995) *Bioremediation*, Springer-Verlag, Berlin.
2. E.P.A. U.S. (1990) *Handbook on in Situ Treatment of Hazardous Waste-Contaminated Soils*, United States Environmental Protection Agency, Washington DC (USA).
3. Marks, P.J., Wujcik, W.J., and Loncar, F.A. (1994) Remediation technologies screening matrix and reference guide, DOD Environmental Technology 02281-012-009.
4. Richards, I.G., Palmer, J.P., and Barratt, P.A. (1993) The reclamation of former coal mines and steelworks, Elsevier Science B.V., Amsterdam.
5. *ANTEA* (1995) Les sites pollués: traitement des sols et des eaux souterraines, Technique et Documentation Lavoisier, Paris.
6. Hrudey, S.E. and Pollard, S.J. (1993) The challenge of contaminated sites: remediation approaches in North America, *Environ. Rev.* 1, 55–72.
7. Bourquin, A.W. (1990) Bioremediation of hazardous waste, *Biofutur* 24–35.
8. Bouwer, E.J. and Zehnder, A.J.B. (1993) Bioremediation of organic compounds — putting microbial metabolism to work, *Tibtech* 11, 360–372.
9. *E.P.A. U.S.* (1996) Bioremediation of hazardous wastes sites: practical approches to implementation, United States Environmental Protection Agency, Seminars EPA/625/K-96/001, Washington (DC).
10. Mueller, J.G., Lin, J.-E., Lantz S.E., and Pritchard, P.H. (1993*a*) Recent developments in cleanup technologies, *Remediation* 369–381.
11. *NTIS* (1990) Bioremediation of contaminated surface soil, National Technical Information Service, NTIS PB90-164047, Sprinfield, VA (USA).
12. Prince, R.C. (1993) Petroleum spill bioremediation in marine environments, *Crit. Rev. Microbiol.* 19, 217–242.
13. Ritter, W.F. and Scarborough, R.W. (1995) A review of bioremediation of contaminated soils and groundwater, *J. Environ. Sci. Health, Part A, Environ. Sci. Eng.* 30, 333–357.
14. Tursman, J.F. and Cork, D.J. (1992) Subsurface contaminant bioremediation engineering, *Crit. Rev. Environ. Cont.* 22, 1–26.
15. Frankenberger, W.T. (1992) The need for a laboratory feasibility study in bioremediation of petroleum hydrocarbons, in E.J. Calabrese, P.T. Kostecki (Eds), *Hydrocarbon Contaminated Soils and Groundwater*, Lewis Publishers, Chelsea, MI (USA), pp. 237–293.
16. Heitzer, A. and Sayler, G.S. (1993) Monitoring the efficacy of bioremediation, *Tibtech* 11, 334–343.
17. Wilson, S.C. and Jones, K.C. (1993) Bioremediation of soil contaminated with polynuclear aromatic hydrocarbons (PAHS): a review, *Environ. Pollut.* 81, 229–249.
18. Blackburn, J.W. and Hafker, W.R. (1993) The impact of biochemistry, bioavailability and bioactivity on the selection of bioremediation techniques, *Tibtech* 11, 328–333.
19. Harms, H. and Bosma, T.N.P. (1996) Mass transfer limitation of microbial growth and pollutant degradation, *J. Ind. Microbiol.* 16.
20. Mueller, J.G., Chapman, P.J., and Pritchard, P.H. (1989) Creosote-contaminated sites, *Environ. Sci. Technol.* 23, 1197–1201.
21. Thomas, A.O. and Lester, J.N. (1993) The microbial remediation of former gasworks sites: a review, *Environ. Technol.* 14, 1–24.
22. Shannon, M.J.R. and Unterman, R. (1993) Evaluating bioremediation: distinguishing fact from fiction, *Ann. Rev. Microbiol.* 47, 715–738.
23. Ellis, B., Harold, P., and Kronberg, H. (1991) Bioremediation of a creosote contaminated site, *Environ. Technol.* 12, 447–459.
24. Warith, M.A., Ferehner, R., and Fernandes, L. (1992) Bioremediation of organic contaminated soil, *Hazard. Waste Hazard. Mater.* 9, 137–147.
25. Hund, K. and Traunspurger, W. (1994) Ecotox-evaluation strategy for soil bioremediation exemplified for a PAH-contaminated site, *Chemosphere* 29, 371–390.
26. Mueller, J.G., Lantz, S.E., Blattmann, B.O., and Chapman, P.J. (1991*a*) Bench-scale evaluation of alternative biological treatment processs for the remediation of pentachlorophenol- and creosote-contaminated materials: solid-phase bioremediation, *Environ. Sci. Technol.* 25, 1045–1055.
27. Brookes, P.C. (1995) The use of microbial parameters in monitoring soil pollution by heavy metals, *Biol. Fertil. Soils* 19, 269–279.

28. Mueller, J.G., Lantz, S.E., Ross, D., Colvin, R.J., Middaugh, D.P., and Pritchard, P.H. (1993b) Strategy using bioreactors and specially selected microorganisms for bioremediation of groundwater contaminated with creosote and pentachlorophenol, *Environ. Sci. Technol.* 27, 691–698.

29. Weissenfels, W.D., Klewer, H.-J., and Langhoff, J. (1992) Adsorption of polycyclic aromatic hydrocarbons (PAHs) by soil particles: influence on biodegradability and biotoxicity, *Appl. Microbiol. Biotechnol.* 36, 689–696.

30. Al-Bashir, B., Cseh, T., Leduc, R., and Samson, R. (1990) Effect of soil/contaminant interactions on the biodegradation of naphthalene in flooded soil under denitrifying conditions, *Appl. Microbiol. Biotechnol.* 34, 414–419.

31. Mihelcic, J.R. and Luthy, R.G. (1988) Microbial degradation of acenaphtene and naphthalene under denitrification conditions in soil–water system, *Appl. Environ. Microbiol.* 54, 1182–1187.

32. Mihelcic, J.R. and Luthy, R.G. (1991) Sorption and microbial degradation of naphthalene in soil–water suspensions under denitrification conditions, *Environ. Sci. Technol.* 25, 169–177.

33. Lee, K., Siron, R., and Tremblay, G.H. (1995) Effectiveness of bioremediation in reducing toxicity in oiled intertidal sediments, in R.E. Hinchee, C.M. Vogel, F.J. Brockman (Eds), *Proc. 3rd Internat. in-Situ and on-Site Bioreclamation Symp.* (San Diego), Battelle Press, Columbus, OH (USA), pp. 117–127.

34. Mueller, J.G., Lantz, S.E., Blattmann, B.O., and Chapman, P.J. (1991b) Bench-scale evaluation of alternative biological treatment processs for the remediation of pentachlorophenol- and creosote-contaminated materials: slurry-phase bioremediation, *Environ. Sci. Technol.* 25, 1055–1061.

35. Bossert, I., Kachel, W.N., and Bartha, R. (1984) Fate of hydrocarbons during oily sludge disposal in soil, *Appl. Environ. Microbiol.* 47, 763–767.

36. Davis, M.W., Glaser, J.A., Evans, J.W., and Lamar, R.T. (1993) Field evaluation of the lignin-degrading fungus *Phanerochaete sordida* to treat creosote-contaminated soil, *Environ. Sci. Technol.* 27, 2572–2576.

37. Lauch, R.P., Herrmann, J.G., Mahaffey, W.R., Jones, A.B., Dosani, M., and Hessling, J. (1992) Removal of creosote from soil by bioslurry reactors, *Environ. Progr.* 11, 265–271.

38. Martens, R. (1982) Concentrations and microbial mineralization of four to six ring polycyclic aromatic hydrocarbons in composted municipal waste, *Chemosphere* 11, 761–770.

39. Qiu, X.J. and McFarland, M.J. (1991) Bound residue formation in PAH contaminated soil composting using *Phanerochaete chrysosporium*, *Hazard. Waste Hazard. Mater.* 8, 115–126.

40. Richnow, H.H., Seifert, R., Hefter, J., Kästner, M., Mahro, B., and Michaelis, W. (1994) Metabolites of xenobiotica and mineral oil constituents linked to macromolecular organic matter in polluted environments, *Org. Geochem.* 22, 671–681.

41. Bollag, J.-M. (1992) Decontaminating soil with enzymes, *Environ. Sci. Technol.* 26, 1876–1881.

42. Huesemann, M.H. (1995) Predictive model for estimating the extent of petroleum hydrocarbon biodegradation in contaminated soils, *Environ. Sci. Technol.* 29, 7–18.

43. Erickson, D.C., Loehr, R., and Neuhauser, E.F. (1993) PAH loss during bioremediation of manufactured gas plant site soils, *Water Res.* 27, 911–919.

44. Lamar, R.T., Evans, J.W., and Glaser, J.A. (1993) Solid-phase treatment of a pentachlorophenol-contaminated soil using lignin-degrading fungi, *Environ. Sci. Technol.* 27, 2566–2571.

45. Cutright, T.J. (1995) Polycyclic aromatic hydrocarbon biodegradation and kinetics using *Cunninghamella echinulata* var. *elegans*, *Int. Biodeterior. Biodegrad.* 35, 397–408.

46. van Veen, J.A., van Overbeek, L.S., and van Elsas, J.D. (1997) Fate and activity of microorganisms introduced to soil, *Microbiol. Mol. Biol. Rev.* 61, 121–135.

PLANT BIOTECHNOLOGY FOR THE REMOVAL OF ORGANIC POLLUTANTS AND TOXIC METALS FROM WASTEWATERS AND CONTAMINATED SITES

Phytoremediation

T. VANĚK[1] AND J.-P. SCHWITZGUÉBEL[2]

[1]*Department of Plant Tissue Cultures, Institute of Organic Chemistry and Biochemistry, Czech Academy of Sciences, Flemingovo namesti 2,166 10 Prague, Czech Republic*

[2]*Laboratory for Environmental Biotechnology (LBE), Swiss Federal Institute of Technology Lausanne (EPFL), CH-1015 Lausanne, Switzerland*

Abstract

The aim of this contribution is to provide some basic knowledge of the phytoremediation methodology, including basic definitions, advantages and potential drawbacks. A list of recent publications, which can give more detailed information about both basic research and applications, is included.

1. Introduction

Plants are thought to be primarily a source of food, fuel and fiber. However, it has been realized recently that plants may serve potentially as environmental counter-balance to industrialization processes, and not only as a sink for increased atmospheric CO_2. Indeed, over the last century, the content of xenobiotic compounds in ecosystems has increased considerably. Many organic synthetic substances, e.g., pesticides, solvents, dyes, by-products of chemical and petrochemical industries, are eventually transported to natural vegetation and cultivated crops, where they can either be harmful to the plant itself, totally or partially degraded, transformed, or accumulated in plant tissues and organs. In the latter case, xenobiotics are concentrated in food chains and finally in human beings, with possible detrimental effects on their health. Such a situation also occurs with heavy metals. Anthropogenic sources of toxic metals in the environment are numerous: metalliferous mining and smelting, electroplating, energy and fuel production, gas exhausts, agriculture or waste disposal. Reports on plants growing in polluted stands without being seriously harmed indicate that it may be possible to detoxicate contaminants using agricultural and biotechnological approaches. Higher plants possess a pronounced ability to metabolize and degrade many recalcitrant xenobiotics and may be considered as "green livers", acting as an important sink for environmentally damaging chemicals. On the other hand, different plant species are able to hyperaccumulate toxic metals in their tissues. It thus appears that crops and cultivated

V. Šašek et al. (eds.),
The Utilization of Bioremediation to Reduce Soil Contamination: Problems and Solutions, 285–293.
© 2003 *Kluwer Academic Publishers. Printed in the Netherlands.*

plants could be developed and used for the removal of hazardous, persistent organic compounds and toxic metals from industrial wastewaters and for phytoremediation purposes.

2. Phytoremediation

Phytoremediation has been defined as the use of green plants and their associated microorganisms, soil amendments and agronomic techniques to remove, contain or render harmless environmental contaminants.

Phytoremediation is expected to be complementary to classical bioremediation techniques, based on the use of microorganisms. It should be particularly useful for the extraction of toxic metals from contaminated sites and the treatment of recalcitrant organic pollutants, such as trinitrotoluene and nitroglycerin. Plant biomass could also be used efficiently for the removal of volatile organic pollutants or different priority pollutants, like pentachlorophenol, other polychlorophenols and anilines.

At present, phytoremediation is still a nascent technology that seeks to exploit the metabolic capabilities and growth habits of higher plants: delivering a cheap, soft and safe biological treatment that is applicable to specific contaminated sites and waste-waters is a relatively recent focus. In such context, there is still a significant need to pursue both fundamental and applied research to provide low-cost, low-impact, visually benign and environmentally sound depollution strategies.

One of the greatest forces driving increased emphasis on research in this area is the potential economic benefit of an agronomy-based technology. Growing a crop can be accomplished at a cost ranging from 2 to 4 orders of magnitude less than the current engineering cost of excavation and reburial. Expected applications will be in the decontamination of polluted soils and groundwater (phytoremediation) or in the clean-up of industrial effluents (plant cells, tissues or biomass immobilized in appropriate containers, whole plants cultivated in constructed wetlands or under hydroponic conditions).

2.1. PHYTOREMEDIATION OF METAL CONTAMINANTS

At sites contaminated with toxic metals, plants can be used either to stabilize or to remove the metals from the soil and groundwater through three mechanisms: phytoextraction, rhizofiltration, and phytostabilization. The same approach can be used for radionuclides (radiophytoremediation).

2.1.1. Phytoextraction

Phytoextraction, also called phytoaccumulation, refers to the uptake and translocation of metal contaminants in the soil by plant roots into the above-ground portions of the plants. Certain plants, called hyperaccumulators, absorb unusually large amounts of metals in comparison to other plants. One or a combination of these plants is selected and planted at a particular site based on the type of metals present and other site conditions. After the plants have been allowed to grow for some time, they are harvested and, if possible, incinerated to recycle the metals. The other possibility is composting, which decreases the amount of metal-containing plant biomass considerably. Of course, because of the high content of metals in the resulting compost, it is not possible to use it for agricultural purposes. This procedure may be repeated as necessary to bring soil contaminant levels down to allowable limits. If plants are incinerated, the ash must be disposed of in a hazardous waste landfill, but the volume of ash will be less than 10 % of the volume that would be created if the contaminated soil itself were dug up for treatment.

2.1.2. Rhizofiltration

Rhizofiltration is the adsorption or precipitation onto plant roots or absorption by roots of contaminants that are in the vicinity of the root zone. Rhizofiltration is similar to phytoextraction, but the plants are used primarily to address contaminated groundwater rather than soil. The plants to be used for cleanup are raised in greenhouses with their roots in water rather than in soil. To acclimate the plants once a large root system has been developed, contaminated water is collected from a waste site and brought to the plants where it is substituted for their original water source. The plants are then planted in the contaminated area where the roots take up the water and the contaminants along with it. As the roots become saturated with contaminants, they are harvested. For example, sunflowers were used successfully to remove radioactive contaminants from pond water in a test at Chernobyl, Ukraine.

2.1.3. Phytostabilization

Phytostabilization is the use of certain plant species to immobilize contaminants in the soil and groundwater through absorption and accumulation by roots, adsorption onto roots, or precipitation within the root zone of plants (rhizosphere). This process reduces the mobility of the contaminant and prevents migration to the ground water or air, and it reduces bioavailability for entry into the food chain. This technique can be used to reestablish a vegetative cover at sites where natural vegetation is lacking due to high metal concentrations in surface soils or physical disturbances to surface materials. Metal-tolerant species can be used to restore vegetation to the sites, thereby decreasing the potential migration of contamination through wind erosion and transport of exposed surface soils and leaching of soil contamination to groundwater.

2.2.   PHYTOREMEDIATION OF ORGANIC CONTAMINANTS

Organic contaminants are common environmental pollutants. There are several ways that plants may be used for the phytoremediation of these contaminants: phytodegradation, rhizodegradation, and phytovolatilization.

2.2.1. Phytodegradation

Phytodegradation, also called phytotransformation, is the breakdown of contaminants taken up by plants through metabolic processes within the plant, or the breakdown of contaminants external to the plant through the effect of compounds (such as enzymes) produced by the plants. Pollutants (complex organic molecules) are degraded into simpler molecules and are incorporated into the plant tissues.

2.2.2. Rhizodegradation

Rhizodegradation, also called enhanced rhizosphere biodegradation, phytostimulation, or plant-assisted bioremediation/degradation, is the breakdown of contaminants in the soil through microbial activity that is enhanced by the presence of the root zone (the rhizosphere) and is a much slower process than phytodegradation. Microorganisms (yeast, fungi, or bacteria) consume and digest organic substances for nutrition and energy. Certain microorganisms can digest organic substances such as fuels or solvents that are hazardous to human beings and break them down into harmless products via a process called biodegradation. Natural substances released by the plant roots (plant exudates) contain organic carbon that provides food for soil microorganisms and the additional nutrients enhance their activity.

2.2.3. Phytovolatilization

Phytovolatilization is the uptake and transpiration of a contaminant by a plant, with release of the contaminant or a modified form of the contaminant to the atmosphere from the plant. Phytovolatilization occurs as growing trees and other plants take up water and the organic contaminants. Some of these contaminants may be transported through the plants to the leaves and evaporate, or volatilize, into the atmosphere. .

2.3. ADVANTAGES AND DRAWBACKS OF PHYTOREMEDIATION

Early research indicates that phytoremediation technology is a promising cleanup solution for a wide variety of pollutants and sites but it has its limitations. The following table (Table 1) reveals that many of the advantages and shortcomings of phytoremediation are consequences of the biological nature of the treatment system. Plant-based remediation systems can function with minimal maintenance once they are established, but they are not always the best solution to a contamination problem. A necessary condition for phytoremediation to be successful is that the pollutant must be bioavailable to a plant and its root system. If a pollutant is located in a deep aquifer, then plant roots cannot reach it. If a soil pollutant is tightly bound to the organic portion of soil, it may not be available to plants or to microorganisms in the rhizosphere. On the other hand, if a pollutant is too water-soluble, it will bypass the root system without any retention (and, therefore, without significant accumulation and/or degradation).

2.4.   PERFORMANCE

A major hurdle for innovative technologies is a lack of performance data, and phytoremediation is no exception. One of the current impediments to gathering performance data is the length of time involved in a phytoremediation project, which is dependent on the rates of plant growth and activity. There is currently a number of pilot-scale projects in existence, but they have not resulted in conclusive performance data at this time. These sites are being monitored and will report results over the next few years. Also, a number of companies have installed phytoremediation systems at polluted sites owned by private clients, so results from those sites are not publicly available. On the other hand, data from basic research are available both from the scientific literature and from a number of Web pages. The readers can obtain recent data about European research in this field, e.g, at   COST   837   WEB page (http://lbewww.epfl.ch/COST837/), where some applications are also described as.

TABLE 1. Advantages and liimitations of phytoremediation

Advantages of Phytoremediation	Limitations of Phytoremediation
in situ	Limited to shallow soils, streams, and groundwater
Passive	High concentrations of hazardous materials can be toxic to plants
Solar driven	Mass transfer limitations associated with other biotreatments
Costs 10% to 20% of mechanical treatments	Slower than mechanical treatments
Transfer is faster than natural attenuation	Only effective for moderately hydrophobic contaminants
High public acceptance	Toxicity and bioavailability of degradation products is very often not known
Fewer air and water emissions	Contaminants may be mobilised into the groundwater
Generate less secondary wastes	Potential for contaminants to enter food chain through animal consumption
Soils remain in place and are usable following treatment	Unfamiliar to many regulators

2.5. COST

In addition to performance data, accurate cost data are often difficult to predict for new technologies. Most lab, pilot, and field scale tests include monitoring procedures far above those expected at a site with a remediation goal. This inflates the costs of monitoring at these test sites. As a result, it is difficult to predict the exact cost of a technology that has not been established through years of use. However, since phytoremediation involves the planting of trees or grasses, then it is by nature a relatively inexpensive technology when compared to technologies that involve the use of large scale, energy consuming equipment.

TABLE 2. Estimates of phytoremediation costs vs. costs of established technologies

Contaminant	Phytoremediation costs	Estimated cost using other technologies
Metals	$80 per cubic yard	$250 per cubic yard
Site contaminated with petroleum hydrocarbons	$70 000	$850 000
10 acres lead contaminated land	$500 000	$12 million
Radionuclides in surface water	$2 to $6 per thousand gallons treated	none listed
1 hectare to a 15 cm depth (various contaminants)	$2500 to $15 000	none listed

Phytoremediation costs will vary, depending on the treatment strategy. For example, harvesting plants that bioaccumulate metals can drive up the cost of treatment when compared to treatments that do not require harvesting. Regardless, of this phytoremediation is often predicted to be cheaper than comparable technologies.

Table 2 presents some estimates of phytoremediation costs in relation to conventional technologies. This table represents some vague and variable estimates due to the current dearth of cost information. It should be kept in mind that costs of phytoremediation are highly site-specific, so that any estimate found in these tables is merely a rough estimate of potential costs. Many of these estimates are speculative, based on laboratory or pilot-scale data.

3. Conclusion

For the efficient utilization of phytoremediation as a general application environment cleaning technology it is still necessary to resolve many problems, at both the level of basic research and practical application.
Some of them are mentioned below:

- Delineation of pathways employed in the uptake and metabolism of organic pollutants by plants.
- Identification of metabolites produced and study of their ecotoxicological behavior.
- Appropriate selection of plants able to hyperaccumulate toxic metals, understanding the physiological and biochemical mechanisms leading to their uptake, translocation and accumulation.
- Production of a databank of genes/enzymes that improve the rate and extent of detoxication of organic pollutants and toxic metals.
- Evaluation of the prospect of using metabolic engineering tools to enhance the capacity of higher plants for phytoremediation and cleanup of industrial effluents.

- Generation/evaluation of plants adapted to the phytoremediation of specifically contaminated sites or wastewaters.
- Execution of pilot studies in the scaleup of selected plants with an increased capacity for biodegradation of xenobiotics and accumulation of toxic metals.

Acknowledgements

This work was supported by COST 837.10 GAAVCR A6055902 and GACR 206/99/1252 grant.

4. Selection of recent relevant publications

1. Bañuelos, G.S., Shannon, M.C., Ajwa, H., Draper, J.H., Jordahl, J., and Licht, L. (1999) Phytoextraction and accumulation of boron and selenium by poplar (*Populus*) hybrid clones, *Internat. J. Phytoremed.* **1**, 81- 105.
2. Bärenwald G., Schneider B. and Schütte H. R. (1994) Metabolism of the bisethylene-glycolesters of 2,4-D and MCPA in excised plants and cell suspension cultures of tomato. *J. Plant Physiol.* **144**: 396-399.
3. Batard Y., Zimmerlin A., Le Ret M., Durst F. and Werck-Reichhart D. (1995) Multiple xenobiotic-inducible P450s are involved in alkoxycoumarin and alkoxyresorufin metabolism in higher plants. *Plant, Cell Environ.* **18**: 523-533.
4. Blake-Kalff M. M. A. and Coleman J. O. D. (1996) Detoxification of xenobiotics by plant cells: characterization of vacuolar amphiphilic organic anion transporters. *Planta* **200**: 426-431.
5. Bockers M., Rivero C., Thiede B., Jankowski T. and Schmidt B. (1994) Uptake, translocation and metabolism of 3,4-dichloroaniline in soybean and wheat plants. *Zeitschr. Naturforsch.* **49c**: 719-726.
6. Bokern M., Nimtz M. and Harms H. (1996) Metabolites of 4-n-nonylphenol in wheat cell suspension cultures. *J. Agric. Food Chem.* **44**: 1123-1127.
7. Bright S. W. J., Greenland A. J., Halpin C. M., Schuch W. W. and Dunwell J. M. (1996) Environmental impact from plant biotechnology. *Ann. N. Y. Acad. Sci.* **792**: 99-105.
8. Burken, J.G., and Schnoor, J.L. (1999) Distribution and volatilization of organic compounds following uptake by hybrid poplar trees, *Internat. J. Phytoremed.*, **1**, 111-125.
9. Chappell J. (1998) Phytoremediation of TCE in Groundwater using *Populus*. Status Report prepared for the U.S. EPA Technology Innovation Office under a National Network of Environmental Management Studies Fellowship.
10. Cole D. J. (1994) Detoxification and activation of agrochemicals in plants. *Pesticide Sci.* 42, 209-222.
11. Coleman J. O. D., Randall R. and Blake-Kalff M. M. A. (1997) Detoxification of xenobiotics in plant cells by glutathione conjugation and vacuolar compartmentalization: a fluorescent assay using monochloro-bimane. *Plant Cell Environ.* **20**, 449-460.
12. Cunningham S. D., Anderson T. A., Schwab A. P. and Hsu F. C. (1996) Phytoremediation of soils cont-aminated with organic pollutants. In: Advances in Agronomy 56, Sparks D.L., ed. Academic Press, 55-114.
13. Davies T. H. and Cottingham P. D. (1994) The use of constructed wetlands for treating industrial effluent (textile dyes). *Water Sci. Technol.* 29, 227-232.
14. Dec J. and Bollag J. M. (1994) Use of plant material for the decontamination of water polluted with phenols. *Biotechnol. Bioeng.* **44**, 1132-1139.
15. Duc, R., Vanek, T., Soudek, P., Schwitzguebel, J.P. (1999) Accumulation and transformation of sulfonated aromatic compounds by rhubarb (*Rheum palmatum*) cells, *Internat. J. Phytoremed.*, **1**, 255-271.
16. Duc, R., Vanek, T., Soudek, P., Schwitzguebel, J.P. (1999) Experiments with rhubarb in Europe, *Soil Groundwater Cleanup* 2/3, 27 – 30.
17. Dushenkov V., Nanda Kumar P. B. A., Motto H. and Raskin I. (1995) Rhizofiltration: the use of plants to remove heavy metals from aqueous streams. *Environ. Sci. Technol.* 29, 1239-1245.
18. Farago M. E. (Ed.) (1994) Plants and the Chemical Elements. Biochemistry, Uptake, Tolerance and Toxicity. VCH. ISBN 3-527-28269-6.

292

19   Fernandez N., Chacin E., Garcia C., Alastre N., Leal F. and Forster C. F. (1996) The use of seed pods from Albizia lebbek for the removal of alkyl benzene sulphonates from aqueous solution. *Process Biochem.* **31**, 383-387.
20   Furukawa K. and Fujita M. (1993) Advanced treatment and food production by hydroponic type wastewater treatment plant. *Water Sci. Technol.* **28**, 219-228.
21   Gareis C., Rivero C., Schuphan I. and Schmidt B. (1992) Plant metabolism of xenobiotics. Comparison of the metabolism of 3,4-dichloroaniline in soybean excised leaves and soybean suspension cultures. *Zeitschr. Naturforsch.* **47 c**: 823-829.
22   Goel A., Kumar G., Payne G. F. and Dube S. K. (1997) Plant cell biodegradation of a xenobiotic nitrate ester, nitroglycerin. *Nature Biotechnol.* **15**, 174-177.
23   Gorsuch J. W., Lower W. R., Lewis M. A. and Wang W. (Eds.) (1991) Plants for Toxicity Assessment. ASTM. ISBN 0-8031-1422-2
24   Greger, M., and Landberg, T. (1999) Use of willow in phytoextraction, *Internat. J. Phytoremed.* **1**, 115-120.
25   Hsu F. C. and Kleier D. A. (1996) Phloem mobility of xenobiotics. A short review. *J. Experiment. Bot.* **47**, 1265-1271.
26   Kreuz K., Tommasini R. and Martinoia E. (1996) Old enzymes for a new job - Herbicide detoxification in plants. *Plant Physiol.* **111**, 349-353.
27   Lakatos G., Kiss M. K., Kiss M. and Juhasz P. (1997) Application of constructed wetlands for wastewater treatment in Hungary. *Water Sci.Technol.* **35**, 331-336.
28   Lasat M. M., Baker A. J. M. and Kochian L. V. (1996) Physiological characterization of root Zn^{2+} absorbtion and translocation to shoots in Zn hypercaccumulator and nonaccumulator species of *Thlaspi. Plant Physiol.* **112**, 1715-1722.
29   Lozano-Rodriguez E., Hernandez L. E., Bonay P. and Carpena-Ruiz R. O. (1997) Distribution of cadmium in shoot and root tissues of maize and pea plants: physiological disturbances. *J. Experiment. Bot.* **48**, 123-128.
30   Marrs K. A. (1996) The functions and regulation of glutathione S-transferases in plants. *Annu. Rev. Plant Physiol. Plant Molec. Biol.* **47**, 127-158.
31   Mattioni C., Gabbrielli R., Vangronsveld J. and Clijsters H. (1997) Nickel and cadmium toxicity and enzymatic activity in Ni-tolerant and non-tolerant populations of *Silene italica* Pers. *J. Plant Physiol.* **150**, 173-177.
32   Mungur A. S., Shutes R. B. E., Revitt D. M. and House M. A. (1997) An assessment of metal removal by a laboratory scale wetland. *Water Sci. Technol.* **35**, 125-133.
33   Nanda Kumar P. B. A., Dushenkov V., Motto H. and Raskin I. (1995) Phytoextraction: the use of plants to remove heavy metals from soils. *Environ. Sci. Technol.* **29**, 1232-1238.
34   Nellessen J. E. and Fletcher J. S. (1993) Assessment of published literature pertaining to the uptake/ accumulation, translocation, adhesion and biotransformation of organic chemicals by vascular plants. *Environ. Toxicol. Chem.* **12**, 2045-2052.
35   Orhan Y. and Büyükgüngör H. (1993) The removal of heavy metals by using agricultural wastes. *Water Sci. Technol.* **28**, 247-255.
36   Raskin I., Nanda Kumar P. B. A., Dushenkov S. and Salt D. E. (1994) Bioconcentration of heavy metals by plants. *Current Opinion Biotechnol.* **5**, 285-290.
37   Roper J. C., Dec J. and Bollag J. M. (1996) Using minced horseradish roots for the treatment of polluted waters. *J. Environ. Qual.* **25**, 1242-1247.
38   Ross S. M. (Ed.) (1994) Toxic Metals in Soil-Plant Systems. Wiley. ISBN 0-471-94279-0
39   Rossini L., Jepson I., Greenland A. J. and Sari Gorla M. (1996) Characterization of glutathione S-transferase isoforms in three maize inbred lines exhibiting differential sensitivity to alachlor. *Plant Physiol.* **112**, 1595-1600.
40   Roy S. and Hänninen O. (1994) Pentachlorophenol: uptake/elimination kinetics and metabolism in an aquatic plant, *Eichhornia crassipes. Environ. Toxicol. Chem.* **13**, 763-773.
41   Roy S. and Hänninen O. (1995) Use of aquatic plants in ecotoxicology monitoring of organic pollutants: bioconcentration and biochemical responses. In: Environmental Toxicology Assessment, Richardson M., ed. Taylor and Francis, pp. 97-109.
42   Salt D. E., Blaylock M., Nanda Kumar P. B. A., Dushenkov V., Ensely B. D., Chet I. and Raskin I. (1995) Phytoremediation: a novel strategy for the removal of toxic metals from the environment using plants. *Bio/Technol.* **13**, 468- 474.
43   Sandermann H. (1992) Plant metabolism of xenobiotics. *Trends Biochem. Sci.* **17**:,82-84.
44   Sandermann H., Schmitt R., Eckey H. and Bauknecht T. (1991) Plant biochemistry of xenobiotics: Isolation and properties of soybean O- and N-glucosyl and O- and N-malonyltransferases for chlorinated phenols and anilines. *Arch. Biochem. Biophys.* **287**, 341-350.

45   Scott C. D. (1992) Removal of dissolved metals by plant tissue. *Biotechnol. Bioeng.* **39**, 1064-1068.

46   Schmidt B., Rivero C., Thiede B. and Schenk T. (1993) Metabolism of 4-nitrophenol in soybean excised leaves and cell suspension culture of soybean and wheat. *J. Plant Physiol.* **141**, 641-646.

47   Schmidt B., Rivero C. and Thiede B. (1995) 3,4-dichloroaniline N-glucosyl- and N-malonyl-transferase activities in cell cultures and plants of soybean and wheat. *Phytochemistry* **39**, 81-84.

48   Schnabel W. E., Dietz A. C., Burken J. G., Schnoor J. L. and Alvarez P. J. (1997) Uptake and transformation of trichloroethylene by edible garden plants. *Water Res.* **31**, 816-824.

49   Schnoor J. L., Licht L. A., McCutcheon S. C., Wolfe N. L. and Carreira L. H. (1995) Phytoremediation of organic and nutrient contaminants. *Environ. Sci. Technol.* **29**, 318A-323A.

50   Schwitzguébel J. P., Vanek T., Mathieu N., Thirot J. L. and Novotny M. (1995) Biotransformation of sulf-onated xenobiotic compounds by plant cells. In: Biosorption and Bioremediation, T. Macek et al., eds. Prague, p. L 2-10.

51   Schwitzguébel J. P. (1996) Biodegradation of xenobiotics by plants. Kurzfassungen der DECHEMA-Jahrestagungen '96, Band I, DECHEMA, Frankfurt am Main, pp. 442-443.

52   Serre A. M., Roby C., Roscher A., Nurit F., Euvrard M. and Tissut M. (1997) Comparative detection of fluorinated xenobiotics and their metabolites through ^{19}F NMR and ^{14}C label in plant cells. *J. Agric. Food Chem.* **45**, 242-248.

53   Shen Z. G., Zhao F. J. and McGrath S. P. (1997) Uptake and transport of zinc in the hyperaccumulator *Thlaspi caerulescens* and the non-hyperaccumulator *Thlaspi ochroleucum*. *Plant Cell Environ.* **20**, 808-906.

54   Shimp J. F., Tracy J. C., Davis L. C., Lee E., Huang W., Erickson L. E. and Schnoor J. L. (1993) Beneficial effects of plants in the remediation of soil and groundwater contaminated with organic materials. *Critic. Rev. Environ. Sci. Technol.* **23**, 41-77.

55   Skipsey M., Andrews C. J., Townson J. K., Jepson I. and Edwards R. (1997) Substrate and thiol specificity of a stress-inducible glutathione transferase from soybean. *FEBS Letts* **409**, 370-374.

56   Srivastav R. K., Gupta S. K., Nigam K. D. P. and Sasudevan P. (1994) Treatment of chromium and nickel in wastewater by using aquatic plants. *Water Res.* **28**, 1631-1638.

57   Trapp S. and Mc Farlane J. C. (Eds.) (1995) Plant Contamination. Modeling and Simulation of Organic Chemical Processes. Lewis Publishers. ISBN 1-56670-078-7.

58   Vanderford M., Shanks J. V. and Hughes J. B. (1997) Phytotransformation of trinitrotoluene (TNT) and distribution of metabolic products in *Myriophyllum aquaticum*. *Biotechnol. Letts.* **19**, 277-280.

59   Wagner G. J. (1993) Accumulation of cadmium in crop plants and its consequences to human health. *Adv. Agronom.* **51**, 173-212.

60   Wayment, D.G., Bhadra, R., Lauritzen, J., J.B. Hughes, J.B., and Shanks, J.V. (1999) A transient study of formation of conjugates during TNT metabolism by plant tissues, *Int. J. Phytoremed.* **1**, 227- 235.

61   Wetzel A. and Sandermann H. (1994) Plant biochemistry of xenobiotics: isolation and characterization of a soybean O-glucosyltransferase of DDT metabolism. *Arch. Biochem. Biophys.* **314**, 323-328.

62   Wright K. M., Horobin R. W. and Oparka K. J. (1996) Phloem mobility of fluorescent xenobiotics in *Arabidopsis* in relation to their physicochemical properties. *J. Exp. Bot.* **47**, 1779-1787.

CONSIDERATION OF PLANT-BASED REMEDIATION AND RESTORATION OF CONTAMINATED SITES CONTAINING HEAVY METALS — THE CANADIAN EXPERIENCE

T. McINTYRE
Environment Canada, Hull, Quebec, Canada

1. Introduction

Although Canada, like many other NATO countries, spends hundreds of millions of dollars annually to remediate contaminated sites, many still remain and continue to represent an economic and environmental liability as well as a technical challenge. Site owners and managers alike are constantly confronted by the need for innovative and cost effective technological solutions to address the more recalcitrant environmental contaminants. One of the more promising environmental technologies on the horizon is that of phytoremediation. Phytoremediation is defined as the *in situ* use of plants to immobilize, remediate, reduce, or recover contaminants from soil, sediment, or water.

The definition applies to all biological, ecological, chemical, and physical processes that affect remediation of contaminated substrates. Although the concept is not new, phytoremediation has only recently been targeted by a number of NATO governments. These governments now recognize the outstanding potential of this technology and they have established dedicated research programs to exploit its potential [1−4].

Environment Canada seeks to protect human health and the environment from risks associated with hazardous waste sites, while encouraging the development of innovative technologies, such as phytoremediation to more effectively clean up and restore these sites to some level of ecosystem health and stability. Given the prevalence of metals at a number of contaminated sites and Brownfield locations across Canada, Environment Canada has recently begun the process to explore with greater interest the potential of plant based technologies for the remediation, recovery, and restoration of contaminated sites containing metals [5].

This paper is intended to highlight key results from both Canadian and global research work and to stimulate discussion on the benefits and limitations of phytoremediation and its potential as an innovative, efficient, and cost-effective environmental technology for treatment of contaminated sites containing metals. It addresses issues that arise and actions that may be required by NATO countries to advance this technology primarily in terms of scientific, technical, regulatory, and envi-

V. Šašek et al. (eds.),
The Utilization of Bioremediation to Reduce Soil Contamination: Problems and Solutions, 295–311.

ronmental research needs. Issues related to intellectual property law, commercialization, biodiversity and public acceptance are also important to the potential application of phytoremediation. However, they are touched on only briefly in this review.

2. The Problem of Contaminated Sites in Canada

The presence of all contaminants, including metals, in Canada's air, surface and groundwater and soil continues to be troublesome to a great many Canadians, prompting governments to initiate environmental laws, policies and programs to address these concerns. Currently there is no comprehensive list of all contaminated sites in Canada as site identification and characterization continues to be an ongoing process by government and industry alike. Federal government departments are establishing inventories of contaminated sites that they have responsibility for with current estimates in excess of 5000 [6]. The estimated number of Brownfield sites range between 2900 and 30 000 [7]. There are approximately 10 000 abandoned mines in Canada and 6 000 abandoned tailing sites (CIELAP 1995, *unpublished data*). The estimated cost for cleanup of abandoned mines alone in Canada has been placed at $ 6 billion (CIELAP 1995, *unpublished data*). There are approximately 875 million tons of mining wastes capable of causing acid mine drainage in Canada (CIELAP 1995, *unpublished data*). Furthermore, it has been estimated that there are approximately 185 million tons of radioactive tailings from uranium mines in Canada.

3. Plant Adaptation to Contaminated Matrices

Botanists have long recognized that some plant species are endemic to soils of high metal and inorganic contamination [4, 8–10]. Indicator plant species of particular metals have proven useful in prospecting for those metals [11]. Considerable research has been conducted on the mechanisms of metal tolerance in plants [12] and the establishment of vegetation on contaminated sites [13]. More recently, the area of forensic ecology is receiving increasing interest as plant ecologists begin to explore the role and origin of plants that have volunteered at selected contaminated sites across North America [2, 14, 15]. Forensic phytoremediation and the associated investigation of plant growth in extreme environments in Canada has resulted in the identification of plants that can both grow under representative contaminated site conditions, and demonstrate capacity to contribute to sequestration/accumulation of metals existing at the site. The increase in numbers of hyperaccumulating plants identified to date that can play a role in cleaning up metals and other pollutants has researchers, regulators and land owners looking at phytoremediation as an effective, innovative, and inexpensive means of remediating contaminated soil, sediment and ground water.

4. Why Do Plants Accumulate Metals?

Explanations for the phenomenon by which plants tolerate and accumulate metals vary but have been generally recognized to be a function of one of six prevailing schools of thought that include: inadvertent uptake, metal tolerance, disposal from plant body, drought resistance, competition strategy, and pathogen/herbivore defense [16]. The rationale for plant tolerance and accumulation of metals as a pathogen/herbivore defense mechanism however, seems to be the most subscribed theory in recognition of the physiological havoc that metal molecules can wreak [17].

5. Putting the "Petal" to the "Metal"

Plants can remediate contaminated sites through a variety of mechanisms. They can directly take up contaminants with subsequent immobilization, volatilization or degradation occurring. Plants have been likened to solar-driven pumps which can extract and concentrate certain elements from their environment [18–20]. Plant root systems represent an enormous surface area that enables plants to absorb and accumulate water and essential nutrients for growth. They have metabolic and adsorption capabilities and possess transport systems that can selectively take up many ions from soils [21]. Plants have evolved a great diversity of genetic adaptations to handle potentially toxic levels of metals that occur in the environment [15]. Most metal-tolerant plants exclude toxic metal ions from uptake, while hyperaccumulators actually tolerate and take up high amounts of toxic metals and other ions, up to several percent of their dry matter [3, 11, 22–24]. Some plant species also can take up organic pollutants and accumulate non-phytotoxic metabolites into plant tissue [20].

Plant roots excrete a variety of compounds that alter the root-soil environment by serving as nutrients and energy sources for soil microorganisms [21, 25–27]. Plant roots can alter pH and transfer oxygen to the root zone for anaerobic mineralization of organics. They stabilize or contain contaminants by forming stable metal chelates. A zone of increased microbial numbers and activity supported by plant roots may contribute to degradation of contaminants in soils [25, 26, 28, 29].

The scientific literature contains numerous descriptions of plants that are hyperaccumulators. Approximately 400 species of metal-hyperaccumulating plants are known, of which about 75 % are endemic to ultramafic soils and accumulate nickel [8, 30]. Globally, over 800 aquatic and terrestrial metal-tolerant and -accumulating plants have been identified by Environment Canada as part of their PHYTOREM data base [31]. Plants identified that could be of potential value in phytoremediation include *Populus deltoides nigra* which takes up atrazine from soils [32], *Alysium* spp. which accumulate nickel [19], *Aeolanthus* sp. which accumulates copper and cobalt [33], *Brassica juncea* which takes up selenium [11], and *Alyxia* sp. which accumulates manganese [34]. *Sebertia acuminata*, a species native to serpentine soils in New

Caledonia is particularly noteworthy as a hyperaccumulator. The latex in this tree contains more than 11 % nickel and is blue because of this content [35].

This battery of remediation functions occurring within plants are made even more attractive by promising and innovative new lines of plant biotechnology research currently being carried out using genetic and protein engineering to create new gene combinations to increase the efficacy of plants used for phytoremediation. The introduction of metal-accumulating traits into fast growing high-biomass plants would be of considerable advantage offering the potential to use commonly cultivated plants such as Brussels sprouts and cauliflower [35]. This approach has been recently shown in the successful transformation of Aradopsis thaliana plants expressing the *merA* gene, a gene that codes for mercury resistance in bacteria [37].

6. Why Phytoremediation / Why Metals / Why Now?

There are a number of compelling reasons in Canada and for NATO countries to consider the enhanced use of plant based remediation and restoration techniques for contaminated sites containing metals. These include:

- the sheer volume of contaminated sites containing metals as the principal constituent or as part of mixed wastes;
- disadvantages associated with traditional physical, chemical, and thermal means for site cleanup ranging from ineffective and variable treatment efficiencies under certain conditions, through site disruption, to high costs [53];
- the triple liability that site owners/managers face (property taxes, clean up liabilities, and alternative use value for some sites) [7];
- the demonstrated potential for plants to sequester, accumulate, and in some instances hyperaccumulate up to 19 different metal species [31];
- volume of metallic waste generated annually by industry;
- human and environmental health concerns associated with a number of existing sites;
- evolving regulatory oversight for specific metals (Hg, Pb, As);
- significant potential for reduced costs associated with plant-based remediation techniques [38].

7. The Promise of Phytoremediation

Numerous benefits have been enumerated for phytoremediation. It is an *in situ*, passive, solar-driven 'green' technology. It has applicability to a wide range of metals, radionuclides and organic substances and to sites that are not readily remediated by other methods. These include large sites that have low concentrations of contaminants that are widely dispersed at a shallow depth. This technology is also appealing in that it may be applied with relative ease. The extensive knowledge developed from agricultural and silvicultural practices can be applied to plant cultivation and management on contaminated sites. Furthermore, the topsoil can be left in usable condition with minimum site disruption [37].

Phytoremediation also offers the advantage of eliminating secondary air- or water-borne wastes. Plants provide ground cover and stabilize soil on contaminated sites to reduce wind-blown dust; an important pathway for human exposure to toxic substances through inhalation of suspended soil particulate matter and ingestion of contaminated food where that suspended matter deposits on food plants [20]. Exposure risks for wildlife may also be reduced. It has been noted that lead-contaminated sites in urban centres pose a particular risk for children who, by playing in the area, can suffer exposure through hand-to-mouth exchange of dirt [35]. Children also may be less inclined to play in areas with substantial plant cover.

The production of contaminated plant biomass offers several advantages. Plants used in phytoremediation may produce recyclable metal and rich plant residues. For example, the tree *Sebertia acuminata* previously mentioned could be tapped for the nickel accumulated in its latex. Biomass produced by hyperaccumulators may be disposed of by incineration resulting in an immense reduction of contaminant mass and volume and the amount of contaminated material to be land-filled. For example, it has been estimated that removing heavy metal contaminated soil from 2 acres to a depth of 18 inches creates about 5000 tons of soil that must be deposited in a hazardous landfill. In contrast, plants that take up the residue and are burned leave a residue of between 25 and 30 tons of ash to be disposed of [35]. Recent developments in microbiology demonstrate even greater promise in reducing these ash quantities significantly through biological leaching techniques which have proven effective in reducing a variety of heavy-metal constituents [39].

Phytoremediation is also seen as a cost-effective means of cleaning up contaminants in soil, groundwater and sediments [38]. For example, the estimated costs for removal of radionuclides from water using sunflowers ranged from $ 2 to $ 6 per thousand gallons including capital and waste disposal costs. Standard microfiltration and precipitation processes cost approximately $ 80 for the same amount [40].

Phytokentics Inc. has considered cost comparisons for phytoremediation and situations involving common cleanup practices where contaminated soil may be removed and isolated, incinerated to destroy or burn off contaminants and the residue returned to the site as sterile material. Reclamation of the site may require replacement of the soil as well. Estimates for this type of cleanup range from $ 200 to $ 600 per ton depending on the characteristics of the site and transportation and landfill costs. Costs for the removal of one meter of soil from a one-acre site can be $ 600 000 to $ 2.5 million, depending on the density of the soil and regional transportation and landfill costs. Costs for phytoremediation of a one acre site including site preparation, planting and harvest of plant material may range from $ 2000 to $ 5000 [38].

The costs of phytoremediation are highly dependent on site characteristics. However, a number of factors contribute to lower costs in this technology. Where site circumstances permit, phytoremediation may use the same equipment and materials common to agricultural practices and costs can be equated to growing crops.

Maintaining a site under cultivation for even ten years is estimated to cost substantially less than conventional methods of site cleanup. Further, in lieu of these favorable costs figures from completed phytoremediation projects, this could allow for more sites to be cleaned up simultaneously.

Phytoremediation offers the additional advantage of making contaminated sites more esthetically appealing, making its use as an innovative environmental technology likely to garner future public interest and support. In fact, a series of public focus group meetings conducted across Canada in 1996 to gauge public perceptions and awareness for environmental applications of biotechnology revealed a high level of public support for the use of plants in a phytoremediation capacity [5].

8. The Limitations of Phytoremediation

A realistic look at phytoremediation points to some serious limitations for this technology. Hyperaccumulating plants often accumulate a specific element only and have not yet been found for all elements of interest or concern to NATO countries. This shortfall requires much more basic herbarium work and field investigations of existing contaminated sites and possibly the establishment of a gene bank to facilitate access to and testing of potential hyperaccumulators [15].

The source of hyperaccumulators could also pose a problem for developing this technology. Many hyperaccumulator plants are relatively rare taxa often occurring in remote areas or are very restricted in distribution [9].

Those associated with metal-rich soils are often threatened by devastation from mining activities [41]. Furthermore, populations may be very small. The collection of propagules from hyperaccumulators may also present problems as native plant seed production is often sporadic. Furthermore, seed from native species may be difficult to harvest, separate and clean. Culturing these species could also be difficult as they may have very specific seed germination requirements and pollination mechanisms as well as susceptibility to diseases and herbivores. It appears that most hyperaccumulators have slow growth rates and produce small amounts of biomass [19]. These traits may also be exacerbated by the colder Canadian climate. Some species may show considerable variation in regard to shoot production. Others may be limited in their efficacy for strongly sorbed (*e.g.* PCB's) or weakly sorbed contaminants. The use of genetic engineering to introduce genes into fast growing cultivars, to regulate root growth, or to increase production of selected plant enzymes may address these problems with this research currently in progress at the *Plant Biotechnology Institute of the National Research Council* in Saskatoon and the *USEPA Athens Georgia Laboratories* [27].

The use of wild plants for phytoremediation raises particular issues of risk assessment. This potential limitation is discussed in greater detail in a later section of this paper. However, much of the information required for the appropriate risk assessment not be readily available with much of the basic research still necessary.

Phytoremediation is frequently slower than traditional physical/chemical techniques, requiring several growing seasons for site cleanup. It is not an appropriate solution to site remediation where the target contaminant presents an eminent danger to human health or the environment. Furthermore, the time requirements may prove disadvantageous, depending on the use that the property is intended for. Property developers, in particular, may not wish to wait out the several growing seasons needed for phytoremediation to effectively clean up the site [20]. These timing constraints may not be of concern where the site is intended for remediation to its natural vegetated state.

The difficulties in characterizing contaminated sites as to the complexity of waste, the selection of appropriate analytical methods and the heterogeneity of site conditions will apply to phytoremediation as with any cleanup technology [42]. Phytoremediation will further require that the site be large enough to accommodate traditional agricultural cropping techniques. The soil texture, contaminant level, pH, salinity, and toxic level must all be within the limits of plant tolerance. Highly soluble contaminants may leach outside the root zone, rendering plant uptake less effective. In general, phytoremediation is appropriate for treatment of shallow contaminated sites [20] although some species like poplar have extensive root systems that can reach deep groundwater contaminants [18].

Difficulties may also arise in relation to the terrain of the site and instability of slopes and surface materials. Microclimates on the site may be inhospitable resulting in low numbers of seed germination and establishment. Phytoremediation could also necessitate considerable input costs such as in pretreating waste material or the sites on which the waste is deposited. Chelating agents, irrigation, soil amendments and insecticides may be required to effectively remediate the site.

The biological interaction of species on the site could be a limiting factor in successful phytoremediation. Plants used for cleanup may be particularly susceptible to herbivores and pests. It may take time and effort to develop a complete understanding of biological mechanisms and to conduct thorough design analyses for biological remediation systems.

The techniques for harvesting/processing and disposal of contaminated biomass must be carefully designed and considered. Further advances in harvesting techniques may yet be required. Sun heat and air drying, ashing or incineration, composting, pressing and compacting and leaching are all methods that are being considered for disposal [19]. The potential environmental impacts of each approach however, must be carefully evaluated.

While phytoremediation appears to be a very cost-effective approach to cleaning up contaminated sites, this still needs to be validated under actual field conditions. Further, the ability to develop cost comparisons and to estimate project cost will also need to be determined on a site-specific basis. The economics of the technology, performance capabilities and comparison to competing technologies are still very much required to determine phytoremediation's potential share of the remediation market as well as its commercial potential [2, 38, 43].

9. Phytoremediation Design

The design of phytoremediation application protocols will vary according to the contaminants, the conditions at the site, the cleanup levels required, and the plants to be utilized [44]. A few common design considerations integral to most phytoremediation applications include the following:

- site characterization
- plant selection
- treatability studies
- irrigation, agronomic inputs, and site maintenance
  contaminant uptake rate, clean up, and harvesting

9.1. CONTAMINANT LEVELS

During the site characterization phase, the concentration level of the contaminants needs to be established. High levels of contamination may eliminate consideration of phytoremediation as a viable option. Plants are not able to treat all metal contaminants and operate ideally in soil conditions that range from pH 4 to pH 7. It is important to also understand both the range of contaminants that can be treated utilizing plants as well as the depth of contaminants in the soil/sediment profile [44].

9.2. PLANT SELECTION

Plants are selected according to the desired application, contaminants of concern, and increasingly, as a result of forensic assessment and functional analysis of existing vegetation already established at the site [15]. A screening test of the literature is recommended to assist the design engineer in the selection process. Candidate plants utilized in the extraction of metals may include such varieties as Indian mustard and sunflowers for lead; pennycress and scented geraniums for zinc, cadmium, and nickel; and sunflowers and a whole range of aquatic plants for radionuclides [31].

9.3. TREATABILITY TESTS

Treatability or plant screening studies are recommended prior to the designing of the phytoremediation plan. Treatability studies assure concerned parties that phytoremediation systems will achieve the desired results. Treatability studies will also reveal data on toxicity and transformation products, as well as the fate of the contaminants in the plant system. Different concentrations of contaminants are tested with different proposed plant species.

9.4. IRRIGATION, AGRONOMIC INPUTS, AND SITE MAINTENANCE

Irrigation of plants ensures a vigorous start to the system even under drought conditions. Hydrological modeling may be required to estimate the rate of percolation to groundwater during irrigation conditions and should be withdrawn if the area receives sufficient rainfall. Agronomic inputs include the nutrients necessary to sustain vigorous growth of the plant cultivars and elected for the phytoremediation project. In addition to the prior soil analysis, a fertilizer regime must be selected to sustain plant growth and ensure plant vitality. Maintenance of the phytoremediation system may include adding fertilizer, agents to bind the metals to the soil, or chelates to assure plant uptake of the metallic contaminants. Replanting may be required due to drought, conditions of phytotoxicity, vandalism, plant pest, and herbivore activity [44]. Many environmental factors, principally abiotic considerations such as wind, temperature, solar radiation, and precipitation, cannot be realistically controlled in a field environment making the presence of regular site maintenance a prerequisite.

9.5. CONTAMINANT UPTAKE, HARVESTING, AND CROP ROTATION

Generic estimates of the uptake of a contaminant, harvesting methods, metal recovery and biomass disposal, and cropping patterns are possible but site specific and will rely extensively on continued collaboration between the site engineer, site owner, appropriate regulatory authority, and agronomist/soil scientist knowledgeable of growing conditions unique to the area.

10. Environmental and Health Risk Assessment Considerations

Considerable attention is now being given to the environmental and health risks associated with the introduction of alien species into new environments [45]. This follows from the severe environmental and economic impacts of alien species that have become invasive of natural and agricultural ecosystems [46, 47]. The following considers how environmental and health risk assessment should be conducted for all plant species introduced into new environments and particular issues for nonindigenous plants used for phytoremediation. These considerations are of utmost importance in recognition of the fact that the majority of those candidate plant cultivars on the Environment Canada PHYTOREM data base were found to be nonnative.

10.1. ELEMENTS OF ENVIRONMENTAL RISK ASSESSMENT FOR PLANTS

Environmental risk assessment of the introduction of plants into new environments has received the most attention for genetically modified crop plants. Plants used for phytoremediation may not be genetically altered but they will likely be species introduced into a new environment. Risk assessment criteria must consider both possibilities as well as risks related to release both on small and large scales. In general, the following elements should be considered in a risk assessment process for the introduction of plant species for phytoremediation.

(i) The characteristics of the plant to be released including information on its taxonomic identification, reproductive biology and sexual compatibility with related species;

(ii) life history characteristics and ecological niche;

(iii) characterization of the ecosystems exposed or potentially exposed to the plant including biotic and abiotic features and ecosystem processes;

(iv) potential environmental impacts of the introduction of the plant species at the genetic, population and ecosystem levels, including:
  - the potential of the plant to interbreed with related species in the ecosystem and nearby ecosystems,
  - the potential of the plant or its hybrids to become established and spread in the ecosystem or nearby ecosystems,
  - the potential of the plant or its hybrids to cause environmental degradation by disrupting ecosystem processes.

10.2. PLANT SYSTEMATICS

A starting point for any risk assessment associated with the introduction of a plant into a new environment is an understanding of plant systematics. For some species used in phytoremediation there may be little issue with taxonomic identification. However, other species, such as crop plants, weeds and ornamentals, may exhibit complex variation and integration, reproductive anomalies and odd distribution patterns which make them difficult to analyze [48]. Furthermore, the classification nomenclature of such species can be inadequate and confusing [48]. If the plant to be introduced contains a deliberately introduced gene, then details of the donor organisms and methods of incorporating the gene into the host seed will be required as well.

10.3. PLANT BIOLOGY

For many plant species, risk assessment may be hampered by fundamental gaps in information on their basic reproductive biology, life history characteristics and ecological niche. This information is used as a basis in risk assessment and, in its absence, data may have to be generated through laboratory and field research.

10.4. THE RECEIVING ENVIRONMENT

The receiving environment for the plant must also be considered in risk assessment. The ecosystem into which a plant is released can be considered on a range of scales from microhabitats of a few square meters to ecozones that range over thousands of hectares. It will be important to adopt a uniform scale of classification that is relevant to the risk assessment process for plants used in phytoremediation. The characterization of ecosystem components presents a considerable challenge and further consideration needs to be given to the depth of information needed and how it can be reasonably obtained.

10.5. IMPACTS ON RARE SPECIES

Rare and endangered species in the ecosystem into which the plant is introduced must be considered. Status reports and data bases for rare species are being developed in Canada [49]. However, it may prove difficult to obtain up-to-date and accurate information on some rare species for risk assessment purposes. Furthermore, it may not be possible to easily obtain information on the biology of those species to determine how they may interact with or be affected by the plant introduced for phytoremediation of a contaminated site.

Phytoremediation also raises an issue in relation to rare species if the source of starting material for phytoremediation is itself a rare plant. Botanists who study rare species do not encourage the collection of voucher specimens unless the population can withstand the loss. An unacceptable loss may be set at more than 4 % of the individuals in the population [49]. Researchers will have to exercise caution that by collecting rare species for use in this area they are not destroying the population in its natural habitat.

10.6. PLANT INVASIVENESS

Risk assessment for phytoremediation must further consider the impacts of the plant once it is released into the environment [47]. The ability of the introduced plant to interbreed with weedy and wild relatives is of particular concern for genetically modified plants but arises as well for those that are not genetically modified. Cross-pollination risks must be considered in terms of the sexuality of the plant, the presence of compatible relatives, the breeding systems of the plants, flowering phenologies, common means of pollination and the potential for cross-pollination, fertilization and viable seed set under field conditions [50]. The challenge of obtaining this information for plant species not used in agriculture or forestry could be formidable.

A further issue to resolve in risk assessment is how to determine if a plant will become established, persist and spread in an ecosystem once it is introduced. There are a number of approaches to predicting the potential for plant species to invade new ranges [51]. These include:

(i) using groupings or lists of species that have become naturalized beyond their range;
(ii) relying on traits of weeds and ruderal species to predict invasiveness;
(iii) predicting a region's future invaders on the basis of their home range climate;
(iv) using models to predict a species rate of spread in a new range;
(v) using controlled growth facilities and greenhouse experiments;
(vi) comparing performance of congeners especially through field trials;

(*vii*) deliberately sowing an alien species beyond its current range; and

(*viii*) deliberate introductions beyond a species' range coupled with simultaneous manipulation of the environment.

There is no one universal approach to predicting invasiveness and accuracy of prediction will be highest when several approaches are combined simultaneously [51]. Further work is also needed to determine the mechanisms whereby plant community diversity may influence invasibility of ecosystems [52]. A further challenge for researchers in this area will be to integrate their research on genetic, population and ecosystem levels [47].

10.7. IMPACTS ON WILDLIFE AND ACCUMULATION IN THE FOOD CHAIN

Of particular concern in relation to ecosystem impacts of introducing plants for phytoremediation purposes lies in impacts on herbivores and pollinator species. While it has been suggested that the metal content in hyperaccumulators would render these plants unpalatable to grazers, this issue requires much more consideration in regard to concentration of toxins in the food chain [2, 20]. Likewise, any ecosystem or health effects from the volatilization of metals or other contaminants by plants should be assessed thoroughly.

10.8. DETERMINATION OF ENVIRONMENTAL FATE AND EFFECTS AND TEST METHOD DEVELOPMENT

The development of models to determine fate, transport and effects (both within and outside of plants) of soil and groundwater contaminants as well as metabolic products resulting from phytoremediation is still in its infancy and will most certainly require special consideration [2, 3]. Screening methods for determination of optimal plant species per site, rates of application of fertilizer and chelating agents, and ideal irrigation practices are also a priority. Toxicity assays for the residual chemicals in soils following phytoremediation will also be required. The bioavailability or mobility of residual chemicals should also be assessed. Concern arises that the residuals may become toxic to target organisms or move off site. This is not an issue for residuals that are tightly bound to soil or are immovable because of their inability to partition to the liquid phase.

11. Status of Phytoremediation Applicability to Metals

Results from demonstration phytoremediation projects in Canada and among a number of NATO countries, coupled with dialogue from representative scientists actively engaged in the area, have identified the commercial readiness of a number of phytoremediation techniques for specific metals of concern [2, 3, 31]. They include the following:

Metal contaminant category	Ni	Co	Se	Pb	Hg	Cd	Zn	As	Cs	Sr	^{3}H	U
Readiness	4	4	4	4	3	2	3	1	3	2	3	3
Uninduced phytoremediation	4	4	4	0	4	4	4	1	2	2	n/a	0
Induced phytoextraction	0	0	0	4	0	0	0	0	0	0	0	3
Regulatory acceptance	Y	Y	N	Y	N	Y	Y	N	Y	Y	Y	N

0	none	1	basic research underway	2 lab scale (ready for field)
3	field scale deployment	4	under commercialization	

12. Research Needs and Priorities

Results from demonstration projects in Canada and among a number of NATO countries also point to a number of research needs and priorities for consideration as a prelude to accelerated phytoremediation applications for remediation and restoration of sites contaminated with metals. A partial listing of these required research activities includes:

- Phytoremediation speciation rates and fate of the volatilized metals with respect to metal speciation
- Comparative risk assessment for different mercury remedial technologies
- Screening of all radionuclides with respect to phytoextraction
- Soil-amendment studies
- Research on selective accumulation and species screening if ^{90}Sr is to be treated by phytoremediation
- Phytoextraction of uranium
- Targeted studies on the basic science of phytoremediation
- Research and development in the area of uptake, removal, fate and transport, and effects of phytoremediation
- Evaluation of crop rotation for enhancement of phytoremediation
- Regulatory acceptance of phytoremediation technologies
- Timeline for deployment of a phytoremediation technology with respect to host countries' regulatory requirements for cleanup
- Risk to ecological receptors from contaminants in plants following uptake
- Plant species research, including plant physiology and the possibility and need for genetic manipulation
- Plant screening on site with respect to contaminant levels
- Basic research on synthetic and organic chelators and inducement of hyperaccumulation
- Fate of chelating agents in the soil
- Individual site testing protocols
- Incineration of contaminated biomass (information exists only on metal recovery)

- Life-cycle costs and regulatory oversight for phytoextraction disposal alternatives
- Cost data on large-scale projects
- Mechanisms to minimize on site access to herbivores
- Improved mechanisms to predict phytotoxicity
- Improved understanding of forensic phytoremediation
- QA/QC, availability, propagation, storage and protection of germplasm of plants from contaminated sites.

13. Conclusion

Phytoremediation is an emerging technology that shows considerable promise for remediating and restoring sites contaminated from a variety of sources. The continued urgency for contaminated site cleanup in NATO countries requires that phytoremediation be given careful, serious, and immediate consideration as a cost-effective, promising and innovative environmental technological solution to this problem. Phytoremediation is expected, in certain situations, to demonstrate superior economic, technical, and environmental advantages over traditional physical, chemical, and thermal remediation techniques. Given the significant strength in academic research, government research facilities and a rapidly evolving environmental biotechnology industry capability, the NATO community is uniquely positioned to draw upon a cadre of silvicultural, agronomic, biological, environmental, and engineering expertise necessary to develop leadership in phytoremediation research and advancement. However, there are still limitations to the technology and numerous outstanding issues that must be addressed before the utility of this technology can be fully exploited. Some of these issues include scientific and technical research needs, environmental and health-risk assessment issues, and the need for regulatory clarification, commercialization, and public acceptance. This paper is intended to catalyze and focus future discussion in these areas.

14. Disclaimer

This manuscript has not been peer-reviewed internally within Environment Canada and reflects only the views of the author. Reference to any particular company or process in this manuscript does not constitute endorsement of that said company or product.

REFERENCES

1. Vangronsveld, J. and Cunningham, S.D. (1999) Metal-contaminated soils. In situ inactivation and phytorestoration. Springer-Verlag, New York.

2. *United States Environmental Protection Agency,* (2000) Introduction to Phytoremediation, EPA/600/R-99/107, p. 104.

3. *United States Department of Energy,* (2000) Proceedings from the workshop on phytoremediation of inorganic contaminants, INEEL/EXT-2000-00207, Idaho National Engineering and Environmental Laboratory, Idaho Falls, Idaho.

4. Fiorenza, S., Oubre, C.L., and Ward, H. (eds.) (2000) *Phytoremediation of Hydrocarbon Contaminated Soil,* Lewis Publishers, New York.

5. McIntyre, T.C., *et al.* (1996) A survey to determine public awareness, perceptions, and concerns over enhanced environmental applications of biotechnology in Canada (*unpublished document*), Creative Research Inc., Ottawa.

6. *National Contaminated Sites Working Group* (NCSWG) (1998), Federal government working group on contaminated sites (*unpublished*), Ottawa, Ontario.

7. *NRTEE* (1998) Brownfields, National Round Table on Environment and Economy, *Ministry of Supply and Services,* Ottawa, Ontario.

8. Baker, A.J.M. and R.R. Brooks (1989) Terrestrial higher plants which hyperaccumulate metallic elements — a review of their distribution, ecology, and phytochemistry, *Biorecovery* 1, 81–126.

9. Brooks, R.R. (Ed.) (1998) *Plants That Hyperaccumulate Heavy Metals ' Their Role in Phytoremediation, Microbiology, Archaeology, Mineral Exploration, and Phytomining,* CAB International, New York.

10. Reilley, *et al.* (1996) Dissipation of polycyclic aromatic hydrocarbons in the rhizosphere. *J. Environ. Qual.* 25, 212–219.

11. Moffat, A.S. (1995) Plants proving their worth in toxic metal clean up. *Science* 269, 302–303.

12. Harrington, C.F., Roberts, D.J., and Nickless, G. (1996) The effect of cadmium, zinc, and copper on the growth, tolerance index, metal uptake, and production of malic acid in two strains of the grass *Festuca rubra. Can. J. Bot.* 74, 1742–1752.

13. Winterhalder, K. (1996) Environmental degradation and rehabilitation of the landscape around Sudbury, a major mining and smelting area. *Environ. Rev.* 4, 185–224.

14. McIntyre, T.C. (1999) Forensic phytoremediation and the Sydney Tar Ponds, presented to the Sydney Tar Ponds Joint Action Group (*unpublished*), Sydney, Nova Scotia.

15. Olson, P. and Fletcher, J. (2000) Ecological recovery at a former industrial sludge basin and its implications to phytoremediation and ecological risk assessment, *Environ. Sci. Pollut. Res.* (*unpublished*).

16. Boyd, R.S. and Martens, S.N. (1992) The raison d'etre for metal hyperaccumulation by plants, in A.J.M. Baker, J. Proctor, and R.D. Reeves (eds.), *The Ecology of Ultramafic (Serpentine Soils)*, Intercept, Andover, pp. 279–289.

17. Borovik, A.S., (1990) Characterization of metal ions in biological systems, in A.J. Shaw (ed.), *Heavy Metal Tolerancein Plants: Evolutionary Aspects,* CRC Press, Boca Raton, pp. 3–5.

18. Gatcliff, E.G. (1994) Vegetative remediation process offers advantages over traditional pump-and-treat technologies. *Remediation* 4, 343–352.

19. Salt, D., *et al.* (1995) Phytoremediation: a novel strategy for the removal of toxic metals from the environment. *Bio/Technology* 13, 468–474.

20. Schnoor, J.J., *et al.* (1995) Phytoremediation of organic and nutrient contaminants. *Environ. Sci. Technol.* 29, 318–323.

21. *United States Department of Energy* (1994) Summary report of a workshop on phytoremediation research needs, Office of Technical Information, Springfield (VA).

22. Chaney, R., *et al.* (1995) Potential use of metal accumulators, *Mining Environ. Manag.* 3, 9–11.

23. Cornish, J.E., *et al.* (1995) Phytoremediation of soils contaminated with toxic elements and radionuclides, in *Bioremediation of Inorganics* (R.E. Hinchee, J.L. Means, and D.R. Burns, eds.), Battelle Press, Columbus (OH).

24. Wang, *et al.* (1995) Bioremoval of toxic elements with aquatic plants and algae, in *Bioremediation of Inorganics* (R.E. Hinchee, J.L. Means, and D. Burris, eds.), Battelle Press, Columbus (OH).

310

25. Sorenson, D.L., *et al.* (1994) Field scale evaluation of grass enhanced bioremediation of PAH contaminated soils, in *20th Annual RREL Research Symposium Abstract Proceedings*, EPA/600/R-94/011, pp. 92–94.

26. Schwab, A.P., Banks M.K., and Arunachalam M. (1995) Biodegradation of polycyclic aromatic hydrocarbons in rhizosphere soil, in *Bioremediation of Recalcitrant Organics* (R.E. Hinchee, D.B. Anderson, and R.E. Hoeppel, eds.), Battelle Press, Columbus (OH).

27. McCutcheon, S. (2000) Contribution of plant enzyme systems to remediation of organic wastes, presented to the USEPA Phytoremediation State of the Science Workshop, Boston (MA).

28. Anderson, T.A., Guthrie, E.A. and Wilson, B.T. (1993) Bioremediation in the rhizosphere. *Environ. Sci. Technol.* 27, 2630–2636.

29. Anderson, T.A., Kruger, E.L. and Coats, J.R. (1994) Enhanced degradation of a mixture of three herbicides in the rhizosphere of a herbicide tolerant plant. *Chemosphere* 28, 1551–1557.

30. Kramer, U., *et al.* (1996) Free histidine as a metal chelator in plants that accumulate nickel. *Nature* 379, 635–638.

31. McIntyre, T.C., (2000) PHYTOREM — a global data base on aquatic and terrestrial plants known to sequester, accumulate, or hyperaccumulate metals in the environment (*under development and to be released*).

32. Burken, J.G. and Schnoor, J.L. (1996) Phytoremediation: plant uptake of atrazine and role of root exudates. *J. Environ. Eng.* 122, 958–963.

33. Brooks, R.R., *et al.* (1978) Copper and cobalt in African species of *Aeolanthus* MART. (*Plectranthinae, Labiatae*). *Plant & Soil* 50, 503–507.

34. Brooks, R.R., *et al.* (1981) The chemical form and physiological function of nickel in some Iberian *Alyssum* species. *Physiol. Plant.* 51, 161–170.

35. Black, H. (1995) Absorbing possibilities: phytoremediation. *Innovations* 103, 12.

36. Coghlan, A. (1996) How plants guzzle heavy metals. *New Scientist*, 17 Feb., p. 17.

37. Rugh, C.L. *et al.* (1996) Mercuric ion reduction and resistance in transgenic *Aribidopsis thaliana* plants expressing a modified *merA* gene. *Proc. Nat. Acad. Sci. USA* 93, 3182–3187.

38. Glass, D. (1998) *The 1998 United States Market for Phytoremediation*, D. Glass and Associates, Inc., Needham (MA).

39. Bosshard, P.B., Reinhard, B., and Brandl H. (1996) Metal leaching of fly ash from municipal waste incineration by *Aspergillus niger*. *Environ. Sci. Technol.* 30, 3066–3070.

40. Phytotech (1996) Sunflowers blossom in tests to remove radioactive metals, in *The Wall Street Journal*, 29 Feb.

41. Newel, J. and Perry, J. (1996) Innovation across the ocean, British research seeks remedial solutions. *Soil & Groundwater Cleanup*, 1–4.

42. Hrudy, S. and Pollard, S.J. (1993) The challenge of contaminated sites: remediation approaches in North America. *Environ. Rev.* 1, 55–72.

43. Yaeck, J. (1997) *An Investigation of the Process Required to Determine the Success of Phytoremediation*. Faculty of Science, University of Waterloo, Waterloo, Ontario (Canada).

44. *Interstate Technology and Regulatory Cooperation Working Group* (ITRC) (1999) Phytoremediation decision tree, http://www.itrcweb.org

45. Ruesink, J., *et al.* (1996) Reducing the risks of nonindigenous species introductions. *BioScience* 45, 465–477.

46. *Office of Technology Assessment* (1993) Harmful non-indigenous species in the United States, OTA-F-565, US Government Printing Office, Washington (DC).

47. Keddy, C. (2000) The role of importation control in protecting native Canadian biodiversity — a discussion paper, prepared for the Canadian Wildlife Service, Environment Canada, Ottawa (Ontario), pp. 9–98.

48. Small, E. (1993) The economic value of plant systematics in Canadian agriculture. *Can. J. Bot.* 71, 1537–1545.

49. Lancaster, J. (ed) (1997) *Alberta Native Plant Council Guidelines for Rare Plant Surveys*, publication of the Alberta Native Plant Council, Alberta.

50. Keeler, K. and Turner, C. (1991) Management of transgenic plants in the environment, in *Risk Assessment in Genetic Engineering: Environmental Release of Organisms* (M. Levin and H. Strauss, eds), McGraw-Hill, New York.

51. Mack, R.N. (1996) Predicting the identity and fate of plant invaders: emergent and emerging approaches. *Biol. Conserv.* **78**, 107–121.

52. Tilman, D. (1997) Community invasibility, recruitment limitation, and grassland biodiversity. *Ecology* **78**, 81–92.

53. Medina, V.F. and McCutcheon, S.C. (1996) Phytoremediation: modelling removal of TNT and its breakdown products. *Remediation*, Winter, pp. 31–45.

COMBINED REMOVAL OF OIL, TOXIC HEAVY METALS AND ARSENIC FROM POLLUTED SOIL

S. N. GROUDEV

Department of Engineering Geoecology,
University of Mining and Geology,
Studentski grad – Durvenitza,
Sofia, Bulgaria

1. Introduction

The bioremediation of soils polluted with crude oil products is largely used on commercial scale in several countries. Different *in situ* and *ex situ* techniques using both indigenous and laboratory-bred microorganisms are used [1-2]. These techniques are also used for bioremediation of soils polluted with toxic heavy metals and arsenic [3-4]. However, the microorganisms involved in the removal of the different pollutants as well as the mechanisms of removal are different. For that reason, the treatment of soils polluted with both organic (oil and oil products) and inorganic (heavy metals and arsenic) pollutants is of special interest.

In the Tulenovo oil deposit, Northeastern Bulgaria, near the Black Sea coast, oil is recovered through numerous wells, which produce fountains of fluid containing brine and oil. The oil content in the fluid recovered from the different wells varies from about 0.1–1 %. The oil is heavy, with a specific gravity of 0.939, rich in asphaltene-resinous substances with a high viscosity. The total ion concentration in the brine is about 2-3 g/L and the pH is in the range of 6.8–7.7. The main components in the brine are sulfates, chlorides, sodium and magnesium but some toxic heavy metals are present in concentrations, which in some cases are higher than the relevant permissible levels.

The fluid from each well is collected in an individual vessel where the oil is separated from the brine as a result of their different specific gravity. The aqueous phase is siphoned off from the relevant vessel and runs into the sea. In most cases a portion of oil escapes together with the water from the vessels. Several ponds having a surface of about 10–30 m^2 each and a depth of about 0.5 m have been constructed to collect the polluted waters, to retain the oil and to prevent its discharge into the sea. A considerable amount of oil has been collected in the ponds during the recovery process in the deposit and the soils in this area, especially those around the ponds, are heavily polluted with oil. Both *in situ* and heap bioremediation techniques were used to clean up the polluted soils [5, 6]. Apart from the oil, some soils are polluted with toxic heavy metals (lead, cadmium, copper, zinc) and arsenic. Pilot-scale operations for a combined removal of oil, toxic heavy metals and arsenic from such polluted soils were applied in the deposit using the heap and *in situ* bioremediation techniques. Some data about these operations are presented in this paper.

313

V. Šašek et al. (eds.),
The Utilization of Bioremediation to Reduce Soil Contamination: Problems and Solutions, 313–318.
© 2003 *Kluwer Academic Publishers. Printed in the Netherlands.*

2. Materials and Methods

Two experimental heaps containing polluted soil were used in the experimental work. The heaps had the shape of a truncated pyramid and were formed on the ground surface covered by a concrete layer. They were 0.7 m high, 20 m long, 5 m wide at the top and 7 m wide at the bottom. The initial oil content in the heaps was in the range of about 20–25 g/kg dry soil and the contents of heavy metals and arsenic were about 1.5–3 times higher than the relevant permissible levels. The initial soil pH was about 6.4.

The polluted soil contained its own indigenous microflora, including different oil-degrading and metal-solubilizing microorganisms. Preliminary experiments had revealed that the inoculation of the soil with a mixed laboratory-bred microbial culture consisting of several very active oil-degrading microorganisms (related to the genera *Bacil-lus*, *Rhodococcus*, *Pseudomonas* and *Corynebacterium*) enhanced considerably the rate of oil degradation and, at the same time, had no negative effect on the rate of metal solubilization.

One of the experimental heaps was inoculated with the above-mentioned microbial culture. After the inoculation, the introduced microorganisms rapidly formed a stable community with the indigenous microflora. The microbial community as a whole demonstrated a well-expressed synergistic action, which produced an efficient degrada-tion of the oil hydrocarbons in the soil.

The second heap was not inoculated with laboratory-bred microorganisms. The degradation of the oil in this heap was carried out only by the indigenous microflora.

The soil treatment was started in March 1997. Both heaps were treated in the same way. Regular cultivation of the soil was used to enhance the natural aeration. The soil was irrigated with water to maintain its water content in the range of about 30–35 %. Zeolite saturated with ammonium phosphate was added to the soil (in amounts in the range of 2–5 kg/ton dry soil) to provide the microorganisms with suitable sources of nitrogen and phosphorus and to improve the physico-mechanical properties of the soil. Periodically, the soil was flushed with water acidified to pH 4.5–5.0 to remove the products from the oil degradation as well as the soluble heavy metals and arsenic. A system to collect the soil effluents was constructed around the heaps. These effluents were treated in a natural wetland located near the heaps.

The experimental plot used in this study for the *in situ* bioremediation was 200 m^2 in size. The soil in this plot was polluted to a depth of about 80–100 cm below the surface where a natural impermeable clay barrier was present. The initial oil content in the soil was in the range of about 5–20 g/kg dry soil. The toxic heavy metals and arsenic were also present at concentrations higher than the relevant permissible levels.

The indigenous microflora of the polluted soil section was very similar to that of the soil heaps used in this study. The soil plot was inoculated with the above-mentioned mixed laboratory-bred culture consisting of several oil-degrading micro-organisms. After the inoculation, the introduced microorganisms formed a stable community with the indigenous microflora.

The treatment of the soil plot was carried out in a way similar to that used for the heap bioremediation. The soil was cultivated down to the clay layer to enhance the natural aeration. Furthermore, air was injected into the soil through ten boreholes installed in the experimental plot. The lower ends of these boreholes also reached the clay layer. The soil effluents after flushing were collected by a system of wells and ditches constructed within and around the experimental plot and were treated by the above-mentioned natural wetland.

The oil hydrocarbons were determined by direct extraction from the relevant sample with 1,1,2–trichlorotrifluoroethane and IR determination. Elemental analyses were done by atomic absorption spectrophotometry and induced coupled plasma spectrophotometry.

The isolation, identification and enumeration of microorganisms were carried out by methods described elsewhere [5, 7].

3. Results and Discussion

Data about the microflora of the heaps are shown in Table 1. The total number of aerobic heterotrophic bacteria as well as the number of oil-degrading microorganisms steadily increased during the treatment and within two months from the start of the experiment exceeded 10^9 and 10^8 cells/g dry soil, respectively. These highest numbers were maintained during the whole treatment period, i.e. until November 1997. Apart from the microorganisms, some higher organisms (protozoa, worms, insects, etc.) inhabited the soil and their numbers increased simultaneously with the diminution of the content of the oil and other pollutants.

The toxic heavy metals and arsenic were solubilized from some easily leachable fractions in the soil mainly as complexes with secreted microbial metabolites (mainly organic acids) and were removed from the heaps by the drainage waters percolating through the soil mass. It must be noted, however, that these pollutants were solubilized also, although at lower rates, even from the relevant sulfide minerals present in the soil. This was due to the activity of some chemolithotrophic bacteria, mainly related to *Thiobacillus thioparus* and *Thiobacillus neapolitanus*. These bacteria enhance the oxidation of sulfide minerals by removing the passivation films of S^o deposited on the mineral surface. Some acidophilic chemolithotrophs (mainly *Thiobacillus ferrooxidans* and *Leptospirillum ferrooxidans*) able to oxidize sulfide minerals were also present in the soil. Their numbers before the treatment were low (less than 100 cells/g dry soil) due to the soil pH values which were unfavorable for these bacteria. However, during the treatment the soil pH slightly decreased and at the end of the experiment was lowered to about 5.9, and there were many microzones with lower pH values (as low as 4.5 – 5.0). The number of acidophilic chemolithotrophs in these microzones was relatively higher.

The oil products from oil degradation and metal solubilization were removed from the heaps by flushing with slightly acidified water. The pregnant heap effluents were treated efficiently by the above-mentioned natural wetland. The pollutants were deposited in the wetland mainly as a result of processes such as the microbial dissimilatory sulfate reduction and biosorption. The contents of heavy metals, arsenic and organic compounds in the wetland effluents were lower than the relevant permissible levels for waters intended for use in agriculture and/or industry.

TABLE 1. Concentrations of microorganisms related to different physiological groups in the soil heaps (2 months after the start of the experiment)

Microorganisms	Cells/g dry soil	
	Heap containing both indigenous and laboratory-bred microorganisms	Heap containing only indigenous microorganisms
Oil-degrading microorganisms	5.10^8	4.10^8
Aerobic heterotrophic bacteria	2.10^9	2.10^9
Nitrifying bacteria	6.10^6	3.10^6
Chemolithotrophs oxidizing Fe^{2+} at pH 7.0	3.10^6	2.10^6
Chemolithotrophs oxidizing Fe^{2+} at pH 72.5	5.10^3	3.10^3
Chemolithotrophs oxidizing S^o at pH 7.0	8.10^5	6.10^5
Anaerobic heterotrophic bacteria	1.10^6	3.10^6
Denitrifying bacteria	5.10^5	5.10^5
Sulfate-reducing bacteria	8.10^5	1.10^6
Fungi	4.10^6	3.10^6
Total cell number	3.10^9	3.10^9

Note: The data shown in the table are average values from analyses of soil samples taken from a depth of 30 cm below the heap surface.

TABLE 2. Contents (in g/kg dry soil) of pollutants in the soil heaps before and after treatment

Pollutants	Heap inoculated with laboratory-bred microorganisms		Heap containing indigenous microorganisms only		Permissible levels of pollutants for soil with pH 5.9
	Before treatment	After treatment	Before treatment	After treatment	
Oil	23	1.9	-	-	-
Lead	0.208	0.062	0.208	0.068	0.070
Cadmium	0.0055	0.0019	0.0055	0.0019	0.002
Copper	0.305	0.107	0.305	0.090	0.120
Zinc	0.284	0.140	0.284	0.125	0.200
Arsenic	0.035	0.015	0.035	0.010	0.025

TABLE 3. Biodegradation of the different hydrocarbon components of oil in the soil heap inoculated with laboratory-bred microorganisms

Hydrocarbon component	Content in the oil, %		Content in the soil, g/kg dry soil		Hydrocarbon degradation, %
	I	II	I	II	
Parafins	20.3	0.5	4.9	0.01	99.8
Naphthenes	58.1	4.4	13.9	0.08	99.4
Aromatics+polars	21.6	95.1	5.2	1.71	67.1
Total	100.0	100.0	24.0	1.80	92.5

Note: I – Before treatment; II – After treatment

The oil content in the heap inoculated with laboratory-bred microorganisms was decreased to less than 2 g/kg soil within 8 months of treatment (from the middle of March to the middle of November 1997), while in the other heap the residual oil content was about 7 g/kg dry soil (Table 2). This residual oil consisted mainly of asphaltenes, resins, and complex aromatics (Table 3). In both heaps the contents of toxic heavy metals and arsenic were decreased below the relevant permissible levels.

The rate of bioremediation markedly depended on the temperature of the soil. The highest rates of oil degradation and metal and arsenic solubilization were achieved during the warmer summer months (June–August) when the soil temperature exceeded 20 °C.

The treatment of the polluted soil by means of the *in situ* technique was also very efficient. About 90 % of the oil was degraded within 8 months (from the middle of March to the middle of November) at temperatures ranging from 3 °C to 35 °C (Table 4). The contents of toxic heavy metals and arsenic at the end of the treatment were lower or very close to the relevant permissible levels (Table 5). The soil pH after the treatment was in the range of 5.7–6.0.

The analyses of the microflora in the experimental plot showed that the total number of aerobic heterotrophic bacteria as well as the number of oil-degrading microorganisms steadily increased during the treatment and in the period from May to October exceeded 10^9 and 10^8 cells/g dry soil, respectively. The numbers of some higher soil inhabitants (protozoa, worms, insects, etc.) increased simultaneously with the diminution of the contents of pollutants in the soil being treated.

TABLE 4. Biodegradtion of the different hydrocarbon components of the oil in the experimental plot treated by means of the *in situ* technique

Hydrocarbon component	Content in the oil, %		Content in the soil, g/kg dry soil		Hydrocarbon degradation, %
	I	II	I	II	
Parafins	21.2	0.7	3.8	0.01	99.7
Naphthenes	59.0	4.8	10.6	0.08	99.3
Aromatics + polars	19.8	94.5	3.6	1.61	55.3
Total	100.0	100.0	18.0	1.70	90.6

Note: I – Before treatment; II – After treatment

TABLE 5. Removal of toxic heavy metals and arsenic from the polluted soil in the experimental plot by means of the *in situ* technique

Pollutants	Contents in the soil, mg/kg dry soil		Pollutant removal, %	Permissible levels for soil with pH<6.0, mg/kg dry soil
	Before treatment	After treatment		
Lead	185	64	65.4	70
Cadmium	5.1	2.3	54.9	2.0
Copper	262	140	46.6	120
Zinc	212	95	55.2	200
Arsenic	32	18	43.7	25

4. Conclusions

The data from this study showed that it was possible to achieve an efficient simultaneous removal of oil, toxic heavy metals and arsenic from heavily polluted soils using bioremediation by heap or *in situ* techniques. It was also possible to enhance the oil biodegradation by inoculating the soil with laboratory-bred microorganisms possessing a high activity toward the most refractory components of the oil. The indigenous and laboratory-bred microorganisms rapidly formed a stable community in the polluted soil. The growth and activity of this microbial community were maintained at high levels by suitable changes in the levels of some essential environmental factors such as water, oxygen and nutrient contents in the soil. The bioremediation markedly depended on the temperature of the soil.

The presence of an impermeable geological barrier below the polluted soil is necessary for the application of the *in situ* technique. Both the heap and *in situ* techniques are connected with the construction of systems to collect the pregnant soil effluents as well as of systems to clean up these effluents.

References

1. Alleman, B.C. and Lesson, A., (eds.) (1999) *In Situ Bioremediation of Petroleum Hydrocarbon and Other Organic Compounds*, Battelle Press, Columbus, Ohio, USA.
2. Alleman, B.C. and Lesson, A., (eds.) (1999) *Bioreactor and Ex Situ Biological Treatment Technologies*, Battelle Press, Columbus, Ohio, USA.
3. Groudev S.N. and Spasova, I.I. (1997) Microbial treatment of soil contaminated with heavy metals and arsenic, Paper presented at the *International Biohydrometallurgy Symposium*, Sydney, August 25-30, 1997.
4. Leeson, A. and Alleman, B.C. (eds.) (1999) *Bioremediation at Metals and Inorganic Compounds*, Battelle Press, Columbus, Ohio, USA.
5. Groudeva, V.I., Groudev, S.N., Uzunov, G.C. and Ivanova, I.A. (1994) Microbial removal of oil from polluted soils in a pilot scale operation in the Tulenovo deposit, Bulgaria, in D.S. Holmes and R.W. Smith (eds.), *Minerals Bioprocessing II*, TMS Minerals, Metals & Materials Society, Warrendale, Pennsylvania, USA, pp. 231-240.
6. Groudeva, V.I., Groudev, S.N., Ivanova, I.A. and Tzarkova, E.I. (1998) Microbial in situ treatment of oil contaminated soils, Paper presented at the *12th Forum for Applied Biotechnology*, Brugge, September 24-25, 1998.
7. Karavaiko, G.I., Rossi, G., Agate, A.D., Groudev, S.N. and Avakyan, Z.A. (1998) *Biogeotechnology of Metals. Manual*, Centre for International Projects GKNT, Moscow.

LANDFARMING FRAMEWORK FOR SUSTAINABLE SOIL BIOREMEDIATION

R.C. SIMS and J.L. SIMS
*Utah Water Research Laboratory, Utah State University,
Logan, UT, USA 84322-8200*

1. Introduction

In this chapter, we address four state-of-the-art aspects of the Landfarming Framework. These include: (*1*) definition and design, (*2*) treatment rate, (*3*) example of full-scale landfarming for soil bioremediation, and (*4*) research and application issues. Landfarming is a technology based on the use of soil microorganisms and agricultural methods in an aerobic environment to reduce soil contamination and associated risk of public exposure through transformation, immobilization, and detoxication processes in order to protect public health and the environment. Landfarming technology accomplishes recycling of organic carbon and nutrients within the biosphere and maintains basic soil characteristics necessary to support plant growth (forestation, vegetation, *etc.*) and thus contributes to natural resource sustainability. The success of landfarming technology depends upon efficient utilization of microorganisms in a soil farm environment to treat soil contamination. Thus the development and availability of tools and methods for effective utilization of microorganisms within a landfarm environment is critical. An improved understanding of chemical−soil−microbe interactions that includes bioavailability, microorganism characterization, and effect of biological formation of soil bound residues on risk assessment and management will result in improved design, management, and monitoring of landfarm systems for sustainable restoration of soil.

2. Definition and Design

Important design factors for each lift (0.15 m depth) of contaminated soil include: (*1*) time required for treatment to risk-based level, (*2*) volume of soil requiring treatment, and (*3*) amount of land required. The components are identified and related as follows:

$$T_L = V/A_r \tag{1}$$

V. Šašek et al. (eds.),
The Utilization of Bioremediation to Reduce Soil Contamination: Problems and Solutions, 319–334.
© 2003 *Kluwer Academic Publishers. Printed in the Netherlands.*

where T_L is time required for lift bioremediation,
      A_r assimilation rate of contaminated soil (m^3/t), and
      V volume of soil lift requiring treatment (m^3).

Total time required is the summation of T_L for all lifts comprising the total soil volume.

2.1. CONTAMINATED SOIL VOLUME

The volume of contaminated soil is determined through a site characterization process that involves three-dimensional (3-D) sampling of the contaminated soil. Discrete samples are recommended rather than composite samples [1–4] because results of composite samples tend toward the mean concentration of the sampled area and do not show the full range and variability of soil contamination. The total mass of contamination is determined as the integration of the concentrations and volumes throughout the site. Since contaminated soil is often blended with other contaminated soil or uncontaminated soil, a mass accounting system is required to inventory the total amount of contaminant initially present and treated through time using landfarming. Hot spots decrease the assimilation rate and therefore increase the time required for remediation (Eq. 1) due to toxicity of parent compounds or metabolized products to soil microorganisms that are responsible for biodegrading the target contaminants, and/or due to mass transfer limitations regarding chemical partitioning to water for biodegradation (bioavailability).

The design volume of the landfarm should be based on the volume of the contaminated soil, with the addition of liners, soil support materials, and water drainage and storage systems where appropriate [2]. The volume of contaminated soil is calculated on the basis of the loose, excavated soil to be placed in a Landfarm, rather than the volume of the compacted *in situ* soil [2, 5]. Environmental conditions appropriate for landfarming, including mass transfer of oxygen, blending contaminated soil with treated soil, and general biological activity, are improved when the soil undergoing bioremediation is maintained in the loose excavated soil density structure rather than in the compacted *in situ* density structure.

For example, consider a soil with the properties shown and the resulting soil volume for landfarming:

in situ density (ϱ_b)	= 1920 kgm^3
excavated soil density (ϱ_{be})	= 1520 kgm^3
volume increase (%)	= $(1920 - 1520)/1920 \times 100 = 21\,\%$ above the volume of undisturbed soil

This can be expressed as:

$$\text{volume increase (\%)} = (\varrho_b - \varrho_{be})/\varrho_b \times 100 \qquad (2)$$

2.2. ASSIMILATION RATE

The assimilation rate of contaminated soil can be determined for each contaminant or class of contaminants of concern, and can be calculated as follows:

$$A_r = T_r \times A \tag{3}$$

where A_r is assimilation rate

T_r treatment rate (m^3 of contaminated soil assimilated per m^2 Landfarm per unit time, and

A amount of land area available for the Landfarm (m^2).

Therefore, the assimilation rate for contaminated soil is directly related to both the area available for treatment and the treatment rate. Using Eqs 1 and 3, if the volume (V) of contaminated soil is 40 000 m^3, the area for treatment (A) is 4 000 m^2, and the rate of treatment is 1 m^3 contaminated soil per 1 m^2 landfarm per month (T_r), the following calculations are appropriate:

$$A_r = 1\ m^3/m^2 \text{ per month} \times 4000\ m^2 \qquad = 4000\ m^3 \text{ per month}$$
$$T = V/A_r = 40\ 000\ m^3/4000\ m^3 \text{ per month} = 10 \text{ months}$$

The assimilation rate is related to transformation, immobilization, and detoxication processes. Measurement, monitoring, and evaluation of these processes for landfarm environments have been addressed before [5–13]. The concept of using risk-based methods for determining soil cleanup levels may be based upon less than complete transformation, immobilization, or detoxication.

Transformation involves the biological destruction of parent compounds and production of metabolites, and transformation rates may refer to the destruction of the parent chemical or the metabolites [10, 14]. Transformation rates of chemicals in soils have been determined primarily for the parent chemicals [15, 16] but also for some metabolites [14, 17, 18]. Mineralization refers to the complete transformation of the parent compound to inorganic end-products (carbon dioxide, water) and nutrients (nitrogen and phosphorus) when they are part of the parent compound.

Immobilization refers to the retardation of a contaminant with respect to the movement of the water phase in subsurface systems. Immobilization occurs through physical sorption and sequestration, chemical reactions, and biological reactions analogous to the soil humification process whereby organic matter detritus is incorporated into soil organic matter [19–21].

Detoxication is measured using general microbial toxicity tests or genotoxicity tests. Detoxication can be evaluated for the aqueous phase within the soil as well as for the soil solid phase. Because many parent compounds and metabolites partition to the soil solid phase, change in toxicity at the soil surface needs to be assessed. Several researchers have identified limitations associated with testing toxicity of only the aqueous phase for toxic chemicals that are highly hydrophobic [22–24]. The Microtox[TM] Aqueous-Phase Assay has been used to evaluate detoxication of the soil solution phase of landfarm environments [7, 13]. The Finland Flash Solid-Phase assay [25] is capable of measuring the toxicity of the soil solid phase, and can be used to measure the extent to which the soil solid phase is detoxicated within a landfarm.

Management of the assimilation rate is important for addressing the 'length of time' design criterion for soil remediation using landfarming technology (Eq. 1). Eq. 3 indicates that treatment rate and treatment area are the relevant variables that affect the assimilation rate in Eq. 1.

2.2.1 Treatment Rate

A critical biological engineering design criterion is the treatment rate of contaminated soil. T_r (Eq. 3) is related to soil microbial activity of those microorganisms that are involved in biological transformation, detoxication, and immobilization of target contaminants. In order to restore contaminated soil to agriculture functions, it may also be necessary to remove non-target contaminants such as oil and grease or non-aqueous phase liquids (NAPLs) to concentrations below risk-based levels, below levels deleterious to plant growth, or to other levels that ensure the sustainability of the soil as a natural resource. Treatment rate is commonly modeled for landfarm systems using a first-order kinetic rate model,

$$\ln(C/C_0) = -kt \tag{4}$$

Use of the first-order model does not take into account mass transfer kinetics, toxicity effects, cometabolism, or effects of environmental variables [26].

The time requirement for remediation of a given volume of contaminated soil is associated with the specific chemical or chemical class requiring the longest time to attain the risk-based soil level, defined as the rate-limiting constituent or constituent class (RLC). The RLC is influenced by the assimilation processes discussed previously and also by the end-point target cleanup level. Therefore, T_r in Eq. (3) is the rate of treatment of the RLC:

$$T_r = T_{r(RLC)} \tag{5}$$

Treatment rate of contaminated soil is affected by four important variables: (1) characteristics of the chemical contaminants and the site soil, (2) mass transfer and toxicity to soil microorganisms, (3) environmental variables, and (4) pretreatment.

Chemical and Soil Characteristics. Chemical characteristics that affect the rate of treatment of organic chemicals within a landfarm include: (1) water solubility, (2) microbial toxicity properties, and (3) degree of halogenation. Contaminants that are more difficult to biodegrade using the landfarming framework have the following properties: (1) low water solubility, (2) biocidal, (3) highly chlorinated. Thus organic wood-preserving contaminants including creosote and pentachlorophenol (PCP) are more difficult to biodegrade [6].

Rate of treatment of these types of contaminated soils can be increased through blending acclimated soil with low volumes of untreated contaminated soil. that is managed to optimize biological activity. Acclimation of indigenous soil microorganisms generally occurs through prolonged exposure to low concentrations of chemicals and gradual incremental increases in concentration [10, 27, 28]. Acclimation of soil generally occurs at the outer boundaries of a contaminated site [28], or within a landfarm that has adapted to the target contaminants.

Incorporation of contaminated soil into treated soil results in an increase in the treatment rate for biocidal and non-biocidal chemicals and for NAPL-containing wastes. This is due to mechanisms that include: (1) increased surface area for contaminant sorption and immobilization, (2) lower dose of chemical (mass of chemical per mass of soil) that reduces toxicity to soil microorganisms, and (3) improved conditions for oxygen transfer and nutrient delivery (oxygen or nutrient consumption per mass of soil). When chemical and soil characteristics limit microbial activity through a shortage of nutrients, nitrogen and phosphorus fertilizers are added.

Soil characteristics that affect rate of treatment are related to characteristics affecting water-holding capacity, mass transfer of oxygen, soil structure, and tilling or blending of soil, i.e., particle size, and nutrient status. While sandy soils have low water-holding capacity that inhibits microbial activity, clayey soils have high water-holding capacity that inhibits aerobic microbial activity through reduced diffusion of oxygen to soil microorganisms. Loamy sands and sandy loam soils, with intermediate water-holding capacity that also allow rapid diffusion of oxygen in the gas phase, have the potential for the highest rate of treatment. Organic matter is generally directly related to microbial activity as well as immobilization of chemicals through sorption, sequestration, and humification. A detailed discussion of soil characterization factors that affect treatment process and treatment rate has been prepared by [29].

Mass Transfer Rate and Toxicity. The rate of mass transfer of target chemicals from NAPL to soil microorganisms plays an important role in affecting the rate at which chemicals become bioavailable and is a function of the "dose" of the NAPL, expressed as weight of NAPL per mass of soil [10, 30]. A reduction in NAPL concentration was demonstrated to result in an increase in the rate of biodegradation in a laboratory landfarm environment [13]. The engineering goal is to decrease the diffusion path length through which the target chemical must travel within the NAPL to arrive at the NAPL/water or NAPL/soil interface where the chemical may become more bioavailable to soil microorganisms. This can be accomplished by decreasing the thickness of the NAPL layer within a contaminated soil matrix, for example, by blending contaminated soil with an uncontaminated soil of the same texture or blending with the soil within the landfarm that has been previously treated to an acceptable assimilation end-point.

Microbial toxicity can also be reduced through decreasing the "dose" of toxic contaminants in order to increase the rate of soil bioremediation [13, 28, 31]. Toxicity reduction can be accomplished by blending the contaminated soil as described above. Toxicity was used as the criterion of waste application dose in a landfarm environment by [11] and resulted in an order of magnitude difference in blending petroleum-based waste (≈ 10 %, W/W) vs. wood-preservative wastes (≈ 1 %, W/W).

Environmental Variables. Environmental variables that affect the rate of soil bioremediation include soil moisture, oxygen, and temperature. Although soil moisture has been shown to have an effect on degradation rate, information is limited, and quantification through mathematical relationships has not been developed [32]. For biological engineering application to landfarms, moisture content should be maintained between 60 and 80 % of the value at the field capacity of the soil for optimum aerobic microbial activity and biodegradation [2, 28, 33].

The effect of oxygen on the rate of soil bioremediation using landfarming technology has been evaluated by (32, 34−36], and has been investigated by others using bioventing technology [37]. For biological engineering of landfarming, a value of 2 % oxygen (V/V) in the soil gas phase is required for non-oxygen limiting conditions. Hurst [34, 35] showed that values of 2, 5, 10 and 21 % (V/V) were approximately equivalent, and as soil oxygen tension decreased from 2 to 0 %, biodegradation rate, using mineralization as the endpoint, decreased to zero. A laboratory landfarm environment was used to evaluate the effect of oxygen tension on the rate of biodegradation of the industrial chemical 7,12-dimethylbenzanthracene (7,12-DMBA) [36]. The relationship of oxygen tension (0, 0.3, 0.8, 2, 7, and 21 % (mol/mol) to rate of biodegradation followed the relationship described by the Michaelis−Menten equation:

$$-dS/dt = k_{max}[O/(K_O + O)] \tag{6}$$

Based on linear transformation of the data using Lineweaver−Burk and Hanes plots, half-saturation constants (K_O) for oxygen were calculated as 2 and 3 %, respectively. Soil oxygen concentrations in full-scale landfarms have been observed to range from 0 to 21 %, depending upon depth and time after application of contaminants [2, 32].

A Michaelis-Menten type relationship between oxygen concentration and biotransformation has been found in biological wastewater treatment systems for nitrification, where K_O values range from 0.1 to 2.0 mg/L. Use of Henry's Law partitioning for oxygen indicates that values of oxygen in water within the above range correspond to values in air in the range identified by others [32, 34−36].

Temperature plays an important role in affecting the rate of treatment in landfarms located in temperate regions where seasonal and diurnal changes in temperature occur. Mathematical relationships for the effect of temperature on biological degradation of chemicals under laboratory landfarm conditions have been evaluated by several researchers [15, 32, 38] and for soil bioremediation within a full-scale landfarm environment by Sims *et al.* [2]. The relationship between temperature and biodegradation follows the relationship described by the Arrhenius equation for the effect of temperature on chemical reactions:

$$d\ln k/dT = E_a/RT^2 \tag{7}$$

Eq. 7 may be approximated by the expression:

$$k_2/k_1 = \Theta^{(T2-T1)} \tag{8}$$

Values for Θ for soil landfarm environments are summarized in Table 1 for the temperature range $10-30$ °C.

TABLE 1. Temperature coefficient values for chemical bio-degradation in soil

Chemical	Θ	Reference
Phenanthrene	1.1	8
Phenanthrene	1.0	38
1-Hydroxy-2-naphthoic acid	1.1	8
Anthracene	1.04	12
Benzo[a]anthracene	1.06	12
Chrysene	1.02	12
Benzo[b]fluoranthene	1.02	12
Benzo[a]pyrene	1.05	12
Benzo[ghi]perylene	1.04	12

Engineering management of soil temperature can be effected through control of irrigation and drainage, as wetter soils have higher heat capacities with less change in soil moisture, and by surface modifications including mulching, residue cover, plastic covers, etc. [28, 39].

While depression of biodegradation at low temperatures has been well demonstrated, indigenous microorganisms adapted to low temperatures (psychrophiles) have been shown to be active. Hinchee [37] observed biodegradation rates in unsaturated soil, under bioventing conditions, to be similar between Eielson Air Force Base, Alaska, and three other sites in Nevada, Utah, and Maryland, thus indicating that hydrocarbon biodegradation can occur in soil at locations continually subjected to cold temperatures.

Engineering Pretreatment. Engineering pretreatment processes including soil blending, physical classification of soil, and organic amendment can be used to increase the rate of soil bioremediation. Soil blending has been discussed previously.

Removal of rock and other debris with diameter of 20 mm or greater allows contaminated soil to be tilled to introduce oxygen and nutrients to be applied in a more uniform treatment process, and also prevents damage to agricultural equipment used to manage the contaminated soil. Removal of material greater than 20 mm in diameter also results in a smaller volume requirement for a landfarm treatment unit [2, 5]. Because the ratio of the surface area (A, $2\pi r^2$) to volume (V, $\frac{4}{3}\pi r^3$) of spherical particles decreases with an increase in radius (r) according to the relationship

$$A/V = 2\pi r^2/\tfrac{4}{3}\pi r^3 = \tfrac{3}{2}r \tag{9}$$

the removed larger material has a low surface area to volume ratio which results in decreased sorption of contaminants compared with smaller soil particles, and therefore is less contaminated. This material can be treated in a "polishing" step, for example, using water irrigation of rocks placed on a concrete or other pad where runoff can be collected and reapplied [2, 5].

Organic amendments can be added to the contaminated soil immediately before or after placement in a landfarm, for example in the form of fresh or aged animal manure, to improve soil organic matter content for microbial activity and contaminant immobilization, improve water retention for stimulating microbial activity, and accomplish nutrient addition. These engineering pretreatment processes are used to improve treatment rate and thus design, management, and performance aspects.

2.2.2 Treatment Area

Contaminated soil in the landfarm treatment area is the proper environment for optimizing biological activity through oxygen transfer, water management (drainage or irrigation), and nutrient addition. It may be possible to optimize treatment rate without blending or incorporation of treated soil through providing the proper environment alone. In other situations it may be necessary to blend untreated contaminated soil into large volumes of previously treated soil that results in an increase in the treatment area, as discussed previously. As illustrated in Eq. (3), an increase in the treatment area results in an increase in the assimilation rate so long as the contaminated soil treatment rate has been optimized.

3. Evaluation of Full-Scale Landfarming for Soil Bioremediation

The *U.S. Environmental Protection Agency* established the Bioremediation Field Initiative to help develop bioremediation as an effective remediation technology. An evaluation of soil bioremediation in a landfarm was conducted as part of the initiative. The landfarm consisted of two one-acre units with liners, berms, and leachate collection systems referred to as prepared beds or land treatment units (LTUs), and was conducted at the Champion International Superfund site in Libby, Montana (USA) [2, 5, 7].

Contamination of soils at the Libby site resulted from wood-preserving operations. Contaminants primarily consisted of residuals from creosote and pentachlorophenol (PCP) wood preservatives. Contaminated surface soils presented a public-health threat through direct contact and ingestion, and release of contaminants to soil and ground water. The purpose of the landfarming system approved was to reduce organic contaminant concentrations to target levels acceptable from a regulatory perspective and to minimize risks to public health and the environment [2]. Risk-based target remediation cleanup levels for the indicator compounds were: 88 mg/kg total carcinogenic polycyclic aromatic hydrocarbons (TCPAHs), 7.3 mg/kg pyrene, and 37 mg/kg PCP [2]. Soil characteristics and landfarm management operations are described in [2].

The objective was to assess: (*1*) treatment effectiveness, (*2*) rates of transformation of target contaminants, and (*3*) detoxication of the contaminated soil. To address (*1*) and (*2*), concentrations of indicator compounds in the LTUs were measured, using discrete soil samples, over time and with depth to measure vertical leaching of contamination. Detoxication was assessed using a battery of assays described in section 3.4.

3.1. DESIGN ASPECTS

Design information for Eqs (1) and (3) was determined as follows [2]. Contaminated soils from three primary source areas (tank farm, butt dip area, and waste pit) were pretreated using screening to remove rocks larger than 20 mm in diameter, stockpiled and pretreated with nutrients before application to the LTUs. For calculation of V in Eq. (1), the following was determined:

Volume = 57 000 m^3 loose excavated soil including rocks >20 mm in diameter
Volume of rock material >20 mm in diameter removed in pretreatment = 23 000 m^3
Volume of contaminated soil after pretreatment to remove rocks = 34 000 m^3 = V

For calculation of A_r in Eq. (3), the following was determined:

Area for treatment = 8 093 m^2
Treatment rate = 0.15 m^3/m^2, was determined as follows. Initial treatability studies conducted by the site owner indicated that 45 days were required for landfarm treatment of a lift of soil equal to 150 mm depth to reach target cleanup levels, giving a treatment rate of [(8093 m^2 × 0.15 m)/8093 m^2] or 0.15 m^3/m^2 for a 45-day treatment period.

Using Eq. (3) the assimilation rate of contaminated soil to reach target cleanup levels is

$$0.15 \text{ m}^3/\text{m}^2 \text{ for 45 days per lift} \times 8093 \text{ m}^2 = 1214 \text{ m}^3/45\text{-day lift.}$$

Due to seasonal changes in temperature, approximately six months allow active treatment, therefore the number of lifts added per year are 180 d/45 d = 4. Using this information, the time required for treatment (*T*) can be calculated using Eq. (1).

$$T = 34\,000 \text{ m}^3/(1\,214 \text{ m}^3/\text{lift} \times 4 \text{ lifts/year}) = 7 \text{ years}$$

3.2. TREATMENT EFFECTIVENESS

Treatment effectiveness of the landfarm was measured in terms of (*1*) attaining target cleanup goals in each lift, and (*2*) preventing downward migration. Target cleanup levels were achieved for all indicator compounds [2]. Downward migration of target indicator compounds was not observed in the LTUs [2].

3.3. TREATMENT RATE

Sampling and evaluation by the site owner indicated that the treatment rate within the landfarm was often lower than initial studies indicated, resulting in lift treatment times significantly longer than 45 days. Average treatment time was 223 days in LTU 1 and 512 days in LTU 2. The extended lift treatment times observed may have been due to several factors impacting treatment rate. These include: (1) the heterogeneous distribution of NAPL and individual contaminants from lift to lift and within a lift of the contaminated soil, (2) mass transfer and microbial toxicity limitations due to the heterogeneous nature of NAPL within the soil, and (3) unfavorable environmental conditions of moisture, oxygen, and temperature within the landfarm. Therefore increases in either the landfarm area or treatment rate, or both, would be required to increase the assimilation rate and meet time requirements for soil remediation of the entire volume of $34\,000$ m^3 requiring treatment [40].

Field-scale rates of treatment for indicator compounds are presented in Table 2. A first-order kinetic model was used to determine the values and to predict treatment time required for the lift. Values presented indicate that pyrene was the rate limiting constituent (RLC), requiring 200 days to attain the target cleanup level of 7.3 mg/kg, and thus controlled the rate at which lifts were added to the landfarm.

TABLE 2. Summary of kinetic information for landfarm (LTU) 2, lift 1 [2]

Indicator chemical	Initial mean concentration mg/kg†	Degradation rate, k 1/d	Half-life d	Predicted treatment time, d
PCP	101.4	−0.0163	43	62
Pyrene	84.9	−0.0125	55	200
TCPAH	204.0	−0.0127	55	66

†n = 20.

Other limiting factors [26] including non-optimal conditions of: (1) moisture, (2) soil oxygen, and (3) temperature at or below 10 °C were potentially important in effecting the longer treatment time required to reach target cleanup levels [2].

Measurements of soil oxygen were taken at depths of 0.3, 0.6, and 1.0 m. Concentrations varied from 0 to 21 %, while carbon dioxide concentration varied from 0.4 to 25 % [2, 24]. Stevens et al. [32] observed landfarm soil oxygen depleted to 0 % following application of petroleum waste to soil, with slow recovery of soil oxygen over weeks. Laboratory tests using the landfarm soil indicated that a critical soil oxygen level necessary for biodegradation was 2 % (V/V) [34, 35]. Biodegradation of target chemicals under anaerobic conditions was minimal, and additions of alternate electron acceptors to landfarm soil under anaerobic conditions did not enhance biodegradation significantly [41]. Oxygen concentrations were less than the critical level during periods of heavy rain when the soil becomes flooded, after the application of a lift of contaminated soil when the oxygen demand rate is greater than the rate of supply of oxygen to the soil, and in "hot spot" areas of the LTU containing high concentrations of chemicals.

The increase in remediation time for a lift of contaminated soil from 45 to 200 days would result in one lift added and treated per year. This would increase the total time for remediation of the total volume (34 000 m^3) to 28 years. To increase the assimilation rate of the RLC the land area for treatment was increased from 2 to 12 acres [40].

3.4. DETOXICATION OF CONTAMINATED SOIL

The third objective, detoxication of landfarm soil, was measured using three assays, the Microtox^{TM} Aqueous-Phase assay, the Microtox^{TM} Solid-Phase assay, and the Finland Flash Solid-Phase Bio Tox^{TM} assay. The Microtox^{TM} Aqueous-Phase assay is applicable to the measurement of reduction in microbial toxicity in the water phase of contaminated soil treated in a landfarm environment [11, 13, 31]. The Microtox^{TM} Solid-Phase assay is a modification of the Aqueous-Phase assay whereby the test organisms are exposed directly to soil particles in an aqueous suspension of the test sample. The Finland Flash Solid-Phase toxicity assay utilizes the same bioluminescent bacteria [25]. The ratio of the 30-s light output value to the initial peak output value is a measure of the toxicity of the soil solid-phase. A sample is characterized as toxic if the toxicity ratio is <0.8.

Microtox^{TM} soil aqueous phase toxicity results showed a reduction from highly toxic values at lift application time (day 1) to nontoxic status after 53 days of treatment [2]. Samples from buried lifts were also nontoxic, thus leaching of contaminants in the downward direction was not observed.

Results of the Microtox^{TM} Solid-Phase assay are presented in Table 3 for untreated soil and LTU 1-treated soil. Results of sequential extractions were interpreted by comparing extracted (aqueous and solvent) soil solid-phase toxicity with unextracted (unwashed) soil solid-phase toxicity. Results for both soils indicate little effect of water washing on changing solid-phase toxicity. Both soils showed greatly reduced toxicity (higher solid-phase dosages required to achieve EC$_{50}$ toxicity) after solvent extraction. These results are expected because the indicator compounds that have toxic characteristics are also highly soluble in the hexane−acetone solvent. Results obtained indicate that both soil solid and aqueous phases undergo detoxication within a landfarm treatment system.

TABLE 3. Microtox^{TM} solid-phase assay toxicity (EC$_{50}$ values, mg/L) for landfarm soil

Soil	Unwashed	Aqueous washed	Solvent extracted
Untreated oil	1 080	930	22 490
LTU 1-treated soil	23 750	14 390	36 550
Clean sand (control)	non-toxic	non-toxic	non-toxic

Table 4 contains results of the Finland Flash Solid-Phase toxicity assay. The toxicity ratio for the LTU 1 soil is similar to the value for the control soil, and both soils show toxicity ratio values greater than 0.8, indicating no solid-phase toxicity. However, the toxicity ratio for the untreated is 0.281, which indicates high toxicity. Results of the three microbial toxicity assays indicated that the time required for soil treatment based on the detoxication (50–75 days) was much shorter than the time required for pyrene to be degraded to 7.3 mg/kg, and thus detoxication of pyrene showed it not to be the rate-limiting compound.

TABLE 4. Finland flash solid-phase toxicity assay ratios for landfarm soil

Soil sample	V_{peak}, mV	V_{30s}, mV	Toxicity ratio, V_{peak}/V_{30s}
Untreated soil	172.3	48.4	0.281
LTU 1-treated soil	260.0	305.5	1.175
Control soil	412.1	527.8	1.281

4. Research and Application Issues

Research is needed to better understand chemical-soil-microbe interactions within landfarm systems. The interactions have implications with regard to rate and extent of treatment, soil residual concentrations, and risk assessment and management. Critical research needs include: (1) PAH degrader characterizaton and bioavailability in soil, (2) biological bound-residue formation, and (3) the role(s) of humic acids in affecting microbe-chemical interactions. Application of research on the effect of oxygen concentration on biodegradation may allow an improved evaluation of the frequency of adding lifts of contaminated soil to a landfarm that may decrease total treatment time.

4.1. PAH DEGRADER CHARACTERIZATION

Laboratory evaluations of the biodegradation of the rate-limiting constituent (RLC), pyrene, in LTU soil under landfarm conditions that simulated the Libby LTUs indicated significant mineralization of spiked ^{14}C-pyrene by indigenous microorganisms

[2, 34]. The microorganisms could be isolated from LTU-treated soil, but not from untreated soil, indicating acclimation and growth of indigenous pyrene degraders within the landfarm soil. Three G$^+$ microorganisms, identified as *Mycobacterium* sp. strains JLS, KMS, and MCS were capable of rapidly mineralizing ^{14}C-pyrene in soil and liquid media. Bioavailability of aged pyrene within the landfarm soil may limit the rate of treatment by these pyrene mineralizers, and current research is underway to evaluate the indigenous pyrene bioavailability. A better understanding of the factors affecting the activity of these mycobacteria has direct implications for developing biological engineering techniques for increasing the treatment rate of the RLC and thereby reducing the time required for treatment of a lift of contaminated soil in the landfarm.

4.2. BIOLOGICAL BOUND RESIDUE FORMATION

Laboratory evaluations of the biodegradation of the RLC, pyrene, as well as PCP in LTU soil under landfarm conditions that simulated the Libby LTUs indicated nonsolvent extractable binding of ^{14}C that resulted from the biodegradation of pyrene and PCP [34, 35, 58]. In an effort to determine the association of ^{14}C-pyrene with specific humic fractions of soil organic matter (SOM), Nieman *et al.* [21] used the methyl isobutyl ketone (MIBK) humic fractionation procedure to evaluate binding of ^{14}C-pyrene to SOM. The affinity of ^{14}C was greatest for the humic acid fraction, increasing to almost 40 % after 294 days of treatment. Bound residue formation rates were calculated based on changes in distribution of ^{14}C among soil fractions using the MIBK method at incubation times of 0, 42, and 294 days in biologically active laboratory soil microcosms. A zero-order rate of 0.022 % per day (8.4 × 10^{-9} mmol pyrene per day) was calculated for the overall biological residue formation process [21]. Bound-residue formation may be an acceptable end point in the remediation of contaminated soil, and management techniques for enhancing bound-residue formation may provide a strategy for increasing the rate of treatment and decreasing the risk to humans and the environment.

4.3. ROLE OF SOIL HUMIC ACID IN MICROBE−CHEMICAL INTERACTIONS

The presence of humic acid has also been observed to increase the rate of biodegradation of pyrene in the presence of the *Mycobacterium* strain JLS, isolated from the Libby landfarm [42]. By utilizing an infrared spectromicroscope and a very bright, nondestructive synchrotron photon source, the influence of humic acid on the progression of degradation of pyrene by strain JLS was monitored *in situ* and over time. The presence of humic acid was observed to shorten the onset time for pyrene biodegradation from 168 to 2 h. In the absence of humic acid, bacteria required 168 h to produce sufficient glycolipids to solubilize pyrene and make it bioavailable for biodegradation. Pyrene biodegradation was observed to result in an increase in the biomass of strain JLS. Additional research is needed to evaluate the role of humic acid augmentation of contaminated soil for increasing the rate of treatment of PAH in landfarm systems.

4.4. MANAGEMENT OF LIFT APPLICATION FREQUENCY AND THICKNESS

Within the landfarm at the Libby site, chemical concentrations of indicator compounds (TCAH, pyrene, and PCP) continued to decrease with time in buried lifts that had reached target cleanup levels and had been covered with another lift. These results indicated an additional option for lift management to decrease total time required for treatment of all of the contaminated soil. If biodegradation continues in buried lifts, then a new lift may be placed on top of a lower lift before the lower lift reaches target cleanup levels for individual compounds, $i.e.$, increased frequency of lift application. A key aspect of continued treatment of the lower lift to reach target cleanup levels is the presence of oxygen at two percent or greater (V/V) in the soil gas phase. The working hypothesis of lift management at landfarm sites is that a buried lift exhibits zero degradation rate, thus a lift must be treated to target cleanup levels before the application of another lift. This represents a constraint on the frequency of lift application. An alternative approach to lift management effectively takes advantage of the non-zero assimilation rate of a buried lift, and thus increases the frequency of application. Increasing application frequency results in decreased time required for bioremediation of the total volume of contaminated soil. In addition, it may be possible to increase the thickness of an applied lift if oxygen is present at or greater than the 2 % critical level.

Acknowledgements

The preparation of this chapter was supported, in part, by a grant from the *Huntsman Environmental Research Center* (HERC) at *Utah State University*.

References

1. Sims, R.C. and Sims, J.L. (1999) Landfarming of petroleum contaminated soils, in D.C. Adriano, J.-M. Bollag, W.T. Frankenberger, Jr., and Sims, R.C. (Eds.), *Bioremediation of Contaminated Soils, Monograph 37*, American Society of Agronomy, Crop Science Society of America, Soil Science Society of America, Madison, WI, Chapter 27, pp. 767–782.
2. Sims, R.C., Sims, J.L., Sorensen, D.L., and McLean, J.E. (1996) *Champion International Superfund Site, Libby, Montana: Bioremediation Field Performance Evaluation of the Prepared Bed Land Treatment System*, Vol. I and II, EPA-600/R-95/156, U.S Environmental Protection Agency, Ada (OK).
3. Barth, D.S., Mason, B.J., Starks, T.H., and Brown, K.W. (1989) *Soil Sampling Quality Assurance User's Guide, Second Edition.* EPA 600/8-89/046, Environmental Monitoring Systems Laboratory, U.S. Environmental Protection Agency, Las Vegas, NV.
4. Mason, B.J. (1983) *Preparation of Soil Sampling protocol: Techniques and Strategies.* EPA-600/4-83/020. Environmental Monitoring Systems Laboratory, U.S. Environmental Protection Agency, Las Vegas, NV.

5. Sims, J.L., Sims, R.C., Sorensen, D.L., and Huling, S.G. (1999a) Prepared bed bioreactors, in D.C. Adriano, J.-M. Bollag, W.T. Frankenberger, Jr., and R.C. Sims (eds.) *Bioremediation of Contaminated Soils, Monograph 37,* American Society of Agronomy, Crop Science Society of America, Soil Science Society of America, Madison (WI), Chapter 21, pp. 559–294.

6. Sims, R.C., Sims, J.L., Zollinger, R.L., and Huling, S.G. (1999b) Bioremediation of soil contaminated with wood preservatives., in D.C. Adriano, J.-M. Bollag, W.T. Frankenberger, Jr., and R.C. Sims (eds.) *Bioremediation of Contaminated Soils, Monograph 37,* American Society of Agronomy, Crop Science Society of America, Soil Science Society of America, Madison, WI, Chapter 25, pp. 719–742.

7. Huling, S.G., Pope, D.R., Matthews, J.E., Sims, J.L, Sims, R.C., and Sorensen, D.L. (1995) Land treatment and the toxicity response of soil contaminated with wood preserving waste, *Remediation,* Spring, 41–55.

8. Ginn, J.S., R.C. Sims, and I.P. Murarka . (1995a) Evaluation of biological treatability of soil contaminated with manufactured gas plant waste, *Hazard. Waste Hazard. Mater.* 12, 221–232.

9. Park, K.S., Sims, R.C., Dupont, R.C., Doucette, W.J., and Matthews, J.E. (1990a) Fate of PAH compounds in two soil types: influence of volatilization, abiotic loss, and biological activity, *J. Environ. Toxicol. Chem.* 9, 187–195.

10. Park, K.S., Sims, R.C., and Dupont, R.C. (1990b) Transformation of PAHs in soil systems, *J. Environ. Enginer.* 116, 632–640.

11. Aprill, W., Sims, R.C., Sims, J.L., and Matthews, J.E. (1990) Assessing detoxification and degradation of wood preserving and petroleum wastes in contaminated soils, *Waste Manag. Res.* 8, 45–65.

12. Sims, J.L., Sims, R.C., and Matthews, J.E. (1990) Approach to bioremediation of contaminated soil, *Hazard. Waste Hazard. Mater.* 72, 117–149.

13. Symons, B.D., and Sims, R.C. (1988) Assessing detoxification of a complex hazardous waste using the Microtox Bioassay. *Arch. Environ. Contam. Toxicol.* 17, 497–505.

14. Ginn, J.S., Doucette, W.J., Smith, D.P., Sorensen, D.L., and Sims R.C. (1996) Aerobic biotransformation of PAH and associated metabolites in soil, *Polycycl. Arom. Comp.* 11, 43–55.

15. Sims, R.C. and Overcash, M.R. (1983) Fate of polynuclear aromatic compounds (PNAs) in soilplant systems, *Resid. Rev.* 88, 1–68.

16. Findlay, M., Fogel, S., Conway, L., and Taddeo, A. (1995) Field treatment of coal tar-contaminated soil based on results of laboratory treatability studies, in *Microbial Transformation and Degradation of Toxic Organic Chemicals,* Chapter 13, Wiley-Liss, New York (NY), pp. 487–513.

17. Howard, P.H., Boethling, R.S., Jarvis, W.F., Meylan, W.M., and Michalenko, E.M. (1991) *Handbook of Environmental Degradation Rates,* Lewis Publishers, Chelsea (MI).

18. Young, L.Y. and Cerniglia, C.E. (1995) *Microbial Transformation and Degradation of Toxic Organic Chemcials,* Wiley-Liss, New York, (NY).

19. Maliszewska-Kordybach, B. (1999) Persistent organic contaminants in the environment: PAHs as a case study, in *Bioavailability of Organic Xenobiotics in the Environment,* Kluwer Academic Publishers, the Netherlands, pp. 3–34.

20. Wild, S.R. and Jones, K.C. (1992) Organic chemicals entering agricultural soils in sewage sludges: screening for their potential to transfer to crop plants and livestock. *Sci. Total Environ.* 119, 85–119.

21. Nieman, J.K.C., Sims, R.C., Sims, J.L., Sorensen, D.L., McLean, J.E., and Rice, J.A. (1999a) [14C]-Pyrene bound residue evaluation using MIBK fractionation method for creosote-contaminated soil, *Environ. Sci. Technol.* 33, 776–781.

22. Kwan, K.K. and Dutka, B.J. (1995) Comparative assessment of two solid-phase toxicity bioassays: the direct sediment toxicity testing procedure (DSTTP) and the Microtox[TM] Solid-Phase Test (SPT), *Bull. Environ. Contam. Toxicol.* 55, 338–346.

23. Benton, M.J., Malot, M.L., Knight, S.S., Cooper, C.M., and Bensen, W.H. (1995) Influence of sediment composition on apparent toxicity in a solid-phase test using bioluminescent bacteria, *Environ. Toxicol. Chem.* 14, 411–414.

24. Ringwood, A.H., DeLorenzo, M.E., Ross, P.E., and Holland, A.F. (1997) Interpretation of Microtox[TM] Solid-Phase Toxicity Tests: the effects of sediment composition, *Environ. Toxicol. Chem.* 16, 135–1140.

25. Lappalainen, J., Juvonen, R., Vaajasaari, J., and Karp, M. (1999) A new flash method for measuring the toxicity of solid and colored samples, *Chemosphere* 38, 1069–1083.

26. Alexander, M. and Scow, K.M. (1989) Kinetics of biodegradation in soil, in B.L. Sawhney and K. Brown (eds.), *Reactions and Movement of Organic Chemicals in Soils,* SSSA Special Publication no. 22, Chapter 10, SSSA, Inc., ASA, Inc., Madison (WI), pp. 243–269.

334

27. Sims, R.C., Sorensen, D.L. Sims, J.L., McLean, J.E., Mahmood, R.H., Dupont, R.R., and Jurinak, J.J. (1986) *Contaminated Surface Soils: In-Place Treatment Techniques*, Noyes Publications, Park Ridge (NJ).
28. Cookson Jr., J.T. (1995) *Bioremediation Engineering: Design and Application*, McGraw-Hill, New York.
29. Burden, D.S. and Sims, J.L. (1999) *Fundamentals of Soil Science as Applicable to Management of Hazardous Wastes*, EPA/540/S-98/500. U.S. Environmental Protection Agency, Office of Research and Development, Washington (DC).
30. Sims, R.C. and Sims, J.L. (1995) Chemical mass balance approach to field-scale evaluation of bioremediation, *Environ. Progr.* **14**, F 2–3.
31. U.S. EPA (1986) *Waste/Soil Treatability Studies on Four Complex Wastes*, EPA/600/6-86/003a,b. U.S. Environmental Protection Agency, Robert S. Kerr Laboratory, Ada (OK).
32. Stevens, D.K., Yan, Z., Sims, R.C., and Grenney, W.J. (1989) *Sensitive Parameter Evaluation for a Vadose Zone Fate and Transport Model*, EPA/600/2-89/039, U.S. Environmental Protection Agency, National Risk Management Research Laboratory, Ada (OK).
33. Baker, K.H. (1994) Bioremediation of surface and subsurface soils, in K.H. Baker and D.S. Herson (eds), *Bioremediation*, McGraw-Hill, New York, pp. 203–259.
34. Hurst, C.J., Sims, R.C., Sims, J.L, Sorensen, D.L., McLean, J.E., and Huling, S.J. (1996) Polycyclic aromatic hydrocarbon biodegradation as a function of oxygen tension in contaminated soil, *J. Hazard. Mater.* **51**, 193–208.
35. Hurst, C.J., Sims, R.C., Sims, J.L., Sorensen, D.L., Mclean, J.E., and Huling, S.G. (1997) Soil gas oxygen tension and pentachlorophenol biodegradation, *J. Environ. Engineer.* **123**, 364–370.
36. Park, H.S., Sims, R.C., Doucette, W.J., and Matthews, J.E. (1988) Biological transformation and detoxification of 7,12-dimethylbenzanthracene in soil, *J. Water Pollut. Contr. Feder.* **60**, 1822–1825.
37. Hinchee, R.E. (1994) Bioventing of petroleum Hhydrocarbons, in R.C. Norris and R.E. Hinchee (Eds), *Handbook of Bioremediation*, CRC Press, Boca Raton (FL) pp. 39–59.
38. Coover, M.P. and Sims, R.C. (1987) The effect of temperature in polycylic aromatic hydrocarbon presistence in an unacclimated agricultural soil, *Hazard. Wastes Hazard. Mater.* **4**, 69–82.
39. Cheng, H.H. and Mulla, D.J. (1999) *The Soil Environment*, in D.C. Adriano, J.-M. Bollag, W.T. Frankenberger, Jr., and R.C. Sims, (eds), *Bioremediation of Contaminated Soils*, ASA, Inc., CSSA, Inc., SSSA, Inc. Madison (WI), pp. 1–13.
40. Nieman, J.K., Sims, R.C., and Cosgriff, D.M. (1999*b*) Regulatory and management essues in prepared bed land treatment: Libby Groundwater Site, in R.E. Hinchee (Ed.), *In Situ and On-Site Bioremediation, Vol. 5, Bioreactor and ex Situ Treatment Technologies*, Battelle Press, Columbus (OH), pp. 97–102.
41. Nieman, J.K.C., Sims, R.C., McLean, J.E., Sims, J.L., and Sorensen, D.L. (2000*a*) Fate of pyrene in contaminated soil amended with alternate electron acceptors. *Chemosphere* **44**, 1265–1271.
42. Holman, H.Y., Nieman, J.K.C., Sorensen, D.L., Miller, C.D., Martin, M.C., Borch, T., McKiney, W.R., and Sims, R.C. (2002) Catalysis of PAH biodegradation by humic acid shown in synchrotron infrared studies, *Environ. Sci. Technol.* **36**, 1276–1280.

SPENT MUSHROOM SUBSTRATE: WHITE-ROT FUNGI IN AGED CREO-SOTE-CONTAMINATED SOIL

T. EGGEN

Jordforsk – Norwegian Center for Soil and Environmental Research
Fredrik A. Dahls gt 20, 1432 Ås, Norway
e-mail: trine.eggen@jordforsk.no

1. Background

The traditional biological treatment methods such as landfarming, biopiles and composting are often applied to the degradation of readily degradable pollutants such as light oils and low-molar-mass polyaromatic hydrocarbons. However, these methods have certain limitations in the application to the more recalcitrant pollutants e.g. high-molar-mass polyaromatic hydrocarbons, polychlorinated biphenyls and some pesticides.

Use of white rot fungi for degradation of recalcitrant pollutants have been studied since the late 1980s. Few of these studies have been performed with soil, and even less with non-sterile soil and aged contaminated soil. An effective, low-cost fungal inoculum is important for the white-rot fungal technique to become a success. Mushroom cultivation is a common method all over the world and also a major income source in different developing countries. In many countries, the spent mushroom substrate is discarded as waste, and usage tends to be limited to soil conditioning and fertilizing. Application of spent fungal substrate from commercial mushroom cultivation in bioremediation, would therefore be an excellent way of recycling agroindustrial by-products.

2. Materials and Methods

This paper summarizes results from experiments which all have in common that they have been performed with aged nonsterile contaminated creosote soil. *Phaerochaete chrysosporium*, the previously most studied white-rot fungus, was not selected as test strain due to its physiological limitations (nitrogen-deficient conditions and low pH) and some less successful application in soil media. The edible mushroom *Pleurotus ostreatus* (oyster mushroom) is produced on a commercial scale and was selected as inoculum in these experiments.

Characteristics of the soils are shown in Table 1. The experiments were performed in polyethylene columns (*d* 12.5 cm, *h* 47 cm) at room temperature and

335

V. Šašek et al. (eds.),
The Utilization of Bioremediation to Reduce Soil Contamination: Problems and Solutions, 335–339.
© 2003 *Kluwer Academic Publishers. Printed in the Netherlands.*

aerated daily from the bottom of the columns. The soil (sieved < 2mm) and fungal substrate ratio was 4:1 (wet substance). The following parameters were tested: pile layout, use of spent mushroom substrate, initial creosote concentration, reinoculation of fungal substrate and mobility of PAHs in soil after fungal inoculation. All PAH analysis are performed in sieved soil (< 2 mm).

TABLE 1. Characteristics of creosote-contaminated soils

Parameters	Soil A	Soil B	Soil C
pH	7.5	6.0	6.2
Organic carbon (g/kg)	24	24	44
Total N (g/kg)	1.0	<0.5	0.7
Total P (g/kg)	658	366	361
2-ring PAHs (mg/kg)	13.7	2.0	1007
3-ring PAHs (mg/kg)	113.6	3632	9892
4-ring PAHs (mg/kg)	696.6	2635	5938
5- and 6-ring PAHs (mg/kg)	59.3	204	378
Σ PAH* (mg/kg)	1909.2	6473	17215
Σ Hetrocyclic** (mg/kg)	na	na	2140

*Σ PAH= 2, 3, 4, 5 and 6-ringPAHs bassed on 16 US.EPA PAHs – not heterovyvlic compounds
** Σ Heterocyclic=dibenzofuran, dibenzothiophene, benzothionaphtene, benzophenanthridin, benzo(d,e,f)dibenzothiophene

3. Results

Pile layout of fungal substrate, layered or mixed, had no significant difference in the degradation rate of the 3-ring PAHs, while homogenized soil and fungal substrate stimulated degradation of the 4- and 5-ring PAHs more than layered (Fig. 1). The used of spent fungal substrate (after cropping) vs. colonized straw before fruit-body formation, was not important for the extent of PAH removal in this case (Fig. 2).

Application of white rot fungi in strongly contaminated creosote soil (18.13 mg/kg Σ PAH) demonstrated that there was a pronounced removal of 3-ring, total PAHs and heterocyclic aromatic compounds within a 4-week incubation period, 75 %,58 % and 86 %, respectively (Fig. 3). However, a less pronounced effect on 4-ring PAH removal was measured (32 %). The reduction in control soil represents the removal due to microbial degradation by the native microflora and by evaporation of the low-molar-mass compounds. Except for bicyclic compounds, the reduction in control soil ranged between 5 and 20 % (not shown). Bicyclic compound removal in control soil was nearly as high as for inoculated soil (96 vs. 99 %, respectively) and was probably caused by evaporation. PAH adsorption to straw was determined at the end of the experiment. Generally, the adsorption of PAH during the experiment period was low, no compounds exceeded 5 %, and adsorption can be eliminated as an important removal pathway (Table 2).

Creosote contaminated soil with an initial concentration of 6.47 mg/kg showed 97 % and 74 % reduction of 3-ring PAHs and total PAH within 12-week incubation, respectively (Fig. 4). A reinoculation with fungal substrate and a following 3-week incubation resulted in a further 90 and 43.5 % reduction for the same PAH groups.

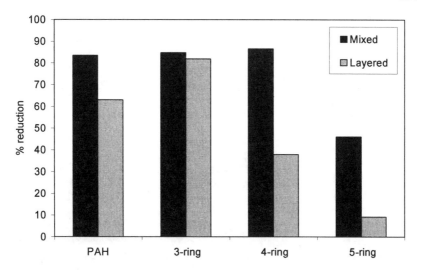

Figure 1. Reduction of PAHs in soil after 7-week incubation with spent mushroom substrate organized in layers or homogenized and mixed with soil (Soil A).

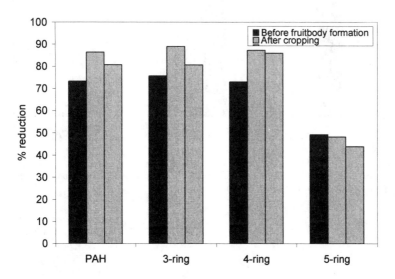

Figure 2. PAH reduction in soil after 7-week incubation with spent fungal substrate (after cropping) or colonized straw before fruit-body formation (Soil A).

TABLE 2. Amount of PAH adsorbed to straw given in percent of total PAH content (Soil C) (n=3).

Σ 2-rings	Σ 3-rings	Σ 4-rings	Σ 5-rings	Σ 6-rings	Σ PAH	Σ HC*
0.01 ± 0.001	0.3 ± 0.03	4.4 ± 0.41	2.5 ± 0.45	1.1 ± 0.16	1.9 ± 0.16	0.4 ± 0.01

* HC= heterocyclic compounds

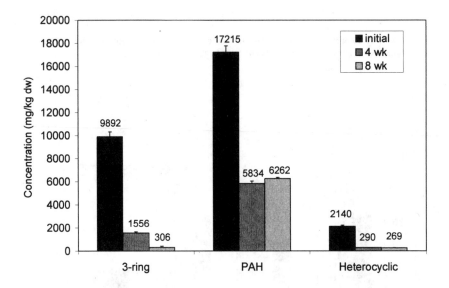

Figure 3. PAH reduction in strongly contaminated soil after 4- and 8-week incubation with spent mushroom substrate (Soil C) (n=3).

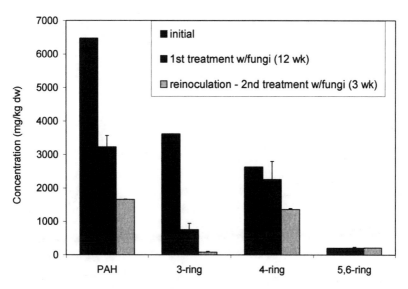

Figure 4. PAH reduction in soil after 12-week incubation followed by reinoculation and 3-week incubation period (Soil B) (*n*=2).

BIOREMEDIATION OF PCBs FROM CONTAMINED SOIL

K. DEMNEROVÁ[1], M. MACKOVÁ[1], P. KUČEROVÁ[1], L. CHROMÁ[1],
H. NOVÁKOVÁ[1], M. B. LEIGH[2], J. BURKHARD[3], J. PAZLAROVÁ[1]
and T. MACEK[4]

[1]Department of Biochemistry and Microbiology, Faculty of Food and
Biochemical Technology, Institute of Chemical Technology, Prague,
Technicka 3, 166 28 Prague, [2]Department of Botany and Microbiology,
University of Oklahoma, 770 Van Vleet Oval, Norman, OK 73019, USA,
[3]Faculty of Environmental Chemistry, Insttitute of Chemical Technology,
Prague, Technicka 3, 166 28 Prague, [4]Instnstitute of Organic Chemistry
and Biochemistry, Academy of Sciences of the Czech Republic,
Flemingovo n. 2, 166 10 Prague 6, Czech Republic.

1. Introduction

It was shown that some biological systems including bacteria, fungi and plants can be
used for the removal of toxic compounds from the environment (1,2,3). In natural
conditions these organisms usually show mutual cooperation and support each other's
abilities. It was shown that certain plant compounds present in root exudates may serve
as natural substrates for the growth and sometimes induction of the genes of the
biodegradative pathway of PCBs in bacteria (4). In our study we grew plants of
different species in an area contaminated with PCBs and we tested the ability of
biological systems to degrade PCBs over 7 months. In pot experiments, we tested the
ability of different plant and bacterial species to degrade PCBs. We also followed
differences in microbial growth between nonvegetated and vegetated soil, and in the
rhizoplane and rhizosphere. This study showed the beneficial effect of plants on
microbial growth. The results from field conditions and pot experiments were compared
with those obtained in laboratory experiments performed with plant tissue cultures of
the same species cultivated *in vitro*.

2. Materials and Methods

2.1. CULTIVATION OF BACTERIA AND PLANT TISSUE CULTURES:

Bacterial consortia were cultivated in 300 ml Erlenmeyer flasks in mineral medium,
always with biphenyl as cosubstrate and inducer of PCB degradation, and 50 ppm of
Delor 103 for 20 d at 25 °C on a rotary shaker. Plant cells cultivated *in vitro* were
incubated with 50 ppm of Delor 103 under the same conditions as bacteria, in

V. Šašek et al. (eds.),
The Utilization of Bioremediation to Reduce Soil Contamination: Problems and Solutions, 341–346.

Murashige-Skoog medium, for 14 d. In laboratory experiments a standard commercial PCB mixture, Delor 103, containing 59 individual congeners substituted with 3 – 5 Cl atoms per biphenyl molecule was used.

2.2. POT EXPERIMENTS

For pot experiments buckets filled with contaminated soil were used. Four different plant species - *Solanum nigrum, Medicago sativa, Nicotiana tabacum* and *Silybum marianum* were cultivated in 10 L metal buckets containing soil contaminated with PCBs. Nonvegetated soil was used as control. Plants were grown for 4 months and then the residual amount of PCBs was analyzed. Also, the total number of microorganisms, pseudomonads and related non-fermenting bacteria were estimated in the rhizosphere and rhizoplane areas of cultivated plants.

2.3. FIELD EXPERIMENTS

Small plants of alfalfa (*Medicago sativa*), thistle (*Silybum marianum*), tobacco (*Nicotiana tabacum)* and black nightshade (*Solanum nigrum*) were planted in contaminated soil. Samples of the nonvegetated soil, rhizoplane and rhizospheric soil vegetated with the above plants were collected for microbial analysis and estimation of residual PCB content after 7 months from the beginning of the experiment. In the samples total number of bacteria, pseudomonads and related nonfermenting species were estimated.

2.4. PCB ANALYSIS – LABORATORY EXPERIMENTS

The analytical approach used for bacterial and plant cells to estimate their ability to degrade PCB was similar. For the use of plant biomass it was slightly modified (5). After homogenization (of plant cultures only) and subsequent sonication, the contents of the flasks were extracted with 10 mL hexane at 20 °C on a rotary shaker for 2 h. Following phase separation, the upper, hexane layer was subjected to GC analysis. Using GC analysis with EC detector, 22 of the Delor 103 congeners were assigned to peaks with areas larger than 0.5 % of the total area of all 59 individual chromatographic peaks. For the calculation of the residual amount of PCBs the above 22 chromatographic peaks were used. These 22 congeners represent 80-90 % of the total sample amount. Controls containing heat-killed cells were included to establish that observed changes in the content of congeners were dependent exclusively on the activity of living cells. Using standard conditions for the preparation of samples for GC analysis, the accuracy of the results obtained was within 15 %. Results were calculated from the residual amounts of each congener peak of the sample, compared to the respective peaks of the controls (dead cells).

2.5. PCB-ANALYSIS – FIELD AND POT EXPERIMENTS

Tested soil samples were dried overnight and sieved through a mesh with 1 mm pore size. Then 1 g of the soil was Soxhlet extracted with hexane for 4 h. The extract was concentrated to approximately 1 mL in volume by nitrogen flow, purified on a Florisil column, diluted with hexane to exactly 10 mL volume, then diluted 1/100 for analysis by GC. Results were calculated from the residual amounts of congener peaks present in the sample, compared to the value recommended by US EPA (Environmental Protection Agency) for expressing the total content of PCBs as a sum of recommended "indicator" congeners.

2.6. IDENTIFICATION AND CHARACTERIZATION OF MICROORGANISMS

10 g samples of nonvegetated soil, rhizospheric soil and roots with attached soil layer (representing rhizoplane bacteria) were extracted with 90 mL of diphosphate solution. Samples were shaken for 2 h with glass beads, then aliquots of the extracts were plated on Petri dishes containing Plate count agar (Oxoid) for estimation of the total number of aerobic bacteria, *Pseudomonas* agar (Oxoid) for detection of nonfermenting bacteria and related species of nonfermentative bacteria. Individual colonies isolated from *Pseudomonas* agar were chosen for identification by Nefermtest (Lachema Brno, Czech Republic).

3. Results

3.1. POT EXPERIMENTS

Total counts of bacteria and pseudomonads and related nonfermenting species present in samples from pot experiments are shown in the Tables. It can be seen that higher numbers of bacteria (Table 1 and 2) were detected in the rhizoplane of all plants in comparison with nonvegetated soil, i.e. the presence of plants and root exudates had a beneficial effect on microbial growth. Species isolated from rhizosphere and rhizoplane area were identified as: *Pseudomonas* sp., *Pseudomonas stutzeri*, *Pseudomonas fluorescens*, *Alcaligenes* sp. Isolated consortia and then individual strains isolated from contaminated soil were tested under laboratory conditions for their ability to grow with biphenyl as the sole carbon source. Unfortunately, growth was very poor. Bacteria were also tested for their ability to degrade PCBs but no degradation was detected.

At the end of pot experiments soils collected from nonvegetated area and from the rhizosphere and rhizoplane of cultivated plants were analyzed for PCB content (see Table 3). Nonvegetated soil originally contaminated by Delor 103 contained an average of about 170 μg/g (ppm) of PCBs, both at the beginning and at the end of experiment. At the end of experiment the soil vegetated by *M. sativa* contained 20 % less PCBs than the nonvegetated control soil, and the soil vegetated by tobacco and black nightshade decreased in PCB concentration by 20 % during four months. The smallest decrease in PCB concentration was analyzed in a soil vegetated by thistle – 15 %.

TABLE 1. Microbial analysis of soil contaminated with PCBs and vegetated with alfalfa and tobacco

Cfu/g soil	Nonvegetated soil	Rhizosphere tobacco	Rhizoplane tobacco	Rhizosphere alfalfa	Rhizoplane alfalfa
Total number of cells	5.5×10^6	4.7×10^7	9.3×10^7	4.3×10^6	6.3×10^7
Nonfermenting bacteria	1.3×10^3	1.4×10^5	4.8×10^6	4.5×10^3	7.1×10^4

TABLE 2. Microbial analysis of soil contaminated with PCBs and vegetated with thistle and black nightshade

Cfu/g soil	Nonvegetated soil	Rhizosphere thistle	Rhizoplane thistle	Rhizosphere nightshade	Rhizoplane nightshade
Total number of cells	5.5×10^6	4.5×10^6	1.2×10^8	4.1×10^6	2.5×10^8
Nonfermenting bacteria	1.3×10^3	9×10^5	7.1×10^4	1.1×10^6	3.2×10^6

TABLE 3. Comparison of PCB content in nonvegetated soil and soil vegetated with various plants during pot experiment; in µg PCB/g soil.

Nonvegetated soil	Alfalfa	Tobacco	Black nightshade	Thistle
178	143 (80 %)*	139 (78 %)*	145 (81 %)*	153 (85 %)

*- Percentage of PCB content related to PCB concentration in nonvegetated soil

3.2. FIELD EXPERIMENTS

Field experiments also showed the beneficial effect of plants on growth of microorganisms in the rhizosphere and especially in the rhizoplane of plants (results not shown). The data in Table 4 show that the final PCB content in soil from field cultivated with tobacco and black nightshade decreased to 76 % and 78 %, respectively, of that in nonvegetated soil. Unfortunately, soil cultivated with alfalfa showed a lower decrease of PCB concentration than was seen in the pot experiment. Thistle was not cultivated due to unsatisfactory results during pot experiment.

Similarly, plant tissue cultures cultivated *in vitro,* using the same species as in pot experiments were followed for their ability to metabolize PCBs. Plant tissue cultures were cultivated as submerged cultures for 15 d with 50 ppm of Delor 103. The results are shown in Table 5.

4. Conclusions

Our experiments proved the beneficial effect of plants on indigenous microflora present in contaminated soil. In the rhizosphere and rhizoplane area of cultivated plants many times higher numbers of bacteria were detected than in nonvegetated soil.

TABLE 4. Comparison of PCB content in nonvegetated soil and soil vegetated with various plants; in µg PCB/g soil.

Nonvegetated soil	Alfalfa	Tobacco	Black nightshade
468	425 (90 %)*	358 (76 %)*	365 (78 %)*

*- Percentage of PCB content related to PCB concentration in nonvegetated soil

TABLE 5. Average PCB content remaining after incubation of *in vitro* plant tissue cultures with 50 ppm of Delor 103 in comparison with dead-cell controls.

Remaining PCB content (%)			
Medicago sativa (ALF)	*N. tabacum* (TO-12)	*S. nigrum* (SNC-9O)	*S. marianum* (OP-35)
70	84	50	90

Unfortunately, bacteria able to degrade PCBs have not been isolated. This fact is probably due to the fact that we were not able to detect right bacteria by classical methods. Generally, using classical isolation and detection methods only 10 % of microorganisms present in the soil can be isolated.

In vitro experiments showed the best results with the hairy root clone of *S. nigrum*. In the pot experiment, soil vegetated with this plant also contained less PCBs than nonvegetated control soil. On the other hand, the results obtained with plant cultures cultivated *in vitro* cannot be simply generalized for the whole plant of the same species, because cultures *in vitro* exhibit much higher spontaneous variability (6). This can be illustrated by our results, e.g. when we examined a dozen *in vitro* cultures of *S. nigrum* for their ability to metabolize PCBs, the results ranged from 0 to 70 % conversion under standard conditions (7, 8). For comparing different species under *in vitro* conditions, trends obtained from analysis of more strains of the same species must be followed. The experiments in contaminated soil showed that plants have a great potential to help remediating soil highly contaminated with PCBs.

346

Acknowledgements

The work was sponsored by grants B6127901 of the Grant Agency of the Czech Academy of Sciences, US National Security Administration NSEP Graduate International Fellowship and GACR No. 203/99/1628.

References

1. Cunningham S.D. and Berti W.R. (1993) Remediation of contaminated soils with green plants: an overview. *In Vitro Cell Dev. Biol.* 29P, 207-212
2. Kás J., Burkhard J., Demnerová K., Kostál J., Macek T., Macková M. and Pazlarová J. (1997) Perspectives in biodegradation of alkanes and PCBs. *Pure and Appl. Chem.* 69, 11, 2357-2369
3. Macek, T., Macková, M., Káš, J. (2000) Exploitation of plants for the removal of organics in environmental remediation. *Biotechnol. Advances* 18 (1), 23-35
4. Fletcher J.S., Donnelly P.K. and Hegde R.S. (1995) Biostimulation of PCB-degrading bacteria by compounds released from plant roots, in R.E. Hinchee, D.B. Anderson and R.E. Hoeppel (eds.), *Bioremediation of recalcitrant organics*, Battelle Press, Columbus, pp. 131-136
5. Burkhard J., Macková M., Macek T., Kucerová P. and Demnerová K. (1997) Analytical procedure for the estimation of PCB transformation by plant tissue cultures. *Anal. Commun.* 34, 10, 287-290
6. Macek T., Macková M., Burkhard J., and Demnerová K. (1998) Introduction of green plants for the control of metals and organics in environmental remediation. In: *Effluents from Alternative Demilitarization Technologies*. (F.W. Holm, Ed), NATO PS Series 1, Vol. 12, Kluwer Academic Publishers, Dordrecht, Boston, London, pp. 71-85
7. Macková M., Macek T., Burkhard J., Ocenásková J., Demnerová K., Pazlarová J. (1997) Biodegradation of polychlorinated biphenyls by plant cells. *International Biodeterioration and Biodegradation*, 39, 4, 317-325
8. Kucerová, P., Macková, M., Poláchová, L., Burkhard, J., Demnerová, K., Pazlarová, J., and Macek, T. : Correlation of PCB transformation by plant tissue cultures with their morphology and peroxidase activity changes. - *Coll. Czech Chem. Commun.* 64 (9), 1497-1509, (1999).

PROBLEMS OF SOIL RENOVATION ON FORMER MILITARY SITES IN LITHUANIA

G. IGNATAVIČIUS, P. BALTRĖNAS

Environmental Protection Department, Vilnius Gediminas Technical University, Lithuania, Saulėtekio al. 11, 2040 Vilnius, Lithuania

1. Introduction

The Soviet army left Lithuania in 1993. At that time they abandoned approximately 500 military installations including 277 Soviet military bases on which 462 military units had been housed [1]. The military sites occupied 67662 ha, or 1.04 % of Lithuania's total land area. Only a fraction of this territory (16.7 %) was needed to satisfy Lithuanian military needs. The rest has been transferred to civilian use.

2. Description of types of pollution

The military sites were installed without any environmental protections. Often they were located on valuable geological formations, such as gravel, sand or sandy loam (Fig. 1) [2].

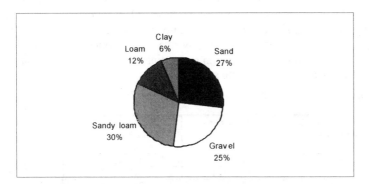

Figure 1. Soil types at military sites.

When Lithuania took charge of the Soviet military bases, an evaluation committee comprised of local experts was established to evaluate the environmental situation at the sites. One of their main tasks was to identify effective measures, which, when applied, could prevent further spreading of pollutants.

The committee identified 2743 sources of pollution on the former military lands (Table 1) [3]. Only 14 % of all the military bases were free of pollution sources (Table 2). The remains of 200 types of toxic chemicals and inflammable substances were found

347

V. Šašek et al. (eds.),
The Utilization of Bioremediation to Reduce Soil Contamination: Problems and Solutions, 347–351.
© 2003 *Kluwer Academic Publishers. Printed in the Netherlands.*

on the remaining 86 % of the sites. Ruins of the former buildings and other sources of potential danger are present at almost every former Soviet military site.

TABLE 1. Types of damage and their distribution (Krüger Consult and Baltic Consulting Group)

Type of fixed damages	Quantity*	Area (ha)
Explosives	12	DC**
Landscape damage	438	7140
Radiation	9	DC
Oil products	566	399
Organic or biological pollution	137	14
Chemicals	56	DC
Destroyed soils	778	11137
Destroyed forest	249	3293
Rocket fuel	20	DC
Rubbish heaps	478	1288
Total	2743	23271

* - Number of pollution sources.
** - Dotted concentration.

As seen in Table 1, damage occurred primarily to soils and landscapes. The major sources of pollution identified were oil products and rubbish heaps prevailing in the military sites as well as the landscape and soil damages. The sources of oil contamination are listed in Table 3. Groundwater contamination by oil products was detected at all the investigated military bases (15) built on sandy and sandy loam soils [4]. At 119 sites the oil–contamination source is located within 200 m of surface water body (river or lake). In 35 of these cases direct evidence of oil contamination on surface streams was observed. A study of the site inventory indicated that environmental contamination from the military bases was usually related to surface run-off. However subsurface infiltration is an important factor in contaminant transport on a region scale where polluted groundwater discharges into surface streams.

TABLE 2. Distribution of pollution sources (Krüger Consult and Baltic Consulting Group)

Number of pollution sources per military site	Number of sites	Total number of pollution sources
0	41 (14%)	0 (0%)
1-5	94 (35%)	268 (10%)
6-10	59 (21%)	447 (16%)
11-20	50 (18%)	750 (27%)
21-50	27 (10%)	840 (31%)
More than 50	6 (2%)	438 (16%)
Total	277 (100%)	2743 (100%)

The 478 rubbish heaps found on the military sites comprise 17 % of all the environmental damage. There is about 333 000 m^3 of waste on 1188 ha of land. Approximately half of the waste originates from military activity The other half is mixed household waste (Table 4). Streams of wastewater and wind-transported heavy metal particles, oil dust, bitumen and other break-up products from the rubbish heaps, to clean sites. Natural ecosystems that have been severely affected by the long-term military activity at the restricted sites cannot be restored easily, and compromise the health of inhabitants in the region. Therefore, the main problem which must be addressed when reusing these military sites is the prevention of pollution being spread further and, specifically, immediate localization and liquidation of those pollution sources that present direct risk to human health or the environment.

TABLE 3. Oil contamination sources in cases where oil contamination is observed (Krüger Consult and Baltic Consulting Group)

Type of source	A. Number of cases
Oil/fuel storage	115
Oil/fuel transfer facility	7
Garages	185
Filling stations	92
Railway ramp	1
Washing stations	58
Other	108
Total	566

Based on the inventory and characterization of the contamination sources at the former Soviet military installations, the sites were grouped according to their need for preventive measures, cleanup and remediation work. Sites within each group were graded from A to F (highest to lowest) based on their degree of damage. Thus, every military site received an index of two letters (e.g. BC). The 1st letter describes the level of investigation, and the necessity of research, or prevention and clean-up work. The 2nd letter indicates the degree of destruction and necessity of restoration work [1].

Preventive activities include estimating the risk originating at military sites such as the potential for contaminant migration or the dispersion of explosive or dangerous materials. The estimates are then used to determine the nature of the protection systems and regimes that must be established at the site [5].

Curative actions include almost all means of prevention and cleaning the territory of pollutants and other dangerous or explosive materials. Also, territory restoration in most cases oversees the arrangement of buildings, full or part recreation of relief, revegetation and restoration of the soil.

The first preventive work conducted at the military sites was the removal of all sources of radioactivity and explosives to the special dumping site. Therefore, radioactive pollution is unlikely at present although radiation might still be detected in areas where radioactive materials had been stored.

According to the prepared classification, military sites and individual objects in which preventive and cleanup works have to be implemented must first be selected. Presently, the Danish company Krüger Int. Consult A/S in cooperation with the "Baltic Consulting Group" Ltd carries out cleaning works by accumulations of free oil from groundwater surface using hydrodynamic method of oil skimming, at the site of the former airport fuel base [5].

TABLE 4. The main types of wastes at military sites (Krüger Consult and Baltic Consulting Group)

Types of wastes	Number of pollution sources	Thousands m^3
Building/metal scraps	157	96
Ruins	50	25
Waste of military/economic activity	182	88
Mixed industrial waste	17	35
Mixed household waste	65	84
Other	7	5
Total	478	333

Cleanup work programs based on the principle that pollution sources must be removed in order to stop further pollution of this site are being developed. The land polluted with oil products is considered to be a secondary source of pollution [6]. According to the existing requirements of directive LAND 9-95, industrial activity in territories polluted with oil products is only permissible when the concentration of oil products in the ground is less than 2000 mg/kg (Table 5). The concentration of oil products at the former military airport ground varies from 2000 to 5000 mg/kg [5].

TABLE 5. Maximal permissible levels of cleaning pollutants in territories mg/kg

Pollutants	I category	II category
Oil products*	300	1000
Pb	100	600
Cd	2	20
Cr	150	800
Cr(VI)	25	100
Co	100	300
Cu	200	600
Ni	200	300
Hg	0,5	10
Zn	500	3000

* - For inhabitants, water protection and sanitary zones – 20 mg /kg.

To date, 10 military sites have been investigated in detail. The sites were selected in different areas so that different types of pollution and typical landscape damages would be represented. Based on the results, new projects related to ecological optimization and renovations for other sites will be designed. The findings also suggest which pollution-preventive works must be carried out immediately in other areas. Newly developed projects will include renovation of those sites and the future management of other former military sites [5].

In all territories where groundwater-cleaning works are being carried out, the treated water is required to meet drinking water standards. This requirement dictates the selection of cleaning methods and technologies.

3. Conclusions

Lithuanian research is steadily improving the environmental situation at the military sites. Scientists from major scientific institutes are collaborating in this direction. We hope that cooperation between researchers from different Lithuanian institutes, business enterprises, and government institutions, and consultation with foreign experts will lead to faster and more effective solutions for solving the renovation problems.

References

1. Ministry of Environmental Protection Lithuania. (1998) *Study - Damage Made by Soviet Military Forces to Lithuanian Environment.* Final Report. Vilnius.
2. P. Baltrėnas, G. Ignatavičius. (1999) Problems of Renovation of Former Military lands in Lithuania. *Reports from Second Conference "Lithuania Without science – Lithuania without future" Environmental Protection Engineering.* Vilnius "Technika". 247 – 251.
3. Krüger Consult and Baltic Consulting Group (1995) *Study - Inventory of Damage and Cost estimate of Remediation of Former military Sites in Lithuania.* Final Report. Vilnius.
4. Geological Survey of Lithuania, Geological Survey of Norway, Canadian Department of National Defence, Norwegian Defence Research Establishment. (1996) *Study - Assessment Methodologies for Soil/Groundwater Contamination at Former Military Bases in Lithuania.* Final Report. NGU.
5. R. Baubinas, J. Taminskas. (1997 – 1998) *Military Nature use in Lithuania in Soviet Years: Ecological Consequences.* Institute of Geography Vilnius.
6. P. Baltrėnas, G. Ignatavičius. (1998). Environmental aspects of renovation of polluted former military lands in Lithuania. *Natural and Nuclear Anomalies and Life Protection. Conference materials.* Vilnius. "Technika", 59-62

THE CAPACITY OF AGRICULTURAL SOILS TO AUTO-REGULATE BIOREMEDIATION OF OLIVE-MILL WASTEWATERS

C. EHALIOTIS*, G. ZERVAKIS, O. ANOLIEFO**,
K. PAPADOPOULOU and A. KARDIMAKI
*National Agricultural Research Foundation, Institute of Kalamata,
Lakonikis 85, 24100 Kalamata, Greece*
*Current address: (*) Dept. of Natural Resources and Agric. Engineering,
Agricultural University of Athens, 75 Iera Odos st., Athens11855,
Greece; (**) Department of Botany, University of Benin, Benin City,
Nigeria*

1. Introduction

Olive-mill wastewaters (OMW) constitute the aqueous phase produced in olive mills after the extraction of olive oil from crushed olives by the addition of water. OMW have a dark color, show high conductivity and exhibit an acidic pH (around 4.8). They present a major environmental problem because of the enormous quantities produced during the winter months, and to their toxic effects and antimicrobial properties. The latter are mainly attributed to the OMW heavy organic load (resulting in BOD and COD values of around 30 and 100 kg/t respectively), and to the presence of recalcitrant polyaromatic compounds (principally phenolics) and perhaps to long-chain fatty acids and volatile acids [1, 2, 3].

In practice the application of OMW to water receptors is prohibited by their high organic load. Previous efforts to reduce the OMW pollutant effect have employed both physico-chemical and biological means [4, 5, 6]. However, technical complexity, high installation and operating costs, and failure to reduce the organic load down to the limits required by law for water receptors have resulted in limited applicability. On the other hand, since OMW are of purely vegetative origin and contain no xenobiotics, added chemicals, or synthetic pollutants, their recycling into agricultural ecosystems, rather than their exhaustive treatment and disposal into domestic sewage or other water receptors, seems environmentally reasonable and financially feasible.

2. Land application of OMW

Studies on the effect of OMW on the physical characteristics of soils demonstrated improved aggregate stability and reduced water evaporation losses [7]. Land

353

V. Šašek et al. (eds.),
The Utilization of Bioremediation to Reduce Soil Contamination: Problems and Solutions, 353–357.
© 2003 Kluwer Academic Publishers. Printed in the Netherlands.

application of OMW to soils leads to a significant increase of available phosphorus and potassium, and of the organic carbon content [8, 9, 10]. Additionally, almost all field trials using sensitive herbage crops and annual weeds showed that phytotoxic phenomena attributed to phenolics lasted for less than 2-3 months following application [8, 11] Therefore, in the case of annual crops, sowing should be carried out at least one month after OMW application, and OMW must not be added when the plants are in their sprouting period [9].

When soil/land characteristics and climatic conditions are favorable, treatment of specifically designated land using high OMW rates (>5000 m^3/ha) could be performed. Indeed, experiments carried out in lysimeters filled with two types of calcareous clay soils showed that a 2-m layer of soil removed almost completely the organic and inorganic components of OMW, when applied at annual rates of 5 000- 10 000 m^3/ha [12]. In field trials, the application of OMW during three successive years at an annual rate of up to 6000 m^3 ha^{-1} resulted in increased concentrations of organic matter, N, soluble NO_3 and available P enhancing soil fertility [12]. Soil electrical conductivity and sodium adsorption ratio also increased, but below the levels representing salinization or sodification hazard for the soil. However, leaching of Na and NO_3 ions below the 1-m layer was detected.

As regards the application of OMW in olive orchards, rates up to 800 m^3/ha showed no significant differences in plant physiology and no modification in the soil microflora [13], whereas applications of 150 m^3/ha showed higher K and P availability and improved plant nutritional status [14], and application rates of 333 m^3/ha resulted in short-term increase in K and Mg availability in a medium texture calcareus soil and long-term increase in organic matter content (sampling five years after application) [15].

Application of OMW in soils was followed by an increase in microbial activity as expressed by either CO_2 evolution measurements, or by the colony-forming units per gram of polluted soil [16, 17]. In both control and polluted soil, standard soil bacteria groups dominated; coryneform bacteria were the most abundant in polluted soil samples, while *Bacillus* spp. occurrence decreased. Once disposal of OMW ceases, soil microflora tends to restore the initial population equilibrium. *Azotobacter* and *Penicillium* strains isolated from soil treated with OMW and from OMW-disposal ponds, respectively, were subsequently used as inocula for the aerobic biodegradation of OMW [18, 19]. Soils already treated with OMW, however, showed lower efficiency to reduce OMW phenolic content following application. This is probably an effect of the wastewater-selecting microorganisms that are resistant to the toxic aromatic compounds rather than selecting those that degrade them [20].

3. Specific problems and experimental approaches on OMW land application in Greece

The potential for land application either directly or following a simple liming procedure has been largely ignored in Greece, perhaps because the national legislative framework does not distinguish OMW from other industrial and urban wastes (Interministerial committee resolution 69728/824/96), and it does not provide appropriate organic load application limits for soils. The recent application of specifically revised limits in Italy [21] had to be withdrawn for not conforming to the EU directives. Lately, however, interest in land application of OMW is growing as the frame for adopting more sustainable agricultural practices, favors recycling of agricultural by-products into land.

Reasons that make land application of OMW suitable for many regions of Greece are: (1) most agricultural soils are poor in organic matter, often shallow, in sloppy terrain and prone to erosion, (2) olive mills are small and widespread in the production areas, close to the appropriate land receptors (olive groves in particular), (3) most olive mill owners are reluctant to shift to two-phase olive oil separators that minimize OMW production (but may slow down olive processing rates and produce instead a hard to treat semisolid sludge).

On the other hand, problems related to OMW application have to be dealt with, e.g. (1) semi-mountainous landscape small-holder farming and limited road access to fields, (2) high winter-rainfall indexes in most olive-production areas of western Greece resulting in restrained distribution of OMW in the fields and increased potential for long-term flooding leading to soil anaerobiosis, (3) soils often developed on crust limestone that could allow substantial leaching under suitable geomorphological conditions.

Our results on Greek soils indicate the following:

- Certain soil bacteria [18] and several fungal species [22] were shown to reduce the phenolic content and to detoxicate the OMW. Phytotoxicity *in situ*, however, is directly related to the presence of phenolics in the soil solution following OMW application. The capacity of clay soils to accept greater volumes of OMW per unit area was observed as they release drastically reduced phenolics in the soil solution (as compared to sandy soils). This property is mainly induced by direct surface adsorption-oxidation phenomena, since it was related to the clay content and to the polyvalent metal oxides of the soils, and it was observed even after immediate extraction of the soils following OMW application [23]. However, short-term incubation (3-6 days) further reduced the extracted phenolics derived from OMW, and this occurred in synchrony with a drastic increase of microbial activity during the first week following OMW application. The pH increase of OMW to 6.8-7.0 using CaO before its application did not affect the availability of phenolics in the soils. Apparently however, other toxic compounds, such as volatile fatty acids, could be neutralized.

- *Lupinus albus* and *Vicia faba*, which are the two legumes mainly used as green manures in acidic and alkaline soils respectively in olive orchards in Greece, were shown to sustain volumes up to 60 m^3/ha of OMW, even when sown after immediate OMW application, without significant growth inhibition. A retardation of seed germination was, however, observed for both plant species. Moreover, an increase was observed in some soils on the nodulation of *Lupinus* plants, which was reflected by their total N content at harvest. This property could be exploited and was probably induced by N immobilization that enhances nodulation following OMW application.

- Nitrogen immobilization may restrict N availability to plants when OMW are applied to agricultural soils. We have observed an N-immobilization phase induced by relatively high rates of OMW in leaching tube experiments that could last for over two months in clay soils. Combined application of fertilizer N with OMW (fertilizer N and essential micronutrients diluted in OMW before application) is now examined as a

356

treatment that would alleviate strong N immobilization, inhibit N losses by leaching during the late winter-early spring period and result in N availability in synchrony with plant demands.

• Inhibitory effects on the growth of soil pathogens (*Phytophthora* spp., *Fusarium* spp.) were observed both *in vitro* and in sandy soils [24], but more work has to be carried out to confirm these effects with different soil types and under different application conditions.

In conclusion, application of OMW to agricultural land may become a sustainable recycling process, improving physical properties and long-term fertility (mainly K, P and organic-matter content) in marginal soils. However, planning of application zones and rates is required, since disadvantages may include leaching risks in sandy and shallow soils, transient phytotoxic and antimicrobial effects and immobilization of soil nitrogen. Difficulties may also arise in handling, storage and field application of the large quantities of OMW produced during winter. Pretreatment to reduce organic load by precipitation of solids or by biological treatment may be beneficial, especially in high-rainfall regions, where addition of materials rich in carbon combined with long-term water saturation could lead to soil anaerobiosis and plant-root asphyxiation.

References

1. Pérez, J., Esteban, E., Gomez, M. and Gallardo-Lara, F. (1986) Effects of wastewater from olive processing on seed germination and early plant growth of different vegetable species, *Journal of Environmental Science and Health* 21, 349-357.
2. Capasso, R., Evidente, A., Schivo, L., Orru, G., Marcialis, M. and Cristinzio, G. (1995) Antibacterial polyphenols from olive mill waste waters, *Journal of Applied Bacteriology* 79, 393-398.
3. Paixao, S.M., Mendoca, E., Picado, A. and Anselmo, A.M. (1999) Acute toxicity evaluation of olive mill wastewaters: a comparative study of three aquatic organisms, *Environmental Toxicology* 14, 263-269.
4. Fiestas Ros de Urcinos, J.A. and Borja Padilla, R. (1992) Use and treatment of olive mill wastewater: current situation and prospects in Spain, *Grasas y Aceites* 43, 101-106.
5. Ramos-Cormenzana, A., Monteoliva-Sanchez, M. and Lopez, M.J. (1995) Bioremediation of alpechin. *International Biodeterioration & Biodegradation* 35, 249-268.
6. Hamdi, M., Bouhamed, H. and Ellouz, R. (1991) Optimization of the fermentation of olive mill waste-waters by *Aspergillus niger*, *Applied Microbiology and Biotechnology* 36, 285-288.
7. Mellouli, H.J., Hartmann, R., Gabriels, D. and Cornelis, W.M. (1998) The use of olive mill effluents ("margines") as soil conditioner mulch to reduce evaporation losses, *Soil & Tillage Research* 38, 85-91.
8. Tomati, U. and Galli, E. (1992) The fertilizer value of waste waters from the olive processing industry, in J. Kubat (ed.), *Humus its structure and role in agriculture and environment*, Elesevier Science Publishers B.V., Amsterdam, pp. 117-126.
9. Fiestas Ros de Urcinos, J. A. (1986) Vegetation water used as fertilizer, in FAO-UNDP (ed.) *Proceedings of the International Symposium on Olive By-products Valorization*, Seville, pp. 321-330.
10. Levi-Minzi, R., Saviozzi, A., Riffaldi, R. and Falzo, L. (1992) Land application of vegetable water: effects on soil properties, *Olivae* 40, 20-25.
11. Bonari, E., Macchia, M., Angelini, L.G. and Ceccarini, L. (1993) The waste water from olive oil extraction: their influence on the germinative characteristics of some cultivated and weed species, *Agricoltura Mediterranea* 123, 273-280.
12. Cabrera, F., Lopez, R., Martinez-Bordiu, A., Dupuy de Lome, E. and Murillo, J.M. (1996) Land treatment of olive oil mill wastewater, *International Biodeterioration & Biodegradation* 38, 215-225.
13. Proietti, P. Cartechini, A. and Tombesi, A. (1988) Influenza delle acque reflue di frantoi oleari su olivi in vaso e in campo, in Atti Tavola Rotonda su *Acque reflue dei frantoi oleari*, Spoleto.
14. Catalano, M., Gomez, T., De Felice, M. and De Leonardis, T. (1985) Smaltimento delle acque di vegetazione dei frantoi oleari: Quali alternative alla depurazione?, *Inquinamento* 27, 87-90.

15. De Monpezat, G. and Denis, J.F. (1999) Fertilisation des sols méditerranéens avec des issues oléicoles, *OCL* 6, 63-68.
16. Paredes, M.J., Monteoliva-Sanchez, M., Moreno, E., Pérez, J., Ramos-Cormenzana A. and Martinez, J. (1986) Effect of waste waters from olive oil extraction plants on the bacterial population of soil, *Chemosphere* 15, 659-664.
17. Moreno, E., Pérez, J., Ramos-Cormenzana, A. and Martinez, J. (1987) Antimicrobial effect of waste water from olive oil extraction plants selecting soil bacteria after incubation with diluted waste, *Microbios* 51, 169-174.
18. Ehaliotis, C., Papadopoulou, K., Kotsou, M., Mari, I. and Balis, C. (1999) Adaptation and population dynamics of *Azotobacter vinelandii* during aerobic biological treatment of olive-mill wastewater, *FEMS Microbiology Ecology* 30, 301-311.
19. Robles, A., Lucas, R., de Cienfuegos, G.A. and Galvez, A (2000) Biomass production and detoxification of wastewaters from the olive oil industry by strains of *Penicillium* isolated from wastewater disposal ponds, *Bioresource Technology* 74, 217-221.
20. Saez, L., Perez, J. and Martinez, J. (1992) Low molecular weight phenolics attenuation during simulated treatment of wastewaters from olive oil mills in evaporation ponds, *Water Research* 26, 1261-1266.
21. Tamburino, V., Zimbone, S.M. and Quattrone, P. (1999) Storage and land application of olive-oil wastewater, *Olivae* 76, 36-45.
22. Antoniou, T., Ehaliotis, C., Panopoulos, N., Lyberatos, G. & Zervakis, G. (2002) Comparative evaluation of white-rot fungi as bioremediation agents of olive-mill waste waters, in *Proceedings of the Regional Symposium on Water Recycling in Mediterranean Region* (*accepted*). Heraklion, Greece.
23. Papaloukopoulou, P., Ehaliotis, C. and Assimakopoulos, J. (2002) Effects from the application of olive-mill wastewaters on the properties of ten different soils, in *Proceedings of the Greek Edaphological Conference*, Greek Edafological Society, Athens, (*in Greek, in press*).
24. Argeiti G., Ehaliotis, C., Katsaris, P., Zervakis, G. and Papadopoulou, K. (2001) Effect of olive mill wastes on soil-borne phytopathogenic fungi, *Phytopathologia Mediterranea* 40, 201 (abstract).

ACCUMULATION OF ORGANIC CONTAMINATION IN PLANT ROOTS AND THE INFLUENCE OF PLANT RHIZOSPHERE ON REMOVAL OF PAH, TPH AND HEAVY OIL FRACTIONS FROM SOIL

A. MALACHOWSKA-JUTSZ and K. MIKSCH

Environmental Biotechnology Department, Silesian Technical University, ul. Akademicka 2, 44-101 Gliwice, Poland

Summary

Contamination of soil with petroleum is an environmental problem affecting many industries, including the petroleum industry, gas stations, car services, cokeries and many others [1].

Petroleum contamination by high amounts of PAHs, TPHs and heavy oil fractions is often treated through landfarming, in which soil is mixed with CaO ($CaCO_3$), fertilizers and microorganisms that can biodegradate organic pollutants. After a few weeks, the soil is ploughed, and then plants are sown [2].

Recent research has demonstrated that plants can enhance the dissipation of organic pollutants in the environment of the roots (rhizosphere). This effect is connected with the growth of microorganisms in this area, due to easy access to organic substances [3].

In the present paper, we described the results of greenhouse experiments in which the use of vegetation to increase the degree of degradation of petroleum contaminants in soils was investigated.

1. Materials and Methods

A petroleum-contaminated soil was collected from the refinery area, from a surface layer not exceeding a depth of 30 cm, and was analysed. Fifteen PAHs were determined by HPLC with fluorescence detection. Concentrations of TPHs were quantified by gas chromatography (GC) with FID. Heavy petroleum fractions were assessed by weighing. Organic carbon, total organic substances, total nitrogen, ammonia nitrogen, nitrate nitrogen, nitrite nitrogen, base exchange complex, phosphorus, and pH were all determined by standard methods.

The testing soil was defined as a clay. The first step in processing soil for the greenhouse experiment was passing samples through a 1-mm sieve to remove large particles and homogenize the soil. Two kg of treated soil was then packed into greenhouse pots having 40 cm in diameter and then the soil was irrigated with distiled water to 25 % moisture by weight and it was allowed to equilibrate in the greenhouse for 7 d.

Plant species were chosen to reflect typical monocotyledonous and dicotyledonous species.

Contaminated soil collected from the refinery area was subjected to different treatments:

V. Šašek et al. (eds.),
The Utilization of Bioremediation to Reduce Soil Contamination: Problems and Solutions, 359–365.
© 2003 *Kluwer Academic Publishers. Printed in the Netherlands.*

I Control E
II With seedlings of grass
III With seedlings of clover
IV With microorganisms
V With microorganisms and seedlings of grass
VI With microorganisms and seedlings of clover

Microbial population was obtained from the top 15-cm layer of petroleum-contaminated soil. Microorganisms were isolated from the polluted soil and multiplied. The
culture was maintained at 20 °C on a mineral salts medium using an extract of the soil
as the sole carbon source. Microorganisms with transmittance equal to 70 % were
added to the selected samples of treated soil at a rate of 100 mL/kg.

2. Results and Discussion

Initial tests of the degree of soil contamination proved that two-ring PAHs represented
4.39 %, three-ring PAHs 30.05 %, four-ring PAHs 53,50 %, five-ring PAHs 6,69 %
and six-ring PAHs 5,36 % (Table 1). Initial content of TPHs was 19,11g/kg dry matter
whereas the content of heavy fractions was 159.11g/kg dry matter.

Table 1. Initial contents of polycyclic aromatic hydrocarbons in investigated soil.

PAH	Content of PAHs [mg/kg dry matter]	Number of rings	Content [%]
Naphthalene	1.411	2	4.39
Acenaphthene	0.714		
Fluorene	0.991	3	30.05
Phenanthrene	6.701		
Anthracene	1.252		
Fluoranthene	8.131		
Pyrene	2.066		
Benzo(a)anthracene	2.160	4	53.50
Chrysene	2.416		
Benzo(b)fluoranthene	1.464		
Benzo(k)fluoranthene	0.957		
Benzo(a)pyrene	1.349	5	6.69
Dibenzo(a,h)anthracene	0.801		
Benzo($g,h,$i)perylen	0.392	6	5.36
Indeno(1,2,3-$cd.$) pyrene	1.330		

In all cases, the highest degradation of PAHs occurred for PAHs with two and three rings. The general trend in the disappearance of naphtalene was typical of landfarming operations. Concentration of naphtalene decreased from 1.41 mg/kg to 0.23 mg /kg during 12 weeks in the sample with microorganisms and clover (Fig. 1). The highest removal of PAHs with three rings was in the soil treated with clover (81.96 % removal), with microorganisms (89.47 %) and with clover and micro-organisms (Fig. 1).

In the cases of PAHs with four, five and six rings, concentrations increased in all samples. The highest increase of PAHs was observed in the control sample (Fig. 1).

The persistence of PAHs in the soil depends on the extent to which they sorb irreversibly to soil constituents. PAHs are multi-ring, nonpolar, hydrophobic organic compounds. They are adsorbed on humic acids and humin [3,4,5,6,7]. Sorption of PAHs onto soil particles makes bioremediation difficult [6,7]. The microorganisms cannot bioremediate PAHs that are insoluble in water [5]. The PAHs with three or more rings have higher molar masses and the greater the number of rings the stronger the compounds are adsorbed to the soil. Strong adsorption coupled with very low water solubility decrease the rate of biodegradation [4,5,8,9,10,11].

It is likely that the higher concentration of PAHs with four, five and six rings and of TPHs after bioremediation in our samples was caused by desorption of the

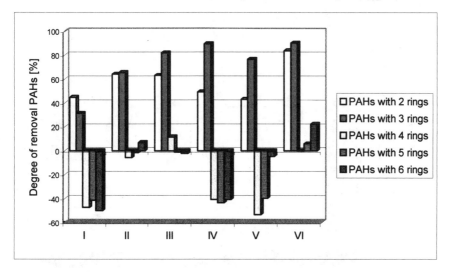

Figure 1. Degree of removal of PAHs after three months of biodegradation

compounds. Microorganisms can exudate biosurfactants during bioremediation. Biosurfactant solutions may solubilize hydrophobic contaminants from soil by reducing the energy of adhesion between the contaminant and the soil, resulting in desorption and incorporation of the organic compound within biosurfactant micelles in the aqueous phase [12,13,14,15].

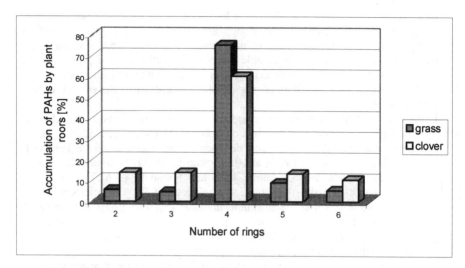

Figure 2. The accumulation of organic contamination by plants roots during 3 months of bioremediation.

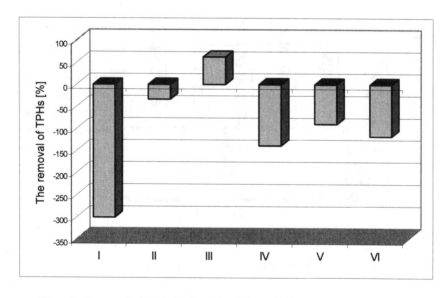

Figure 3. The removal of TPHs after 3 months of bioremediation

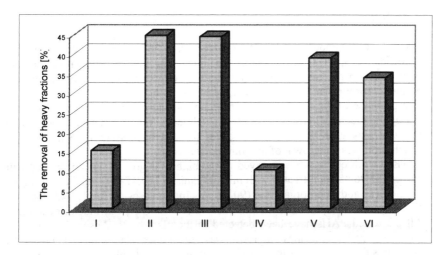

Figure 4. The removal of heavy fractions after 3 months of bioremediation

The composition of the organic matter in soils has a significant influence on the sorption of organic compounds. Humins have a higher sorption capacity for PAHs, compared to humic and fulvic acids.

All PAHs have been detected in vegetation. Among the PAHs, the highest accumulation by plants roots during a period of three months was for PAHs with four rings – 75.3 % by grass roots and 60.2 % by roots of clover (Fig. 2).

Recent research indicates that adsorption of PAHs onto roots may be significant but translocation of PAHs with four and more rings from roots to foliar portions of the plants is negligible [3,16], probably because these PAHs are not soluble in water and have a high molar mass.

The removal of TPHs was highest in sample III (soil treated with clover), *viz.* 62.64 %. In other samples, some desorption of TPHs from soil particles probably took place. The highest concentration of TPHs was in the control soil, *viz.* 301 % of the initial concentration, a lower concentration of TPH was in sample II (soil treated with grass) (Fig. 3).

The removal of the heavy fractions was significantly greater in vegetated soil than in unvegetated soil I and IV after 12 weeks of incubation (Fig. 4).

In this experiment, it was not possible to evaluate the contribution of individual plant species and of their associated rhizosphere, although it is likely that each species may have a unique influence on the rhizosphere. In vegetated soil characteristic phenomena for areas contaminated with petroleum have been observed. Plant growth was initially slow followed by a phase of enhanced development and at 3-4 weeks after germination, the plants rapidly died. In the case of clover, foliar portions of the plants had small leaves and thin short stalks.

3. Conclusions

1. The 12-week period of experiment is too short to confirm a very significant influence of applied agrotechnical treatments on the biodegradation process, particularly in the case of soils with large cation-exchange capacity (during this period, desorption of organic pollutants from soil prevailed).
2. The presence of vegetation and microorganisms enhances the removal (probably through biodegradation) of TPHs, PAHs and heavy fractions. This is most likely a result of exudation of organic substances from plant roots into the rhizosphere, which promotes the growth of microbial population.
3. PAHs with three rings are removed to the highest degree in the soil treated with clover (81.96 %), soil treated with microorganisms (89.47 %), and the soil treated with clover and microorganisms (90.11 %). In the cases of PAHs with four and five rings and of TPHs in all samples, desorption from soil constituents occurred.
4. All the tested PAHs have been detected in the plant roots. Adsorption of PAHs with four and more rings on roots was the highest, but translocation of the pollutants from plant roots to the foliar portion was negligible.
5. The use of vegetation is attractive because it is inexpensive and requires little care, and could be superior to many alternative cleanup technologies.

References

1. Bąkowski W., Bodzek D. (1988) Wielopierścieniowe węglowodory aromatyczne w naturalnym środowisku człowieka – pochodzenie, występowanie, toksyczność, oszacowanie emisji w Polsce., Archiwum Ochrony Środowiska tom **3-4** .
2. Twarda B, Miksch K. (1995) Ekologia mikroorganizmów ryzosfery występujących w warunkach skażenia gleby WWA, III Sympozjum Naukowo – Techniczne „Biotechnologia Środowiskowa", Ustroń – Jaszowiec.
3. Reilley K. A., Banks M. K., Schwab A. P. (1996) Dissipation of policyclic aromatic hydrocarbons in the rhizosphere, J. Environ. Qual., **25**, 212-219.
4. Kosinkiewicz B., Mokrzycka M. (1988) Transformacja i wykorzystanie antracenu przez mikroorganizmy glebowe, Archiwum Ochrony Środowiska, tom **1-2**.
5. Maliszewska-Kordybach B. (1992) Wpływ nawożenia organicznego na trwałość wielopierścieniowych węglowodorów aromatycznych w glebach, Archiwum Ochrony Środowiska, **2**, 153-162.
6. Łebkowska M. (1996) Wykorzystanie mikroorganizmów do biodegradacji produktów naftowych w środowisku glebowym, Gaz, Woda i Technika Sanitarna, **3**.
7. Maliszewska – Kordybach B. (1995) The persistence of pollutants in soil is related – among other factors – to their sorption. Hydrophobic Xenobiotics, e.g. highly carcinogenic and mutagenic polycyclic aromatic hydrocarbons (PAH), are sorbed mainly on the organic fraction of soil, Archiwum Ochrony Środowiska, **2**, 183-190.
8. Sztompka E. (1995) Biodegradacja paliwa Diesla w glebie, materiały III Ogólnopolskiego Sympozjum Naukowo – Technicznego „Biotechnologia Środowiskowa", Ustroń – Jaszowiec.
9. Maliszewska –Kordybach B. (1991) Wpływ wapnowania na trwałość wielopierścieniowych węglowodorów aromatycznych w glebach., Archiwum Ochrony Środowiska, **3-4**, 69-78.
10. Fudryn G., Kawala Z. 1996 Odnowa zanieczyszczonych gruntów metodami in situ., Ochrona Środowiska, tom **2**, 61.
11. Lisowska K., Fijałkowska S., Długoński J. (1997) Mikrobiologiczna degradacja związków ropopochodnych w nieobecności i obecności metali ciężkich, materiały V Ogólnopolskiego Sympozjum Naukowo – Technicznego „Biotechnologia Środowiskowa", Ustroń -Jaszowiec.
12. Aroinstein Boris N., Calvillo Y., Alexander M. (1991) Effect of surfactans at low concentrations on the desorption and biodegradation of sorbed aromatic compounds in soil, Environ. Sci. Technol., **25**, 1728-1731.

13. Manilal V., Alexander M. (1991) Factors affecting the microbial degradation of phenantrene in soil, Appl. Microbiol. Biotechnol., **35,** 401-405.
14. Obermremer A., Muller-Hurtig R., Wagner F. (1990) Effect of addition of microbial surfactants on hydrocarbon degradation in a soil population in a stirred reactor, Appl. Microbiol. Biotechnol., **32,** 485-489.
15. Laha S., Luthy R. (1992) Effects of nonionic surfactans on the solubilization and mineralization of phenantrene in soil-water systems, Biotech. and bioengineering, **40,** 1367-1380.
16. Jones K. C., Stratford J. A., Tidridge P., Waterhouse K. S., Johnston A. E. (1989) Polynuclear aromatic hydrocarbons in an agricultural soil: Long – term changes in profile distribution, Environmental Pollution **56,** 337 – 351.

Perspectives of Stable Isotope Approaches in Bioremediation Research

H. H. RICHNOW[1], R. U. MECKENSTOCK[2], AND M. KÄSTNER[1]

[1] *Center for Environmental Research Leipzig-Halle (UFZ), Permoserstr. 15, 04318 Leipzig, Germany*
[2] *Center for Applied Geosciences, Eberhard-Karls-University of Tübingen, Willhelmstr. 56, 72076 Tübingen, Germany.*

Abstract

Stable isotope signature of organic compounds (natural abundance and enriched tracer compounds) can be used to investigate the distribution, transformation, and fate of organic substances in environmental compartments and organisms. The degradation of organic substances by microorganisms often leads to a relative enrichment of the heavier carbon isotopes in the residual substrate fraction. Based on the isotopic signature of a detected pollutant concentration and the analysis of the fractionation processes, the microbial *in situ* degradation in contaminated aquifers can be calculated. In addition, tracer compounds of organic xenobiotics labeled with stable isotopes can be used for the calculation of mass balances. The concept has been employed to characterise the transformation of ^{13}C-labeled polycyclic aromatic hydrocarbons (PAH) in closed soil bioreactors.

1. Introduction

In the last decade, stable isotope chemistry (natural abundance and enriched tracer compounds) has received increasing attention in environmental science [2]. The stable isotope signature of organic substances yields useful information to decipher the distribution, transformation and final fate of organic substances in the environment. The major elements of organic substances are carbon, hydrogen and nitrogen. Hydrogen (^{1}H) has a natural abundance of 99.9844, whereas deuterium (^{2}D) was found to represent 0.0156 % [4]. Carbon was found with a natural abundance of 98.89 % ^{12}C and 1.11 % ^{13}C. The abundance of nitrogen in the atmosphere is 99.64 % ^{14}N and 0.36 % ^{15}N. The changes of isotopic composition of these elements during biological activity are of interest to solve crucial questions of modern environmental and biological research. Recent developments in IRM-GC-MS enable the exploration of the large potential of isotope chemistry in remediation research. Moreover, tracers labeled with stable isotopes may be applied in structural studies with conventional GC-MS and

367

NMR spectroscopy. In this contribution, two concepts of applying stable isotope approaches in environmental and remediation research are presented.

2. Analysis of isotopic fractionation to characterize *in situ* biodegradation processes

The assessment of *in situ* biodegradation of pollutants is required for a precise prediction of the fate of contaminants in aquifers. A decrease of the concentration of a pollutant downstream a contamination plume may have many causes, such as dilution, adsorption to soil, or biodegradation by the indigenous microflora. To characterise the *in situ* microbial degradation of aromatic hydrocarbons, we applied a concept based on isotope fractionation of contaminants during biodegradation [12,15]. The degradation of organic substances by micro-organisms often leads to the relative enrichment of the heavier carbon isotopes in the residual substrate fraction. Carbon isotope fractionation during microbial degradation was proven for BTEX compounds and halogenated solvents in numerous studies [1,3,5,10,17,18]. For implementation of the concept, compound specific isotopic fractionation factors ($\alpha^{13}C$, αD) are determined from microbial degradation of contaminants in laboratory experiments. Using these factors, the *in situ* biodegradation processes can be calculated based on the differences of the isotopic signatures of detected pollutant concentrations downstream of the contamination plume within the aquifer [16].

The $^{13}C/^{12}C$ isotope ratio of aromatic hydrocarbons was studied in soil percolation columns with toluene as sole carbon and energy source and with sulfate as electron acceptor [9]. After 2 months of incubation, the soil microbial community was able to degrade 32 mg/L toluene to less than 0.05 mg/L, generating a stable concentration gradient within the column. The $^{13}C/^{12}C$ isotope composition of the residual non-degraded fraction of toluene showed a significant increase in the $^{13}C/^{12}C$ ratio corresponding to an isotope fractionation factor of αC 1.0017.

In a field study, a contaminated anoxic aquifer was analyzed for BTEX and PAH contaminants. Toluene revealed a significant concentration gradient from 160 to 1.9 µg/l along a contamination plume of 800 m [16]. A distinct increase of the $\delta^{13}C$ values was observed for the residual non-degraded toluene (7.2 ‰). The contribution of microbial degradation to the total contaminant removal was calculated on the basis of laboratory-derived $^{13}C/^{12}C$ isotope fractionation factors for toluene, indicating that more than 98 % of the compounds were already degraded by microbial activity (Fig. 1). Isotopic fractionation may be a promising tool to estimate the degradation process in the subsurface or in biotechnological processes independent of other concentration diminishing processes such as adsorption and dilution.

3. Application of stable isotope labels to trace the fate of organic chemicals

This approach is used for the determination of the fate of a specific compound in environmental compartments and organisms. Stable carbon isotopic measurements

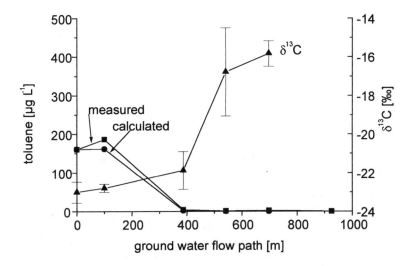

Figure 1: Toluene concentrations (squares) and $^{13}C/^{12}C$ isotope ratios (triangles) along the monitoring transect of a contaminated aquifer. Circles depict the theoretical toluene concentration as calculated based on the measured isotope ratios. Data from [16]

have been employed to characterize the transformation of the 9-[^{13}C]-labeled polycyclicaromatic hydrocarbon (PAH) anthracene in closed soil bioreactors. Beside the application of labeled substances for structural studies with GC-MS, the ^{13}C-label was applied to calculate the carbon mass balance including mineralization, transformation, and formation of nonextractable soil-bound residues (Fig. 2). The comparison of mass balance with ^{13}C- and ^{14}C-labeled anthracene in separate batch experiments revealed similar results [8,11,13]. The transformation of ^{13}C-labeled PAH into nonextractable residues depends on the metabolic activity of the soil microflora and occurs during the early phase of biodegradation. The extent of residue formation is controlled by the capability of the microflora to degrade the contaminant [6,7]. Periodical additions of anthracene over a period of 300 d leads to the adaptation of the soil microflora to mineralize the added compound. Results of long-term experiments indicate that non-extractable residues are relatively stable over time [14].

The sensitivity of the ^{13}C-tracer method meets the requirements of classical radio tracer experiments in concentration ranges typical of real PAH-contaminated sites. Thus, the presented balancing method based on stable isotope labeled chemicals may substitute radiotracer experiments in several cases.

With the improvement of sensitivity of Isotope-Ratio-Monitoring-GC-MS systems, studies of microbial degradation of specific organic compounds in environmental systems became feasible. The application of stable isotope labeled substances may allow to trace the speciation of the xenobiotic carbon into specific fractions such as

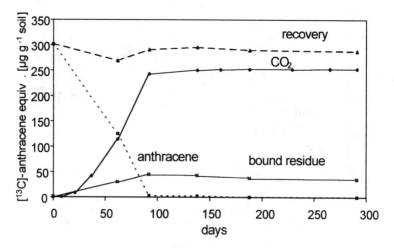

Figure 2: Mass balance of the ^{13}C-label during microbial degradation of 9-[^{13}C]-anthracene in soil (anthracene equiv. = recovery of the labeled ^{13}C-carbon calculated as anthracene equivalents) [13].

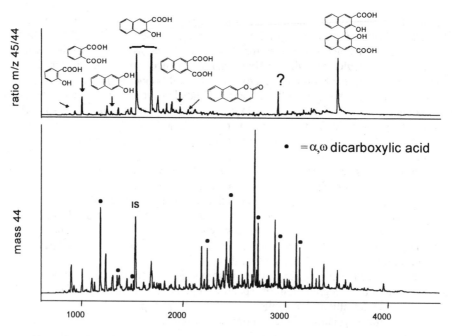

Figure 3: IRM-GC-MS of alkaline hydrolysis products of the bound residue fraction after degradation of 9-[^{13}C]-anthracene in soil. The upper trace (45/44) depicts the variation of mass 45 ([13]CO_2) to mass 44 ([12]CO_2). ^{13}C-labeled metabolites can be easily identified by their characteristic isotope composition even if the absolute concentration is low. (dots = α,ω-dicarboxylic acids with an isotopic composition between −27 and −33 ‰ which is typical of long-chain carboxylic acids in soil).

metabolites or biomass components quantitatively. This includes, e. g., the detection of xenobiotic carbon incorporated into bacterial fatty acid or amino acid fractions as was found with several compounds. However, the incorporation of ^{13}C-labeled carbon into the biomass was not observed from the 9-[^{13}C]-position of anthracene (Fig. 3). Stable-isotope-labeled substrates can be applied to open systems and thus may be an interesting tool in environmental and remediation studies.

References

1. Ahad, J.M.E., Sherwood Lollar, B., Edwards, E.A., Slater, G.F., and Sleep, B.E. (2000) Carbon isotope fractionation during anaerobic biodegradation of toluene: implication for intrinsic biodegradation, *Environ. Sci. Technol.* **34**, 892-896.

2. Grossman, E.L. (1997) Stable carbon isotopes as indicators of microbial activity in aquifers, in: Hurst, C.J., Knudsen, G.R., McInerney, M.J., Stetzenbach, L.D., and Walter, M.V. (eds.), *Manual of environmental microbiology*. ASM Press, Washington DC, pp 565-576.

3. Heraty, L.J., Fuller, M.E., Huang, L., Abrajano, T.A., and Struchio, N.C. (1999) Isotopic fractionation of carbon and chlorine by microbial degradation of dichloromethane, *Org. Geochem.* **30**, 793-799.

4. Hoefs, J. (1997) *Stable isotope geochemistry,* Springer Verlag, Berlin.

5. Hunkeler, D., Aravena, R., and Butler, B.J. (1999) Monitoring microbial dehalogenation of tetrachloroethene (PCE) in groundwater using compound specific stable carbon isotope ratios: microorganism and field studies, *Environ. Sci. Technol.* **33**, 2733-2738.

6. Kästner, M., Streibich, S., Beyrer, M., Richnow, H.H., and Fritsche, W. (1999) Formation of bound residues during microbial degradation of [^{14}C]-anthracene in soil, *Appl. Environ. Microbiol.*, **65**, 1834-1842.

7. Kästner, M. (2000) Degradation of aromatic and polyaromatic compounds, in H.-J. Rehm, G. Reed, A. Pühler, and P. Stadler (eds.). *Biotechnology,* 2nd. Edition, Vol. 11b; Environmental Processes. Wiley-VCH, Weinheim, pp 211-239.

8. Kästner, M., Richnow, H.H. (2001) Formation of Residues of Organic Pollutants within the Soil Matrix – Mechanisms and Stability, in: Stegmann, R., Brunner, G., Calmano, W., and Matz, G. (eds.), *Treatment of Contaminated Soil.* Springer. Berlin, Heidelberg, New York, pp 219-251.

9. Meckenstock, R.U., Morasch, B., Warthmann, R., Schink, B., Annweiler, E., Michaelis, W., and Richnow, H.H. (1999) ^{13}C/^{12}C isotope fractionation of aromatic hydrocarbons during microbial degradation, *Environ. Microbiol.* **1**, 409-412.

10. Morasch, B., Richnow, H.H., Schink, B., Meckenstock, R.U. (2001), Stable hydrogen and carbon isotope fractionation during microbial toluene degradation: mechanistic and environmental aspects, *Appl. Environm. Microbiol.* **67**, 4842-4849.

11. Richnow, H.H., Eschenbach, A., Seifert, R., Wehrung, P., Albrecht, P., and Michaelis, W. (1998), The use of 13-C-labeled polycyclic aromatic hydrocarbons for the analysis of their transformation in soils, *Chemosphere* **36**, 2211-2224.

12. Richnow, H.H., and Meckenstock, R.U. (1999a) Isotopen-geochemisches Konzept zur *in-situ*-Erfassung des biologischen Abbaus in kontaminiertem Grundwasser - Biodegradation in kontaminierten Grundwasserleitern,. *TerraTech* **5**, 38-41.

13. Richnow, H.H., Eschenbach, A., Mahro, B., Kästner, M., Annweiler, E., Seifert, R. and Michaelis, W. (1999b) Formation of nonextractable residues - a stable isotope approach, *Environ. Sci. Technol.* **33**, 3761-3767.

14. Richnow, H.H., Annweiler, E., Fritsche, W., and Kästner, M. (1999c) Organic pollutants associated with macromolecular soil organic matter and the formation of bound residues, in J.C. Block, Ph Baveye, and V.V. Goncharuk (eds.) *Bioavailability of organic xenobiotics in the environment.* NATO ASI Series, Volume XX, Kluwer Academic Publishers, Netherland, pp 297-326.

15. Richnow, H.H., Vieth, A., Kästner, M., Gehre, M., and Meckenstock, R.U. (2002), Isotope Fractionation of Toluene: A Perspective to Characterise Microbial In Situ Degradation, *TheScientificWorldJOURNAL ISSN#1537-744X, 2002,* **2**, 1227-1234 (www.thescientificworld.com).

16.   Richnow H.H., Annweiler E., Michaelis W., Meckenstock R. (2002) Microbial *in situ* degradation of aromatic hydrocarbons in a contaminated aquifer monitored by carbon isotope fractionation, *Journal of Contaminant Hydrology* (in press).

17.   Sherwood Lollar, B., Slater, G.F., Ahad, J., Sleep, B., Spivack, J., Brennan, M., and MacKenzie, P. (1999) Contrasting carbon isotope fractionation during biodegradation of trichloroethylene and toluene: Implications for intrinsic bioremediation, *Org. Geochem.* **30**, 813.

18.   Stehmeier, L.G., Francis, G.F., Jack, T.R., Diegor, E., Winsor, L., and Abrajano, T.A. (1999) Field and in vitro evidence for in-situ bioremediation using compound-specific $^{13}C/^{12}C$ ratio monitoring, *Org. Geochem.*, **30**, 821.

SCREENING OF LITTER DECOMPOSING FUNGI FOR DEGRADATION OF POLYCYCLIC AROMATIC HYDROCARBONS (PAH) PHENANTHRENE AND BENZO[A]PYRENE

A. MAJCHERCZYK, A. BRAUN-LÜLLEMANN, A. HÜTTERMANN
Institute of Forest Botany, University of Göttingen
D-37077 Göttingen, Büsgenweg 2, Germany

1. Introduction

A long-lasting, high level contamination of soil with polycyclic aromatic hydrocarbons (PAH) has in most cases resulted from the accidental disposal of wastes by the chemical industry. Large regions surrounding these centers also display a significant contamination resulting from diffusion and leakage of PAH. Transport of PAH *via* aerosols (e.g. on coal dust) has resulted in a low-level contamination but in deposits over large areas. These can not be detoxicated in an economical way by on-site decontamination techniques. *In-situ* detoxication which utilizes endogenous soil micro-organisms is limited by the inability of such organisms to degrade highly condensed PAH effectively; the organisms which have overcome this problem, wood-degrading white-rot fungi cannot be maintained in the soil *in situ* for the length of time necessary for decontamination. Recently, the ability of numerous ectomycorrhizal fungi to metabolize PAH has been demonstrated [1]. In the present preliminary study we report on the ability of litter-decomposing fungi (48 isolates from Central European soil and litter) belonging to 12 species (24 strains) to degrade phenanthrene and benzo[a]pyrene in a pure liquid culture.

2. Material and Methods

Stock cultures of fungi were grown on modified Melin-Norkrans medium [7] with 2 % agar. The degradation of PAH was studied in liquid batch cultures (five replicates) using three different media (15 mL medium in 500-mL Erlenmeyer flasks). Medium A was modified Moser medium [6, 10]. Medium B was the same as the former but without peptone, and Medium C was a modified Melin-Norkrans medium.

Acetone stock solution of phenanthrene and benzo[a]pyrene was mixed with the desired medium containing Tween 80. Sterile aliquots of these solutions were added to 3–4-week-old fungal cultures resulting in final concentrations of each PAH of 20 ppm. After 4 weeks of incubation cultures were extracted by isooctane containing 5 ppm of fluoranthene and perylene as internal standards. Samples were analyzed by gas chromatography and mass spectrometry (GC-MS, single ion monitoring).

V. Šašek et al. (eds.),
The Utilization of Bioremediation to Reduce Soil Contamination: Problems and Solutions, 373–376.
© 2003 *Kluwer Academic Publishers. Printed in the Netherlands.*

3. Results

The selected semi-defined media were previously reported on within the context of mycorrhizal and litter decomposing fungi. Medium A was a C- and N-rich medium and Medium B was simply a modified version of A with A depletion. Medium C was a C-rich medium with additional glucose and only an inorganic A source. Many of the fungi grew very slowly; therefore, a preincubation period of three to four weeks was selected to obtain a sufficient amount of biomass. Since no optimization of media for single fungus could be established in this screening, it was not surprising that the growth of the organisms differed among the media with no clear general preference. Several isolates of the same fungus even displayed a strong difference in growth with respect to the culture medium; these were tested if a visible growth of fungus was detected. Still, the degradation ability did not correspond to the visible growth of mycelia in all cases.

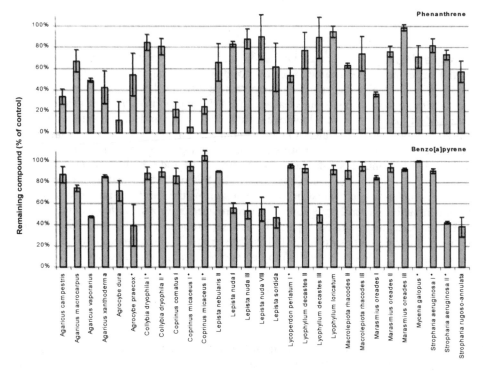

Figure 1. Degradation of phenanthrene and benzo[a]pyrene by litter decomposing fungi – Medium A.
* Fungi growing on litter and wood.

Most of the isolates grow well on a complex Medium A with peptone and yeast extract, but the best overall metabolization of PAH was obtained on Medium B without peptone (Fig. 1 and 2). All of the fungi tested in liquid cultures with Medium A and B were able to degrade phenanthrene. Approximately 38 % and 80 % of fungi metabolised this compound to over 50% in Medium A and B, respectively. *Agrocybe dura, Coprinus micaceus*, strains of *Lepiota nauciana, Lepista nebularis, Lepista sordida*, and

Stropharia rugoso-annulata removed approximately 80 % of the phenanthrene. The degradation ability on Medium C was much less expressed: none of the fungi removed more than 40 % of phenanthrene and almost 60 % of fungi did not metabolized this PAH.

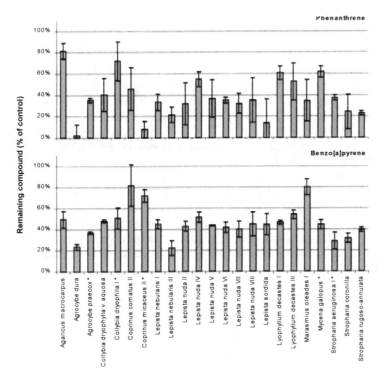

Figure 2. Degradation of phenanthrene and benzo[*a*]pyrene by litter decomposing fungi – Medium B.
* Fungi growing on litter and wood.

A similar pattern of metabolism with regard to the culture medium was also observed in the degradation of benzo[*a*]pyrene. More than 50 % and 100 % of the fungi metabolized this compound growing on Media A and B, respectively. No significant degradation or by less than 10 % was detected in cultures with Medium C. The best degraders of benzo[*a*]pyrene – *Agrocybe dura, Lepista nebularis, Stropharia aeruginosa,* and *Stropharia coronilla* – removed approximately 70 % of this highly cancerogenic and the most relevant PAH.

4. Discussion

Dead wood litter forms a large reservoir of mineral nutrients, which are unavailable for primary biomass producers until they are released by decomposer organisms, mostly litter degrading fungi (basidiomycetes and ascomycetes). Degradation of the lignin aromatic moiety by a number of litter decomposers (e.g. [9]) has led to the inference that such organisms should also be able to degrade condensed aromatic structures. This was previously demonstrated for single fungi, e.g. *Marasmius*, in the metabolism of pyrene [4]. Our screening demonstrates that numerous litter decomposing fungi are able to metabolize such polycyclic aromatic hydrocarbons as phenanthrene and benzo[*a*]pyrene; compounds that differ in condensation grade and ionization (oxidation) potential. Removal of the PAH tested depends strongly on the culture medium; however, no general or significant correlation between A depletion and degradation was found. The comparatively slow yet very efficient metabolism of phenanthrene and benzo[*a*]pyrene was comparable to results obtained in experiments with white rot fungi.

Numerous litter decomposing fungi grown in pure culture secrete oxidative enzymes such as polyphenol oxidases, laccases, tyrosinases or peroxidases (e.g. [2, 5, 8]) and the involvement of comparable extracellular oxidative enzymes in the degradation of PAH by white rot fungi has been studied extensively. The ability of phenol oxidases to metabolise PAH in the presence of low molar mass phenols, aromatic amines or sulfhydryl compounds – secreted by fungi or produced in degradation processes – has been also demonstrated [3].

The potential of the litter decomposing fungi to degrade PAH and their ability to inhabit soil may prove to be very promising in the recultivation and bioremediation of contaminated areas *in situ*. After the primary degradation step performed by fungi, the mineralisation of pollutants such as PAH is expected to proceed even more readily in complex soil systems which prevent an accumulation of intermediate or dead-end metabolites.

ACKNOWLEDGEMENTS

This study was supported by a grant from the Volkswagen Stiftung, Germany. We would like to thank Prof. W. Fritsche and his co-workers in the Institute of Microbiology, University of Jena (Germany) for providing some of the fungal cultures.

References

1. Braun-Lüllemann, A., Hüttermann, A., and Majcherczyk, A. (1999) Screening of ectomycorrhizal fungi for degradation of polycyclic aromatic hydrocarbons, *Appl. Microbiol. Biotechnol.* **53**, 127-132.
2. Gramss, G., Günther, T., and Fritsche, W. (1998) Spot tests for oxidative enzymes in ectomycorrhizal, wood-, and litter decaying fungi, *Mycol. Res.* **102**, 67-72.
3. Johannes, C. and Majcherczyk, A. (2000) Natural mediators in the oxidation of polycyclic aromatic hydrocarbons by laccase mediator systems, *Appl. Environ. Microbiol.* **66**, 524-528.
4. Lange, B., Kremer, S., Sterner, O., and Anke, H. (1996) Metabolism of pyrene by basidiomycetous fungi of the genera *Crinipellis*, *Marasmius*, and *Marasmiellus*, *Can. J. Microbiol.* **42**, 1179-1183.
5. Lindeberg, G. and Holm, G. (1952) Occurence of tyrosinase and laccase in fruit bodies and mycelia of some hymenomycetes, *Physiol. Plantarum* **5**, 100-114.
6. Moser, M. (1958) Die künstliche Mykorrhizaimpfung an Forstpflanzen, *Forstw. Centrbl.* **77**, 32-40.
7. Norkrans, B. (1949) Some mycorrhiza-forming *Tricholoma* species, *Svensk Botan. Tidskr.* **43**, 485-490.
8. Tagger, S., Perissol, C., Gil, G., Vogt, G., and LePetit, J. (1998) Phenoloxidases of the white-rot fungus *Marasmius quercophilus* isolated from an evergreen oak litter (*Quercus ilex* L.), *Enzyme Microbial. Technology* **23**, 372-379.
9. Tanesaka, E., Masuda, H., and Kinugawa, K. (1993) Wood degrading ability of Basidiomycetes that are wood decomposers, litter decomposers, or mycorrhizal symbionts, *Mycologia* **85**, 347-354.
10. Trojanowski, J., Haider, K., and Hüttermann, A. (1984) Decomposition of carbon-14 labelled lignin, holocellulose and lignocellulose by mycorrhizal fungi, *Arch. Microbiol.* **139**, 202-206.

PHOTOSYSTEM II-BASED BIOSENSORS FOR PHYTOREMEDIATION

M.T. GIARDI, D. ESPOSITO, E. PACE, S. ALESSANDRELLI, A. MARGONELLI, G. ANGELINI, P. GIARDI.

IBAF-ICN, CNR, Area della ricerca di Roma, 00016 Monterotondo Scalo. Italy EM

1. Introduction

Phenylurea, triazine and diazine represent economically very important compounds since they are used in chemical, pharmaceutical and agricultural industries. Although new herbicides are now available, they still represent the basic products for weed control. All the compounds of these classes constitute about 40 % of all herbicides used at present in agriculture, amounting to thousands of tons all over the world. These compounds are absorbed through the roots and then translocated via the xylem to the leaves. Some of these herbicides are directly absorbed by the leaves. They act by inhibiting photosynthesis at the level of the photosystem II-dependent electron transfer and further block production of ATP and NADPH. In soils, these compounds are quite persistent and they are adsorbed on soil colloids and on organic substances in proportion to the cation exchange capacity of these soil constituents.

Bioremediation by photosynthetic plants (phytoremediation) is a well-established method to recover herbicide-polluted soils. The ideal plant for phytoremediation should exhibit the following characteristics: (i) be resistant to the herbicides, (ii) to allow herbicide translocation to the leaf: (iii) to be able to degrade the herbicide. We can distinguish naturally resistant plants and mutants of sensitive species. In naturally resistant plants the mechanisms of resistance include a slow translocation into the chloroplast, high protein repair turnover, immobilization and detoxication by endogenous enzymes through conjugation and/or degradation. Glutathione S-transferases are involved in the detoxication of herbicides. Often, more than one mechanism is active at the same time. Moreover, prolonged agronomic use of PSII inhibitors, specifically triazine herbicides, has resulted in the evolution of mutants of sensitive species. In this case, the predominant basis for resistance is a single nucleotide substitution in the chloroplast *psb*A gene encoding the D1 protein, which precludes the binding of herbicide to the protein. However, resistance of mutated biotypes has also been attributed to modifications in the activities of the glutathione enzyme [1].

The best results in phytoremediaton can be obtained using mutant plants that are resistant to the herbicide with increased enzymic activities able to degrade the herbicide. Mutant plants with amino acid modifications in the D1 target protein are less effective, whereas plants with slow herbicide translocation are not useful [1]. Usually enzymic analyses and radioactive tracers are necessary to reveal the mechanism of resistance of these plants. Chl fluorescence is frequently used as a potential indicator of environmental and chemical stresses and a screening method of resistant plants. It allows nondestructive, simple, and rapid measurements both in the laboratory and in the

V. Šašek et al. (eds.),
The Utilization of Bioremediation to Reduce Soil Contamination: Problems and Solutions, 377–380.

field. Photosynthetic herbicides cause changes in fast fluorescence induction; when a herbicide binds to the Q-B pocket of D1 protein, electron transfer is blocked and no re-oxidation of the secondary quinone acceptor can occur. Thus, fluorescence induction increases rapidly between minimal fluorescence yield and maximal fluorescence yield compared to untreated control. As a consequence, the area over the fluorescence curve and the maximal fluorescence are particularly reduced in the presence of a herbicide. Therefore, this technique could be used to rapidly determine the herbicide action site at the level of the D1 protein.

Based on fluorescence induction changes we previously developed some biosensors to measure herbicide concentration in water [2;3]. We built some biosensors based on the measurement of PSII fluorescence, to select the photosynthetic plant for phytoremediation of a specific herbicide. The biosensor methodology should avoid laborious field plant testing. A biosensor incorporates a biologically-active element, able to reveal selectively and reversibly the activity of a chemical compound in any type of sample. A biosensor consists of two elements: a biological component (biomediator) interfaced to a sensor, a chemical or physical signal transduction system. The transduction system measures changes in biomediator activity that reflect the effects of an environmental pollutant.

2. Materials and Methods

The following species were used: *Amaranthus retroflexus, Avena sativa, Pisum sativum, Poa annua, Spinacea oleracea, Solanum nigrum, Senecio vulgaris, Triticum durum, Vicia faba, Zea mays,* and the atrazine resistant mutants *Amaranthus retroflexus, Poa annua, Solanum nigrum* and *Senecio vulgaris* obtained from the wild type by plant selection after long-term field treatment with atrazine. Seedlings were grown in soil or in vermiculite in plastic containers and watered with Hoagland's solution. Greenhouse temperature was maintained at 25 °C with a growth irradiance of 450 µmol photons per m^2 per s. Two four-week-old plants were used.

3. Results and Discussion

Two sets of analyses are necessary for the determination of herbicide resistance: fluorescence analysis of intact plants and of thylakoids both treated with herbicide. Two devices were built for this purpose.

The first device is composed of several vials in series, with plant petioles kept in bubbled solutions containing the herbicide under light (Fig. 1). It uses a sensor that provides excited light and detects the fluorescent signal emitted by plants. An optic fiber coil connects the sensor to a device that measures the signal. A computer then processes the data. Each part is linked to a timer that activates the fluorescent sensor at established time intervals. The optical fibre moves along a mechanical rail made of 22 switches and the fluorescence is measured automatically. The measurement is recorded after 10 min of dark adaptation, every 20 min for each plant.

The F_v/F_m is affected by herbicide plant treatment *in vivo* and we determined the following I_{50} for the tested plants kept in herbicide solution under air current and light for 10 h. We can see from the biosensor data (Table 1) that resistant mutants of sensitive plants can easily be selected. Data represent the mean of four experiments.

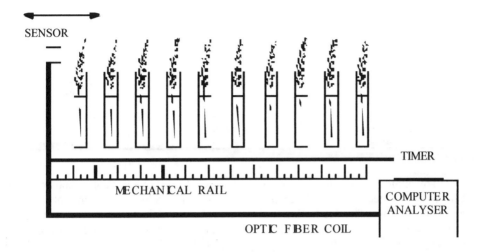

Figure 1. Schematic illustration of the sensor system

The second biosensor uses a sensor that provides excited light and detects the fluorescent signal emitted by a biomediator that is set in a chamber where herbicide solution is allowed to flow. As with the previously described biosensor, an optic fiber coil connects the sensor to a device that measures the signal, and a computer processes the data. Each part is linked to a timer that activates the fluorescent sensor at established time intervals [2, 3].

TABLE 1. Herbicide concentrations resulting in a 50% inhibition of plant activity (I 50 values) expressed as Fv/Fm fluorescence ratio.

Plant	I_{50} (μM)
Avena sativa	2.4 ± 0.06
Amaranthus retroflexus.	2.9 ± 0.07
Amaranthus retroflexus. mutant	> 4
Pisum sativum	2.7 ± 0.05
Poa annua	3.0 ± 0.10
Poa annua mutant	> 4
Solanum nigrum	2.0 ± 0.10
Solanum nigrum. mutant	> 4
Triticum durum	2.1 ± 0.08
Vicia faba	1.9 ± 0.08

380

TABLE 2. Biomediator: thylakoids from *Amaranthus retroflexus*; herbicide concentration: 10^{-6} M, herbicide flow time 20 min.Ddata are an average of 3 analyses, SE was about 16%.

Herbicide	Percentage of fluorescence area reduction in the presence of herbicide	
	Wild type	Mutant
Diuron	99	96
Bromoxynil	97	95
Dinoseb	99	93
Atrazine	96	15
Simazin	95	14
Ametryn	99	21
Desmetryn	92	10

Thylakoids were isolated from the mutants reported above in Table 1 and 50 µg of Chl were trapped in the biosensor flow cell. The herbicide solution at various concentrations was allowed to flow and the area above the fluorescence induction curve was measured. This experiment reveals the binding of herbicides to the thylakoids. In the case of resistance at the level of D1 protein no herbicide binding can occur and the fluorescence activity is not affected. In the case of resistance at the level of the content of the enzymic degradative system, herbicide can normally bind thylakoids and its presence reduces the induction curve area in a dose-dependent manner.

Among the tested plants a reduction of the fluorescence induction area was not observed for *Poa annua* mutant but was observed for *Amaranthus retroflexus* mutant (Table 2).

In conclusion, the use of two biosensors allows the choice of photosynthetic plants suitable for phytoremediation purpose, and avoids laborious field plant testing.

References

1. Mattoo, A., Giardi, M.T., Raskind, A. and Edelman, M. (1999). Dynamic metabolism of photosystem II reaction center proteins and pigments. a review. *Physiol. Plant.* **107**, 454-461.
2. Koblizek, M., Masojidek, J., Komenda, J., Kucera, T., Pilloton, R., Mattoo, A. and Giardi, M.T. (1998). An ultrasensitive PSII-based biosensor for monitoring a class of photosynthetic herbicides. *Biotechnol. Bioeng.* **60**, 664-669
3. Giardi, M.T., Esposito, D., Leonardi, C., Mattoo, A., Margonelli, A. and Angelini, G. Sistema portatile per il monitoraggio ambientale selettivo di erbicidi inquinanti basato sulla applicazione di un biosesnore da Fotosistema II (PSII). Italian Patent, 112, 2000

DEGRADATION OF POLYCHLORINATED BIPHENYLS (PCBs) IN CONTAMINATED SOIL WITH HORSREDISH PEROXIDASE AND PEROXIDASE FROM WHITE REDISH *RAPHANUS SATIVUS*

G. KÖLLER AND M. MÖDER

Analytical Department, Center for Environmental Research Ltd., PermoserStr.15, D-04318 Leipzig, Germany

Abstract

White-rot fungi have been found to be useful organisms for the decontamination of toxic materials like polychlorinated biphenyls (PCBs). These microorganisms use certain peroxidases to degrade chlorinated as well as nonchlorinated biphenyl systems. Not only fungal peroxidases but also plant peroxidase are able to degrade PCBs. In our experiments we investigated horseradish peroxidase as well radish juice as peroxidase source for the degradation of PCBs in a highly polluted soil (12 ppm PCBs).

In both cases degradation of PCBs was observed. The PCB content decreased to 25 % of the initial concentration. Dichloro-, trichloro-, tetrachloro- and pentachloro-biphenyls were degraded to different low and medium concentrations, whereas the content of hexachloro-, and heptachlorobiphenyls were not decreased.

1. Introduction

The thermal and electric properties of PCBs have led to their widespread use for industrial flame retardands, hydraulic oils and dielectrics. Due to their high toxicity and persistence in the environment, much research is currently underway to develop methods for complete degradation of PCBs. White-rot fungi namely, *Phanerochaete chrysosporium* and *Trametes multicolor* were extensively investigated.

The ability of white-rot fungi to metabolize cross-linked aromatic polymeric substances is well known [1-3]. Further investigations have shown that these fungal systems can also degrade DDT [1], chlorinated phenols [2] chlorinated dibenzo-dioxines (PCDDs) and dibezofuranes (PCDFs) [3] as well as PCBs [4]. The white-rot fungus *P. chrysosporium* has been widely used in PCB degradation studies. Yadav *et al.* [5] reported the degradation of Arochlor mixtures, showing equal selectivity towards ortho-, meta- and para-substituted congeners.

Due to the similarities between fungal peroxidase systems and horseradish peroxidase (HRP) [6] we have suggested that HRP could be an appropriate model

V. Šašek et al. (eds.),
The Utilization of Bioremediation to Reduce Soil Contamination: Problems and Solutions, 381–384.
© 2003 *Kluwer Academic Publishers. Printed in the Netherlands.*

system for the degradation of PCBs. The HRP-H_2O_2-system is successfully used in the removal of chlorinated phenols from wastewater [7]. Dec *et al.* [8] reported the removal of phenols from wastewater with cut radish.

Horseradish peroxidase is one of the most studied peroxidases. The enzyme is endowed with similar properties and capabilities like LiP from *P. chrysosporium*. Like LiP, HRP is also involved in the metabolism of lignin. Both, LiP and HRP, have the same cofactor, protoporphin IX. The small differences in the oxidation potentials between these two enzymes are caused by differences in the protein structure [9]. Despite this, the enzymes are comparable in their mechanisms.

That is why we investigated the degradation of PCBs in a contaminated soil with commercially available horseradish peroxidase and juice from white radish *Raphatus sativus*.

2. Experimental

Two types of experiments were made. In both cases, soil contaminated with PCBs was used. One type of experiment, commercially available horseradish peroxidase and hydrogen peroxide were applied together to treat the PCB-contaminated soil. In the other type of experiment we used hydrogen peroxide and freshly prepared radish juice as peroxidase source instead the pure enzyme. Both the enzyme solution and the radish juice were added in different batches. Hydrogen peroxide was added thrice daily. Weekly one sample was taken to analyze the PCB content. The enzyme activity was monitored several times daily. The experiments were aborted after three weeks. The enzyme tests were performed according to Arnao [10]. The PCB were analyzed with a GCMS-Iontrap system (GCQ, Finnigan MAT, San Jose).

3. Results and Discussion

3.1. MONITORING THE ENZYME ACTIVITY

The behavior of peroxidase activity during the treatment with HRP in the first week is shown in Figure 1A. Owing to the adsorption of the solved enzyme to the soil, after 20 h the activity of solved enzyme decreased dramatically.

To maintain the enzyme activity, freshly prepared peroxidase solution was added every 24 h during the full period of the experiment. The adsorbed enzyme might be nevertheless active but is not assayed in the enzyme test.

Due to the PCB immobilized on the humic matter of the soil, only solved enzyme can attack these substances efficiently. Therefore, it is important to monitor and maintain the enzyme activity of the solved enzyme through the entire run time.

Figure 1B shows the peroxidase activity in the experiments with radish juice. After adding the radish juice to the contaminated soil, the enzyme activity decreased to 50 % of the initial activity. This decrease is caused by soil adsorption of the enzyme.

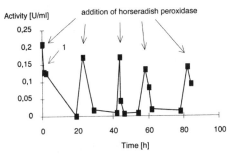

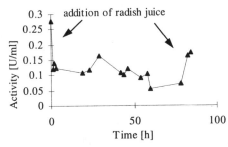

1 = addition of hydrogen peroxide

Figure 1. Enzyme activity of HRP (A) and radish juice (B) during the degradation experiments

Until 80 h only a small decrease of the enzyme activity were observed followed by a longer stable phase of lower activity. It seems that another process sets of the loss of peroxidase activity, owing to adsorption. Therefore, we suggest that peroxidase is delivered furthermore by shreds and cells of the radish root containing in the juice, because it was not filtered prior use. After 80 h the same amount of freshly prepared radish juice was added.

3.2. MONITORING THE PCB CONTENT

The PCB concentration in the soil during the experiments is shown in Figure 2. Both, the experiments with horseradish peroxidase as well as with radish juice, show a decrease in PCB content to 25 % of the initial PCB concentration.

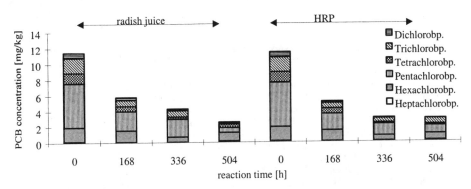

Figure 2. Content of PCBs in the soil and the corresponding profile of homologues

Figure 2 shows similar behavior, because the activities of the solved enzyme in both types of experiments, horseradish peroxidase and radish juice, were set equally. The initial PCB concentration of about 12 mg/kg in the soil was reduced to 3 mg/kg and 5 mg/kg respectively. Originally, PCB pattern ranged from dichlorobiphenyls to heptachlorobiphenyls. Tetra-, penta- and hexachlorobiphenyls are the major contaminants in the examined soil. Figure 2 shows also the degradation of each homologue group for the experiments with radish juice and HRP respectively. The lower chlorinated PCBs, like di-, tri- and tetrachlorobiphenyls were degraded, whereas the content of hexa- and heptachlorobiphenyls remained constant. Earlier experiments with single hexachlorobiphenyl (PCB 138) showed that higher chlorinated PCBs were not degraded by horseradish peroxidase. The soil contained 2mg/kg of hepta- and hexachlorobiphenyls. Therefore, it is not surprising, that 25 % of the initial amount remained after degradation within three weeks. The reduction of the PCB concen-tration owing to the degradation of especially tetra- and pentachlorobiphenyls plays an important role in the detoxification of contaminated soil, because these congeners have the highest iTE-factors.

It could be shown, that PCBs in complex soil matrices can be degraded with plant peroxidases. Owing to adsorption processes, the use of isolated horseradish peroxidase requires a continuous supply on fresh enzyme to maintain the activity of the solved enzyme. Freshly prepared radish juice is well suited and not expensive source for plant peroxidases. The loss of enzyme activity is compensated probably by the biosynthesis of peroxidases. However, no differences were observed in both experiments with respect to PCB degradation. Except for hepta- and hexachlorobiphenyls, the initial amount of all PCBs was reduced. The final PCB concentration was 3 mg/kg in the experiments with radish juice and 5 mg/kg in experiments with isolated HRP, respectively.

References

1. Bumpus, J.A., Tien, M., Wright, D., and Aust, S.D. (1985) Oxidation of persistent environmental pollutants by a white rot fungus, *Science* **228**, 1434-1436.
2. Joshi, D.K. and Gold, M.H. (1993) Degradation of 2,4,5-trichlorphenol by the lignin-degrading basidiomycete *Phanerochaete chrysosporium, Appl. Environ. Microbiol.* **59**, 1779-1785.
3. Valli, K., Brock, B. J., Joshi, D. K., and Gold, M. H. (1992) Degradation of 2,7-dichlorodibenzo-p-dioxin by the lignin degrading basidiomycete *Phanerochaete chrysosporium, J .Bacteriol.* **174**, 213 - 2137.
4. Thomas, D.R., Carswell, K. S., and Georgiou, G. (1992) Mineralization of biphenyl and PCBs by white rot fungus *Phanerochaete chrysosporium, Biotechnol. Bioeng.* **40**, 1395-1402.
5. Yadav, J.S., Quensen III, J. F., Tiedje, J. M., and Reddy, C.A. (1995) Degradation of polychlorinated biphenyl mixtures (Arochlors 1242, 1254, 1260) by the white rot fungus *Phanerochaete chrysosporium* as evidenced by congener-specific analysis, *Appl. Environ. Microbiol.* **61**, 2560-2565.
6. Smith, A.T., Sanders, S.A., Thornley R.N., Burke J.F., and Bray, R.R.. (1992) Characterisation of a haem active-site mutant of horseradish peroxidase, Phe41 ->Val, with altered reactivity towards hydrogen peroxide and reducing substrates, *Eur. J. Biochem.* **207**, 507-519.
7. Tatsumi, K., Wada, S., and Ichikawa, H. (1996) Removal of chlorophenols from wastewater by immobilized horseradish peroxidase, *Biotechnol. Bioeng.* **51**, 126-130.
8. Dec, J. and Bollag, J.-M. (1994) Use of plant material for the decontamination of water polluted with phenols, *Biotechnol. Bioeng.* **44**, 1132-1139.
9. Bumpus, J.A., Mileski, G., Brock, B., Ashbaugh, W., and Aust, S. D. (1991) Biological oxidation of organic compounds by enzymes from a white rot fungus, *Innovative Hazard. Waste Treat. Technol. Ser.* **3**, 47-54.
10. Arnao, M.B., Acosta, M., del Rio, J.A., Varón, R., and Garcia-Cánovas, F. (1990) A kinetic study on the suicide inactivation of peroxidase by hydrogen peroxide, *Biochim. Biophys. Acta* **1041**, 43-47.

PART V

BIOREMEDIATION MARKET
ANALYSIS

SUCCESSFUL CHALLENGES DURING THE DEVELOPMENT AND APPLICATION OF INNOVATIVE PROCESSES FOR THE BIOREMEDIATION OF SOILS

D. E. JERGER
Shaw Environmental & Infrastructure, Inc.
Technology Applications Group
312 Directors Drive, Knoxville, TN 37923, USA

1. Introduction

Remediation technology is selected for its ability to meet the desired cleanup goal. The conventional approach to soil cleanup has been incineration, solidification and excavation and disposal in a hazardous-waste landfill. Incineration has met public opposition and is costly. Solidification and excavation are not destructive treatment processes but simply immobilize the contamination or move it to a new location.

The limitations of conventional technologies were important in the development and application of innovative technologies. The use of innovative technologies increased for soil treatment at Superfund sites and underground storage tank sites [1]. One of the key reasons for choosing an innovative technology is to control environmental exposures and to reduce future risks in a cost efficient manner [2]. However, environmental statute and federal regulations have not encouraged the development of innovative technical solutions.

The successful demonstration of an innovative biological treatment process does not imply commercialization. Other barriers or challenges must be overcome which may impede the transition from full-scale demonstration to implementation and commercialization.

These common barriers or challenges include:

- the lack of convincing and reliable full-scale cost and performance data
- the lack of industry and regulatory involvement in process evaluation
- a lack of knowledge about the site-specific applicability of the process
- the lack of funding to conduct site-specific bench and pilot testing,
- fear of penalties for the failure of the technology to achieve site-specific treatment goals

EPA's Technology Innovation Office has attempted to reduce these barriers and the risks associated with the application of innovative technologies and accelerate their implementation [3]. One of the most important issues for the development of innovative technologies is to generate and disseminate quality, cost and performance information [4].

V. Šašek et al. (eds.),
The Utilization of Bioremediation to Reduce Soil Contamination: Problems and Solutions, 387–404.
© 2003 *Kluwer Academic Publishers. Printed in the Netherlands.*

A recent publication by EPA summarizes some key publications with respect to an in-depth understanding of what the barriers are, how they affect development and use of innovative technologies, and provide suggestions for stakeholders to coordinate initiatives to remove or reduce these barriers. The barriers were grouped into four categories including institutional, regulatory and legislative, technical, and economic and financial. Initiatives were also presented that have been undertaken to help mitigate the effects of these barriers. In addition, EPA has published The Innovative Treatment Technology Developer's Guide to Support Services, Fourth Edition which is available online at http://clu-in.org. and tabulated by EPA [2].

Public and regulatory acceptance of innovative technologies has been a secondary concern during the development and demonstration of innovative technologies. The principal focus is usually on performance. This approach may result in the development of technologies that are difficult to deploy because of regulatory obstacle or public concern. Stakeholder involvement during the development of a technology has proven successful due to the early identification of issues and concerns [5]. The basic approach was to identify people and organizations with a stake in the remediation process and hence in the development and demonstration of the technology. The stakeholders included public interest groups, regulators, technology users [5].

2. Marketing and Marketing Trends

The remediation market in the United States has been historically driven by federal, state and local regulation that has compelled the cleanup of contaminated sites. In recent years economic forces have begun to impact the regulatory forces, which has led to a change in the marketplace. Most experts agree that the U.S. market has matured and is expected to decline in upcoming years. The estimated U.S. market in 1998 ranged from $6–$8 billion of which 66–70 % was attributed to the federal sector, 25–30 % to private spending and less than 5–10 % to state government. The market for explosive-contaminated soils was estimated at $100 million per year in 1998 [6].

Innovative treatment technologies are gaining acceptance in site cleanup and are estimated to comprise 30–40 % of the market share. The US market for bioremediation technologies in 2000 was $400 to $500 million. The shift from a regulation-driven market to an economics-driven market was cited as a factor in slowing the U.S. market due to reduced enforcement of environmental laws and the advent of economic decision making has removed many of the incentives for cleanup [1]. The lack of Superfund reform has impacted the market. The US Congress has been discussing reform for years and the opinions are divided between sweeping reform and modest change, such as liability reduction. However, relaxation of environmental regulations through the adoption of risk-based decision-making activity has led to an increase in remediation activity. Risk-based remediation, where endpoints take into account potential human exposure and the future use of contaminated sites is being accepted in state programs and by EPA. In addition, the application of natural attenuation processes may be selected as the preferred remedial alternative when shown to attain site-specific remediation goals.

The international remediation market is overall at an earlier stage of the life cycle for remediation technologies. These markets may, however, be affected by a poor or nonexistent regulatory framework and the lack of investment capital. The unified European market was estimated at $2–$4 billion (US dollars) in 1998. Including the central and eastern European nations, Glass (2000) [6] estimated an annual market growth rate ranging from 15–20 %. Information on contaminated sites in 30–40 European nations has been assembled by the European Top Center on Soil of the European Environment Agency. These publications contain legislative and regulatory framework, government programs, cleanup methods and site-remediation costs. The information covers all nations of the European Union, the European Free Trade Association, and the Baltic and Central and Eastern European nations [6].

3. Technology Development

Technology development is a product of experience to define and overcome the limitations of existing conventional technologies, establish better performance and cost information and expand the process application to new contamination problems. Technology development involves the initial application during which acceptance increases due to successful application and communication, establishing design practices which is typically the rapid growth stage, and the mature phase in which the focus shifts from benefits to limitations and the technology is vulnerable to replacement. The use of the technology may also increase if these process limitations are overcome or the technology becomes applicable to a new contaminant. The use of the technology can also increase, based on showing improvements in effectiveness or a decrease in costs compared to proven technologies [3]. The acceptance of the technology is based in part on the accessibility of performance documentation that can be used to evaluate the benefits and limitations. Some frequently encountered problems are incomplete technology reports, lack of a critical scientific evaluation, lack of reliable cost data, and proprietary information.

The remediation market is also fragmented by client type and site type. The market categories include both private and public sectors of which both sectors offer a wide variety of clients, sites and responsible agencies. Consulting companies also represent many clients. Due to the variety of site-specific parameters significant technical expertise is needed for a site remediation; therefore it is difficult for technology developers to present a turnkey solution. Therefore technology vendors must either diversify or sell directly to consultants.

The EPA Site Program is the leading program in the U.S. for funding the evaluation, development and implementation of innovative treatment technologies. The program was established by OSWER and ORD in response to SARA [2]. The EPA SITE Demonstration Program evaluates and verifies cost and performance to provide reliable information. The EPA SITE Emerging Technology Program, which has been discontinued provided research funding at bench and pilot level. The EPA technology transfer program disseminates technical information including engineering performance and cost data on innovative technologies to remove impediments.

The development of biological treatment technologies has been slow due primarily to a lack of field experience for larger *ex situ* treatment processes. Some technical problems unique to bioremediation include the difficulties encountered in simulating field conditions in the laboratory. Technology developers should supply more information on process science, understand the process variables and field scale-up issues, and should provide detailed results on economics, process implementation and operation. The technology appliers generally need to minimize risk, maximize revenues and margins, capital and operating costs, and want to choose the most appropriate technology in terms of cost performance and reliability to meet the desired treatment goals. The technology appliers do not want cost overruns, poor performance, and scale-up issues. In other words the developers should try to better understand how end users select and apply the technologies. These issues are also influenced by other stakeholders such as the public and regulators.

Environmental remediation business is challenging and very competitive where added value can be very important. A successful situation includes revenue, profit margins and a market advantage for continued business development. Innovative technologies provide a means to add value to the customer if employed early in the process. There must be a good understanding of the process for site application. The cost and process information must be made available by the developers to the application firms. Conventional technologies have extensive design databases that allow estimate of performance, time and cost. Engineers responsible for technology application must have comparable information for the selection of innovative technologies.

Key factors which are necessary for the continued development of innovative processes for bioremediation of contaminated soils include an:

- increase in awareness of successfully completed projects
- acceptance by regulators, customers and environmental consultants
- improvement in process operation and performance
- application to an increased range of contaminants
- improvement in process economics
- and a favorable regulatory environment [7].

Bioremediation of soils is a scientifically intensive process that must be tailored to a site-specific application. The initial costs for site assessments, characterization and treatability testing may be higher in comparison to conventional approaches. Extensive monitoring is typically necessary during process operation to obtain the necessary data to evaluate the impact of operating conditions on process performance. Microbiological monitoring may be necessary in addition to the standard physico-chemical monitoring associated with nonbiological processes. In addition, projects in the U.S. must meet all of the requirements of the environmental regulations applicable to the site and contaminants. The regulations may include RCRA, CERCLA, as well as applicable state and local issues.

4. Regulation and Permitting

Laws and regulations drive the supply and demand for cleanup technologies including bioremediation. In the US several environmental statutes as well as the regulations and technologies are being considered as the concept of risk based cleanup standards gain prominence on a site specific basis.

CERCLA was designed to remedy mistakes made in past hazardous waste management practices while RCRA focuses on proper management practices in the present and the future. A CERCLA response is triggered by the release or substantial threat of release of a dangerous substance. CERCLA authorizes removal and remedial actions. The CERCLA (Superfund) process consists of several steps resulting in a feasibility study (FS) that identifies a preferred remedy from a list of remedies in a record of decision (ROD). The cleanup is conducted under the remedial design and action stage. Under the Superfund Amendments and Reauthorization Act (SARA) CERCLA response actions must attain all applicable or relevant and appropriate requirements (ARAR's) of other federal laws and more stringent State environmental laws. ARARs can raise the threshold for cleanup to exclude bioremediation in favor of conventional technologies. Similar to the RCRA program there is a predisposition to familiar technologies such as thermal treatment and stabilization. In addition, the focus of the regulatory agencies seems to be on contaminant mass reduction versus risk-based criteria such as contaminant mobility reduction. US EPA issued technology versus risk-based standards after comments from Congress. These standards required a method of treatment or a specific concentration level for the target contaminant in the waste, which were based on the best-demonstrated technology or BDAT. "Environmentally acceptable endpoints" is a concept that has been proposed to increase the number of cleanups at a substantial cost savings [8]. Development of site-specific risk-based approach and/or generic remediation goals using reasonable maximum exposure limits are approaches that would benefit the application of innovative technologies such as bioremediation.

5. Financing

Innovative markets attract public and private investment to support basic research, technology growth and development, and initial commercialization. The most critical stage for funding has been identified as the stage between the R&D phase and successful commercialization. Investors typically refer to this phase as the "Valley of Death" because so many startup firms fail during this phase (Fig. 1) [9].

The "Valley of Death" shows the relationship between technology developers and the relative availability of investment capital as the idea is proven and graduates to bench scale testing. Developmental funds can usually be found (i.e. SBIR, EPA) in the US. Technical skills are the most important during early stages of development. As the technology further evolves, financing needs to grow during the next stages of development but financing becomes more difficult. The inability of a private inventor to get funding has been identified as the famous valley of death that lasts until revenue or positive cash flow is generated. Leadership skills must change from a technical to market-oriented business approach [1].

or positive cash flow is generated. Leadership skills must change from a technical to market-oriented business approach [1].

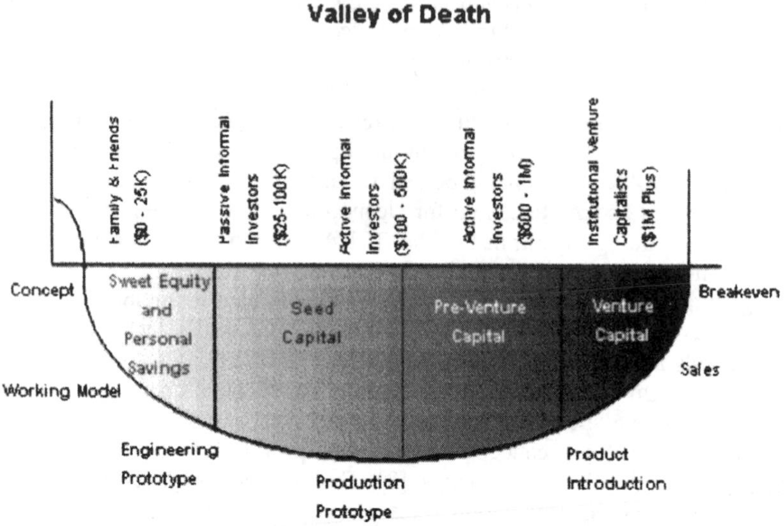

Valley of Death

Figure 1. The "Valley of Death" stage between development of a new concept and the successful commercialization of the concept. Adapted from Reference 9.

6. Innovative Technology Development for Explosive-Contaminated Soils: Case Histories

6.1 PROCESS APPLICATIONS

The biological treatment processes typically utilized for the *ex situ* soil treatment are land treatment, biopile, composting, fungal treatment and bioslurry treatment. The typical soil contaminant classes for *ex situ* soil treatment include petroleum hydrocarbons, polycyclic aromatic hydrocarbons, PCBs, pesticides, and explosives. The technologies which have been developed for explosive-contaminated soils include windrow composting, land treatment, white-rot fungal process, and bioslurry treatment. The measures of success for the development of these processes include technical performance, commercial characteristics, and public and regulatory acceptance.

The bases for technical performance criteria includes:

- risk reduction
- operating ranges and sensitivity
- ease of implementation
- maintenance
- predictability
- and any emissions or process residuals.

The commercial characteristics include:

- capital costs
- operating costs
- copyright and patent restrictions
- profitability
- and accessibility.

Public and regulatory acceptance includes:

- potential community disruption
- site disruption
- safety
- regulatory challenges
- future land use.

The occurrence of explosive-contaminated soils has presented an environmental challenge at numerous military installations. The major explosives of concern include TNT, DNT, RDX and HMX. Various biotreatment technologies have been developed and demonstrated for the biodegradation of these contaminants [10]. The biological treatment processes currently employed in the field have been developed and commercialized over the past 10–15 years and have met the general criteria for successful application in the field, i.e. implementability, performance and cost. These processes include windrow composting, Grace Daramend land treatment and the Simplot Anaerobic Biological Remediation (SABRE) slurry-phase treatment process. Windrow composting was the only process developed specifically for explosive-contaminated soils. The commercial applications for these technologies have been discussed in a previous publication [10]. In addition to these processes, other technologies have been or are under development for explosives-contaminated soils either as the primary contaminant or a secondary contaminant.

Successful field application of these processes typically requires a project management plan and a detailed work plan. These plans are typically required by the client and by the federal, state and local regulatory agencies prior to the initiation of work. Site-specific cleanup goals are developed for each project based on risk-based action levels for the constituents of concern. These levels integrate current EPA toxicity and carcinogenicity values with health-protective exposure assumptions to evaluate chemical concentrations in environmental media. The levels are derived to reflect the potential risk from exposure to the constituents of concern based on specific land use.

6.2. REGULATORY FRAMEWORK

In the U.S. RCRA regulations apply primarily to active sites and hazardous wastes disposed after November 1980. Soils contaminated with explosives can be considered hazardous by RCRA guidelines under two conditions: the characteristics of the material (i.e., ignitable, corrosive, reactive, or toxic) or the specific process that generated the material. Once the contaminated soil is classified as a hazardous material, RCRA has established minimum guidelines for the management of the material. The guidelines govern the procedures for all facilities that treat, store or dispose of hazardous wastes, including provisions for groundwater monitoring; corrective actions programs; record keeping; and the design, installation and operation of the treatment systems. In general for the onsite biological treatment of explosive-contaminated soils the guidelines will include containment liners and/or buildings for the treatment process, sampling and analytical requirements, and the ultimate disposition of the treated materials.

The Department of Defense (DOD) in the U.S. is the primary purchaser of services for the cleanup of explosive-contaminated soils. The DOD was responsible for the installations at which explosives were manufactured, and munitions were loaded, stored and tested. The investigation and cleanup of DOD sites is governed by CERCLA as amended by SARA, RCRA and state and local environmental laws and regulations. While there may be significant differences among the regulatory requirements that apply to the investigation and cleanup processes at DOD installations, the sequence of events required in each case can be generalized to consist of four major components: investigation, interim action, design and cleanup.

6.3. MARKET

The U.S. Army Environmental Center has identified at least 50 installations that have explosive-contaminated soils and are in various stages of remediation [11]. These have been segregated into three categories: Cleanup Completed; Cleanup in Progress and Sites with Known or Suspected Contamination. Incineration was the technology used at the installations to treat over 320 000 yards of soil where the cleanup is complete between 1990 and 1998. Windrow composting was the process used to treat over 15 000 yards of soil at the Umatilla Depot Army. The project was initiated in 1992 and completed in 1996. Windrow composting is the primary technology employed at sites where cleanup is in progress. These sites represent between 600 000 and 1 000 000 yards of contaminated soil. The majority of these facilities have completed the Remedial Investigation/Feasibility Study (RI/FS) process, the Engineering Evaluation/Cost Analysis (EE/CA) and the Remedial Design (RD) Stages and are involved in the Remedial Action for Cleanup Stage.

In Germany many explosive-manufacturing facilities were demolished at the end of World War II and used for industry and housing. These facilities produced approximately 20 000 tons of explosives per month during the peak production period. Limited site characterization has been completed at a few undeveloped facilities. A technology guideline handbook is being prepared by the German government to provide a basis for the use of biotreatment processes for cleanup of contaminated soils at these sites (12). The author also indicated that some site characterization has been done in the UK, Canada, and Australia but little cleanup is in progress. Regarding other countries the author states "In the rest of the world the extent of contamination with explosives is either undetermined or not available to the public. Anecdotal information suggests that the scope of the problem is significant" [12].

6.4. INNOVATIVE TECHNOLOGY DEVELOPMENT CASE HISTORIES

In an effort to better understand the challenges faced by developers of these technologies with respect to the technical, financial and regulatory challenges the questions below were submitted to key individuals involved in the respective process development and implementation.

1. Who invented the process and what was their involvement in the developmental stages?
2. What soil contaminants were originally targeted for treatment? Was the list of contaminants expanded and why?
3. What was the funding history of the R&D and commercialization of the process? What types of public and/or private funding were applied for and awarded and not awarded?
4. What types of expertise do (did) the people have who were involved in various stages of R&D and commercialization? Was the right expertise involved at the right time? Was a critical scientific evaluation conducted? Were complete technology reports issued?
5. What were the perceptions of the technology developers in terms of what was needed for development and application and was this perception realistic?
6. In terms of marketing the technology, was marketing started during the R&D phase to assist in further funding of R&D or was a critical scientific evaluation completed prior to marketing?
7. Does the current market match the one that was originally conceived during technology R&D?
8. Did the technology developers participate in any programs such as the EPA SITE program and was the participation beneficial?
9. How was the technology marketed and was this successful? Was the market target private or public sector?
10. What impact have regulatory standards had on technology development?
11. How many demonstration projects have been completed? What were the challenges in performing the field demonstration? Was cost information developed from the demonstration? How did this compare with previous cost data?
12. Is this process proprietary and/or patented? How was the proprietary nature of the technology maintained during development? Do you feel this hampered development?
13. What commercial projects have been completed? Is further commercial application easier or is every project as challenging as the first and why?

The information received in response to these questions was summarized and is presented below for each technology.

6.5. TECHNOLOGY DEVELOPER CASE HISTORIES

6.5.1. *U.S. Army Environmental Center - Windrow composting*

Composting of explosive-contaminated soils began in 1980. Three composting technologies including aerated static pile, agitated in vessel and windrow composting, were screened at the bench scale to determine process efficiency and economics. The initial testing was conducted by the U.S. Army Environmental Center (USAEC), Environmental Technology Division, Aberdeen, MD. Further testing at the field scale indicated that aerated static pile compost could not achieve the desired treatment goals. Mechanically agitated in vessel composting was able to achieve the desired treatment goals but the process was too costly. The windrow composting process was able to achieve the desired treatment goals at a cost comparable to the aerated static pile process.

Prior to 1980 some laboratory testing was completed with 2,4,6, trinitrotoluene (TNT) and hexahydro-1,3,5,-trinitro-1,3,5-triazine (RDX) to determine if these compounds were biodegradable. When compost testing began, the list of contaminants included all of the conventional explosives used by the Department of Defense including, TNT, RDX, 1,3,5,7-tetranitro-1,3,5,7-tetrazacyclooctane (HMX), 2,6-dinitrotoluene and 2,6-dinitrotoluene (DNT), 2,4,6-tetranitro-N-methylamine (Tetryl), and 1,3,5-trinitrobenzene (TNB) because the vast majority of sites contained most of these compounds. There were fewer than 5 installations that had single explosive contaminants. The U.S. Army provided all funding for the development and testing of composting of explosive-contaminated soils. Commercialization began with the cleanup of explosive contaminated soils at Umatilla Depot Army in 1992. This was the first explosive-contaminated soil site to be remediated using biological treatment. The process is not patented and available to the public.

Numerous professional disciplines and expertise were used throughout the development, testing and commercialization of the compost process. A critical scientific and engineering evaluation was conducted throughout the development of the process producing numerous technical reports. Engineering design and cost analyses were conducted throughout the process development stages. Off-the-shelf equipment was utilized when applicable.

The conceptual application of the composting process was:

- to develop a treatment process that would require minimal soil preparation and handling, treat any soil type with a wide range of moisture contents
- to treat soils with an explosives concentration up to 10 % (safety constraints dictate that the soil be blended in place to less than 10 % explosive concentrations before transport to the treatment area)
- to use off the shelf equipment
- to operate using a wide range of compost amendments found locally
- to require minimal on-site personnel and utilities
- to be accepted by public and regulators
- not to produce any secondary waste streams.

The development of the process met these expectations. The soil is typically screened to remove debris and rocks (over 2 inches) to prevent damage to the windrow compost turner. Process operation consumes water therefore an adequate water supply is required. Process operation during a 12-month period in a cold climate requires a treatment building.

Incineration was the only commercially available process for treatment of explosive-contaminated soils in the early 1980s. The biggest obstacle to obtain resources for the initial development of this process was to convince the funding agencies that composting would be much cheaper than incineration. Based on limited available funding a stepwise approach was implemented that built upon each successful demonstration. The success stories were presented at technical conferences and symposia that in turn generated more interest in the process. The successful application of composting at Umatilla Depot Army provided full-scale process economics that showed a 50 % reduction in costs compared to incineration.

The technology is currently applied at numerous sites, probably more sites than originally anticipated. Issues with the production of dioxins and furans by incineration have resulted in a decline in the use of this process due to negative perception by both the regulatory agencies and the public.

The technology has been successfully marketed primarily by verbal communication. USAEC personnel have spent much time discussing the process with regulators, construction and operation personnel, and domestic and foreign contractors assisting in planning, startup and operation. USAEC personnel have also visited most of the composting operations to assist in bench and pilot tests, and commercial operation.

Regulatory standards have had little impact on the development and application of explosive-contaminated soils windrow composting. The treatment standards for each site are generally established using risk-based criteria. These cleanup goals cover a wide range of explosive contaminant concentrations in soils.

The composting process is operated until these soil cleanup goals are achieved.

Approximately 15–20 project demonstrations have been completed since the onset of process development. Regulatory involvement was more pronounced in the early process developmental stages however as more successful projects were completed their involvement lessened. The process was adaptable to a wide variety of conditions due to its simplicity and cost information was always generated during the demonstration. During the latter stages of development, demonstrations were conducted to assess a variety of amendment mixtures to obtain the most cost-effective amendment with local availability. Process costs have also been reduced over time, as contractors have become more efficient at implementation and operation. In addition competition in the marketplace has greatly influenced treatment costs.

The windrow composting process is not a proprietary process and the information from each demonstration had been readily available to the public. The USAEC believes that if the technology was proprietary a much slower rate of development would have occurred (Wayne Sisk, personal communication).

Soils from four DOD facilities have been treated using windrow composting process while seven DOD facilities are currently treating explosive-contaminated soils. Windrow composting is in the planning stages at four installations. Two private sector sites have used windrow composting to treat contaminated soils. Success builds upon success as information from each application is passed onto the remediation and regulatory communities. In addition the process enjoys favor with the public and the regulatory personnel because it is a "green' technology.

6.5.2. AstraZeneca - Xenorem Process

The Xenorem process was invented by a research team of scientists and engineers at the Sheridan Park Environmental Laboratory of Astra Zeneca in Mississauga Ontario. The original targeted compounds were DDT, DDD and DDE. The list of chlorinated pesticides was expanded as the pesticide-contaminated soils typically contained many different compounds. The technology has been shown to be effective for treating explosive-contaminated soils. All of the funding for development and demonstration was obtained internally. Each phase of the program was funded separately by the organization that had site-specific responsibilities. External funding was never obtained. Multi-disciplinary teams were involved throughout the process development and commercialization. The research team that performed the actual work consisted of chemists, biologists, microbiologists and engineers. The team was accountable to a business group that consisted of personnel with business, law and engineering backgrounds. The research team was directed by a small group of biologists, chemists and chemical engineers.

The team approach assured that the science and technology were usable and economical for process application. The team composition varied as the process moved from the research and development stage towards commercialization. The scientific validation of the process during the early stages of development was very important. Complete technical reports were prepared and submitted to the US EPA and state agencies since the actual application of the process at former SMC sites needed final approval by these agencies.

Marketing of the technology began early in the overall program. The actual market has not changed since the early stages of the program because the research group was focussed on solving real environmental problems in a cost-effective manner and under direction of market-oriented personnel.

Federal and state agencies provided program oversight during the development of the technology. The technology was demonstrated at the commercial scale with this oversight prior to permission to proceed with actual site cleanup.

The technology is patented and used by others under a license. The technology was licensed to a small number of major environmental companies within the U.S. Outside of the U.S. the technology is licensed to any qualified organization. During process development existing provisions of EPA confidential policy and procedure protected information. Disclosure to non-agency parties was protected by secrecy agreements. The developer did not feel that the proprietary nature of the process hampered commercialization.

The rigorous risk-based cleanup standards for chlorinated pesticides, sampling and analytical QA/QC and a valid statistical sampling plan caused the developmental process to require more time and to cost more than originally planned.

6.5.3. Gas Technology Institute (GTI) The Chemical/Biotreatment (Chem-Bio) Technology

The Chem-Bio technology was invented by scientists and engineers at GTI, formerly the Institute of Gas Technology. The process involves either chemical treatment followed by biological treatment, biological treatment followed by chemical treatment or concurrent chemical–biological treatment. Two patents have been granted (U.S. Patent Nos. 5,610,065 and 5,955,350).

The process was originally developed for the treatment of PAH-contaminated soils from former MGP sites and later expanded to treat PCB and explosive-contaminated soils.

The funding for development of the process was obtained primarily from GTI's Sustaining Membership Program(SMP) and the Gas Research Institute(GRI). Additional funding was obtained from various gas industries, DOD, DOE and EPA.

GTI determined that a diverse group of experts was needed for successful process development. The professional disciplines involved during the developmental stages included microbiologists, soil chemists, environmental engineers, geotechnical engineers, and chemical engineers. An internal review committee conducted a critical review of the process during the developmental stages and several internal technology reports were issued.

The marketing of the technology was developed during the R&D stage in an effort to obtain ongoing funding. The current market trend is different than the originally planned because the application has expanded from *ex situ* to both *in situ* and *ex situ* processes. GTI is also investigating the use of risk-based endpoints in place of or in addition to traditional cleanup goals. In addition GTI has made environmentally acceptable endpoints a major thrust in both R&D and field applications. GTI has an internal marketing and commercialization group in addition to SMP program guidance. The marketing efforts were coordinated among these groups. The recent merger of IGT and GRI to form GTI will provide a greater in-house market focus and capability.

Five pilot scale projects and more than 40 bench-scale projects have been completed. Cost information was developed from the pilot-scale tests and was within 20 % of the projected costs. A full-scale commercial project has not been completed but is under development.

The process is a proprietary, patented process that is licensed on a site-specific basis. The technology development was not impacted by the use of invention disclosures.

6.5.4. *Grace DARAMEND Process*

The DARAMEND Process was developed by the current director of Grace Bioremediation Technologies (GBT) who participated in all the stages of process development.

The early stages of the laboratory process development were funded by the Canadian government between 1988 and 1992. The objective of the research was to develop an effective, reliable, low-cost technology for treatment of soil contaminated with pentachlorophenol and polycyclic aromatic hydrocarbons from wood-preserving sites. The list of treatable contaminants was expanded through an ongoing R&D program to respond to market conditions including chlorinated pesticides and explosives.

The early research was funded by the Canadian government and Grace. GBT was unsuccessful in obtaining funding through various R&D programs supported by the DOD including SERDP, ESTCP and FCT. GBT also applied for participation in the BAA contracting program conducted by the US Navy which led to a contract vehicle for a project at a Navy site but no funding was supplied by the Navy. The main benefit was to provide the Navy with a contract mechanism with which to access the DARAMEND technology.

The technology developer holds degrees in soil science and microbiology, a combination of physical and biological training suited for the development of the process. During the commercialization stages individuals with expertise in process engineering, finance, intellectual property, licensing, and sales and marketing were involved.

R&D and marketing were conducted concurrently after the process was developed to the level sufficient to support field demonstration. Performing commercial projects provided additional R&D funding and a greater understanding of the market to better focus additional R&D efforts. The technology was originally developed to treat soils at wood- treating sites. However, this market was limited and has since been surpassed by opportunities in the treatment of pesticides, phthalates and explosives.

The Grace DARAMEND Process was demonstrated at a wood-treating site under the US EPA SITE program in 1993. The developers indicated that the EPA SITE report provided an independent confirmation of the performance, cost and credibility of the technology. Successful field implementation with independent verification by EPA provided a level of confidence to prospective customers considering commercial application.

The technology was marketed through direct contact with federal and industrial clients, and by indirect marketing through engineering, consulting and remediation companies. Licensing of the technology for fixed facility and site-specific applications has also occurred. Bioremediation processes have historically been more acceptable to regulatory agencies and the public; therefore, approval for site-specific projects has been less cumbersome. However, delayed enforcement actions and the acceptance of nontreatment or institutional controls has presented increased competition. Also the Land Disposal Regulations (LDRs) and the related low treatment standards has made bioremediation of many compounds impractical.

The DARAMEND process has been demonstrated successfully at 6 sites. Some of the challenges for performing the field demonstrations included securing sufficient funding, and forming and maintaining an experienced, internal project team to stay within budget. Cost information developed following each demonstration project indicated that the DARAMEND process offered a substantial cost saving in comparison to alternative technologies. The developers feel this was a key element in the successful scale-up of the process.

The DARAMEND process is proprietary and protected by patents which has not impeded development. GBT indicates that the DARAMEND process has developed faster than any of the other innovative bioremediation technologies. A more substantial barrier in the federal marketplace has been competition against processes developed with internal funding.

The DARAMEND process has been used on over 20 sites commercially.

6.5.5. EarthFax Development Corporation – White-Rot Fungal Process

The application of white-rot fungi (WRF) for the degradation of hazardous compounds in soils was developed from work to decolorize and detoxicate Kraft process bleach effluents. The original investigators developed a process for the decolorization and detoxication of Kraft bleach effluents and other aqueous wastes containing chlorinated organic compounds. Several patents were granted that have been licensed but the process has not been applied at the commercial scale. EPA provided funds to study the application of the WRF for soil remediation at the laboratory scale in 1985. The USDA Forest Service Forest Products Laboratory undertook further development of the process in 1990. The EPA primarily funded process development during the 1990s.

The initial laboratory testing was done on low-molar-mass chlorinated organics that expanded to PCBs, chlorinated dioxins and furans, chlorinated pesticides and explosives such as TNT. The applied testing focussed primarily on the treatment of soils contaminated with wood- preserving chemicals such as PCP and creosote-range PAHs due to the work being done at the FPL. One of the initial challenges was the scale-up of the process from aqueous to soil systems. Many of the contaminants that were degradable in aqueous systems were more recalcitrant in soil systems due to bioavailability. Therefore the technology was initially marketed based on results from aqueous microcosms with very little process information from soil systems. The soil contaminants such as PCP and TNT were very biodegradable in soil systems while many of the other contaminants were more difficult. Many of these compounds such as the PAHs associated with MGP and creosote-contaminated soils can now be effectively treated using surfactants with the process.

The US EPA provided the primary funding for the development of the WRF technology. The DOD also contributed funding for laboratory testing and process development through a SBIR. Funding for laboratory research was also obtained through NIEHS and EPRI. Commercialization of the technology was attempted by three companies, Mycotech, Intech 180 and Earthfax Development. In addition to private funding these companies relied heavily on SBIR for process development and commercialization.

The initial research was conducted by biochemists. Earlier involvement of applied mycologists and engineers would have assisted in the development by providing technology for large-scale cultivation of the fungi, and construction and maintenance of biopiles. Process failures during the early scale-up phases of process development were due to the inability to produce large quantities of a viable inoculum or overheating of the biopiles. The companies participating in the early technology development and commercial application stages had little experience in the environmental market and may not have had a strong business plan outlining market areas and development costs. During the developmental stages the only technical report that was prepared was the EPA SITE report which described a land farming process versus the current biopile process. The final version of the report is under review.

The perception of the technology developers was that one successful field demonstration would be sufficient for commercial applications. In addition, the market was identified as any contaminant that had been successfully degraded in aqueous culture. The marketing was initiated during the R&D phase to obtain further funding. A critical scientific evaluation of the process was not completed prior to marketing the process/technology. A technical protocol was also not developed. The lack of understanding of the potential markets and process scale-up caused early failures in the field that created a negative perception of the technology.

The original market for the WRF technology included almost all of the organic contaminants that were successfully degraded using cultures grown under aqueous conditions in the laboratory. It appears that the current market is soils contaminated with chemicals such as dioxins, furans, and chlorinated pesticides that are difficult to treat using other biological processes. The technology was marketed to environmental consulting and remediation companies (somewhat successful), federal agencies such as DOD and EPA, and state regulatory agencies (unsuccessful), and industry (unsuccessful). The regulatory standards did not appear to have any impact on the technology development.

Participation in the EPA SITE program was not very beneficial in terms of further development of the technology.

7. Conclusions

In summary the processes presented have been developed and applied for the full-scale treatment of explosive contaminated soils and soils contaminated with other recalcitrant compounds. The developers themselves identified the challenges faced and overcome during the development and application of these technologies. Categories of barriers identified by EPA (2000) which included

- institutional – barriers imposed by entities that regulate, develop or select innovative treatment technologies for use
- regulatory and legislative – barriers imposed by government agencies
- technical – barriers associated with lack of cost and performance data
- economic and financial – barriers that reduced or eliminated financial incentives were compared to the challenges identified by the technology developers (Table 1). The institutional barrier appeared to be the greatest challenge. The regulatory barriers were site-specific as the developers performed site-specific demonstrations. The technical barrier was seen by the fungal technology developers as a key challenge during process development. Much of the funding for the technologies presented here was internal, therefore, the economic barrier was not frequently mentioned. A summary of the comments is provided.

TABLE 1. Barriers identified by technology developers

TECHNOLOGY	INSITUTIONAL	REGULATORY	TECHNICAL	ECONOMIC
Windrow Composting	X			
Xenorem		X		
Chemical/Biotreatment	X		X	
Daramend	X	X		X
White-rot fungi	X		X	X

Windrow composting is being applied at more sites than originally anticipated by the developers. The developer was the US Army and the process was developed to treat explosive-contaminated soils at DOD sites. The major barrier was institutional, that is trying to convince the US Army funding agency that windrow composing would be cheaper than incineration. Once the initial funding was in place any potential technical and economic barriers were overcome by successful case histories.

Regulatory standards were not a barrier since each site applied risk-based cleanup standards that were achieved by the process.

The Xenorem process was developed through internal funding to provide a less costly process in comparison to incineration for the treatment of pesticide-contaminated soils at sites owned by the parent company. Due to the rigorous risk-based cleanup standards for chlorinated pesticides the technology development took longer and was more costly than originally planned. The Xenorem process appears to be cost-competitive with other biological treatment processes for explosive-contaminated soils.

The Chem-Bio technology was developed for treatment of PAH-contaminated soils at former MGP sites and later expanded to treat PCB- and explosive-contaminated soils. A full-scale project has not been undertaken for explosive-contaminated soils, probably due to institutional and technical challenges.

The DARAMEND process was developed for pentachlorophenol- and PAH-contaminated soils at wood-preserving sites and later expanded to include explosives-contaminated soils. The challenges encountered during field demonstrations included securing sufficient funding and competition from windrow composting. The DARAMEND technology was successful in recent tests conducted by the U.S. Army at the Joliet Army Ammunition Plant [13].

The white-rot-fungal technology was initially developed to treat contaminants typically encountered at wood-preserving sites. One of the early technical challenges was scale-up of the process from aqueous to soil systems. Early funding was obtained through the EPA SITE program. The technology developers faced all of the barriers discussed. Although the technology was successful in the U.S. Army tests [13] the process has not been used at full scale for explosive-contaminated soils. There is much interest, however, in application of the white-rot-fungal technology for soil treatment in western and eastern European countries.

Acknowledgements

The author would like to thank the following individuals for providing information included in this paper. Wayne Sisk, USAEC, Aberdeen, MD, Windrow composting; Frank Peter, AstraZeneca, Wilmington, DE, Xenorem; Robert Paterak, Gas Technology Institute, Des Plaines, IL, Chemical/Biotreatment; Robert Ferguson, Grace Bioremediation, Philadelphia, PA, Daramend; and Richard Lamar, EarthFax Development Corporation, Logan, UT, White Rot Fungal Technology. The information provided by these technology developers made this paper possible.

References

1. National Research Council (1997) *Innovations in Groundwater and Soil Cleanup: From Concept to Commercialization.* National Academy Press, Washington, DC.
2. EPA (Environmental Protection Agency). (2000) An Analysis of Barriers to Innovative Treatment Technologies: Summary of Existing Studies and Current Initiatives, EPA 542-B-00-003, Cincinnati, OH
3. Ryan, M.J. and G. Ganapathi. (1997) Moving Innovative Technologies to the Field: An Engineering Perspective, in *Bioremediation of Surface and Subsurface Contamination,* R. Bajpai and M. Zappi (eds.), New York Academy of Science, New York, NY, Volume 829, pp 6-15.

404

4.   Mandalas, G.C., J. A. Christ, G.D. Hopkins, P.L. McCarty and M.N. Goltz (1998) Remediation Technology Transfer from Full-Scale Demonstration to Implementation: A Case Study of Trichlorethylene Bioremediation, in *Risk, Resource, and Regulatory Issues, Remediation of Chlorinated and Recalcitrant Compounds,*. G. Wickramanayake and R. Hinchee (eds.), Battelle Press, Columbus, OH, pp 235-240.

5.   Hund, G.(1998) Stakeholder Involvement-Free Consulting that Results in Enduring Decisions, in *Risk, Resource and Regulatory Issues, Remediation of Chlorinated and Recalcitrant Compounds.*, G. Wickramanayake and R. Hinchee (eds.), Battelle Press, Columbus, OH pp 211-216.

6.   Glass, D. (2000) International Remediation Markets: Perspectives and Trends, in *Risk, Regulatory and Monitoring Considerations, Remediation of Chlorinated and Recalcitrant Compounds,*. G. Wickramanayake, in A. Gavaskar, M.E. Kelley, and K. Nehring (eds.), Battelle Press, Columbus, OH, pp 33-40.

7.   Devine, K. (1995) U.S. Bioremediation Market: Yesterday, Today and Tomorrow, in *Applied Bioremediation of Petroleum Hydrocarbons,* R. Hinchee, J. Kittel, and J. Reisinger (eds.), Bioremediation 3(6) pp 53-59. Battelle Press, Columbus, OH.

8.   Devine, K. and P. LaGoy, (1998) Regulatory Issues Applied to Bioremediation as a Risk Reduction Technology, in *Bioremediation: Principles and Practices. Volume III Bioremediation Technologies.* S. Sikdar and R. Irvine (eds.), Technomic Publishing Co., Lancaster, PA, pp 195-222.

9.   SBA (Small Business Administration). Bridging the Valley of Death: Financing Technology for a Sustainable Future. Washington, DC.

10.   Jerger, D. and P. Woodhull (2000) Applications and Costs for the Biological Treatment of Explosives-Contaminated Soils in the U.S , in *Biodegradation of Nitroaromatic Compounds and Explosives,* J. Spain, J. Hughes, H. Knackmuss (eds.), Lewis Publishers, Boca Raton, FL pp 395-424.

11.   Broder, M.F. and R.F. Westmoreland (1998) An Estimate of Soils Contaminated with Secondary Explosives. Report No. SFIM-AEC-ET-CR-98002. U.S. Army Environmental Center, Aberdeen Proving Ground, MD.

12.   Spain, J. (2000) Introduction, in *Biodegradation of Nitroaromatic Compounds and Explosives,* J. Spain, J. Hughes, H. Knackmuss eds. Lewis Publishers, Boca Raton, FL . pp 1-5.

13.   PSC (Plexus Scientific Corporation) (1999) U.S. Army Environmental Center Biotechnology Demonstration" Final Report, SFIM-AEC-ET-CR-99012. U.S. Army Environmental Center, Aberdeen, MD.

CURRENT STATE OF BIOREMEDIATION IN THE CZECH REPUBLIC

V. MATĚJŮ
ENVISAN-GEM, Inc., Biotechnological Division,
Radiova 7, CZ-102 31 Prague 10, Czech Republic

1. Introduction

The research into bioremediation as well as the application of bioremediation methods in full-scale operations have a long time history in former Czechoslovakia and more recently in the Czech Republic. Whereas basic research into bioremediation and biodegradation is undertaken at universities and research institutes (Table 1), applied research is principally carried out by companies actively engaged in the bioremediation business.

Bioremediation is defined in many ways [1-3], but all definitions imply that it is a use of living organisms to eliminate hazardous compounds in the environment or in hazardous wastes. This definition means that the term „bioremediation" can comprise all living organisms. Nevertheless, the processes studied and used in the Czech Republic employ bacteria, white-rot fungi and plants only. Even if phytoremediation should be included among bioremediation methods, by definition, we omit this technique in the following discussion, mainly because of its presently very limited significance in the Czech context.

2. Brief Overview of Bioremediation Research in the Czech Republic

Research in bioremediation is very intensive in the Czech Republic at a number of institutions (Table 1). The key players are the Institute of Microbiology of the Czech Academy of Sciences, University of Chemical Technology in Prague, and Masaryk University in Brno.

2.1 INSTITUTE OF MICROBIOLOGY

Research at this institute is aimed at the possibility to use bacterial strains and white-rot fungi or their combination for bioremediation. The pollutants studied are polychlorinated biphenyls (PCBs), polyaromatic hydrocarbons (PAHs), azodyes and some other recalcitrant pesticides, herbicides, etc. Research of bacterial degradation of

V. Šašek et al. (eds.),
The Utilization of Bioremediation to Reduce Soil Contamination: Problems and Solutions, 405–412.
© 2003 *Kluwer Academic Publishers. Printed in the Netherlands.*

PAHs utilizes strains isolated from contaminated sites and model compounds such as benzo[a]pyrene and tries to enhance the rate and efficiency of the biodegradation.

TABLE 1. Research institutes and universities engaged in bioremediation research in the Czech Republic

Institution	Type of Research
University of Chemical Technology, Prague	Biodegradation and bioremediation of PCBs, PAHs, oil hydrocarbons, halogenated hydrocarbons, phenols Ecotoxicity of bioremediated materials
Institute of Microbiology, Czech Academy of Sciences, Prague	Biodegradation and bioremediation using bacteria and white-rot fungi Target pollutants: PCBs, PAHs, dyes, recalcitrant pollutants Ecotoxicity of bioremediated materials
University of Ostrava, Faculty of Science, Ostrava	Mutagenicity of bioremediated materials
Masaryk´s University, Faculty of Science, Brno	Biological dehalogenation, Biological catalysts for bioremediation
Charles University, Faculty of Science, Prague	Enzymes for Biodegradation
Institute of Organic Chemistry and Biochemistry, Czech Academy of Sciences, Prague	Phytoremediation of PCBs, PAHs, explosives Bioaccumulation of heavy metals
University of Pardubice, Dept. of Theory and Technology of Explosives, Pardubice	Phytoremediation of explosives

process via optimization of environmental conditions and by bioaugmentation using specialized bacteria isolated mostly from the contaminated site [4]. PCB degradation is studied using genetically modified bacteria in the Laboratory of Physiology of Microorganisms.

The Laboratory of Experimental Mycology carries out very intensive research on bioremediation using composting, white-rot fungi, bacteria and a two-step process employing white-rot fungi and bacteria. Pollutants of interest include PAHs, azodyes, PCBs and some other recalcitrant pollutants [5,6].

2.2 UNIVERSITY OF CHEMICAL TECHNOLOGY

At the Institute of Biochemistry and Microbiology, research on the bioremediation of PAHs in suspension proceed with various bacteria (*Pseudomonas putida* 8368, *Sphingomonas* sp.) and, in cooperation with the Laboratory of Experimental Mycology, also with white-rot fungi (*Irpex lacteus, Pleurotus ostreatus, Phanerochaete chrysosporium*). As model compounds, various PAHs and their mixture (phenanthrene, fluoranthene, anthracene) are used [7]. Another research activity deals with bacterial transformations of PCBs and the ecotoxicity and mutagenity of biotransformation products. At the Institute of Fermentation and Bioengineering, the research on bioremediation focusses on biodegradation pathways of acetone, acetonitril, and phenols using the bacteria *Comamonas acidovorans, Rhodococcus equi* and the fungi *Candida maltosa* and *Fusarium proliferatum,* sorbed to solid particles. The emphasis is put on the formation of an active biofilm [8]. The Institute of Water and Environmental Technology studies intensely the biodegradability of various compounds and relates this biodegradability with possible biological treatment [9].

2.3 MASARYK´S UNIVERSITY

The intensive research on the biodegradation of halogenated hydrocarbons and enzymatic systems involved in this process is carried out in the Laboratory of Structure and Dynamics of Biomolecules. Haloalkane dehydrogenases are the enzymes of interest due to their broad range of substrate specificity [10-12]. The main aim of the research is the modification of enzymes via mutation to increase their reaction rates [13].

3. Bioremediation in the field

3.1 HISTORY

Pollution by organic compounds is a very significant environmental problem in the Czech Republic. It is a consequence of intensive industrial activities without regard to the environment. The other source of pollution are military bases operated from 1968 to 1991 by the Soviet Army.

Soviet troops operated more than 80 military bases in the Czech Republic. Most of these were heavily polluted with oil hydrocarbons (like jet fuel, diesel fuel, gasoline and lubricants), various solvents (dichloro- and trichloroethylene, perchloroethylene, etc.) and other pollutants. The total amount of contaminated soil has been estimated to be in the range of 1.5 to 2.0 million tons [14]. The largest pollution was found at the Hradčany and Milovice bases (about 500000 to 600000 tons of contaminated soil at each site). The concentration of petroleum hydrocarbons in the contaminated soils ranges from 1000 to 20000 mg/kg. Occasionally it was found to be as high as 350000 mg/kg.

Contamination of groundwater is also widespread. For example, polluted groundwater area around the air force base in Milovice occupies 58 hectares. The layer of free-phase jet fuel above the water table is up to 2.6 m thick. At another site, Hradačany, the polluted groundwater area is smaller than in Milovice but the thickness of the layer of free-phase jet fuel reaches 6.0 m [15].

There exist approximately 250 industrial polluted sites in the Czech Republic. Some of these polluted sites experienced pollution for more than 100 years. The most abundant pollutants are petroleum hydrocarbons, PAHs, raw oil, halogenated hydrocarbons, phenols, coal tar, creosote and many more. Heavily polluted industrial areas are in some cases very large. For example, in Northern Moravia, the industrial agglomeration around Ostrava comprises hundreds of hectares of polluted sites.

Such huge pollution was neglected for a long time by the state authorities of the communist regime. The first cleanup projects started in the mid-eighties when drinking water sources in villages in the neighbourhood of Soviet air-force bases became polluted with jet fuel and were no longer usable leaving the whole villages without a drinking water source. In the late eighties, the first bioremediation company emerged on the remediation market in the former Czechoslovakia. After the „velvet revolution", the new government took responsibility for the decontamination of the environment and started funding remediation projects. The boom for remediation companies started in the nineties. During two years (1990-1992) six bioremediation companies emerged on the Czech market. Thereafter, the number of bioremediation companies steadily increased, reaching about 25 at the beginning of the year 2000.

In the late eighties and early nineties, all bioremediation companies used only a single technique. They employed landfarming with bio-augmentation and they treated only excavated soil polluted with oil hydrocarbons. These technologies were very simple, had low efficiency, needed large areas for treatment and displayed a number of other disadvantages. The general scheme was as follows: Excavated soil was dumped on a decontamination plateau as a 0.2 to 0.3 m thick layer, and a bacterial suspension was used for inoculation, nutrients were added and, during the treatment, the soil was turned, ploughed or mixed. This process offers little opportunity for control. During the years 1990 to 1992 some 50000 tons of soil contaminated with oil hydrocarbons were treated using bioremediation techniques.

Because of the large extent of polluted sites and limited funds for remediation the priorities are related to risks arising from a contaminated site and the use of low-input technologies is stressed. Biodegradation techniques belong to lower input technologies and are cheaper in comparison with other remediation techniques [16,17].

This success of bioremediation, the huge market for bioremediation in the Czech Republic, the state funding of cleanup projects and the nearly absent legislation to control the use and full-scale application of bioremediation technologies contributed to encouraging many companies (having initially nothing to do with remediation business at all) to start bioremediation operations in the Czech Republic during the early nineties. The quality of services offered by some companies was very low, at best.

In the course of the year 1993 to 1995 some new bioremediation companies appeared on the scene, characterized by very good scientific and technological support. These companies brought innovative technologies to the market and started application of bioremediation *in situ*. They used bioremediation for treatment of various organic pollutants different from oil hydrocarbons. They currently are the leaders in the bioremediation business in the Czech Republic. During these years, also, many of bioremediation companies disappeared from the market.

3.2 CURRENT SITUATION

3.1.1 Legislation

At present, in the Czech Republic there is nearly no regulatory legislation for full-scale application of bioremediation techniques. The only mandatory condition to be fulfilled is the obtaining an approval of hygienic safety of a bioremediation technology issued by the Institute of Public Health. This approval is issued in accordance with the general act of public health protection. The other legislation that could be applied to bioremediation is a regulation for manipulation, treatment and disposal of hazardous waste.

There is no official obligation to test the efficacy of a given bioremediation technology before its practical use in the Czech Republic. More interestingly, the leading investors of remediation projects (The Fund of National Property - the organization that funds the remediation projects of contaminated industrial sites - and the Ministry of the Environment, that funds remediation projects at former Soviet army military bases) are not interested in the quality, efficiency and reliability of the technologies that are used for treatment of contaminated sites. This is the reason for many failures in application of bioremediation in the Czech Republic in the late eighties and early nineties. In some cases, very peculiar approaches to bioremediation were adopted.

3.1.2 Former experience

The first bioremediation technology on the market in the former Czechoslovakia was Russian technology that used the bacterial strain *Pseudomonas putida* as a biodegradation-enhancing agent. This first bioremediation technology was on the market from the year 1988 until 1997, and it was used to treat excavated soil contaminated with oil hydrocarbons *ex situ*. This bioremediation technology had a number of positive results.

In the early nineties, an American bioremediation technology was introduced into the Czech Republic. This technology used a „confidential" biocatalyst that, according to company claims, supplied oxygen for the biological oxidation of organic pollutants in soil and groundwater, lowered bacterial biomass production, increased overall efficiency of the biodegradation process and enhanced the biodegradation rate. Because of failures of this bioremediation technology in several applications, a scientific evaluation of the process was funded by one unsatisfied client. It came out from the scientific evaluation made by three independent research institutes that the technology is absolutely inefficient and that the „confidential" biocatalyst was a freeze-dried dung imported from the United States.

Another American company started its bioremediation activity in former Czechoslovakia by supplying its bacterial preparation to a Czech company that had no specialist on bioremediation. This fact caused problems at the very beginning. The lack of qualified personnel at this company caused difficulties from time to time. Sometimes, as a result, the company had to withdraw from remediation.

More recently a French company tried to promote on the market in the Czech Republic a bioremediation technology that involved white-rot fungi for the biodegradation of PAHs, as claimed by the company. Because the company declared that every detail of the technology was confidential, pilot tests were ordered by the Institute of Public Health to verify the hygienic safety of the process. During the pilot tests it was found that white-rot fungi, even if inoculated into the treated soil, survived only a few days after seeding, without any significant change of the concentration of PAHs. The results of the tests proved that the inoculation of contaminated soils with white-rot fungi according to the protocol of the French company was inefficient for PAH biodegradation in soil.

A subsidiary of a German bioremediation company introduced its bioremediation technology on the Czech market in 1991. This technology was based on a very careful homogenization of soil to be treated, separation of stones and building debris and amendment of nutrients and selected bacterial strains produced in a bioreactor. Sometimes this technology employed lignocelullosic materials to improve soil structure. During treatment the soil was turned and if necessary water was supplied. The results of this technology were very good but the company withdrew from the market in the year 1998.

Czech companies started their activities in the bioremediation business in 1991; Some companies met with success, others with resounding failure. It took three to four years to introduce innovative bioremediation technologies, to improve efficiency and to start new applications of bioremediation as *in situ* treatment, soil flushing with biological polishing, enhancement of availability using surface active compounds etc.

3.1.3 Current status of bioremediation in the Czech Republic

As mentioned above, about 25 companies are engaged in the bioremediation business in the Czech Republic these days. Only two companies advertise bioremediation as their main activity. Only these two companies possess research and development divisions and apply innovative and efficient technologies. The rest of the companies started bioremediation as a minor business along their main activities in hydrogeology. These companies apply bioremediation technologies very often without well-trained staff, as part of their cleanup strategies based mainly on physical and physicochemical treatment techniques. It is worth noting that among foreign companies once active in the bioremediation business only one French company is still operating in the Czech Republic. The following discussion concerns the two Czech companies whose activities focus on bioremediation.

Technologies used. The use of on-site treatment of excavated soil is applied in various systems with forced aeration, mechanical turning etc. When very low target limits are to be achieved in the final stage a nonionic surfactant is added to enhance the bioavailability of pollutant molecules adsorbed to solid particles. This technology was successfully applied to the treatment of soils contaminated with the following pollutants: jet fuel, heating oil, diesel oil, phenols, PAHs, butylacetate, perchloroethylene, trichloroethylene, dichloroethylenes and various sludges with high concentration of oil hydrocarbons and raw oil. Techniques that are commonly used include bioventing and biosparging in combination with vapor extraction and treatment of extracted soil air in biofilters or by adsorption on activated carbon. Many variations of these methods have been employed.

One company uses specially selected bacteria with enhanced production of biosurfactants for *in situ* treatment of sites contaminated with creosote or coal tar. In this case it was demonstrated that most of the present PAHs is not completely degraded but is only biotransformed. Analytical determination showed important decreases of total PAH concentration but subsequent study of toxicity and mutagenicity of treated excavated soils shows an increase in mutagenicity of treated soils compared with the untreated one [18,19]. This case shows the importance of evaluating the toxicity, ecotoxicity and mutagenicity of bioremediated soils, because chemical analysis alone is not sufficient to determine the quality of any given bioremediated soil.

Cooperation between a bioremediation company and a chemical company producing surface-active compounds yielded a new composition of nonionic surface- active compound which is nontoxic for biodegraders in concentration up to 5 %, possesses very low critical micelle concentration and has high biodegradability. The new technology for enhancement of bioavailability of sorbed pollutant molecules has been designed and successfully applied. The new surfactant is biodegraded very easily and there is no danger of its accumulation in the environment. Combined soil flushing with biological polishing of freed pollutant molecules and surfactant proved to be a powerful mean for treatment of contaminated sites.

Bioremediation was successfully applied for *in situ* treatment at a site contaminated by various fire extinguishers. These foaming agents were degraded in soil as well as in groundwater by enhancement of the activity of autochthonous bacteria and via amendment with nutrients and terminal electron acceptor. The process was laboratory tested before implementation at the site.

Natural attenuation was applied at an industrial site contaminated with chlorinated ethenes. Biological degradation governs natural attenuation at the site. It was used after physical and physico-chemical treatments lost their efficiency. After two years of monitoring of natural attenuation it is clear that this process proceeds at a rate that ensures protection of recipient surface water [20].

4. Conclusion

The use of bioremediation in the Czech Republic has a long history. This method has been applied at some 50 sites *in situ*. The amount of excavated soil treated by bioremediation is of the order of 500000 to 750000 tons [12]. There exist about 50 sites where bioremediation was employed. It has never been employed as a sole remediation technique. It has generally been combined with physical or physico-chemical treatment technologies.

References

1. Alexander, M. (1994) *Bioremediation and Biodegradation,* Academic Press, San Diego, CA.
2. Dasappa, S.M. and Loehr, R.C. (1991) Toxicity reduction in contaminated soil remediation processes, *Wat. Res.* **25**, 1121-1130.
3. Gibson, D.T. and Saylor, G.S. (1992): *Scientific Foundation of Bioremediation: Current Status and Future Needs,* American Academy of Microbiology, Washington, D.C.
4. Pavlů, L., Stuchliková, K., Brenner, V. (2000) Biodegradation capacity of autochtonous microflora at a PAH-contaminated site in O. Halouškova (ed.) *Proceedings of the Symposium Bioremediation IV*, March 8-9, Seč, Vodni Zdroje Ekomonitor, ISBN 80-7080-374-6, pp. 8-13.
5. Bhatt, M., Cajthaml, T., and Šašek, V. (2002) Mycoremediation of PAH-contaminated soil, *Folia Microbiol.* **47**, 255-258.
6. Kubátová, A., Erbanová P., Eichlerová I., Homolka L., Nerud F., and Šašek V. (2001) PCB congener selectivie biodegradation by the white rot fungus *Pleurotus ostreatus* in contaminated soil, *Chemosphere* **43**, 207-215.
7. Zrotalová, K., Cajthaml, T., Demnerová, K., and Šašek, V. (2000) Possibilities for biodegradation ofpolyaromatic hydrocarbons in biological suspension system in O. Haloušková (ed.) *Proceedings of the Symposium Bioremediation IV*, March 8-9, Seč, Vodni Zdroje Ekomonitor, ISBN 80-7080-374-6, p. 144.
8. Siglová, M.,. Masák, J., Čejková, A., Jirků, V., and Kotrba, D. (2000) Stimulation of adhesion of single cell organisms capable to degrade toxic compounds in O. Haloušková (ed.) *Proceedings of the Symposium Bioremediation IV*, March 8-9, Seč, Vodni Zdroje Ekomonitor, ISBN 80-7080-374-6, p. 137.
9. Sýkora, V. and Pitter, P. (2000) Biodegradability of ethyleneamine derivatives in O. Halouskova (ed.) *Proceedings of the Symposium Bioremediation IV*, March 8-9, Seč, Vodni Zdroje Ekomonitor, ISBN 80-7080-374-6, p. 144.
10. Jasenská, A., Sedláček, I., and Damborský, J. (2000) Dehalogenation of haloalkanes by *Mycobacterium tuberculosis* H37Rv and other mycobacteria, *Appl. Environ. Microbiol.* **66**, 219-222.
11. Damborský, J., Kuty, M., Němec, M.,and Koca, J. (1997) A molecular modelling study of the catalytic mechanism of haloalkane dehalogenase: I. quantum chemical study of the first reaction step, *J. Chem. Inf. Comp. Sci.* **37**, 562-568.
12. Damborský, J., Boháč, M., Prokop, M., Kuty, M., and Koca, J. (1998) Computational site-directed mutagenesisof haloalkane dehalogenase in position 172, *Prot. Eng.* **11**, 901-907.

412

13.   Damborský, J., Jesenská, J., Hynková, K., Kmuníček, J., Boháč, M., Kuta-Smetanová, I., Marek, J., and Koca, J. (2000) Screening and construction of biocatalysts in O. Halouskova (ed.) *Proceedings of the Symposium Bioremediation IV*, March 8-9, Seč, Vodni Zdroje Ekomonitor, ISBN 80-7080-374-6, p. 14-18.

14.   Ružicka, J. (1993) Decontamination of former Soviet military bases in the Czech Republic, *Planeta* 1 (8), 23-249.

15.   Svoma, J. (1993) Decontamination of Soviet military bases, *Planeta*, 1(9), 25-28

16.   Thacker, B.K., Ford, C.G. (1999): In situ bioremediation technique for sites underlain by silt and clay, *J. Environ.Engrg.*, ASCE **125** (12), 1169-1172.

17.   Robertiella, A., Lucchese, G., Leo, C. di, Boni, R., and Carren, P. (1994) In situ bioremediation of a gasoline and diesel fuel contaminated site with integrated laboratory simulation experiments in Hinchee, R.E., Alleman, B.C., Hoeppel, R.E., Miller, R.N. (eds.), *Hydrocarbon Bioremediation*, Lewis Publishers, Boca Raton, CA, pp.133-139.

18.   Malachová, K. (1999) Using short-term mutagenicity tests for the evaluation of genotoxicity of contaminated soils, *J. Soil Contam.* **8** (6), 667-680.

19.   Malachová, K., Lednická, D., and Dobiáš, L. (1998) Bacterial assays of mutagenicity in estimating the effectivness of biological decontamination of soils, *Biologia* (Bratislava) **53**, 699-704.

20.   Matějů, V. and Kyclt, R. (2000): Natural biological attenuation of chlorinated hydrocarbons, Screening and construction of biocatalysts in O. Haloušková (ed.) *Proceedings of the Symposium Bioremediation IV*, March 8-9, Seč, Vodni Zdroje Ekomonitor, ISBN 80-7080-374-6, p. 65-68.

SUBJECT INDEX

SUBJECT INDEX

A

Acinetobacter   102, 103
Alcaligenes   105
Ames test   211-214
anthracene   53, 62, 63, 276, 325, 350, 359, 360
Arthrobacter   79
Aspergillus   7, 115, 116, 143, 145
assimilation rate   322

B

bacteria   3, 7, 15, 32, 39, 40, 51, 57, 76, 81, 96-102, 106, 116, 314, 316, 331, 345, 405, 406, 411
benzo[*a*]anthracene   53, 62, 275, 325
benzo[*a*]pyrene   53, 65, 130-132, 229, 231, 257, 273, 275, 276, 280, 325, 350, 363, 365, 366, 405, 406
benzo[*e*]pyrene   53, 62
bioassays   157, 161 165, 168, 169, 205, 212, 329, 330'
bioavailability   53, 245, 256, 275, 319, 330
bioluminiscence   32, 169, 171, 173, 205-208, 218, 329
biomass   3, 29, 31, 33, 35, 36, 38, 39, 79, 115, 117, 121
biopiles   229, 230, 233-245, 269
biomarker   3, 12,
biosensor   377-380
biosorption   78, 115, 117, 120, 121
Bjerkandera   144-146, 252, 254, 257
bound residue formation   331
Burkholderia   51, 58

C

carcinogenicity   54, 56
chlorophenols   249, 255
chrysene   53, 326, 350
Comamonas   51, 58, 100, 407
compost(ing)   267, 269, 273-275, 278, 392, 396, 402
creosote   52, 271, 325, 326
Cunninghamella elegans   57
cyanide   21-26, 269-272
     cyanide hydratase   21-26

D

DDT (dichlorodiphenyl trichlorethene)   252, 381, 388
decolorization   127-129, 135-140,143-147
dioxins   249, 255, 401, 402
dyes   135-141, 143-147, 250, 255

E

earthworms   169, 170, 199-202,
ecosystem health   29, 30,
ecotoxicity   167, 199, 205, 217, 218, 329, 406, 410, 411
Eisenia fetida   169, 170, 199, 200
enzymes   35-37, 41, 43,
environmental probes   22,
ergosterol   38,
ex situ treatment   229, 230, 267, 375, 392, 399, 409

F

fluoranthene   53, 62, 229, 230, 350
fluorene   53, 280, 350
fungal technology (see mycoremediation)
fungi   33, 57, 77, 116, 117, 143, 213, 247, 316, 325, 331, 345, 364, 365, 400
Fusarium   23-26,

G

genetically modified organisms   101, 119
genotoxicity   54, 211, 321

H

herbicide   367-370, 405
horseradish peroxidase   381-384

I

immobilization   78, 120
in situ treatment   3, 13, 21, 22, 41, 105, 108, 188, 267, 271, 289, 314, 315, 317, 318, 320, 358, 363, 408, 409
Irpex lacteus   143-147, 256, 406